Essential Computational Modeling in Chemistry

Essential Computational Modeling in Chemistry
A derivative of Handbook of Numerical Analysis Special Volume: Computational Chemistry, Vol 10

General Editor

P.G. Ciarlet

Laboratoire Jacques-Louis Lions
Université Pierre et Marie Curie
4 Place Jussieu
75005 PARIS, France

and

Department of Mathematics
City University of Hong Kong
Tat Chee Avenue, KOWLOON, Hong Kong

Guest Editor

C. Le Bris

CERMICS
Ecole Nationale des Ponts et Chaussées
77455 Marne La Vallée, France

Amsterdam • Boston • Heidelberg • London • New York • Oxford • Paris
San Diego • San Francisco • Singapore • Sydney • Tokyo

ELSEVIER

Elsevier
30 Corporate Drive, Suite 400, Burlington, MA 01803, USA
Linacre House, Jordan Hill, Oxford OX2 8DP, UK
Radarweg 29, PO Box 211, 1000 AE Amsterdam, The Netherlands

First edition **2011**

Notice
No responsibility is assumed by the publisher for any injury and/or damage to persons or property as a matter of products liability, negligence or otherwise, or from any use or operation of any methods, products, instructions or ideas contained in the material herein. Because of rapid advances in the medical sciences, in particular, independent verification of diagnoses and drug dosages should be made.

British Library Cataloguing in Publication Data
A catalogue record for this book is available from the British Library

Library of Congress Cataloging-in-Publication Data
A catalog record for this book is available from the Library of Congress

For information on all **North Holland** publications
visit our web site at books.elsevier.com

Printed and bound in Great Britain
11 12 13 10 9 8 7 6 5 4 3 2 1

ISBN: 978-0-444-53754-6

General Preface

In the early eighties, when Jacques-Louis Lions and I considered the idea of a *Handbook of Numerical Analysis*, we carefully laid out specific objectives, outlined in the following excerpts from the "General Preface", which has appeared at the beginning of each of the volumes published so far:

> During the past decades, giant needs for ever more sophisticated mathematical models and increasingly complex and extensive computer simulations have arisen. In this fashion, two indissociable activities, *mathematical modeling* and *computer simulation*, have gained a major status in all aspects of science, technology and industry.
>
> In order that these two sciences be established on the safest possible grounds, mathematical rigor is indispensable. For this reason, two companion sciences, *Numerical Analysis* and *Scientific Software*, have emerged as essential steps for validating the mathematical models and the computer simulations that are based on them.
>
> *Numerical Analysis* is here understood as the part of *Mathematics* that describes and analyzes all the numerical schemes that are used on computers; its objective consists in obtaining a clear, precise and faithful representation of all the "information" contained in a mathematical model; as such, it is the natural extension of more classical tools, such as analytic solutions, special transforms, functional analysis, as well as stability and asymptotic analysis.
>
> The various volumes comprising the *Handbook of Numerical Analysis* thoroughly cover all the major aspects of Numerical Analysis, by presenting accessible and in-depth surveys, which include the most recent trends.
>
> More precisely, the Handbook covers the *basic methods of Numerical Analysis*, gathered under the following general headings:
>
> – Solution of Equations in $\mathbb{R}^n$,
> – Finite Difference Methods,
> – Finite Element Methods,
> – Techniques of Scientific Computing.

It also covers the *numerical solution of actual problems of contemporary interest in Applied Mathematics*, gathered under the following general headings:

 - Numerical Methods for Fluids,
 - Numerical Methods for Solids.

In retrospect, it can be safely asserted that Volumes I to IX, which were edited by both of us, fulfilled most of these objectives, thanks to the eminence of the authors and the quality of their contributions.

After Jacques-Louis Lions' tragic loss in 2001, it became clear that Volume IX would be the last one of the type published hitherto, that is, edited by both of us and devoted to some of the general headings defined above. It was then decided, in consultation with the publisher, that each future volume will instead be devoted to a single *"specific application"* and called for this reason a *"Special Volume"*. *"Specific applications"* includes Mathematical Finance, Meteorology, Celestial Mechanics, Computational Chemistry, Living Systems, Electromagnetism, Computational Mathematics, etc. It is worth noting that the inclusion of such "specific applications" in the *Handbook of Numerical Analysis* was part of our initial project.

To ensure the continuity of this enterprise, I will continue to act as Editor of each Special Volume, whose conception will be jointly coordinated and supervised by a Guest Editor.

P.G. CIARLET
July 2002

Contributors

Alán Aspuru-Guzik, William A. Lester, Jr.
Department of Chemistry, University of California, Berkeley, California, USA

André D. Bandrauk, Hui-Zhong Lu
Laboratoire de Chimie Théorique, Faculté des Sciences, Université de Sherbrooke, Canada

Eric Brown
Program in Applied and Computational Mathematics, Princeton University, Princeton, NJ, USA. (E. Brown)
E-mail: ebrown@princeton.edu

J.P. Desclaux
Sassenage, France.
E-mail: jean-paul.desclaux@wanadoo.fr

J. Dolbeault
CEREMADE, Unité Mixte de Recherche du CNRS no. 7534 et Université Paris IX-Dauphine, Paris cedex, France.
E-mail: dolbeaul@ceremade.dauphine.fr

M.J. Esteban
CEREMADE, Unité Mixte de Recherche du CNRS no. 7534 et Université Paris IX-Dauphine, Paris cedex, France.
E-mail: esteban@ceremade.dauphine.fr

Jean-Luc Fattebert
Center for Applied Scientific Computing (CASC), Lawrence Livermore National Laboratory, P.O. Box 808, L-561, Livermore, CA, 94551, USA
E-mail: fattebert1@llnl.gov

M.J. Field
Laboratoire de Dynamique Moléculaire, Institut de Biologie Structurale – Jean-Pierre Ebel, 41 rue Jules Horowitz, Grenoble cedex 1, France

Wilhelm Huisinga
Institute of Mathematics II, Department of Mathematics and Computer Science, Free University (FU) Berlin, Germany
E-mail: huisinga@math.fu-berlin.de

P. Indelicato
Laboratoire Kastler-Brossel, Unité Mixte de Recherche du CNRS no. C8552, École Normale Supérieure et Université Pierre et Marie Curie, Paris cedex, France.
E-mail: paul.indelicato@spectro.jussieu.fr

Patrick Laug
Dipartimento di Chimica e Chim. Indus., Via Risorgimento 35, University of Pisa, Pisa, Italy

INRIA Rocquencourt, Gamma project, BP 105, Le Chesnay cedex, France

B. Mennucci
Dipartimento di Chimica e Chim. Indus., Via Risorgimento 35, University of Pisa, Pisa, Italy

INRIA Rocquencourt, Gamma project, BP 105, Le Chesnay cedex, France

Marco Buongiorno Nardelli
Department of Physics, North Carolina State University, Raleigh, NC, and Center for Computational Sciences (CCS) and Computational Science and Mathematics Division, Oak Ridge National Laboratory, Oak Ridge, TN, 37830, USA

Herschel Rabitz
Department of Chemistry, Princeton University, Princeton, NJ, USA.
E-mail: hrabitz@princeton.edu

Christof Schütte
Institute of Mathematics II, Department of Mathematics and Computer Science, Free University (FU) Berlin, Germany
E-mail: schuette@math.fu-berlin.de

E. Séré
CEREMADE, Unité Mixte de Recherche du CNRS no. 7534 et Université Paris IX-Dauphine, Paris cedex, France.
E-mail: sere@ceremade.dauphine.fr

J. Tomasi
Dipartimento di Chimica e Chim. Indus., Via Risorgimento 35, University of Pisa, Pisa, Italy

INRIA Rocquencourt, Gamma project, BP 105, Le Chesnay cedex, France

Gabriel Turinici
INRIA Rocquencourt, B.P. 105, Le Chesnay cedex, France
E-mail: Gabriel.Turinici@inria.fr

CERMICS–ENPC, Champs sur Marne, Marne la Vallée cedex, France. (G. Turinici)
E-mail: Gabriel.Turinici@inria.fr

Contents

The Modeling and Simulation of the Liquid Phase

J. Tomasi

Dipartimento di Chimica e Chim. Indus., Via Risorgimento 35,
University of Pisa, Pisa, Italy
INRIA Rocquencourt, Gamma project, BP 105, Le Chesnay cedex,
France

B. Mennucci

Dipartimento di Chimica e Chim. Indus., Via Risorgimento 35,
University of Pisa, Pisa, Italy
INRIA Rocquencourt, Gamma project, BP 105, Le Chesnay cedex,
France

Patrick Laug

Dipartimento di Chimica e Chim. Indus., Via Risorgimento 35,
University of Pisa, Pisa, Italy
INRIA Rocquencourt, Gamma project, BP 105, Le Chesnay cedex,
France

Computational Chemistry
Special Volume (C. Le Bris, Guest Editor) of
HANDBOOK OF NUMERICAL ANALYSIS
P.G. Ciarlet (Editor)

1

Contents

An overview on the theoretical and computational methodologies developed so far to study liquids and solutions is presented. The main characteristics of the different methods are outlined and the advantages and shortcomings of the computational approaches are discussed. Particular attention is focused on a specific class of methods, known as continuum solvation models, for which a more detailed description of theoretical and computational aspects is presented. For this class of methods, the concept of molecular cavity is introduced together with a description of the numerical techniques developed to mesh the corresponding surface. For selected methods representing those of larger use, an overview on their applications to the evaluation of energies and properties of liquid systems and molecules in solution is also presented.

CHAPTER I

Introduction to Liquids

For very long tradition chemistry has to deal with liquids. The chain of chemical manipulations performed in laboratories and in factories are mostly performed in a liquid medium and one of the first questions asked when a new chemical is introduced in the use is what is its solubility in solvents. There is a large number of solvents in use in the chemical practice, more than two thousand, and this number is not sufficient: often it is convenient to use solvent mixtures. Chemical reactions, the core of chemistry, are very sensitive to the solvent used to put in intimate contact the reagent species: often a delicate reaction fails if the solvent was not properly chosen. Important areas of chemical research are inherently tied to a solvent. An important example is that of biological systems. All the complex machinery of biological systems works only if the system is in water: without this solvent event the intimate structures of biological molecules collapse and loose their activity.

Chemists are so accustomed to consider many properties of liquids, from the basic physical properties to others, more specific and playing a role in specific cases in the large variety of problems that chemistry has to face. We cannot give here an overview of the properties of interest for liquids, many among which could be the object of computer modeling and we limit ourselves to indicate a few points more directly related to what will be exposed in the following of this chapter.

Liquids often occur in the chemical practice in large quantities. The effects due to the spatial limitations of the liquid sample have not great influence on the phenomena occurring at the interior, which in the current practice is called the bulk solvent. On large scale the bulk pure liquids and solutions exhibit isotropy. Liquids can be however dispersed. A typical example are the water droplets dispersed in the atmosphere, as for example in the fog. Other examples occur in liquid mixtures that present a miscibility lacuna. In all cases the ratio "surface region/bulk" can be relatively large and phenomena occurring at the interface may present properties different with respect to the same occurring in the bulk.

The region of separation between a bulk liquid and a solid body has in some cases a decisive importance. One example is the separation between a salt solution and a metal electrode. All the electrochemical phenomena are influenced by the behavior of this thin portion of the liquid. Another example are the phenomena leading to dissolution of solid bodies in the liquid phase, and to the deposition of material components of the liquid into solid particles. The surface of separation of a liquid and a solid surface presents features of different type when the liquid is in a limited amount.

The liquid phase can be present under the form of separate drops, or of a thin liquid film on the solid surface: both occurrences have a great importance in some chemical problems.

Surfaces of bulk liquids can be covered by a thin layer of an immiscible liquid. This is a phenomenon frequently observed in everyday life. Such thin layers can be organized into a structure that maintains some aspects of normal liquids but presents a high local order. With an appropriate selection of the chemical composition of this second phase, it is easy to form ordered layers of well-defined molecular thickness and membranes. Membranes, on their part, preserves fluid-like properties at their interior which can be exploited, for example, in the machinery of biological systems. A liquid can contain small solid particles within the bulk. Many phenomena and many practical applications are based on this particular type of liquid systems. A well-known phenomenon that has played an important role in the development of science is the Brownian motion. The solvent eases the dispersion in the bulk of such small particles, that are however subjected to gravity forces leading to sedimentation. Gravity, one of the basic forces in the universe, plays a little role in the microscopic approach to the study of material systems, but this example shows that there are reasons to forecast that in the future also gravity will be included into some computational models. A liquid can also contain at its interior portion of a second phase organized with specific shapes. This is the domain of micelles, vesicles and other similar structures, all subject of intense research both of basic type and addressing practical applications.

The ordering of specific chemical systems into layers and vesicles has a counterpart in bulk liquids. A special category of liquids are called liquid crystals. They combine macroscopic properties shared by all liquids, in particular the properties of assuming the shape of the vessel in which they are put and the capability of dissolving other chemicals, with a long range order making them similar to crystals. This long range order may assume different forms, to which corresponds specific names for the liquid crystal phase, smectic, nematic, cholesteric. In general the liquid crystal are specific phases of a liquid that can also have an isotropic phase without long range order. For many substances the changes in physical parameters temperature and pressure (T, P) rule the transformation from a solid, to a liquid and then to a gaseous phase (in the order on increasing T and decreasing P), while in the substances giving origin to liquid crystals there are some additional phases between solid and isotropic liquid. The liquid crystals are a subject of intensive study, because of their quite peculiar optical and electric properties.

1. Physical approaches to the study of liquids

The distinction we introduce here between physical and chemical approaches to the study of liquid systems is mainly justified by exposition reasons. The basic principles are the same in both approaches and there is a mutual interchange of methods and procedures: both approaches grow up in harmony. Actually there is a basic difference in the motivations of the two approaches: chemistry is directly interested to the details due to differences in the chemical composition of the fluid (the study of pure liquids could be viewed in this context as an extreme case of solutions), physics arrives to consider effects due to chemical composition (and of specific solutions in particular) at the end of a longer route. The physical approaches we shall consider here do not play much attention to the chemical composition of the system (but some aspects have still to be considered).

1.1. Macroscopic approaches

Historically, and also logically, the first contribution of the physical understanding of liquids was obtained with macroscopic approaches. We shall not consider here the mechanical studies that come first in the physical enquiry about fluids, but we directly shift to the thermodynamic approach elaborated in the nineteenth century. The science of heat, thermodynamics, regards all the matter in general, and considers liquids as a specific state of aggregation of the matter, to be treated on the same footing as the others. This emphasis on the uniformity of thermodynamic laws actually is at the basis of our understanding of the phenomena of phase transformation we have quoted above, and of the transfer of models (that we shall quote in the following) elaborated for the gas phase, being simpler to study, to liquids.

Thermodynamics is a rigorous discipline, and the definition of thermodynamic functions (at the equilibrium and out of the equilibrium) must be always reminded in performing studies on liquids with theoretical tools. Even when the model is reduced to a molecular model with attention paid to the details of quantum mechanical (QM) calculations on a reduced portion of the liquid, it is wise (often necessary) to keep in mind what is the thermodynamic status of the system, and to what selection of fixed macroscopic variables (pressure, volume, temperature, energy, chemical potential) is made in assessing the models.

A second macroscopic approach of interest for us regards the electric properties of the liquid. Here again we have to go back to the nineteenth century to find the elaboration of the macroscopic theory. A large portion of liquids are poor conductors of electricity (while there are notable exceptions, the whole electrochemistry is based on the conducting properties of ionic solutions), and for this reason the dielectric behavior plays the prominent role. It is convenient to give a short summary because we shall need it later. An external electric field induces a polarization of continuum dielectric media. In the standard version of the Maxwell elaboration, the attention is focused on the dipole polarization with respect to a homogeneous electric field. A vector field, $\mathbf{P}$, is defined giving the value of the dipole density; this vector is related to the two other vector fields defined by Maxwell to satisfy the two basic constitutive relations for electrostatic fields in vacuo: the electric field $\mathbf{E}$ and the displacement vector $\mathbf{D}$ (by definition $\mathbf{E} - \mathbf{D} = 4\pi\mathbf{P}$).

In the simplest case (homogeneous linear dielectrics, constant electric field) the relationship $\mathbf{P} = \chi\mathbf{E}$ with $\chi = (1 - \varepsilon)/4\pi$ holds. The macroscopic dielectric description of the liquids is not limited to the basic homogeneous and isotropic description we have here recalled. For anisotropic fluids (as liquid crystals) a tensorial definition of χ and of the dielectric constant ε must be introduced. There are cases in which the linear regime is not sufficient and the polarization must be described with the aid of higher order terms in the electric field. Models for specific cases call for specific modifications of the basic dielectric equation (a notable example of wide interest in chemistry is given by the Debye–Hückel model for ionic solutions). Other modifications are used for liquids spread on a metallic body.

All the examples we have given here refer to specific applications that shall be considered more in detail later, but of wider occurrence are the dynamic aspects introduced by the time dependence of the external field $\mathbf{E}(t)$, which gives rise to a large number of important phenomena. Relations paralleling those defined for all the static cases we have mentioned can be derived when changes in the electric field are not excessively fast (for the definition of such limits one has to make reference to a microscopic picture of matter). At a basic level, the dynamic case can be studied with the help of sinusoidally varying electric fields: $E(t) = E^0 \cos \omega t$ giving so origin to a complex dielectric response function that can be split into a real part, $\varepsilon'(\omega)$, and an imaginary part, $\varepsilon''(\omega)$. The first is a generalization of the static dielectric constant, $\varepsilon'(0) = \varepsilon$, and the second is called the *loss factor* (BÖTTCHER [1973], BÖTTCHER and BORDEWIJK [1978]). Both functions play an important role in the following of this chapter.

1.2. Microscopic approaches

The use of microscopic models based on molecules to describe liquids was introduced at the beginning of the past century. The impact has been enormous pervading all the research fields on liquids. The theory of liquid is at present a molecular theory. Two are the reasons of a so large impact: first, microscopic approaches have opened the way to the use of statistical mechanics, second, they have introduced explicit consideration of molecular interaction potentials, a subject on which chemists have a lot to say. For more than fifty years the research has been led by physicists, following the strategy to which physicists are more inclined, namely discarding the details of the molecular interaction potentials, spending more effort to establish models based on statistical mechanics and taking the model for ideal gases as starting point. Only in the late sixties of the past century aspects of more direct interest for chemistry were taken into consideration. In this chapter, we are more interested to the chemical way of describing liquids and so we shall deserve only few sentences to the physical approach, especially to introduce terms that will be used in the following chapters (among the vast literature on the argument, we quote here few titles: BALESCU [1991], REICHL [1980], MCQUARRIE [1976]).

The basic ingredient in this approach is the probability density P in the $6N$-dimensional phase space of a liquid system which satisfies the invariance of a selected set of three macroscopic quantities (the number of particles N, the temperature T, the energy E, or the chemical potential μ). In the formulation of the theory of statistical thermodynamics use is made of another concept, that of ensemble of systems, an ideal collection of numerous systems, all defined by the same invariants, in close contact and exchanging among them what permitted by the invariants. These ensemble are usually indicated as *microcanonical, canonical* and *grand canonical*. We report in the table names and invariants of the systems to which are associated. In the following we shall indicate single systems with the name of the ensemble, to avoid proliferation of names. The systems that have been selected correspond, respectively, to closed and isolated, to closed but not isolated and to open systems. Several other systems may be defined, according to the experimental setting used for the study of the properties shown in Table 1.1.

The time evolution of the probability density $P(t)$ can be described in terms of a Liouville equation. The description given by the full P function is too detailed as in practice we only need the probability density to find expectation values or correlation functions for various observables. The observables generally corresponds to one- and two-body operators and so it is sufficient to use reduced densities limited at these two orders: P_1 and P_2, formally obtainable by integration of P on the other variables. The equations of motion for reduced densities must be however expressed in terms of a hierarchy of equations, called BBGKY, where P_1 is given in terms of P_2, P_2 in terms of P_3, and so on, until the so-called thermodynamic limit (quite large). To use the BBGKY hierarchy there is so the need of defining a closure relation, and the various physical methods differ in the choice of it. The BBGKY hierarchy was defined within the classical picture of the fluid; in quantum systems, where the Hamiltonian replaces the Liouvillian operator, use is made of the Wigner functions which lead to a similar hierarchy.

Another approach is available, based on the correlation functions. The two approaches are connected. For brevity we shall limit ourselves to stationary states of the Liouville equation, namely to systems in equilibrium. P can be expressed as a product of uncorrelated one-molecule distribution functions $P_1(i)$ modified by a correlation function g_N:

$$P_N(1, 2, \ldots, N) = P_1(1)P_1(2) \cdots P_1(N)g_N(1, 2, \ldots, N).$$

We have indicated with i the set of coordinates of the ith molecule (remark that implicitly, we have here discarded the grand canonical system, that can be recovered later, and that we have not paid attention to the possible differences among the N molecules of the system). For isotropic and homogeneous fluids each $P_1(i)$ is independent on the coordinates of i, so we may express it in terms of the numeral density ρ

$$P_N(1, 2, \ldots, N) = \rho g_N(1, 2, \ldots, N).$$

The correlation function is then subjected to a cluster expansion:

$$g_N(1, 2, \ldots, N) = \sum_{i<j} g_2(i,j) + \sum_{i<j<k} g_3(i,j,k) + \cdots.$$

TABLE 1.1

Name of the system	Physical invariants
Microcanonical	N, V, E
Canonical	N, V, T
Grand canonical	V, T, μ

Let us consider the two-body correlation function $g_2(i, j)$. It can be divided into two parts, the first called *direct correlation function* $c(i, j)$, describing a direct effect only depending on the interaction potential $V(i, j)$, and the second describing many-body effects, in which i influences j through other molecules. The Ornstein–Zernicke equation, set in 1914 for spherical rigid molecules (nonadditive repulsive potential), and generalized in 1972 to rigid nonspherical molecules, is used as an approximation to connect $g_2(i, j)$ to the direct correlation function. After some manipulations (use is made of an intermediate function indicated as $h(i, j)$) one obtains an expansion into powers of the numeral density, in which the coefficients are given by the integration of appropriate products of direct correlation functions:

$$g_2(i,j) = 1 + c(i,j) + \rho \int c(i,k)c(j,k)\mathrm{d}r_k + \rho^2 \int \int c(i,k)c(j,k)c(k,l)\mathrm{d}r_k\mathrm{d}r_l + \cdots.$$

Clearly this approach presents the same problems of the *equation of motion* method. The same set of possible closure relations are in fact used in the two approaches. It is worth to remark that the use of these approaches are limited, with a very few exceptions regarding very simple model fluids (low pressure gases), to a calculation of $g_2(i,j)$ discarding higher terms in the cluster expansion. The same limitation is used in the computer simulations on liquids we shall present in the following of this chapter.

To close this chapter on the physical approach we consider some formal aspects of the calculation of properties. The number of properties that can be treated with physical approaches is large, but does not include many properties of chemical interest (mostly related to the behavior of specific molecules in the liquid). In parallel several properties computed with the physical approach are of little interest in chemistry. It may be worth recalling that properties object of scrutiny in the physical approach can be divided into three categories. First, the properties having a microscopic counterpart and that be defined as averages on the microscopic dynamical functions (they are often called mechanical quantities), examples are the internal energy E and the pressure. Second, the properties that have no microscopical meaning and must be defined in terms of the whole probability density (called thermal quantities); examples are the temperature T and the entropy S. Third, quantities that are fixed in value by the experiment without reference to the internal state of the system; they are called external parameters and typical examples are the number of particles N, the volume V, and the strength of external fields.

The method we are sketching here below can be applied to properties of the first and second type. We limit ourselves to consider time-independent properties in equilibrium systems. The value of a property is given as the average of the values the property has for each distribution, each multiplied by the appropriate statistical weight. The normalization factor of this sum is denoted as the *partition function Z*. Each kind of systems has its own partition function. There is no complete uniformity in the literature about the symbols used to indicate the corresponding partition function, here we shall use Ω, Q, and Ξ for the microcanonical, canonical and grand canonical systems, respectively. The system of reference is the canonical one. The steps leading from these systems to the other will be here omitted. In classical thermodynamics the complex (and complete) set of differential relations connecting the various thermodynamical functions is expressed taking one among them as the leading term (characteristic function). In the (N, V, T) systems the characteristic function is the Helmholtz free energy A. Its expression in statistical thermodynamics is quite simple:

$$A(N, V, T) = -kT \ln Q.$$

The most important thermodynamic functions for canonical systems are reported here below:

Pressure	$P = kT(\partial \ln Q / \partial V)_T,$
Entropy	$S = kT(\partial \ln Q / \partial T)_V + k \ln Q,$
Internal energy	$E = kT^2(\partial \ln Q / \partial T)_V,$
Enthalpy	$H = -(\partial \ln Q / \partial b)_V + kTV(\partial \ln Q / \partial V)_T,$
Free energy	$A(\text{or} G) = -kT \ln Q.$

Among the five energy functions reported in the table the most important in chemistry is G, also called Gibbs free energy, or free energy tout court. G is the characteristic function for another type of systems: (*NPT*) called isobaric–isothermal, and it also plays the leading role in the definition of grand canonical systems, through a related thermodynamic function, the chemical potential (in the chemical literature the term chemical potential is often used instead of Gibbs free energy). The function A plays a limited role in chemistry, it has been introduced because the analysis in terms of the canonical ensemble is easier than using other invariants. Remark, however, that when the attention is limited to liquid systems near the standard conditions there is no difference in practice between A and G.

2. Chemical approaches to the study of liquids

The simple model of liquids used in the chemical approach considers liquids as composed by an assembly of molecules at close contact, undergoing incessant collisions. Macroscopic conditions (T and P) rule the kinetic aspects of these collisions, that are however strongly modulated by molecular interactions depending on the chemical nature of the molecules. These interactions are responsible of rare events (with respect to the number of collisions) which give origin to reactions and other phenomena of chemical interest. The same interactions are also responsible of the creation of a partial ordering of local nature, which fades away at larger distances. Chemists tend to mentally average the thermal collisions and to pay more attention to the effects of chemical interactions, both for reactions and for the partial local ordering, which affects the properties of the composing molecules, the solute in particular.

2.1. Molecular interactions

The mutual molecular interactions are of different type, some are of very short range, others exert their effects at a longer distance. In general they are nonadditive, but with some exceptions. Chemical practice has permitted to learn a lot about the effects of these forces in the various solvents and obtained some hint about their nature. A more precise classification and definition came from the study with theoretical tools of the behavior of small molecular clusters, composed by two, three or little more molecules (MARGENAU and KESTNER [1971], HOBZA and ZAHRADNIK [1988]). The classification in use for the decomposition of these interactions is reported in Table 2.1.

The whole set of interactions can be described at the QM level using the usual electrostatic Hamiltonian, other terms not present in it, like spin-orbit effects or relativistic corrections play a little role in the study of liquid systems.

The complete interaction energy within a given cluster can be easily computed with standard QM tools. It is sufficient to compute the energy of the whole system, considered as a *supermolecule*, and then subtract the energies of the separate partners. This interaction energy can be decomposed with the help of some additional QM calculations performed with the same procedure but with deletion of some terms and/or of some steps in the solution of the resulting computational problem. The details are not reported here, suffice it to say that the analysis of such decompositions performed on small clusters at different geometries has permitted to gain a good confidence on the relative importance of these terms, and on their spatial anisotropy.

There is another approach that do not introduce the calculation of the energy of the supermolecule, it was the only approach in use before the advances in electronic computers made possible calculations on supermolecules. Use is made in this approach of the perturbation theory (ARRIGHINI [1981]). The unperturbed Hamiltonian is defined as the sum of the Hamiltonians of the isolated monomers, and the perturbation is just given by the interaction. The perturbative corrections to the energy are expressed order by order and within each order a separation in additive terms corresponding to interactions of different physical origin is performed. There are problems, however, due to the fact that the full Hamiltonian of the cluster has a higher permutational symmetry than the unperturbed one due to the exchange of electrons among partners: the wavefunction of the cluster has to be fully antisymmetrized with respect to all the electrons. This lack of antisymmetry in the exchange of electrons among partners greatly complicates the formulation of the method, and the quality of the results. Here again we do not give details about the procedures elaborated in the years to partly overcome these problems. Suffice it to say that the results obtained for lower orders of the perturbation expansion on small clusters are in fairly good agreement with

TABLE 2.1

Component	Acronym	Physical meaning
Coulombian	COUL	Electrostatic interactions between rigid charge distributions
Induction	IND	Electrostatic deformation of the charge distributions
Exchange	EXC	Quantum repulsion effect due to the Pauli exclusion principle
Dispersion	DISP	Interactions between fluctuations in the charge distributions
Charge transfer	CHTR	Transfer of electronic charge among partners

those obtained with the decomposition of the supermolecule description. The basic point is that neither approaches can be directly used to get a description of the liquid around a given solute.

There are two reasons, strictly connected. The first is that experience has learned that models composed by small size clusters give a very poor description of local solvent effects (and almost nothing on the properties of the whole solution). The second is that the larger aggregates that should be used in modeling require an accurate description of the thermal motions, properly averaged to reach a consistent thermo-dynamic status. QM supermolecule treatments are ruled out mostly for the difficulty of getting the thermal average. QM calculations at single geometries of a large cluster are feasible with some efforts, but the spanning of the portion of interest of the conformational space, and the following determination and use of the system partition function are out of question. The device often used in the study of isolated chemical systems of looking only at minima and other topological critical points on the poten-tial energy surface (PES) cannot be applied here. The PES presents an exceedingly number of local minima, separated by very low barriers and with flat portions of the surface, hard to treat with statistical mechanics. In practice, the concept of PES is of little utility in treating large molecular aggregates. The direct use of the explicit expressions of the various components of molecular interactions obtained either via decomposition of the supermolecule energy or perturbation theory analysis are out of question for similar reasons. The computational cost is nowadays similar to that of getting the whole supermolecule energy (in addition, the extension of such formulas to the case of many body clusters presents serious difficulties).

There is so the need of reconsidering the problem, paying attention to the physical effects produced by the various types of interactions. We shall give here a short summary of this analysis (TOMASI, MENNUCCI and CAPPELLI [2001]).

COUL. It is the interaction having the larger long range effect. It is strictly nonad-ditive (i.e., limited to two-body contributions) and strongly anisotropic. In most cases it is the more important contribution (for example in water solutions). It determines the most favorable orientation among partners having asymmetric charge distributions (e.g., dipoles). The formal expression for a two body interaction

$$E_{\text{COUL}} = \iint \rho_A(r_1) \frac{1}{r_{12}} \rho_B(r_2) \mathrm{d}r_1 \mathrm{d}r_2 = \int \rho_A(r_1) V_B(r_1) \mathrm{d}r_1$$

suggests that simplifications of this expression can be searched with an opportune modeling of the total charge distribution $\rho(r)$ or of the electrostatic molecular poten-tial $V(r)$. This last presents, as a rule for neutral molecules, positive and negative regions, and the same holds for the integrand giving origin to COUL. Acceptable simplifications must preserve the anisotropy of the property, which, as already said, is essential. One-center multipole expansions work badly if the molecule has a com-plex shape: better it is to pass to many center expansions. One formulation of large use in chemical computations (especially for solutions of large biological systems) consist in keeping only the first term (a point charge) of multipole expansions cen-tered on all the atoms of the molecule. This approximation is rather grossly, and should be avoided, if possible.

IND. This term is decidedly less anisotropic than the preceding one, but it exhibits a strong nonadditivity. The integrand giving origin to IND is everywhere negative. The modeling of the interaction is generally based on multipole expansions of the electric molecular polarizability. The effect of the other molecules on the charge distribution is in general far from that of a uniform electric field assumed in the definition of the first order polarizability tensor. In spite of it, a formulation often used in computations consists in placing a single isotropic value for the polarizability placed at the center of the molecule. It is a poor approximation, partially justified for a very common solvent, water. The nonadditivity of the contribution is active also at relatively long distances: the electric field of distant molecules must be considered in the contribution at each polarizable center. In turn polarization effects, even when reduced to a single contribution, affects the total polarization. IND must be computed in an iterative way, when these models are used.

EXCH. To satisfy antisymmetry, closed electronic shells have to repel each other. This contribution is everywhere positive. It is nonadditive, but only the effect of nearby molecules play a role, being the contribution considerably short-ranged. Exchange forces can be relatively well described by a set of repulsive stiff potentials, centered on the various nuclei of the molecule, each with a spherical symmetry (typically with a R^{-12} decay law). The exchange forces are the interaction terms left in the oversimplified physical models, in which the molecule is replaced by a sphere, hard or with a soft repulsive potential.

DISP. This term is nonadditive, with quite moderate anisotropy. It may be formally treated as IND, making use here of dynamic multipole polarizabilities. The simple approximation used in chemical computations uses the first term in the development of the first polarizability in terms of the inverse powers of the distance (it is a R^{-6} term) applied again to each atom of the molecule. Dispersion forces are relatively weak and have a moderate long range effect, by far smaller that of COUL and IND. It happens however that the long range effects of the two "classical" electrostatic terms, coupled to their anisotropy, tend to orient molecules in the liquid in the more appropriate orientation, producing so a screening of the global effect. The screening is not active for dispersion terms, and so happens that for several of the examples of liquids we have done in the introduction, the long range interactions are almost solely ruled by dispersion forces. This effect is more evident in the presence of massive bodies.

CHTR. The effects due to the transfer of electronic charge during the molecular encounter are harder to model than the preceding ones. Of course these effects belong to the category of "rare events" but a large part of chemistry is based just on these "rare events". In the approximated formulation of interaction effects we have outlined, electron transfer effects have not been inserted. This is a limitation in the computational procedures using this modelistic elaboration. To consider the effect of charge transfer other methods must be devised.

We have outlined a reasonable way of computing molecular interactions without passing through QM calculations. This procedure can be inserted as a basic element into other procedures performing the systematic scan of the conformational space on the desired thermodynamic ensemble, to get the properties of the liquid. This scan is given via molecular simulations, Monte Carlo (MC; according to a Gibbs picture of thermodynamic averages) or to Molecular Dynamics (MD; according to a Boltzmann picture). These topics will be considered later in this chapter. Such simulations, largely used in the last 20 years, have given a considerable wealth of information about the properties of liquids, by far more detailed and more precise than those obtainable with the physical approach. All the types of liquids we mentioned in the introduction have been at least partially examined with these approaches, and it may be said that what we know today about liquids derives from, or has been confirmed by, molecular simulations. It has been a big effort, quite rewarding. The magnitude of this effort, that has been at a good extent of methodological type, can be appreciated by looking at the final computational costs. To perform a simulation on a liquid system there is to repeat many times the computation of the geometry and of the interactions of a relatively large cluster. For this reason people is compelled to use descriptions of the molecular interactions as simple as possible. In addition, this approach does not give the essential information for which chemists undertook studies on liquids: the effect of a solvent on a chemical reaction and on the molecular properties of the solute. To have them a QM description at high level is in fact necessary.

An approach alternative to the simulation on simplified semiclassical interaction potentials exists, and it is able to reach the requested high quality in the description of the molecular units of interest (TOMASI and PERSICO [1994]). This approach is the main subject of this chapter. It is still based on the analysis of the various interaction terms, but in a different way. The basic consideration is that the thermal average performed as the final step in the above described methods can be avoided by replacing the molecular discreteness of the molecular distribution with a continuous distribution function. It is possible to formally define continuous solvent response functions corresponding to the various interaction terms, and satisfying the macroscopic thermodynamic conditions: density, temperature, pressure. It is also possible to formulate equilibrium and nonequilibrium expressions, and to introduce boundary conditions to treat systems different from the bulk liquid. These response functions are then applied to a small molecular system, including the molecule (or molecules) of direct interest, using a QM approach that can be, when desired, of high accuracy. We shall call it the continuum approach, in contrast to the discrete approach of which we have given in the preceding pages a short summary of its basic definition.

CHAPTER II

Continuum Models

As said above, continuum models tend to simplify the problem by introducing solvent response functions describing its interaction with the focused model (we shall call it for brevity the "solute"). The main advantage of this approach consists in a very large reduction of the internal degrees of freedom of the system one has to consider. In passing from the whole solution to the solute + continuum model, a detailed monitoring of all the degrees of freedom of the solvent molecules is no more necessary. The solvent response functions are specific for the various types of interactions occurring in liquids. Among them, that carrying more information is the electrostatic response. For several years also modern continuum methods were limited to the electrostatic interactions only. We shall consider now the origin and evolution of this electrostatic model starting from the seminal paper by ONSAGER [1936] in which he presented several concepts that are the basis of modern continuum methods.

The model used in Onsager's paper was quite simple. A solute, reduced to a point dipole, μ, but provided of polarizability α, is placed into a spherical cavity of appropriate radius; the solvent, placed out of the cavity is described as a continuous isotropic dielectric. The solute charge distribution induces a polarization of the dielectric, and in turns, this polarized medium polarizes the solute charge distribution, via an electric field, called by Onsager the *solvent reaction field* **R**. As a consequence the solute dipole changes from μ to μ^*, this last depending on μ, its polarizability α, the dielectric constant ε, and the radius a of the cavity. The elaboration of this formula, quite simple, was done on the basis of classical electrostatics.

We have anticipate that this simple model introduces some basic concepts, among them we first quote the cavity. Onsager's definition of cavity is a physical entity: it corresponds to a portion of the physical space in which the solvent is not allowed, because already occupied by the solute molecule. Kirkwood, another eminent figure in the study of liquid systems, remarked a short time after that this was an epoch-making innovation: all the cavities used before (e.g., Maxwell and Lorentz cavities, BÖTTCHER [1973], BÖTTCHER and BORDEWIJK [1978]) were just mathematical devices. Onsager was well aware of it. In its quoted papers he paid attention to the shape of the cavity, to its dependence on the thermally induced volume changes, to the problems related to the possible occurrence of hydrogen bonds. These are problems amply treated in the more recent versions of continuum models.

The second concept introduced by Onsager is that of the reaction field. He spoke of the reaction field because its solute model was a dipole, now we are speaking more generally of a reaction potential. Here again he paid attention to the physical content of this concept: he analyzed some aspects of the nonideal behavior the homogeneous continuum dielectric may have in real systems, like the phenomena of nonlinearity that can be phenomenologically related to dielectric saturation and to electrostriction effects.

Onsager also introduced a third concept, that of the cavity field G, occurring when the liquid sample is subjected to an external electric field E. G is related to E and to the geometrical factors defining size and shape of the cavity. The field R is the father of the electrostatic solute–solvent interaction potentials we are using in QM continuum models, G plays an important role in the very recent extensions of the continuum model to compute molecular properties. The development of these basic ideas will be discussed in the following pages. Here we remark that the original Onsager model continues to be amply employed, especially to get a rationale of experimentally observed trends in chemical properties. Some modifications of this models introduced new concepts. We quote its application to the description of solvent shifts in electronic spectra.

An important development was the translation of this model in a QM language. In the models called Onsager–SCRF (or simply SCRF, were the acronym stays for self-consistent reaction field) the solute is described at the QM level, put in a spherical cavity and subjected to the action of a R field having as its origin the dipolar contribution of the charge distribution of the solute. This model is quite simple to implement and to use when a code for the calculation of QM molecular wavefunctions is available. For this reason the first implementation of the Onsager–SCRF model (TAPIA and GOSCINSKI [1975]) was done forty years after the original Onsager paper. Actually few years before (1973) a proposal was made by Rivail's group in Nancy to introduce in the model other terms of the multipole expansion of the solute charge distribution. The proposal was expressed in a better way and documented by results in RIVAIL and RINALDI [1976]. This innovation eliminates at a good extent a defect of the too simple description of the electrostatic terms given in the original version: the use of the dipole only may produce serious deformations in the description of the solvent effects for molecules with a complex shape. A typical example is that of solutes having two or more identical polar groups, which are spatially arranged in the molecule to give a net value of the total dipole equal to zero: in this case the Onsager–SCRF model gives zero solvent effects. The original SCRF model has been continuously refined: we quote here the extension to ellipsoidal cavities and to cavities with a molecular shape. SCRF models are easy to implement and to use when the cavity has a constant curvature (sphere, ellipsoid). At present, they may also be used for more realistic cavities with a shape modeled on that of the molecule, but the elaboration of the model and in particular its use is a bit more delicate. The multipole expansion still presents some limits for complex molecules, that could be partially eliminated by using segmental local expansions, with expansion centers placed at opportune sites of the molecule. There exists programs able to do it, but they are rarely employed. The SCRF method has been further generalized to high-level QM approaches (like MCSCF and Coupled-Cluster) by MIKKELSEN, CESAR, ÅGREN and JENSEN [1995] who implemented the spherical version of the SCRF model in the Dalton QM code (HELGAKER ET AL. [2001]).

An alternative approach to SCRF-like continuum models was proposed by our group. The first paper is of MIERTUŠ, SCROCCO and TOMASI [1981]. In this approach the multipole expansion used in all the previous models was replaced by another way of solving the electrostatic problem posed by the model. The solution of the Laplace and Poisson equations requested by the model was expressed in term of an apparent charge distribution, defined by applying theorems of classical electrostatics, and spread on the cavity surface. The solvent reaction potential was expressed in terms of this apparent charge properly discretized into point charges and introduced in the Hamiltonian of the solute as a solute–solvent interaction potential. In such a way the multipole expansions of both the molecular and the reaction potentials, rather cumbersome in the case of cavities of molecular shape, was avoided. The method was presented within a QM formalism for computing molecular wavefunctions of *ab initio* type (the previously quoted SCRF methods were at that time all at the semi-empirical level, probably because it was not easy to compute high multipole integrals with *ab initio* codes). The procedure we devised was the application to a QM problem of the boundary element method (BEM). We realized it later; actually at that time the name was not yet diffused in the literature. Afterwards it was called Polarizable Continuum Model (PCM) and it continues to keep this name even if many important improvements have been introduced in the years with respect to the original version (CAMMI and TOMASI [1995b], CANCÈS and MENNUCCI [1998b], CANCÈS, MENNUCCI and TOMASI [1997], MENNUCCI, CANCÈS and TOMASI [1997]).

We shall consider later more details and extensions of PCM. Other methods sharing some similarity with PCM appeared in the nineties: among them we quote the so-called COSMO model and the methods making explicit use of other mathematical techniques, as the Finite Element method (FEM) and the finite difference (FD) approach. In the last two cases the solvent response function is obtained by sampling the solvent electric potential at a relatively large number of points inside the bulk dielectric. One among these methods has gained wide popularity in a semi-classical version (the solute charge distributions is reduced to a set of point charges on the nuclei) to study large biomolecules (CHEN and HONIG [1997]). In the method known with the acronym COSMO, the model is different as the screening effects in the dielectric are replaced by the screening effects in a conductor (KLAMT and SCHÜÜR-MANN [1993], TRUONG and STEFANOVICH [1995]). In other words, COSMO method is a solution of the Poisson equation designed for the case of very high ε, and it takes advantage of the analytic solution for the limit case of a conductor ($\varepsilon = \infty$), for which the boundary condition reduces to $V = 0$ on the surface. Anyway, apart from these differences in the theoretical background, COSMO can be described exactly in the same way as PCM.

The last solvation approach we are here quoting adopts a different strategy. It is an extension of a simple model Born employed many years ago (1921) to describe the solvation energy of a simple atomic ion. The approach is often called GB (generalized Born) approach (CRAMER and TRUHLAR [1991]). The charge distribution of the molecule is reduced to atomic point charges; the Born formula is generalized taking into account that at each atom a sphere is assigned and that such spheres partially overlap each other. The procedures following this approach contain a number of parameters that have to be empirically fixed. A version of this approach, with a line of versions progressively improving the method, has been accurately parametrized and it is widely used, first in semi-empirical versions, and now also with *ab initio* codes (ZHU, LI, HAWKINS, CRAMER and TRUHLAR [1998]). The QM part of the procedure is necessary to update during the calculations the solute atomic charges, modified by the reaction potential. We have summarized the methods we consider more important for the applications the have had in the chemical domain, but several other exists, reference is made to a detailed review by TOMASI and PERSICO [1994] that can be partially adjourned by looking at CRAMER and TRUHLAR [1999].

In the following we shall made reference to the PCM-like methods only, because they are the methods that have had more extensions. We have had until now considered electrostatic models for molecules in the bulk of homogeneous liquids. The structure of PCM makes easy to introduce further boundary conditions similar to the cavity surface. Without important changes in the general structure, it is easy to divide the whole space into nonoverlapping portions, each having a given dielectric function and containing, or not, an electric charge distribution to be treated classically or at the desired QM level. This opens the possibility to study important aspects of many phenomena of chemical interest, as the description of molecular encounters giving origin to associates or to new compounds, to follow the decomposition of a molecule into separate portions, to describe the interactions among molecules separated by the solvent.

We have here recalled other phenomena occurring in the bulk of the solvent, but the extension to other types of liquids is also immediate. We quote as examples the description of the partition of a solute between two immiscible liquid phase (a procedure which has very old traditions in chemistry and many applications in drug design), the chemical processes occurring within a droplet dispersed in another liquid or suspended in the atmosphere, the behavior of molecules within a membrane, a vesicle or a molecular host (great attention is now paid to the technological exploitation of that change in the molecular properties induced by these restricted systems). The formal transformation of PCM conceived for bulk is easy, but to have a proper modelistic description of the related phenomena there is to consider other aspects of the basic solvation models we have not yet introduced. Another subject of interest is the behavior of liquids, and of solutes at the liquid surface. A lot of chemistry is given at the surfaces: liquid/gas, liquid/liquid, liquid/solid. It is an enormous field in which important chapters of basic science are intermingled with very important practical problems. It is not possible to give a correct impression just quoting few names of phenomena or of sub-fields, so we prefer to limit ourselves to this indication, adding the remark that the work in this field is at its very beginning, and that continuum electrostatic methods have to be strongly supplemented to give reliable results.

We may add here at the list of extensions of PCM, those regarding salt solutions and liquid crystals. Both have introduced in PCM by a renewed version, called Integral Equation Formalism (IEF; (CANCÈS and MENNUCCI [1998b], CANCÈS, MENNUCCI and TOMASI [1997], MENNUCCI, CANCÈS and TOMASI [1997]) on which more will be said in the following. Here, we add only few remarks. Both slat solution and liquid crystals are of considerable scientific and practical importance. Biological processes solely occur in saline solutions, and the chemical physics of salt solutions has been the object of continuous studies and modelizations since almost 150 years. The liquid crystals are relatively newcomers in the realm of chemistry, but have rapidly gained popularity because their very peculiar properties, exploited in many ways: optical display, to give an example are almost all composed by liquid crystals subjected to an external electric field.

3. The energy of the system and the solvation energy

The energy of dissolution of a chemical into a solvent is one of the basic quantities in chemistry. Precise measurements are required in many fields of chemistry and chemical engineering. Actually to get accurate experimental values is a very hard task and, after more than one hundred years of efforts, intrinsic solvation energies are known for a very limited number of substances in a few solvents, and often with large error bars. According to the experimental setup solvation free energies or enthalpies are obtained. We recall that the former has a wider interest, because it governs the chemical transformations, the change of liquid phase, the electrochemical processes, and all the other processes in which chemistry is involved: enthalpies are not sufficient to such purposes. We also recall that thermodynamic quantities always require reference values. To formulate the solvation energy problem in computational methods it is convenient to make use of canonical ensembles. We shall consider now how this problem has been formulated in the several versions of the continuum PCM methods.

The PCM continuum model considered in this chapter uses *ab initio* QM methods to describe the solute. In this framework the reference energy (the zero value in the final outcome) is that of a system composed by the pure unperturbed liquid at the given (N, P, T) conditions (alternatively to the N, V, T conditions, according to the already expressed remarks) supplemented by the opportune number of electrons and nuclei, necessary to describe the solute, not interacting and at zero kinetic energy. We shall now consider two different ways of defining the total energy of the system and the energy difference corresponding to the solvation energy.

(A) The solvation process can be split into two parts: (1) the formation of a cavity of suitable shape and size within the homogeneous liquid, and (2) a charging process of formation the solute, starting from the electrons and nuclei, within the cavity. During this process the solute–solvent interactions play an active role contributing to give the final result in which both solvent and solute have reached an equilibrium including mutual interaction effects. The energy related to the first process must be computed apart.

In this scheme we have implicitly made use of one of the physical approaches to the statistical description of liquids that we have introduced before. This method, called Scaled Particle Theory (SPT), gives exact analytical expressions for the formation of a cavity of spherical shape in a liquid composed by spherical units (PIEROTTI [1976]). For coherence with the other terms this process is quantified in terms of the corresponding cavitation free energy, G_{cav}. In the SPT model, the solvent distribution function is modified only in the region corresponding to the cavity without disturbing the remainder, as requested by our two-step process. Actually, the procedure requires some checks, because in realistic applications neither the solute nor the solvent has a spherical shape. Checks have been done via MC and MD simulations, with satisfactory results, but others would be advisable.

The main part of process (2) is given by the *ab initio* PCM calculation in which the electrostatic, dispersion and repulsion interaction terms are coupled to the inner electrostatic interactions giving rise to the molecule. In doing this, a problem appears. In the various replicas of the canonical ensemble, the point where the cavity has been formed and the molecule has been built, occurs at different positions, being not fixed by the model. This fact gives rise to an additional entropic contribution, called communal entropy; it corresponds to the fact that the solute molecule is free to wander within the whole volume V of the system. Actually, it is a not complete freedom, because V is occupied at a large extent by solvent molecules. From here it arises the concept of free volume, on which fifty years have been spent in discussions. The solution we consider to be definitive, has been given by BEN-NAIM [1974], BEN-NAIM [1987] with the introduction of the concept of *liberation free energy*, the free energy spent to allow to solute to pass from a given position to others within V. No details will be given on the elaboration of this idea, which has been accepted in PCM. The free energy G given by PCM actually corresponds to a molecule with nuclei frozen at a given geometry (the Born–Oppenheimer approximation is used in the *ab initio* calculations). The contribution of nuclear motions are computed with statistical mechanics techniques and may be divided into vibrational and rotational contributions: G_{vib} and G_{rot}. To do it, formally one has to factorize the canonical partition function into molecular components q. This factorization is not exact; there are couplings, especially in the rotational component, a fact which prompts to improve further the model.

(B) A second possible approach is divided into three steps: (1) cavity formation, (2) formation in vacuo of the molecule with a charging process, and (3) insertion of the molecule in the cavity. With this second approach we may give an expression of the solvation energy based on the passage of the solute molecules from the gas to the liquid phase:

$$G_{sol} = \Delta G + G_{cav} + \Delta G_{vib} + \Delta G_{rot},$$

where $\Delta G = G - G^0$ is the difference between the free energies given by the two QM calculations (in the BO approximation), G^0 in the ideal gas phase and G in solution. The translational contribution to the free energy has been omitted, because of the use of the Ben-Naim treatment, based on a definition of the standard concentration in terms of the volume, and not of the pressure as it is usual for the gas phase.

The definition of the energy becomes a bit more complex when one studies (i) an association process, where two solutes A and B, merges into a unique entity AB with the consequence that six degrees of rotational and vibrational type of the two molecules become internal degrees of vibrations; (ii) in the case of liquid crystal, in which G depends on the orientation of the solute with respect to the axes of the dielectric tensor; (iii) in the case of molecules near a boundary, where G depends on the distance (and orientation) with respect to the boundary. We will not dwell on these problems and others similar.

The whole charging process done to compute G can be partitioned into three separate charging processes when the solute–solvent interaction potential is separated into three parts: $V = V_1 + V_2 + V_3$, where 1, 2, 3 stay for electrostatic, repulsion, and dispersion. The charging processes can be done separately, if one neglects couplings. The expression to use is based on the fact that $\ln Q = -kT(G + \Delta G)$, where G is the free energy of the system in absence of solute-solvent interactions and:

$$\Delta G = \sum_{m=1}^{3} \int_0^1 d\lambda_m \int dr \rho V_m g'(r; \lambda_m)$$

in which λ_m are three charging parameters (assimilated respectively to charges, overlaps and dynamic polarizabilities) going from 0 to 1, and $g'(r; \lambda_m)$ is the reduced solute-solvent distribution function when the interaction potential is $\lambda_m V_m$. Actually, this last condition is satisfied only for the electrostatic term, being the others computed at constant distribution function. The formula is of course approximate, but comparison with solvation energies computed with the approach B (in which couplings are at least in part considered), and the analysis of a few computer simulations performed ad hoc, indicate that in normal cases the errors are quite limited.

3.1. Dispersion and repulsion contributions

In general, the modeling of dispersion and repulsion interactions in solution is based either on a discrete molecular description of the liquid or on a continuum model.

The discrete approach is generally based on the use of pair potentials related to atoms or groups of atoms of the solvent S (here indicated with l) and the solute M (here indicated with m):

$$V_{\text{dis–rep}} = \sum_{m \in M} \sum_{l \in S} V_{ml}(r_{ml}) \quad \Leftrightarrow \quad V_{ml}(r_{ml}) = \sum_n \frac{d_{ml}^{(n)}}{r_{ml}^n}$$

the dispersion ($n = 6, 8, 10$) and the repulsion ($n = 12$) coefficients are taken from the literature. Often an alternative exponential expression, more related to the physical interpretation of the interaction, is used for the repulsion term: $c_{ml} \exp(-\gamma_{ml} r_{ml})$. The related approximate expression for the dispersion–repulsion contribution to G is derived in terms of continuous distribution functions $\rho_{ml}(\mathbf{r}_{ml}) = N_l n_S g_{ml}(\mathbf{r}_{ml})$, where N_l is the number of groups of type l in each solvent molecule, n_S the solvent macroscopic density, and $g_{ml}(\mathbf{r}_{ml})$ a correlation function depending on the position of l with respect to m (we use here the bold character to indicate vectors). As a final result we can write:

$$G_{\text{dis–rep}} = \sum_{l \in S} \sum_{m \in M} \int \rho_{ml}(\mathbf{r}_{ml}) V_{ml}(\mathbf{r}_{ml}) d\mathbf{r}_{ml}. \tag{3.1}$$

It should be clear that to go further other approximations are needed; in particular the integral operation can be simplified by defining for each l an appropriate portion of space in which there are no l centers, and $g_{ml}(\mathbf{r}_{ml})$ is always zero. This portion of space can be identified with the cavity C_l related to the Van der Waals spheres centered on the nuclei of the solute and enlarged to take into account the correspondent radius of the solvent. When the set of C_l cavities is known, we may replace the volume integrals in Eq. (3.1) with surface integrals over the surface Σ_l of the cavities C_l (FLORIS, TOMASI and PASCUAL-AHUIR [1991]) by introducing auxiliary vector functions defined in terms of the pair potential and the correlation functions. Simple analytic expressions of these functions can be obtained by reducing each $g_{ml}(\mathbf{r}_{ml})$ to a step function 0 inside and 1 outside C_l (this is often called the "uniform approximation", and is congruent with the properties of solutions at infinite dilution) and by performing a simple one-dimensional integration over r. It has to be noted that such methodology is completely independent on solute charge distribution; it thus does not involve any QM description of the system, but it only leads to an additional term in the total free energy.

The second possible approach we have mentioned to get $G_{\text{dis-rep}}$ is based on a continuum model; in this case the two contributions are treated separately. The starting point is the general expression derived from the intermolecular forces. In the previous discrete approach however, this equation were applied to calculated or estimated potentials V_{rep}^{AB} available from the literature. We now substitute V_{rep}^{AB} for a suitable expression taken directly from the theory considering that, as originated from the Pauli exclusion principle, the repulsion forces between two interacting molecules increase with the overlap of the two distributions and are strictly related to the density of electrons with the same spin (AMOVILLI and MENNUCCI [1997])

$$G_{\text{rep}} = \frac{1}{2}\rho_B \int g_{AB}(\mathbf{r})d\mathbf{r} \int \frac{d\mathbf{r}_1 d\mathbf{r}_2}{r_{12}} P_A(\mathbf{r}_1;\mathbf{r}_2)P_B(\mathbf{r}|\mathbf{r}_2;\mathbf{r}_1).$$

Here, the label A refers to the solute, B to the solvent, $\mathbf{r}$ is an appropriate set of coordinates defining the internal geometry of the complex AB, ρ_B is the number density and g_{AB} is a correlation function which is 0 inside the solute cavity ($\mathbf{r} \in C$) and 1 outside ($\mathbf{r} \notin C$). At this point, because in the continuum approach here exploited the electron density of the solvent is not given, it is useful to make the following two assumptions: (i) each valence electron pair of the solvent molecules can be localized in bond and lone pair regions and (ii) each pair, owing to the thermal motion of the solvent molecules, will have the same probability to be found at any point of the solution not occupied by the solute. The resulting solvent density $P_{\text{pair}}(\mathbf{R}|\mathbf{r}_2;\mathbf{r}_1)$ (here $\mathbf{R}$ is the coordinate of the centroid of the localized orbital containing the pair in a reference frame fixed on the solute molecule) can be represented in terms of a Gaussian representation of localized orbitals; this yields the simple expression

$$G_{\text{rep}} = \alpha \int_{\mathbf{r} \notin C} d\mathbf{r} P_A(\mathbf{r}), \tag{3.2}$$

where α is a suitable constant defined by some selected properties of the solvent.

We note that in Eq. (3.2) the repulsion energy is proportional to the fraction of solute electrons outside the cavity; as a consequence the inclusion of this contribution provides an automatic confinement of the electronic cloud of the solute.

The continuum approach to the dispersion term has a much longer history than for the repulsion, and different procedures have been developed.

The original formal theory is expressed in terms of quantum electrodynamics and the continuum medium characterized by its spectrum of complex dielectrics frequencies. A successive formulation derived from this theory, is based on the extension of the reaction field concept to a dipole subject to fluctuations exclusively electric in origin. In this framework, the dispersion free energy of a molecule placed in a cavity immersed in the solvent is related to the molecular polarizability and the complex dielectric constant. A couple of methods now in use (RINALDI, COSTA-CABRAL and RIVAIL [1986], AGUILAR and OLIVARES DEL VALLE [1986]), have extended this treatment, in origin limited to the dipolar approximation and a spherical cavity, to a full multipolar expansion and to any cavity shape. Actually, the two methods are quite different either in the solvation model they exploit or in the practical implementation; anyway an important aspect is common to both: the inclusion of the dispersive term in the solute effective Hamiltonian which allows a real influence of these interaction forces on the electronic structure of the solvated molecule (i.e., on the description of its wavefunction). In 1997 an alternative procedure (AMOVILLI and MENNUCCI [1997]) has been formulated starting, as for the repulsion contribution, from the theory of intermolecular forces. In this framework, the expression of the dispersion energy between two molecular systems A and B is given in terms of generalized frequency dependent polarizabilities. Following a scheme commonly exploited to derive the electrostatic contribution to the interaction energy, the molecule B is substituted by a continuum medium, the solvent, described by a surface charge density $\sigma_B[\varepsilon_B(i\omega),$ $P_A(0K|\mathbf{r})]$ induced by the electric field of the solute A and spreading on the cavity surface. In defining this surface charge density the transition density $P_A(0K|\mathbf{r})$ for the solute A (for transition to state K) has to be computed as well as the solvent dielectric constant calculated at imaginary frequencies, ε_B $(i\omega)$ has to be known. The reduction of the general expression to an equation containing only terms related the solute ground state can be obtained through some simplifications on the form of σ_B and the nature of the excited states to be considered.

3.2. Electrostatic contributions

The electrostatic interactions between the charge distributions which compose the molecular system (point nuclei and electronic cloud in quantum chemistry, points charges and multipoles in molecular mechanics) are affected by the presence of the dielectric. Indeed, in solvation continuum models in which the molecular system is placed in a volume (called molecular cavity) surrounded by a continuum dielectric, the interaction energy between the two charge distributions ρ_1 and ρ_2 reads

$$E_s(\rho_1, \rho_2) = \int_{\mathbb{R}^3} \rho_1 V_2 = \int_{\mathbb{R}^3} \rho_2 V_1 = \int_{\mathbb{R}^3} \varepsilon \nabla V_1 \cdot \nabla V_2,$$

where the electrostatic potential V_k generated by ρ_k satisfies

$$-\mathrm{div}(\varepsilon(x) \nabla V_k(x)) = 4\pi \rho_k(x) \tag{3.3}$$

with $\varepsilon(x) = 1$ inside the cavity Ω and $\varepsilon(x) = \varepsilon$ outside (ε denotes the macroscopic dielectric constant of the solvent).

It is useful to decompose the potential V in the sum of the electrostatic potential

$$\phi := \rho \star \frac{1}{|x|}$$

generated by the charge distribution ρ in vacuo and the "reaction potential"

$$V^r := V - \phi.$$

If we indicate as $G(x,y) = 1/|x-y|$ the Green kernel of the operator $-(1/4\pi)\Delta$, $G^s(x,y)$ the Green kernel of the operator $-(1/4\pi)\text{div}(\varepsilon \nabla \cdot)$ and $G^r(x,y) := G^s(x,y) - G(x,y)$, the following relations yield

$$V(x) = \int_{\mathbb{R}^3} G^s(x,y)\rho(y)dy,$$

$$\phi(x) = \int_{\mathbb{R}^3} G(x,y)\rho(y)dy,$$

$$V^r(x) = \int_{\mathbb{R}^3} G^r(x,y)\rho(y)dy.$$

Following this scheme the total interaction energy E_s (ρ_1, ρ_2) is usually split into two terms

$$E_s(\rho_1, \rho_2) = \mathcal{D}(\rho_1, \rho_2) + E_r(\rho_1, \rho_2),$$

where $\mathcal{D}(\rho_1, \rho_2)$ and $E_r(\rho_1, \rho_2)$, respectively, denote the interaction energy in the vacuum, and the so-called reaction-field contribution to the energy

$$\mathcal{D}(\rho_1, \rho_2) = \int_{\mathbb{R}^3}\int_{\mathbb{R}^3} \frac{\rho_1(x)\rho_2(y)}{|x-y|}\,dxdy,$$

$$E^r(\rho_1, \rho_2) := \int_{\mathbb{R}^3}\rho_1 V_2^r = \int_{\mathbb{R}^3}\rho_2 V_1^r = \int_{\mathbb{R}^3}\int_{\mathbb{R}^3}\rho_1(x)G^r(x,y)\rho_2(y)dxdy.$$

3.3. Beyond the standard continuum model

The electrostatic problem of charge distributions embedded in a cavity surrounded by a continuum dielectric can be extended from the standard homogeneous isotropic dielectrics, characterized by a constant scalar permittivity, ε to more complex systems such as homogeneous anisotropic dielectrics, characterized by a constant tensorial permittivity, ε, systems formed by two or more different dielectrics separated by well-defined boundaries, and solutions in which the dissolved electrolyte charges are free to move in the surrounding medium.

3.3.1. Anisotropic solvent

To extend the continuum model defined by Eq. (3.3) to anisotropic solvents in which the dielectric permittivity is represented by a tensor, we have to solve the anisotropic Poisson equation

$$-\text{div}(\varepsilon(x)\nabla V(x)) = 4\pi\rho(x). \tag{3.4}$$

The dielectric permittivity $\varepsilon(x)$ is no more a scalar quantity but a 3×3 tensor of the type

$$\varepsilon(x) = \begin{cases} \mathbf{I}_3 & \text{if } x \in \Omega, \\ \varepsilon_s & \text{if } x \in \mathbb{R}^3 \backslash \bar{\Omega}. \end{cases}$$

($\mathbf{I}_3$ is here the unity 3×3 tensor.) For physical reasons the tensor ε is symmetric. The model of anisotropic continuum well represent liquid crystalline phases and crystal matrices

3.3.2. Ionic solutions

The ionic solutions (i.e., solutions in which electrolyte charges are dissolved) well represent biological environments and in particular physiological solutions. An ionic solution can be described through a continuum model if the Poisson equation (3.3) is replaced by the (nonlinear) Poisson–Boltzmann equation:

$$-\text{div}(\varepsilon(x)\nabla V(x)) + \varepsilon(x)\kappa^2(x)kT \sinh(\mathbf{V}(x)/kT) = 4\pi\rho(\mathbf{x}) \tag{3.5}$$

with

$$\varepsilon(x) = \begin{cases} 1 & \text{if } x \in \Omega, \\ \varepsilon_s & \text{if } x \in \mathbb{R}^3 \backslash \bar{\Omega}, \end{cases}$$

and

$$\kappa(x) = \begin{cases} 0 & \text{if } x \in \Omega, \\ \kappa_s & \text{if } x \in \mathbb{R}^3 \backslash \bar{\Omega}. \end{cases}$$

The constant κ accounts for the ion screening: its inverse is known as the Debye length.

The hyperbolic sine function in Eq. (3.5) derives from a treatment of statistical physics in the thermodynamical equilibrium approximation, namely it represents the distribution of the mobile ions in the field of the nonlinear electrostatic potential V. If we represents such function in terms of a Taylor expansion and we take the linear term only, we obtain the so-called linearized PB equation (LPB).

$$-\text{div}(\varepsilon(x)\nabla V(x)) + \varepsilon(x)\kappa^2(x)V(x) = 4\pi\rho(x).$$

The LPB equation is a result of the Debye–Hückel approximation applicable in the case of low potentials, a condition approached at low concentrations. For electrolytes of high-valence type and for electrolytes in media of low dielectric constant, the same approximation becomes not good.

4. Computation of the electrostatic contribution

Let us now focus on the calculation of the reaction field interaction energy

$$E^r(\rho, \rho') = \int_{\mathbb{R}^3} \rho' V^r$$

with $V^r = V - \phi$ and $\phi = \rho \star \frac{1}{|x|}$, V denoting the unique solution to Eq. (3.3) zeroing at infinity. The problem whose V^r is solution has the following characteristics: (i) it is posed on $\mathbb{R}^3$; (ii) it exhibits an interface; (iii) inside (resp. outside) the cavity the partial differential equation is linear and the differential operator has constant coefficients. Those three characteristics make it natural to resort to an integral method: the original three-dimensional problem (3.3) posed on an unbounded domain ($\mathbb{R}^3$) can thus be replaced by a two-dimensional problem posed on a bounded manifold (the interface $\Gamma = \partial\Omega$).

4.1. Basics on integral equations

For the reader convenience, we state here some basic results on integral equations which are used below. For more details, the reader is referred to HACKBUSH [1995], NÉDÉLEC and PLANCHARD [1973], NÉDÉLEC [1994]. Let us consider a function V satisfying

$$\begin{cases} -\Delta V = 0 & \text{in } \Omega, \\ -\Delta V = 0 & \text{in } \mathbb{R}^3 \backslash \bar{\Omega}, \\ V \to 0 & \text{at infinity,} \end{cases}$$

and whose interior $(V_i, \frac{\partial V}{\partial n}|_i)$ and exterior $(V_e, \frac{\partial V}{\partial n}|_e)$ traces on $\Gamma = \partial\Omega$ are well-defined and continuous. Denoting by

$$[V] := V_i - V_e \quad \text{and} \quad \left[\frac{\partial V}{\partial n}\right] := \frac{\partial V}{\partial n}\Big|_i - \frac{\partial V}{\partial n}\Big|_e,$$

the following *representation formulae* can be stated: the function V satisfies for any $x \notin \Gamma$,

$$V(x) = \int_\Gamma \frac{1}{4\pi|x-y|} \left[\frac{\partial V}{\partial n}\right](y)\mathrm{d}y - \int_\Gamma \frac{\partial}{\partial n_y}\left(\frac{1}{4\pi|x-y|}\right)[V](y)\mathrm{d}y \tag{4.1}$$

and for any $x \in \Gamma$,

$$\frac{V_i(x) + V_e(x)}{2} = \int_\Gamma \frac{1}{4\pi|x-y|}\left[\frac{\partial V}{\partial n}\right](y)\mathrm{d}y - \int_\Gamma \frac{\partial}{\partial n_y}\left(\frac{1}{4\pi|x-y|}\right)[V](y)\mathrm{d}y. \tag{4.2}$$

For $x \in \Gamma$, one has in addition

$$\frac{1}{2}\left(\frac{\partial V}{\partial n}\Big|_i + \frac{\partial V}{\partial n}|_e\right)(x) = \int_\Gamma \frac{\partial}{\partial n_x}\left(\frac{1}{4\pi|x-y|}\right)\left[\frac{\partial V}{\partial n}\right](y)\mathrm{d}y \tag{4.3}$$

$$-\int_\Gamma \frac{\partial^2}{\partial n_x \partial n_y}\left(\frac{1}{4\pi|x-y|}\right)[V](y)\mathrm{d}y. \tag{4.4}$$

The two last relations suggest to introduce the operators S, D, D^* and N formally defined for $\sigma : \Gamma \to \mathbb{R}$ and $x \in \Gamma$ by

$$(S\cdot\sigma)(x) = \int_\Gamma \frac{1}{|x-y|}\sigma(y)\mathrm{d}y, \tag{4.5}$$

$$(D \cdot \sigma)(x) = \int_\Gamma \frac{\partial}{\partial n_y} \left(\frac{1}{|x-y|} \right) \sigma(y) dy, \tag{4.6}$$

$$(D \cdot \sigma)(x) = \int_\Gamma \frac{\partial}{\partial n_x} \left(\frac{1}{|x-y|} \right) \sigma(y) dy, \tag{4.7}$$

$$(N \cdot \sigma)(x) = \int_\Gamma \frac{\partial^2}{\partial n_x \partial n_y} \left(\frac{1}{|x-y|} \right) \sigma(y) dy. \tag{4.8}$$

When the surface Γ is regular (C^1 at least), the Green kernels of the operators S, D and D^* present integrable singularities on the surface Γ; it is easy to check that they behave as $1/|x-y|$ when y goes to x (then $|(x-y) \cdot n_x| \sim |(x-y) \cdot n_y| \sim |x-y|^2$ when y is close to x). On the other hand, the Green kernel of the operator N is hyper-singular (it behaves as $1/|x-y|^3$ when y is close to x) so that the notation (4.8) is only formal: even for regular σ, the integral $\int_\Gamma \frac{\partial^2}{\partial n_x \partial n_y} \left(\frac{1}{|x-y|} \right) \sigma(y) dy$ has to be given a sense of Cauchy principal value.

The operators S, D, D^* and N satisfy the following relations (which follow from the properties of Calderon operators (HACKBUSH [1995])): (i) the operators S and N are self-adjoint on $L^2(\Gamma)$ and D^* is the adjoint of D; (ii) one has $DS = SD^*$, $DN = ND^*$, $D^2 - SN = 4\pi^2$, $D^{*2} - NS = 4\pi^2$.

Let us end up this section with the definition of the so-called single-layer and double-layer potentials.

A *single-layer potential* is a function V which can be written as

$$V(x) = \int_\Gamma \frac{\sigma(y)}{|x-y|} dy, \quad \forall x \in \mathbb{R}^3,$$

with $\sigma \in H^{-1/2}(\mathbb{R}^3)$. A single-layer potential is well-defined and continuous on $\mathbb{R}^3$ (in particular $[V] = 0$). Its normal derivative presents a discontinuity at the crossing of the interface Γ given by the formula

$$\left[\frac{\partial V}{\partial n} \right] = \frac{\partial V}{\partial n} \Big|_i - \frac{\partial V}{\partial n} \Big|_e = 4\pi \sigma.$$

The density σ is solution to the integral equation on Γ

$$S \cdot \sigma = V.$$

A *double-layer potential* is a function V which can be written as

$$V(x) = \int_\Gamma \frac{\partial}{\partial n_y} \left(\frac{1}{|x-y|} \right) \sigma(y) dy, \quad \forall x \in \mathbb{R}^3,$$

with $\sigma \in H^{-1/2}(\mathbb{R}^3)$. A double layer potential is continuous on $\mathbb{R}^3$ Γ but presents a discontinuity at the crossing of the interface Γ given by

$$[V] = V|_i - V|_e = 4\pi \sigma.$$

On the other hand, its normal derivative is continuous at the crossing of Γ. The density σ is solution to the integral equation on Γ

$$N{\cdot}\sigma = -\frac{\partial V}{\partial n}.$$

4.2. Numerical aspects

Usually, integral equations arising in potential theory are numerically solved by either a *collocation* or a *Galerkin* method; in the latter case, boundary elements are most often used. Let us detail both methods on the example of a linear integral equation

$$A{\cdot}\sigma = g, \tag{4.9}$$

where the unknown σ belongs to $H^s(\Gamma)$, the right hand side g is in $H^{s\prime}(\Gamma)$, and the integral operator $A \in \mathcal{L}(H^s(\Gamma), H^s(\Gamma))$ is characterized by the Green kernel $a(x, y)$:

$$(A{\cdot}\sigma)(x) = \int_\Gamma a(x, y)\sigma(y)\mathrm{d}y, \quad \forall x \in \Gamma.$$

Let us consider a mesh $(T_i)_{1 \leqslant i \leqslant n}$ on Γ, that will be considered as drawn on the curved surface Γ; let us denote by x_i a representative point of the element T_i (e.g., its "center"). Theoretical and numerical details on the methodologies developed to determine surface meshes will be given in Section 6.

The P_0 collocation and Galerkin methods for solving Eq. (4.9) provide two approximations of σ in the space V_h of piecewise constant functions whose restriction to each element T_i is constant: in the collocation method, σ^c is the solution to

$$\int_\Gamma a(x_i, y)\sigma^c(y)\mathrm{d}y = g(x_i), \quad \forall 1 \leqslant i \leqslant n;$$

whereas in the Galerkin method, σ^g satisfies

$$\forall_\tau \in V_h, \quad \langle A{\cdot}\sigma^g, \tau \rangle_\Gamma = \langle g, \tau \rangle_\Gamma.$$

These two methods, respectively, lead to the matrix equations

$$[A]^c{\cdot}[\sigma]^c = [g]^c \quad \text{and} \quad [A]^g{\cdot}[\sigma]^g = [g]^g,$$

where

$$[A]_{ij}^c = \int_{T_j} a(x_i, y)\mathrm{d}y, \quad [g]_i^c = g(x_i),$$

$$[A]_{ij}^g = \int_{T_j} \int_{T_j} a(x, y)\mathrm{d}x\,\mathrm{d}y, \quad [g]_i^g = \int_{T_i} g,$$

$[\sigma]_i^c$ and $[\sigma]_i^g$ denoting, respectively, the values of σ on T_i under the collocation and Galerkin approximations. The collocation method is more natural and easier to implement (at least at first sight); for these reasons, it is often used by Chemists; on the other hand, the Galerkin method leads to a *symmetric* linear system when the operator A is itself symmetric, which may appreciably simplify the numerical resolution of the linear system (QUARTERONI and VALLI [1997]).

The BEM follows the standard FEM. The only difference proceeds from the fact that in usual applications of the FEM, the operator is *local* (typically a Laplacian) whereas it is *nonlocal* in most applications of the BEM. Consequently, the stiffness matrix [A] is generally sparse for FEM and full for BEM.

In many applications, the surface Γ is partitioned in small patches called tesserae T_i. This approximation renders easier the computation of the coefficients of the stiffness matrices

$$[S]_{ij} = \int_{T_i}\int_{T_j} \frac{1}{|x-y|}\,\mathrm{d}x\mathrm{d}y \quad \text{and} \quad [D]_{ij} = \int_{T_i}\left(\int_{T_j}\frac{\partial}{\partial n_y}\left(\frac{1}{|x-y|}\right)\mathrm{d}y\right)\mathrm{d}x.$$

In the Galerkin approximation, the exterior integration can be performed with an adaptive Gaussian integration method, the number of integration points depending on the distance and on the relative orientation of the elements T_i and T_j.

4.3. The case of an interior charge

Let us recall that in the standard case, the external medium is modelled by a homogeneous and isotropic dielectric whose dielectric constant ε_s equals the macroscopic permittivity of the solvent. Let us denote by ρ and ρ' two charge distributions of the generic form $\sum_{k=1}^{M} q_k \delta_{\bar{x}_k} - \rho_{\mathcal{D}}$ with $\bar{x}_k \in \Omega$ and $\rho_{\mathcal{D}} \in C^{0,1}(\bar{\Omega})$. By extension, the duality brackets for which the integrals below make sense are also denoted by the symbol $\int$.

Our goal is to compute the energy

$$E^r(\rho,\rho') = \int_{\mathbb{R}^3} \rho' V^r,$$

where the reaction potential V^r generated by ρ is uniquely defined by

$$V^r := V - \phi, \quad -\nabla(\varepsilon(x)\nabla V(x)) = 4\pi\rho(x), \quad -\Delta\phi = 4\pi\rho,$$

with $\varepsilon(x) = 1$ inside the cavity Ω and $\varepsilon(x) = \varepsilon_s$ in the external domain $\mathbb{R}^3\backslash\bar{\Omega}$. It is easy to check that V^r is C^2 in $\bar{\Omega}$ and in $\mathbb{R}^3\backslash\Omega$ and satisfies

$$\begin{cases} -\Delta V^r = 0 & \text{in}\,\Omega, \\ -\Delta V^r = 0 & \text{in}\,\mathbb{R}^3\backslash\bar{\Omega}, \\ [V^r] = 0 & \text{on}\,\Gamma, \\ V^r \to 0 & \text{at infinity.} \end{cases}$$

The representation formulae (4.1) and (4.2) therefore enable ones to write down the reaction potential V^r as a single layer potential

$$V^r(x) = \int_{\Gamma} \frac{\sigma(y)}{|x-y|}\,\mathrm{d}y, \quad \forall x \in \mathbb{R}^3,$$

with $\sigma = \frac{1}{4\pi}\left[\frac{\partial V^r}{\partial n}\right]$. In order to get the apparent surface charge distribution σ, suffices it to use the relations

$$\left.\frac{\partial V^r}{\partial n}\right|_i - \left.\frac{\partial V^r}{\partial n}\right|_e = 4\pi\sigma,$$

$$\frac{1}{2}\left[\left.\frac{\partial V^r}{\partial n}\right|_i + \left.\frac{\partial V^r}{\partial n}\right|_e\right] = D^* \cdot \sigma$$

and the jump condition on the interface Γ

$$0 = \left.\frac{\partial V^s}{\partial n}\right|_i - \varepsilon_s\left.\frac{\partial V^s}{\partial n}\right|_e \tag{4.10}$$

$$= \left.\frac{\partial V^r}{\partial n}\right|_i - \varepsilon\left.\frac{\partial V^r}{\partial n}\right|_e + (1-\varepsilon)\frac{\partial \phi}{\partial n}, \tag{4.11}$$

which lead by simple algebraic manipulations to the integral equation

$$\left(2\pi\frac{\varepsilon_s + 1}{\varepsilon_s - 1} - D^*\right) \cdot \sigma = \frac{\partial \phi}{\partial n}. \tag{4.12}$$

It can be shown that a solution σ of Eq. (4.12) exists and it is unique.

This technique for calculating the reaction potential, based on integral equations, is referred in chemistry by the acronym ASC (apparent surface charge). It is by far the most often used method. In particular, the expression (4.12) is the one used in the original version of the PCM method introduced in the previous sections. The energy $E^r(\rho, \rho')$ can then be obtained in the following way

$$E^r(\rho, \rho') = \int_{\mathbb{R}^3}\rho' V^r = \int_{\mathbb{R}^3}\rho(x)\left(\int_\Gamma \frac{\sigma(y)}{|x-y|}\,dy\right)dx$$

$$= \int_\Gamma \sigma(y)\left(\int_{\mathbb{R}^3}\frac{\rho'(x)}{|x-y|}\,dx\right)dy = \int_\Gamma \sigma\phi'$$

with $\phi' = \rho' \star \frac{1}{|x|}$.

This method presents a difficulty as the electronic charge distribution is not entirely supported in Ω (the electrons are delocalized in the whole space $\mathbb{R}^3$). Rigorously speaking, it is not possible to used the technique detailed above to compute this term. However, as the part of the electronic cloud which lays outside the cavity is generally small, an accurate approximation of this interaction term can be obtained by a slightly more sophisticated integral method.

4.4. The escaped charge problem

In the derivation of Eq. (4.12) we have assumed that ρ is supported inside the cavity, in this case the interaction energy $E^r(\rho, \rho')$ equals the exact energy; otherwise the two quantities differ. Besides, if ρ (or ρ') is not supported inside the cavity, the symmetry property $E^r(\rho, \rho') = E^r(\rho', \rho)$ is broken: this is an important problems of ASC continuum models when coupled to QM calculations. In QM in fact, although most of the electronic density ρ_{el} lays inside the cavity, electronic tails always spread outside the cavity: this is the so-called *escaped charge*. When this happens, integral equations methods like PCM can only provide an approximation of the reaction-field energy; an exact computation requires a 3D calculation. Whereas the escaped charge is usually small (a fraction of atomic units), it can nevertheless affect the evaluation of energies and, even more significantly, of properties of the solute. For this reason, it is important to take into account the escaped charge in quantum chemistry calculations.

As reported in the previous sections, the PCM method has been reformulated in a new version known as IEF (CANCÈS and MENNUCCI [1998b], CANCÈS, MENNUCCI and TOMASI [1997], MENNUCCI, CANCÈS and TOMASI [1997]). IEF method can be seen as an improvement of the standard PCM method; in particular it appears to be more efficient in the way the escaped charge is accounted for. The IEF energy reads

$$E_r^{IEF}(\rho, \rho') = \int_\Gamma \sigma^{IEF} \phi', \tag{4.13}$$

where σ^{IEF} is the solution to the integral equation

$$\left(2\pi \frac{\varepsilon+1}{\varepsilon-1} - D\right) S\sigma^{IEF} = -(2\pi - D)\phi.$$

Contrary to the PCM energy, the IEF energy E_r^{IEF} is actually symmetric in the two arguments ρ and ρ' since, denoting by $\langle \cdot, \cdot \rangle_\Gamma$ the scalar product on Γ

$$\int_\Gamma \sigma^{IEF} \phi' = \langle L\phi, \phi' \rangle_\Gamma = \langle \phi, L\phi' \rangle_\Gamma = \int_\Gamma \sigma'^{IEF} \phi$$

the self-adjointness of the linear operator

$$L = \left[\left(2\pi \frac{\varepsilon+1}{\varepsilon-1} - D\right) S\right]^{-1} (-2\pi + D)$$

being a consequence of the commutation property $DS = SD^*$.

4.5. Anisotropic dielectrics and ionic solutions

In Section 3 we have presented the extension of the standard electrostatic problem of a charge distribution immersed in a continuum dielectric to more complex systems such as anisotropic dielectrics and ionic solutions; here we present the solutions of the corresponding problems. In these cases the Green kernels to be defined are, respectively:

$$G_e(x, y) = \begin{cases} \left(\sqrt{\det\varepsilon}\right)^{-1} \left[(\varepsilon^{-1}(x-y))\cdot(x-y)\right]^{-1/2} & \text{(anisotropic)} \\ \exp(-\kappa|x-y|)(\varepsilon|x-y|)^{-1} & \text{(ionic solution)} \end{cases} \quad \text{if } x \notin C,$$

where C indicates the cavity.

In a parallel way as that we have used to define the operators (4.5)–(4.8) we can define two other operators, S_e and D_e, by replacing G with the corresponding function G_e defined in the outer space. In this case the derivative operator means ∂_y $G_e\,(x,\,y) = (\varepsilon \cdot \nabla_y G_e(x,\,y)) \cdot n(y)$, where the dielectric matrix reduces to the scalar dielectric constant for ionic solutions.

For brevity's sake, here we do not report the formal derivation that, exploiting specific characteristics of these operators, leads to the definition of the surface charge σ, but we only say the latter is the unique solution to the equation:

$$A \cdot \sigma = -g \tag{4.14}$$

where:

$$A = (2\pi - D_e)S_i + S_e(2\pi + D_i), \tag{4.15}$$

$$g = (2\pi - D_e)\phi + S_e\frac{\partial \phi}{\partial n} \tag{4.16}$$

being I the unit operator.

5. Geometry optimization

As shown in the previous section the electrostatic interactions between the charge distributions which compose the molecular system are affected by the presence of the dielectric. Indeed, in solvation continuum models, the interaction energy between two charge distributions ρ_1 and ρ_2 can be redefined as the interaction energy we obtain in vacuo for a new system composed by ρ_1, ρ_2, and the further apparent charge $\rho_1{}^a$ whose density is the surface density σ we have defined in the previous section; namely we have:

$$E(\rho_1(\lambda), \rho_2(\lambda)) = \mathcal{D}(\rho_1, \rho_2) + \mathcal{D}(\rho_1^a, \rho_2), \tag{5.1}$$

where

$$\mathcal{D}(\rho, \rho') = \iint\limits_{\mathbb{R}^3 \times \mathbb{R}^3} \left[\frac{\rho(x)\rho'(y)}{|x - y|}\right] dxdy$$

denotes the interaction energy between any charge distributions ρ and ρ' in vacuo. To preserve all generality, we have considered that the charge distributions ρ_1, ρ_2, depend on n real parameters $(\lambda_1, \ldots, \lambda_n)$ and that the cavity also depends on the λ_i. We note in a geometry optimization algorithm the set of parameters λ_i represent the nuclear coordinates of the atoms which constitute the molecular system.

Passing now to the partial derivatives $\partial E/\partial \lambda_i$, in standard solvation methods the usual way to proceed is the following. It consists in differentiating the interaction energy expression (5.1), which leads to

$$\frac{\partial}{\partial \lambda_j}(E(\rho_1(\lambda), \rho_2(\lambda))) = \frac{\partial}{\partial \lambda_j}\mathcal{D}(\rho_1, \rho_2) + \mathcal{D}\left(\rho_1^a, \frac{\partial \rho_2}{\partial \lambda_j}\right) + \mathcal{D}\left(\frac{\partial \rho_1^a}{\partial \lambda_j}, \rho_2\right)$$

and in computing the derivative $\partial\rho_1^a(\lambda)/\partial\lambda_j$ (the only term for which problems may arise) by differentiating the basic integral equation (4.12) after approximation by the BEM quoted above. Recently we have suggested an alternative way to proceed (CANCÈS and MENNUCCI [1998a], CANCÈS, MENNUCCI and TOMASI [1998]), which in particular avoids computing any of the derivatives $\partial\rho_k^a(\lambda)/\partial\lambda_j$. Globally speaking, the new approach consists in deriving first the electrostatic equation so as to write $\partial V_k/\partial\lambda$ as a solution of the differentiated equation, and next inserting it in the derivative of E.

In this scheme, we obtain the following derivative formula:

$$\frac{\partial E}{\partial\lambda_j}(\lambda) = \left\langle \frac{\partial\rho_1}{\partial\lambda_j}(\lambda), V_2(\lambda) \right\rangle + \left\langle \frac{\partial\rho_2}{\partial\lambda_j}(\lambda), V_1(\lambda) \right\rangle + \frac{1}{4\pi}\int_{\Gamma(\lambda)} \tau(\lambda)(U_{\Sigma(\lambda)}^j \cdot n_{\Sigma(\lambda)}) \quad (5.2)$$

with

$$\tau(\lambda) = \frac{16\pi^2\varepsilon}{\varepsilon - 1}\sigma_1(\lambda)\sigma_2(\lambda) + (\varepsilon - 1)(\nabla V_1(\lambda))_{\|}(\nabla V_2(\lambda))_{\|}. \quad (5.3)$$

In Eq. (5.3) $(\nabla V_k(\lambda))_{\|}$ is the projection of $\nabla V_k(\lambda)$ on Γ, whereas in Eq. (5.2) we denote with U_Γ^j the derivative of the interface Γ when λ varies. This quantity assumes a very simple form for standard cavities given as union of spheres, each of them centered on a solute nucleus.

Each of the three terms on the r.h.s. of (5.2) has a clear meaning: the first two are due to the variations of the charges ρ_1 and ρ_2, respectively, while the third one comes from the deformation of the cavity. From a numerical point of view, the second term of (5.3) is not easy to deal with. Fortunately, its contribution appears to be small in practical cases and can therefore be neglected. This behavior can be formally proved only for smooth cavities and for dielectrics with high permittivities, but numerical tests performed so far have shown that this is still true for all practical cases encountered in chemistry, even for low dielectric permittivity ($\varepsilon \simeq 2$). In view of these arguments, Eq. (5.2) can be always reduced to the simplified formula

$$\frac{\partial E}{\partial\lambda_j} \simeq \left\langle \frac{\partial\rho_1}{\partial\lambda_j}, V_2 \right\rangle + \left\langle \frac{\partial\rho_2}{\partial\lambda_j}, V_1 \right\rangle + \frac{4\pi\varepsilon}{\varepsilon - 1}\int_{\Gamma} \sigma_1\sigma_2(U_\Gamma^j \cdot n_\Sigma).$$

6. Molecular surface meshing

In the solvation continuum methods presented in this chapter, the solute is modeled by a domain Ω, called *molecular cavity* or *molecular volume*, in the tridimensional space $\mathbb{R}^3$. Its boundary $\Gamma = \partial\Omega$ is referred to as a *molecular surface*. As shown before, physical problems can be formulated in terms of partial differential equations (PDE) and solved using for instance the FEM or the BEM. These methods are based on a spatial discretization, or *mesh*, of the surface Γ (and also the volume Ω in some cases). It is thus necessary to construct a partition of the molecular surface Γ in geometrically simple elements $\{T_i\}$ such as triangles or quadrilaterals, sometimes called herein *tesserae*.

The generation of such a mesh raises several issues that are considered in this section. First of all, a geometric definition of a molecular surface must be given precisely, in a way which makes sense in computational chemistry. This is detailed in the first subsection which shows that, in fact, different kinds of surfaces can be defined for a given molecule: VWS, SAS and SES. Then, a boundary representation of this molecular surface must be obtained, so that a mesh generator can use this model as a geometric support (see the second subsection for VWS and SAS, and the third subsection for SES). Finally, a surface mesh must be generated to simulate the chemical phenomena numerically. In such a context of numerical simulation, it is now clearly established that mesh quality has a strong influence on solution accuracy, as also convergence and speed of the computing scheme. Consequently, it is essential that the mesh elements are as regular as possible (i.e., almost equilateral), while conforming to a user-specifiable size map and closely approximating the surface (see the fourth subsection). To show some applications of the methods presented, several examples of molecular surface meshes are provided in the fifth subsection, and a brief conclusion is given at the end.

6.1. Geometric definitions of molecular surfaces: VWS, SAS, SES

To model a molecule, the basic idea is to assimilate each constituting atom to a ball B_i whose size is determined by its Van der Waals radius. Let $\mathcal{B} = \bigcup B_i$ be the union of all these possibly overlapping balls. Then, it is possible to define different kinds of surfaces, the most commonly used being the Van der Waals surface (VWS), the solvent-accessible surface (SAS) and the solvent-excluded surface (SES), as explained in a review by CONNOLLY [1996].

The VWS is simply the boundary of the union $\mathcal{B}$ (see Fig. 6.1).

The SAS, as introduced by LEE and RICHARDS [1971], involves a sphere of radius r_p, called the *probe*, which represents a single solvent molecule, for instance a water molecule. When the probe sphere rolls on the VWS, the locus of its center defines the SAS. In fact, this definition amounts to the previous one, by increasing the radius of each atom by the constant value r_p (see Fig. 6.2).

FIG. 6.1 Van der Waals surface (VWS).

FIG. 6.2 Solvent-accessible surface (SAS).

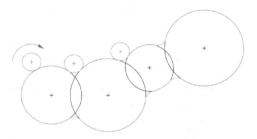

FIG. 6.3 Bidimensional diagram showing the SES of a simplified 4-atom molecule.

The SES is made up of two parts, the "contact surface" and the "reentrant surface".
Figure 6.3 shows a bidimensional diagram where the probe (red circle) rolls on a
simplified 4-atom molecule, tracing the contact surface (green arcs) and the reentrant
surface (cyan arcs). In three dimensions, the contact surface is the part of the VWS
that can be touched by the probe sphere, following an idea which was introduced
by RICHMOND and RICHARDS [1978]. The reentrant surface, as introduced by
RICHARDS [1977], is obtained when the probe sphere is in contact with two atoms
or more. As shown on Fig. 6.4, the SES consists of spherical patches lying either
on the atoms (red) or on the probe (green), and toroidal patches defined when the
probe rolls over a pair of atoms (cyan). As we will see later, the SES is G^1 continuous
(i.e., not folded) in most cases, depending on the probe size, and this is an important
property for many computation models. In the literature, the SES is also called
"smooth molecular surface" or "Connolly surface".

At present, a molecular surface being given, it is necessary to obtain its boundary
representation (B-rep), that is, a set of patches and the topological relations between
them. The next subsections deal firstly with VWS and SAS (containing spherical
patches only), and secondly with SES. (These descriptions, as well as mesh generation
methods, can also be found in LAUG and BOROUCHAKI [2001].)

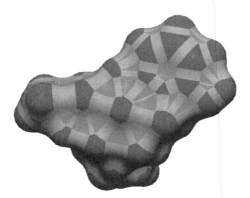

FIG. 6.4 Solvent-excluded surface (SES).

6.2. Boundary representation of VWS and SAS

In the case of *Van der Waals* or *solvent-accessible* surfaces (VWS or SAS), the problem can be mathematically stated as follows:

Given a set of arbitrary spheres $\{S_i\}$ *(several spheres may intersect), where each sphere* S_i *is defined by its center* (x_i, y_i, z_i) *and its radius* r_i, *determine the envelope E of the union of these spheres.* By definition, the envelope E is the topological boundary of the union of balls $\mathcal{B} = \bigcup B_i$, where each ball B_i is bounded by the sphere S_i.

To allow for meshing, the envelope E will be determined in the form of a composite parametric surface, i.e., the union of several patches in the 3D space. Two different patches may have a common boundary curve, called an *interface curve*, or may be totally disjoint. Each patch is the image of a planar domain, called a *parametric space*, which is defined from its boundary curves. A single 3D interface curve may be the image of several planar boundary curves.

To give a parametric representation of the envelope E, *first* its interface curves can be obtained by using the following algorithm:

(1) Compute the intersections of spheres $\{S_i\}$ two by two, giving a set of circles $\{C_j\}$.

(2) Compute the intersections of circles $\{C_j\}$ two by two, giving a set of points $\{P_k\}$.

(3) The different points $\{P_k\}$ on a circle C_j define a partition of the circle into a set of arcs. Extract the set of arcs that are outside the spheres, giving the desired interface curves.

Second, mapping functions $\{\sigma_i\}$ on parametric spaces $\{\Omega_i\}$, whose images form the envelope E, are defined. An efficient solution is to use the inverse function of a projection sometimes used in cartography.

Third, the projections of the interface curves of E give the boundaries of the parametric domains $\{\Omega_i\}$ in a 2D space. Here, the curve discretization plays an important role for the validity and shape quality of the planar mesh.

All the above steps are detailed in the following paragraphs.

6.3. Computing the intersections of spheres

We want to determine the circle representing the intersection of two spheres (S_1, S_2) with centers (C_1, C_2) and radii (r_1, r_2), if this circle exists.

Let us consider a plane containing the vertical axis C_1Z and passing through point C_2. In this plane, it is easy to find points A and B at the intersection of two circles with centers (C_1, C_2) and radii (r_1, r_2). Actually, if $AB < \varepsilon$, where ε is a given small value, it is generally preferable to ignore the intersection because of the floating point errors and to avoid a mesh which would be locally too fine.

The intersection of spheres S_1 and S_2 in $\mathbb{R}^3$ is a circle with diameter $[AB]$. The circle lies in a plane defined by $ax + by + cz + d = 0$, where $\overrightarrow{C_1C_2} = (a, b, c)$ is a normal vector. The molecule surface does not contain the part of sphere S_1 (resp. S_2) satisfying the inequality $ax + by + cz + d > 0$ (resp. <0), and this property will be used later (paragraph entitled "determining the external arcs").

6.4. Computing the intersections of circles

Let us consider two circles drawn on the same sphere with center O and radius r. Each circle lies in a plane whose normal vector is known. Let $\overrightarrow{v_1}$ and $\overrightarrow{v_2}$ be these normal vectors. Let us determine the intersection points of these two circles, if they exist.

The idea is to consider a plane whose normal vector is perpendicular to both $\overrightarrow{v_1}$ and $\overrightarrow{v_2}$. The projection of each circle on this plane reduces to a straight segment. It is then easy to compute their intersection. In practice, if the two circles are nearly intersecting, or if the intersection points are very close, only one intersection point can be retained, to avoid again a locally too fine mesh.

While computing the intersections of the circles, the same points should not be computed several times. Actually, if three spheres S_1, S_2 and S_3 intersect each other, their intersections circles are $C_1 = S_1 \cap S_2$, $C_2 = S_2 \cap S_3$ and $C_3 = S_3 \cap S_1$. These three circles define only two intersections points, for $C_1 \cap C_2 = C_2 \cap C_3 = C_3 \cap C_1 = S_1 \cap S_2 \cap S_3$ (by associativity).

Now, let us consider a given circle C_j and the set of intersection points $\{P_k\}$ belonging to it. Some close points can be eliminated directly because of the above remarks, but others may remain. For instance, Fig. 6.5 represents the envelope of four slightly shifted spheres with radius 1 and coplanar centers $(0, 0)$, $(1, 0)$, $(1, 1)$ and $(0, 0.999)$, showing (left) many triangles, generally small and distorted, at the center of the mesh. To avoid this problem, it is sufficient to sort the points on a given circle with respect to their angles. If the difference between two consecutive angles is less than a certain threshold (for instance, $0.3°$), the corresponding points are merged provided they remain close to their defining circles (see Fig. 6.5, right).

6.5. Determining the external arcs

Let us consider again a circle C_j and the set of intersection points $\{P_k\}$ belonging to it. This set defines a partition of circle C_j into several arcs. The set of the arcs that are not inside a sphere form the interface of all the spherical surfaces (see Fig. 6.6). To determine if an arc is inside or outside, the equations of the planes containing the circles are used (see above).

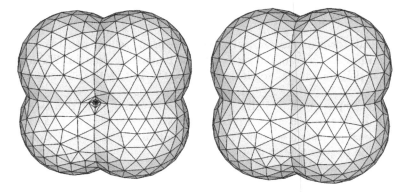

FIG. 6.5 Mesh without (*left*) and with (*right*) merging the closest points.

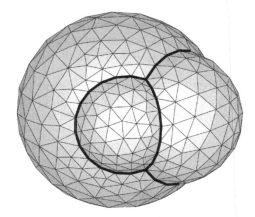

FIG. 6.6 External arcs at the intersection of three spheres.

More precisely, let us denote by $\{A_l\}$ the set of all the arcs found. To each arc A_l is associated a flag $A_l.ext$ with value T (true) if the arc is external and F (false) otherwise. To this end, the following algorithm written in pseudo-code can be used:

For each arc A_l
  $A_l.ext = T$
End
For each sphere S_i
  For each circle C_j lying on the sphere S_i
    (a, b, c, d) = coefficients of the plane containing C_j
    For each arc A_l
      If A_l is a part of C_j **then** cycle
      (x, y, z) = middle point of arc A_l
      If $ax + by + cz + d > 0$ **then** $A_l.ext = F$
    End
  End
End

6.6. *Parameterizing a spherical surface*

Each spherical surface must now be parameterized, which consists conversely in defining a projection from the surface to the parametric domain. We will obtain here rational polynomials, as for instance in BAJAJ, LEE, MERKERT and PASCUCCI [1997].

One of the simplest method derives from the *orthogonal projection* (see Fig. 6.7, point P_0). If $P = (x, y, z)$ is a point on a sphere with center $C = (0, 0, R)$ and radius R, its orthogonal projection is point $P_0 = (u, v)$ with $u = x$ and $v = y$. Conversely, a hemisphere can be parameterized by:

$$\sigma_0(u, v) = \begin{bmatrix} u \\ v \\ \sqrt{R^2 - u^2 - v^2} \end{bmatrix}.$$

The main drawback of this parameterization is that it can be highly unstable near the equator, for the partial derivatives become infinite. As an illustration, Fig. 6.8 (left) shows the image by the mapping function σ_0 of a uniform mesh on a plane disk Ω.

To avoid degenerate derivatives, it is more convenient to consider the *stereographic projection* (see Fig. 6.7, point P_s). If $N = (0, 0, 2R)$ is the sphere's "North Pole", the stereographic projection of point P is defined as the intersection P_s of the

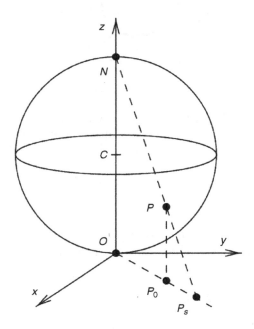

FIG. 6.7 Diagram of an orthogonal (P_0) and stereographic (P_s) projection.

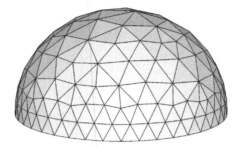

FIG. 6.8 *Left*: inverse orthogonal projection of a uniform mesh. *Right*: inverse stereographic projection of the same mesh.

straight line (NP) with the plane $z = 0$. We now have $P_s = (u, v)$ with $u = \frac{2Rx}{2R-z}$ and $v = \frac{2Ry}{2R-z}$. Conversely, the sphere without point N can be parameterized by:

$$\sigma_s(u, v) = \frac{2R}{u^2 + v^2 + 4R^2} \begin{bmatrix} 2Ru \\ 2Rv \\ u^2 + v^2 \end{bmatrix}.$$

This parameterization is stable near the equator, continuously differentiable, and preserves angles (but not distances). As a consequence, if the triangulation represents the geometry of the sphere accurately, any equilateral triangle in the parametric domain Ω is almost equilateral on the sphere, giving directly a tridimensional surface mesh with a good shape quality (see Fig. 6.8; right).

Since it is impossible to parameterize a whole sphere with only one domain, it can be divided into a pair of hemispheres when this case occurs. However, as far as molecular surfaces are concerned, only some parts of a sphere generally remain, hence only one domain is necessary (by choosing the sphere's "North Pole" outside these parts).

6.7. Defining and discretizing the bidimensional boundaries

We have already shown that the interface curves in 3D are circle arcs. Each arc is discretized conforming to a pre-specified size map. Finally, using a stereographic projection, a discretization of the bidimensional boundaries is obtained. However, this discretization may lead to an invalid definition of the bidimensional parametric domain, or give bad quality elements. It is thus necessary to check the crossing edges, the close edges and the adjacent edge lengths of the domain boundaries, as explained below.

    (a) *Checking crossing edges*. Consider the example of Fig. 6.9. The geometric definition of the boundary is represented by dashed curves, and the initial discretization of this support by its vertices (small circles) and its edges (thin solid lines). Some edges are intersecting each other, meaning that the discretized

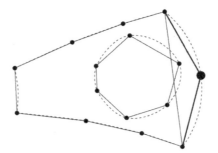

FIG. 6.9 Crossing edges.

boundary of the domain is not well defined. To rectify this, the longest crossing edges can be recursively subdivided (here, one larger circle and two thick solid lines). In fact, we present just below an algorithm which is more strict and produces a better mesh, at the cost of more calculations.

(b) *Checking close edges.* Figure 6.10 shows a bidimensional boundary whose edges are very close. In this case, the domain mesh, which must comply with the boundary discretization, would contain very flat triangles. To avoid this, the idea is to associate to each edge a rectangle (dashed lines on the figure). Its length is equal to the edge length, and its width is proportional to it. If the rectangle intersects other edges, then the corresponding edge is subdivided. This process is repeated until there is no intersection. In the case where two edges make a sharp angle, this algorithm must be adapted by defining a rectangle with a smaller width.

(c) *Checking adjacent edge lengths.* Once the boundaries have been discretized (initially or after the above correction), the lengths of two adjacent edges may be very different, leading to a bad mesh quality. For instance, Fig. 6.11 presents a partial mesh in its initial state (left) and after the previous correction (middle). Then, an iterative algorithm which limits the lengths ratio of adjacent edges has been used, and the final mesh can be seen on the same figure (right).

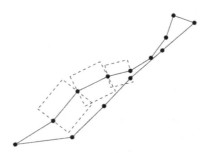

FIG. 6.10 Close edges.

FIG. 6.11 Initial mesh (left), after checking close edges (middle) and after checking edge lengths (right), thus improving the shape quality of triangles.

6.8. *Boundary representation of SES*

So far, we have considered *Van der Waals* or *solvent-accessible* surfaces (VWS or SAS), which are made up of several spherical patches. If we now focus on *solvent-excluded* surfaces (SES), also called Connolly surfaces (cf. first subsection), different kinds of patches may be encountered:

- When the *probe sphere* (PS) is in contact with only one atom, it defines a spherical surface, which is in fact a part of the classical VWS.
- When it rolls while touching simultaneously two atoms, the part facing the molecule traces a toroidal patch. In the usual case, this patch is bounded by four arcs. However, if the two atoms become more distant compared to the probe diameter, we obtain a pair of patches bounded by three arcs each. These two 3-sided patches share a curve along which the surface is not G^1 continuous.
- When it touches three atoms at the same time, it cannot roll anymore and defines a spherical surface which is now reentrant.

Several algorithms to compute analytical models of the SES are referenced in an article by SANNER, OLSON and SPEHNER [1996], who developed an efficient program called MSMS which determines the constituting patches (spherical contact, spherical reentrant, toroidal-3 or toroidal-4). This program can either give a Connolly surface mesh suitable for visualization, or provide an intermediate description of the patches in terms of centers, radii, bounding arcs, etc. To generate a mesh from such a boundary representation, a parameterization of the obtained patches is necessary.

To parameterize a *spherical* patch (contact or reentrant), a stereographic projection can be used as before. To parameterize a *toroidal* patch, a basic idea is to consider an arc on a rotating plane. The center of this arc is defined by a radius r_1 and an angle $\varphi \in [\varphi_1, \varphi_2]$. For a given φ, a point on the arc is defined by a radius r_2 and an angle $\theta \in [\theta_1, \theta_2]$. Then, if r_1 and r_2 are known, any point on the torus is defined by the two angles φ and θ. However, in general, the parametric domain cannot be defined as the rectangle $[\varphi_1, \varphi_2] \times [\theta_1, \theta_2]$. A problem occurs when $r_1 \leqslant r_2$, i.e., when the arc intersects the axis. In this case, a toroidal patch is bounded by only three arcs (see Fig. 6.12). Then, instead of considering the two angles φ and θ, it is preferable

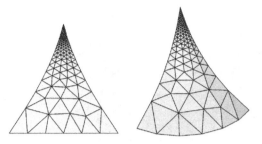

FIG. 6.12 Toroidal patch bounded by three arcs (planar and surface meshes).

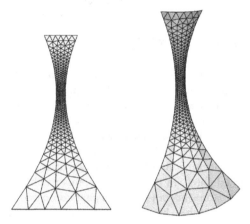

FIG. 6.13 Thin toroidal patch bounded by four arcs (planar and surface meshes).

to use curvilinear abscissae. This also gives better results when $r_1 \simeq r_2$ (but $r_1 > r_2$), for the shapes of the 2D and 3D domains become similar and the 2D triangles are not too distorted (see Fig. 6.13).

6.9. Surface meshing

Having a boundary representation and a parameterization of a molecular surface, a surface mesh must now be generated. As explained before, the size and shape of the mesh elements are a crucial point in a computational scheme. The problem of meshing, in particular surface meshing, for numerical simulation purposes is a broad and interdisciplinary field, and it is beyond the scope of this Handbook to describe the various existing algorithms suitable to produce such meshes (a comprehensive survey about mesh generation techniques for surfaces and volumes can be found in FREY and GEORGE [2000]). In short, there are two approaches to meshing parametric surfaces: direct and indirect. In the *direct* approach, the mesh is generated over the surface directly in $\mathbb{R}^3$. Among the direct approaches we can cite the octree-based method, the advancing-front-based method and the paving-based method. A direct approach, conceived mainly for molecular surface visualization, is described in AKKIRAJU and EDELSBRUNNER [1996]. The *indirect* approach consists of meshing the parametric domain and mapping the resulting mesh onto the surface. It is conceptually straight-forward, as a two-dimensional mesh is generated in the parametric domain, and thus it is expected to be faster than the direct approach. However, one problem with this method is the generation of a mesh which conforms to the metric of the surface. It is briefly presented below (for more details, see BOROUCHAKI, LAUG and GEORGE [2000]).

Let Σ be such a surface parameterized by:

$$\sigma : \Omega \to \Sigma, \quad (u, v) \mapsto \sigma(u, v) = (x, y, z),$$

where Ω denotes the parametric domain. First, from the size specifications, a Riemannian metric $\mathcal{M}_3 = \frac{1}{h^2} \mathcal{I}_3$ (h being the specified size function and $\mathcal{I}_3$ the identity matrix) is defined so that the desired mesh has unit length edges with respect to the related Riemannian space (such meshes being referred to as "unit" meshes). Then, based on the intrinsic properties of the surface, namely the first fundamental form:

$$\mathcal{M}_\sigma = \begin{pmatrix} {}^t\sigma\sigma'_u & {}^t\sigma\sigma'_v \\ {}^t\sigma\sigma'_u & {}^t\sigma\sigma'_v \end{pmatrix},$$

the Riemannian structure $\mathcal{M}_3$ is induced into the parametric space as follows:

$$\widetilde{\mathcal{M}}_2 = \frac{1}{h^2} \mathcal{M}_\sigma.$$

The initial size specification is isotropic while the induced metric in parametric space is in general anisotropic, due to the variation of the tangent plane along the surface. Finally, a unit mesh is generated completely inside the parametric space such that it conforms to the induced metric $\mathcal{M}_2$. This mesh is constructed using a combined advancing-front – Delaunay approach applied within a Riemannian context: the field points are defined after an advancing-front method and are connected using a generalized Delaunay type method.

One can control explicitly the accuracy of a generated element with respect to the geometry of the surface if careful attention is paid. Indeed, a mesh of a parametric patch whose element vertices belong to the surface is "geometrically" suitable if the two following properties hold:

- each mesh element is close to the surface, and
- each mesh element is close to the tangent planes related to its vertices.

A mesh satisfying these properties is called a *geometric mesh*. The first property allows us to bound the gap between the elements and the surface. This gap measures the greatest distance between an element and the surface. The second property ensures that the surface is locally of order G^1 in terms of continuity. To obtain this, the angular gap between the element and the tangent plane at its vertices must be bounded. A sufficient condition is that the element size is locally proportional to the minimal radius of curvature. Here, we deal with two kinds of surface, namely spheres and tori. For a sphere, the minimal radius of curvature is simply its radius.

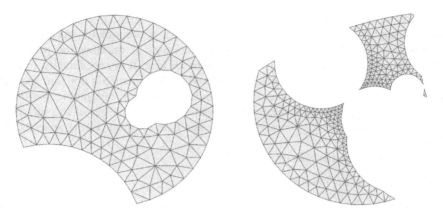

FIG. 6.14 Two examples of parametric domain meshes.

In the case of a torus, this radius can be easily computed from the radii of the two main defining circles.

Note that if a given size map is specified, the two above properties can be locally violated. In fact, it is more useful to find a compromise between the geometric approximation of the surface and the size map conformity.

To give an idea of possible shapes of parametric domains, Fig. 6.14 shows two examples where several topological sides, trimming curves and multiple loops can be noticed. The first one (left) has one external and one internal boundary, each boundary being made up of several arcs. The second one (right) is constituted by three different sub-domains.

6.10. Application examples

Several sample meshes are presented in this subsection to illustrate the above methods. They all have been generated by a software package called BLMOL (BAUG and BOROUCHAKI [2002]), except in the second example (ice SES). Each molecule is represented by a list of atoms, and each atom is described by its chemical symbol, its center coordinates and its radius length. Such descriptions can be obtained for instance from the MathMol library (NEW YORK UNIVERSITY [2002]) in a PDB (Protein Data Bank) file. The following examples involve molecular surfaces of fullerene (C_{60}), ice, pentane, DNA and echistatin.

C_{60}. The C_{60} molecule can be modeled by 60 identical spheres whose centers are at the vertices of pentagons and hexagons. Figure 6.15 (top) shows the default mesh generated by BLMOL. It meets the most common requirements for computational chemistry. It is a geometric mesh with a tolerance angle of $3°$ for the curves and $9°$ for the surfaces, using a gradation parameter of 1.5. The total CPU time is 5 s on a HP 9000/785 400 MHz for a mesh without gradation, and 12 s for the mesh shown (three iterations are necessary when a gradation is required). The number of vertices is 7,863 and the number of triangles is 15,718. Figure 6.15 (bottom) shows a mesh

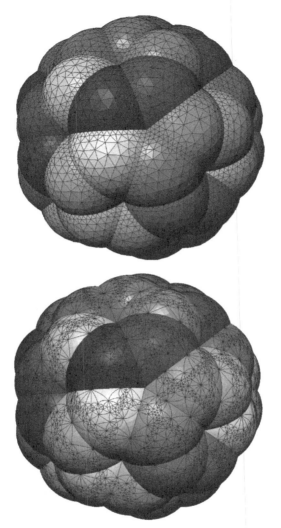

FIG. 6.15 Carbon 60. *Top*: P^1 geometric mesh. *Bottom*: P^2 mesh (curved triangles) with a given
analytical field.

which conforms to a given analytical field. Here, curved quadratic triangles (P^2) are
used instead of the usual ones (P^1), thus decreasing the number of elements for a
given accuracy. In this example, the mesh contains 40,776 nodes and 20,386 six-node
triangles.

Ice. Figure 6.16 shows two meshes of a solvent-excluded surface (SES) of a network of hydrogen bonded water molecules. Initially, its boundary representation has
been generated by the MSMS program (SANNER, OLSON and SPEHNER [1996]), as
explained in the subsection entitled "Boundary representation of SES" (a very small

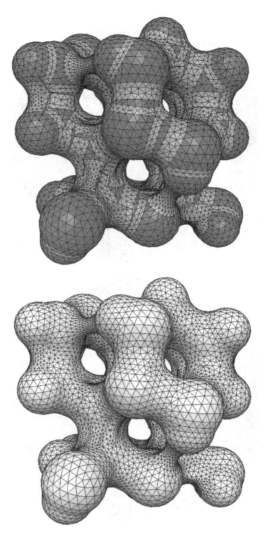

FIG. 6.16 Solvent-excluded surface (SES) of a network of hydrogen-bonded waters. *Top*: patch-dependent mesh. *Bottom:* simplified mesh.

probe has been used to have holes in the network). Then, the BLSURF software package (LAUG and BOROUCHAKI [1999]) was used to create surface meshes. The first mesh (on the bottom) has been constructed while preserving the contours of each patch. It contains 17,446 vertices and 35,020 triangles. However, within the context of numerical simulation, it is more convenient to enhance the quality of such a mesh by merging the extremities of small edges and moving points. In the example shown on the top of the figure, the counts reduce to 16,365 vertices and 32,858 triangles with a better shape quality.

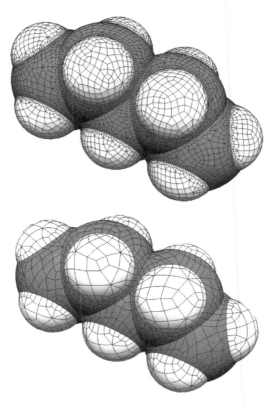

FIG. 6.17 Quadrilateral meshes of pentane. *Top*: linear (Q^1). *Bottom*: quadratic (Q^2).

n-Pentane. The *n*-pentane molecule (C_5H_{12}) is made up of 17 atoms. The two meshes shown on Fig. 6.17 have quadrilateral elements that are, respectively, linear (Q^1) and quadratic (Q^2). They have been obtained by pairing adjacent triangles. Let us remark that this process have been applied to an "optimal" triangular mesh, which cannot in any case generate an optimal quadrilateral mesh. The Q^1 mesh is a geometric mesh with a tolerance angle of $3°$ for curves and $5°$ for surfaces, and a gradation parameter of 1.2. It contains 8,520 vertices and 8,518 quadrilaterals. For the Q^2 mesh, we have $10°$ for curves and $20°$ for surfaces without gradation, producing a mesh with 5330 nodes and 1776 eight-node quadrilaterals.

DNA. Figure 6.18 shows a mesh of a fragment of DNA strand, whose structure is a double helix. It has 637 atoms, 154,220 vertices and 309,004 triangles.

Echistatin. Finally, Fig. 6.19 shows a mesh of a sphere of water containing an echistatin molecule with 12,596 atoms, 1,329,865 vertices and 2,673,604 triangles.

6.11. Conclusion

To conclude briefly this section, some methods for generating a mesh on any molecular surface (Van der Waals, solvent-accessible or solvent-excluded surface) have

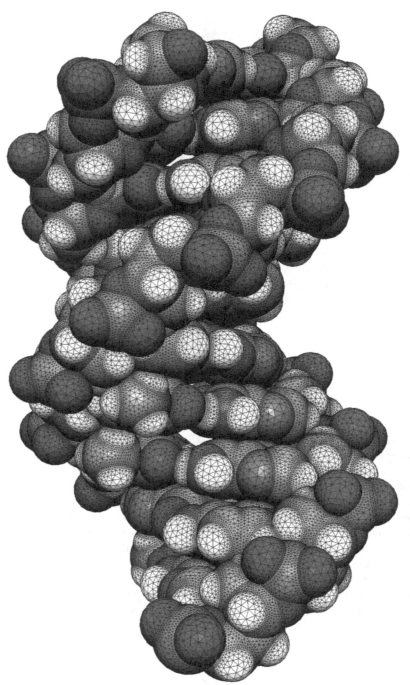

FIG. 6.18 Surface mesh of DNA.

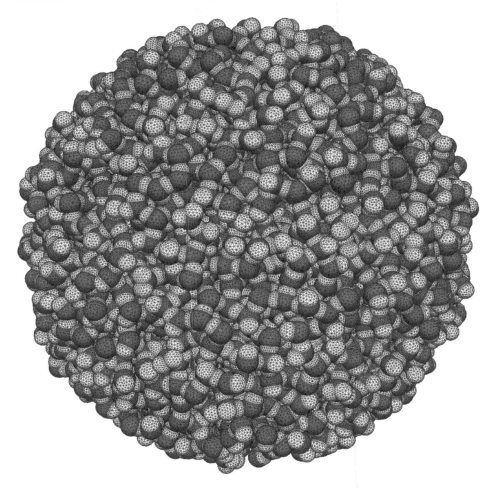

FIG. 6.19 Sphere of water containing an echistatin molecule.

been presented and examples have been shown. In computational chemistry, like in any field of computing science, mesh quality is an important issue to obtain accurate results. The choice of elements may also be taken into consideration (P^1 or P^2 triangles, Q^1 or Q^2 quadrilaterals). At present, though molecules with several thousand atoms can be addressed, a remaining challenge is to find out and implement even more efficient methods for very large molecules and dynamic problems.

CHAPTER III

Computer Simulations

Computer simulations are considered a valuable alternative to experiments to get information on the liquid state. The literature on this subject is large, with a sizeable number of detailed monographs (HANSEN and MCDONALD [1986], GRAY and GUBBINS [1984], ALLEN and TILDESLEY [1987], FRENKEL and SMIT [1996]). Here, we shall be concise, limiting our attention to the two basic approaches, MC and MD, and giving for both a very schematic outline.

7. Monte Carlo methods

MC technically is a procedure to compute an integral with a random sampling of integrand values. It may so be used to compute the averages over the ensemble distributions necessary to get the properties in the Gibbs approach. Actually MC methods are generally limited to properties without an explicit dependence on time.

The basic formalism is simple. The value of a property X depending on the spatial $\mathbf{r}^N$ and angular ω^N set of coordinates, averaged on time, is given in the Gibbs approach by

$$\langle X \rangle = \int P(\mathbf{r}^N \omega^N) X(\mathbf{r}^N \omega^N) d\mathbf{r}^N d\omega^N = \frac{\int X(\mathbf{r}^N \omega^N) \exp[-V/kT]}{Q_N}, \qquad (7.1)$$

where V indicate the potential energy of the N interacting particles. We have here used the canonical formulation but the method can be applied to other ensembles as well. In the MC approach the integrals appearing in the numerator as well as in the denominator of Eq. (7.1) are replaced by summations. Each term of the sum refers to a specific choice of the values of the coordinates, which we now collectively indicate with $\mathbf{R}$. Expression (7.1) may be so approximated by

$$\langle X \rangle = \sum_k X(\mathbf{R}_k) P_k = \frac{\sum_k X(\mathbf{R}_k) \exp[-V(\mathbf{R}_k)/kT]}{\sum_k \exp[-V(\mathbf{R}_k)/kT]}. \qquad (7.2)$$

To evaluate Eq. (7.2) with the MC procedure one has to:

(1) Specify the initial coordinates of atoms $\mathbf{R}_0^N$.
(2) Generate new coordinates $\mathbf{R}_a^N$ by changing the initial coordinates at random.
(3) Compute the transition probability $W(0, a)$.
(4) Generate a uniform random number t in the range $[0,1]$.
(5) If $W(0, a) < t$, then take the old coordinates as the new coordinates and go to step 2.

(6) Otherwise accept the new coordinates and go to step 2.

The most popular realization of the MC method (though not the only one) for molecular systems is the Metropolis method (METROPOLIS, A.W. ROSENBLUTH, M.N. ROSENBLUTH, A.N. TELLER and E. TELLER [1953]) which consists in: (i) specifying the initial atom coordinates (e.g., from molecular mechanics geometry optimization), (ii) selecting some atom i randomly and move it by random displacement: ΔR_i; (iii) calculating the change of potential energy ΔV corresponding to this displacement. If $\Delta V < 0$, the new coordinates are accepted and the cycle is repeated. Otherwise, if $\Delta V \geqslant 0$, a random number γ in the range [0,1] is selected and, if $e^{-\Delta V/kT} < \gamma$, the new coordinates are accepted, while if $e^{-\Delta V/kT} \geqslant \gamma$ the original coordinates are kept unchanged. In both cases the process goes back to point (iii) and iterated.

Note that the iterations are independent of one another (i.e., the system does not contain any "memory"). The possibility that the system might revert to its previous state is as probable as choosing any other state. This condition has to be satisfied if the system is to behave as a Markov process, for which the methods of calculating ensemble (i.e., statistical) averages are known. Another important condition to be satisfied is the continuity of the potential energy function in order for the system to be ergodic. In this context, ergodicity means that any state of the system can be reached from any other state.

There are several problems related to the use of this procedure. The formal elements of the Markov theory are not accompanied by stringent rules on the representativity of the chain in spanning the whole ensemble. If the random selection of new configurations is not well calibrated, the sampling will be only performed in a limited region of the conformational space, giving so a value of the property with a small statistical error, but with a large systematic error.

The move generally regards a single coordinate of a single molecule, under the form

$$q_{im}(R_{i+1}) = q_{im}(R_i) + \lambda \Delta q_m,$$

where q_m is one of the six translational and rotational coordinates of the molecule i, Δq_m is the maximal change permitted for that coordinate and λ is a random number between 0 and 1. Experience has learned what are acceptable Δq_m values for small molecules, but for large molecules as well as for cases in which many molecules exhibit positive correlation in their position (e.g., strong hydrogen networks, polymers) this definition of the moves surely is inadequate. There are now "smarter" definitions of the moves. To alleviate problems of inefficient sampling the length of the Markov chain has been greatly increased in the years: standard calculations include now several millions of accepted moves, after some extra million moves of "equilibration" used to find an initial R_i conformation not too severely biased.

The ensemble on which the simulation is performed cannot be too large for computational reasons. Generally it is defined in terms of a square box of appropriate dimensions containing a given number N of molecules. To reduce border effects the box is surrounded by other similar boxes all containing the N molecules in the same R_i conformation (translational symmetry). So when a move brings a molecule out of the box a similar molecule on the opposite side replaces it. The number N must be as large as possible, but there are limitations due to the finite computer power (the maximum for

N is now a few thousands). This fact introduces new problems, especially for ionic solutions in which long range unscreened Coulomb interactions are active. The collection of boxes introduces in these cases a spurious periodic pattern of charges that severely modifies the ensemble. In any case, simulations on solutions are more sensitive to errors than simulations on pure liquids, because the low mole ratio does not permit to average local properties on many molecules.

Actually to perform calculations on a given property the general formula (7.2) is rewritten in a computationally more suitable form. For example the energy V is immediately reduced to a weighted average of the energies of the various conformations, the heat capacity at constant volume C_V to an average of the fluctuations of V around its mean value $\langle V \rangle$, that can be recovered during a single simulation run (even if with a larger error than for $\langle V \rangle$). Things are not so simple for the free energy A for which there is the need of averaging over the $\exp[-V(R)/kT]$ function. This means that to have a meaningful result for $\langle A \rangle$ one has to sample regions in the conformational space high in energy, generally less represented in the MC chain.

Generally, MC simulations work well for the canonical ensemble; one could pass from the canonical to another ensemble giving a simpler expression for the desired property. The performances of MC simulations are however worse in passing from the canonical one to other ensembles, in particular the use of the grand canonical ensemble (that it would be the best for free energy) is quite difficult.

We have dwelt upon problems and limits in the use of MC simulations, and have now to stress the great merits of this method. It is quite flexible, open to the researcher's ingenuity to derive information not limited to the thermodynamic functions: a complete listing of the successful applications of MC to condensed systems (of every type) would be very instructive and stimulating. Unfortunately we cannot review here these performances and we limit ourselves to remark a point. There are no formal limits in the use of potential functions to describe the collective or local behavior of molecules in the condensed systems. The limitations we have signaled in the use of physical methods (which are progressively reduced at the cost of great efforts) do not exist here. The only real limitation is due to the computational cost, a limitation more and more reduced by the impressive increase of computer performances. There are for example MC methods which compute the interaction potentials on the spot at the QM level, as well as MC strategies to follow the evolution of a chemical reaction along its reaction path. The limitations due to the cost are still present, however: a large part of recent studies continues to use empirical two-body potentials (now more and more expressed at the level of nonrigid molecules and with the introduction of contributions describing modifications to the potential due to nearby molecules); the use of explicit three body potentials still is a rarity.

As a result of a stochastic simulation, the large number of configurations (geometries) are accumulated and the potential energy function is calculated for each of them. These data are used to calculate thermodynamic properties of the system. The MC method is accepted more by physicists than by chemists, probably because MC is not a deterministic method and does not offer time evolution of the system in a form suitable for viewing. It does not mean however, that for deriving the thermodynamic properties of systems MD is better. In fact many chemical problems in statistical mechanics are approached more efficiently with MC, and some (e.g., simulations of polymers chains on a lattice) can only be done efficiently with MC. Also, for Markov chains, there are efficient methods for deriving time related quantities such as relaxation times. Currently, the stronghold of MC in chemistry is in the area of simulations of liquids and solvation processes.

8. Molecular Dynamics

The deterministic approach, called MD, actually simulates the time evolution of the molecular system and provides the actual trajectory of the system. The information generated from simulation methods can in principle be used to fully characterize the thermodynamic state of the system. In practice, the simulations are interrupted long before there is enough information to derive absolute values of thermodynamic functions, however the differences between thermodynamic functions corresponding to different states of the system are usually computed quite reliably. In MD, the evolution of the molecular system is studied as a series of snapshots taken at close time intervals (usually of the order of femtoseconds). For large molecular systems the computational complexity is enormous and supercomputers or special attached processors have to be used to perform simulations spanning long enough periods of time to be meaningful. Typical simulations of small proteins including surrounding solvent cover the range of tens to hundreds of picoseconds, i.e., they incorporate many thousands of elementary time steps.

Based on the potential energy function V, we can find components F_i of the force acting on an atom as:

$$F_i- = -\partial V / \partial x_i.$$

This force results in an acceleration according to Newton's equation of motion. By knowing acceleration, we can calculate the velocity of an atom in the next time step. From atom positions, velocities, and accelerations at any moment in time, we can calculate atom positions and velocities at the next time step. Integrating these infinitesimal steps yields the trajectory of the system for any desired time range. There are efficient methods for integrating these elementary steps with Verlet and leapfrog algorithms being the most commonly used.

To start the MD simulation we need an initial set of atom positions (i.e., geometry) and atom velocities. In practice, the acceptable starting state of the system is achieved by "equilibration" and "heating" runs prior to the "production" run. The initial positions of atoms are most often accepted from the prior geometry optimization with molecular mechanics. Formally, such positions correspond to the absolute zero temperature. The velocities are assigned randomly to each atom from the Maxwell distribution for some low temperature (say 20 K). The random assignment does not allocate correct velocities and the system is not at thermodynamic equilibrium. To approach the equilibrium the "equilibration" run is performed and the total kinetic energy (or temperature) of the system is monitored until it is constant. The velocities are then rescaled to correspond to some higher temperature, i.e., the heating is performed. Then the next equilibration run follows. The absolute temperature, T, and atom velocities are related through the mean kinetic energy of the system:

$$T = \frac{2}{3Nk} \sum_{i=3}^{N} \frac{m_i |v_i|^2}{2},$$

where N denotes the number of atoms in the system, m_i represents the mass of the ith atom, and k is the Boltzmann constant. By multiplying all velocities by $\sqrt{T_{\text{desired}}/T_{\text{current}}}$ we can effectively "heat" the system. Heating can also be realized by immersing the system in a "heat bath" which stochastically (i.e., randomly) accelerates the atoms of the molecular system. These cycles are repeated until the desired temperature is achieved and at this point a "production" run can start. In the actual software, the "heating" and "equilibration" stages can be introduced in a more efficient way by assigning velocities in such a way that "hot spots" (i.e., spots in which the neighboring atoms are assigned high velocities) are avoided.

MD for larger molecules or systems in which solvent molecules are explicitly taken into account, is a computationally intensive task even for the most powerful supercomputers, and approximations are frequently made. The most popular is the SHAKE method (RYCKAERT, CICCOTTI and BERENDSEN [1977]) which in effect freezes vibrations along covalent bonds. This method is also applied sometimes to valence angles. The major advantage of this method is not the removal of a number of degrees of freedom (i.e., independent variables) from the system, but the elimination of high frequency vibrations corresponding to "hard" bond stretching interactions. In simulations of biological molecules, these modes are usually of least interest, but their extraction allows us to increase the size of the time step, and in effect achieve a longer time range for simulations. Another approximation is the united-atom approach where hydrogen atoms which do not participate in hydrogen bonding are lumped into a heavier atom to form a pseudo-atom of larger size and mass (e.g., a CH group).

Even supercomputers have their limitations and there is always some practical limit on the size (i.e., number of atoms) of the simulated system. For situations involving solvent, the small volume of the box in which the macromolecule and solvent are contained introduces undesirable boundary effects. In fact, the results may depend sometimes more on the size and shape of the box than on the molecules involved. To circumvent this limited box size difficulty, periodic boundary conditions are used. In this approach, the original box containing a solute and solvent molecules is surrounded with identical images of itself, i.e., the positions and velocities of corresponding particles in all of the boxes are identical. The common approach is to use a cubic or rectangular parallelepiped box, but other shapes are also possible (e.g., truncated octahedron). By using this approach, we obtain what is in effect an infinite sized system. The particle (usually a solvent molecule) which escapes the box on the right side, enters it on the left side, due to periodicity. Since MD simulations are usually performed as an *NVE* (microcanonical) ensemble (i.e., at constant number of particles, constant volume, and constant total energy) or an *NVT* (canonical) ensemble, the volume of the boxes does not change during simulation, and the constancy in the number of particles is enforced by the periodicity of the lattice, e.g., a particle leaving the box on left side, enters it on the right side.

There are also techniques for performing simulations in a *NPT* (isothermal–isobaric), and *NPH* (isobaric–isoenthalpic) ensembles, where the pressure constancy during simulation is achieved by squeezing or expanding box sizes. The constant temperature is usually maintained by "coupling the system to a heat bath", i.e., by adding dissipative forces (usually Langevin, friction type forces) to the atoms of the system which as a consequence affects their velocities. However, each approximation has its price. In the case of periodic boundary conditions we are actually simulating a crystal comprised of boxes with ideally correlated atom movements. Longer simulations will be contaminated with these artificially correlated motions. The maximum length for the simulation, before artifacts start to show up, can be estimated by considering the speed of sound in water (15 Å/ps at normal conditions). This means that for a cubic cell with a side of 60 Å, simulations longer than 4 ps will incorporate artifacts due to the presence of images.

In the other popular approach, stochastic boundary conditions allow us to reduce the size of the system by partitioning the system into essentially two portions: a reaction zone and a reservoir region. The reaction zone is that portion of the system which we want to study and the reservoir region contains the portion which is inert and uninteresting. For example, in the case of an enzyme, the reaction zone should include the proximity of the active center, i.e., portions of protein, substrate, and molecules of solvent adjacent to the active center. The reservoir region is excluded from MD calculations and is replaced by random forces whose mean corresponds to the temperature and pressure in the system. The reaction zone is then subdivided into a reaction region and a buffer region. The stochastic forces are only applied to atoms of the buffer region, in other words, the buffer region acts as a "heat buffer". There are several other approximations whose description can be found in the MD monographs quoted here. The straightforward result of a MD simulation, a movie showing changing atom positions as a function of time, contains a wealth of information in itself. Viewing it may shed light on the molecular mechanisms behind the biological function of the system under study.

MD, contrary to energy minimization in molecular mechanics, is able to climb over small energy barriers and can drive the system towards a deeper energy minimum. The use of MD allows probing of the PES for deeper minima in the vicinity of the starting geometry. It is exploited in a *simulated annealing method* (KIRKPATRICK, GELATT JR. and VECCHI [1983]), where the molecular system is first heated to an artificially high temperature and then snapshots of the trajectory at high temperature are taken as a starting state for cooling runs. The cooled states obtained from hot frames correspond frequently to much deeper potential energy minima than the original structure taken for dynamics. It is particularly suitable for imposing geometrical constraints on the system. Such constraints may be available from experimental results. MD will try to satisfy these artificially imposed constraints by jumping over shallow potential energy minima – behavior which is not possible in molecular mechanics.

CHAPTER IV

Hybrid Methods

9. Quantum mechanics/molecular mechanics (QM/MM)

Each computational method has strengths and weaknesses. Molecular mechanics (MM) can model very large compounds quickly. Quantum mechanics (QM) is able to compute many properties and model chemical reactions. It is possible to combine these two methods into one calculation, which models a very large compound using molecular mechanics and one crucial portion of the molecule with quantum mechanics (QM/MM). This is designed to give results that have very good speed where only one region needs to be modeled quantum mechanically. This can also be used to model a molecule surrounded by solvent molecules. The literature on QM/MM methods is vast and rapidly updating; here we only quote a general review: GAO [1995].

The basic idea to describe molecules in solution is to build an hybrid QM/MM potential and to introduce the solvent field into QM calculations of the solute molecule on the fly during a computer simulation such that the molecular wavefunction of the solute will be polarized by the dynamic change of the surrounding solvent molecules. The method was first described by WARSHEL and LEVITT [1976] and a detailed prescription for MO calculations was presented by FIELD, BASH and KARPLUS [1990]. The idea was to divide a condensed phase system into a QM region and a MM region, plus an appropriate boundary treatment to mimic the bulk effects. Consequently the effective Hamiltonian of the system is written as follows:

$$H_{\text{eff}} = H_{\text{QM}}^0 + H_{\text{MM}} + H_{\text{QM/MM}}^{\text{elec}} + H_{\text{QM/MM}}^{\text{VdW}} \tag{9.1}$$

where H_{QM}^0 describes the completely QM part of the system, H_{MM} the purely molecular mechanics and the last two terms the QM-MM electrostatic $\left(H_{\text{QM/MM}}^{\text{elec}}\right)$ and van der Waals $(H)_{\text{QM/MM}}^{\text{VdW}}$ interactions. The van der Waals term $(H)_{\text{QM/MM}}^{\text{VdW}}$ is to ensure that the QM and MM systems will not get too close because of a lack of electronic structural description of the solvent MM system. It turns out that the most significant part is to account for the short range electron repulsions; however, for convenience as well as for inclusion of some dispersion interactions, a Lennard-Jones term is typically used for H^{VdW}.

$$H_{QM/MM}^{VdW} = \sum_{s=1}^{S} \sum_{m=1}^{M} 4\varepsilon_{sm} \left[\left(\frac{\sigma_{sm}}{R_{sm}} \right)^{12} - \left(\frac{\sigma_{sm}}{R_{sm}} \right)^{6} \right],$$

where S and M are the number of solvent (MM) and solute (QM) atoms, and σ and ε are empirical parameters. It should be emphasized that these parameters must be carefully examined for a given QM (model, basis set, and level of theory) and MM (force fields) combination. However, on the other hand, it gives us the opportunity to optimize the performance of the hybrid QM/MM potential.

In the hybrid QM/MM scheme, solute-solvent (or QM/MM) interactions are given by

$$E_{QM/MM} = \langle \Psi | H_{QM/MM} | \Psi \rangle + E_{QM/MM}^{VdW},$$

where Ψ is the wavefunction of the solute molecule in solution, which minimizes the energy of the Hamiltonian of Eq. (9.1).

Commonly it is implicitly assumed that only point charges $\{q_s\}$ in the MM region contribute to the modified Hamiltonian in the hybrid QM/MM method through the electrostatic $H_{QM/MM}^{elec}$ term, namely:

$$H_{QM/MM}^{elec} = -\sum_{s=1}^{S} \sum_{i=1}^{2N} \frac{q_s}{r_{is}} + \sum_{s=1}^{S} \sum_{m=1}^{M} \frac{q_s Z_m}{R_{ms}},$$

where i indicate electrons, r and R are the distances of the QM electrons and nuclei from the solvent sites, respectively. This allows the wavefunction and charge distribution of the solute molecule to be polarized. However, the solvent charges are kept fixed, without responding to the solute charge reorganization as well as the solvent dipole reorientations. To improve this description and to allow both the solute and solvent polarizations on the same footing. The MM solvent molecules are represented by a set of Van der Waals parameters, a set of charges, and a set of atomic polarizabilities. Therefore, an additional term is added to the system effective Hamiltonian:

$$H_{eff} = H_{QM}^{0} + H_{MM} + H_{QM/MM}^{elec} + H_{QM/MM}^{VdW} + H_{QM/MM}^{pol},$$

where the last term represents the interaction between the QM region and the induced dipoles of the solvent MM region. It is given below.

$$H_{QM/MM}^{pol} = \sum_{i=1}^{N} \sum_{s=1}^{S} \frac{R_{is} \cdot \mu_s}{R_{is}^{3}} + \sum_{a=1}^{M} \sum_{s=1}^{S} \frac{Z_a R_{as} \cdot \mu_s}{R_{as}^{3}},$$

where μ_s are MM induced dipole vectors depending on the atomic polarizabilities. The solution of the corresponding Schrödinger equations and the convergence of MM induced dipoles are coupled, and must be solved iteratively.

If the QM and MM regions are separate molecules, having nonbonded interactions only, this scheme is sufficient. On the contrary, if the two regions are parts of the same molecule, it is necessary to describe the bond connecting the two sections. In most cases, this is done using the bonding terms in the MM method being used. This is usually done by keeping every bond, angle or torsion term that incorporates one atom from the QM region.

It is sometimes desirable to include the effect of the rest of the system, outside of the QM and MM regions. One way to do this is using periodic boundary conditions, as is done in liquid state simulations. Some researchers have defined a potential, which is intended to reproduce the effect of the bulk solvent. This solvent potential may be defined just for this type of calculation, or it may be a continuum solvation model. To represent a solid continuum, a set of point charges, called a Madelung potential, is often used.

10. Layered methods: ONIOM

An alternative formulation of QM/MM is the *energy subtraction method*. In this method, calculations are done on various regions of the molecule with various levels of theory. Then the energies are added and subtracted, to give suitable corrections. This results in computing an energy for the correct number of atoms and bonds, analogous to an isodesmic strategy. Three such methods have been proposed by FROESE and MOROKUMA [1998]. The integrated MO + MM (IMOMM) method combines an orbital based technique with an MM technique. The integrated MO + MO method (IMOMO) integrates two different orbital based techniques. The "our own *n*-layered integrated MO and MM" method (ONIOM) allows for three or more different techniques to be used in successive layers. The acronym ONIOM is often used to refer to all three of these methods since it is a generalization of the technique. This technique can be used to model a complete system as a small model system and the complete system. The complete (or *Real*) system would be computed using only the lower level of theory. The *model* system would be computed with both levels of theory. The energy for the complete system, combining both levels of theory, would then be

$$E = E_{\text{real}}^{\text{low}} + E_{\text{model}}^{\text{high}} - E_{\text{model}}^{\text{low}}.$$

Likewise a three layer system, could be broken into small, medium, and large regions, to be computed with low, medium and high levels of theory (L,M,H, respectively). The energy expression would then be

$$E = E_{\text{small}}^{H} + E_{\text{medium}}^{M} - E_{\text{small}}^{M} + E_{\text{large}}^{L} - E_{\text{medium}}^{L}.$$

This method has the advantage of not requiring a parameterized expression to describe the interaction of various regions. Any systematic errors in the way that the lower levels of theory describe the inner regions will be canceled out.

The definition of the model system is rather straightforward if there is no covalent bond between the layers, and is then identical to the high level layer. When covalent bond exists between the layers, the dangling bonds are saturated with link atoms, which is in fact the method of choice in many QM/MM schemes. One chooses link atoms that best mimic the substituents, and usually hydrogen atoms yield good results when carbon–carbon bonds are broken. The atoms that exist both in the model system and in the real system will have the same coordinates both systems. C_{model}–H_{link}

bonds are assigned the same angular and dihedral values as the C_{model}–C_{real} bonds in the real system, with the bond lengths adjusted by scaling the C–C bond length in such way that a reasonable C_{model}–C_{real} distance yields a reasonable C_{model}–H_{link} distance. Because the geometry of the model system is a function of the geometry of the real system, the number of degrees of freedom remains 3N–6 (or 3N–5), which ensures that any method for the investigation of PESs available for conventional methods, can be used for ONIOM as well. The derivatives of the ONIOM energy with respect to the geometrical parameters can be obtained in a similar fashion as the energy.

In addition to the energy and its geometrical derivatives, other properties are available in the ONIOM framework as well. For example, the integrated density is expressed as

$$\rho = \rho_{real}^{low} + \rho_{model}^{high} - \rho_{model}^{low}.$$

Properties related to the density, such as potentials or electric field gradients, can be expressed in the same way. Higher order or mixed properties are available as well, but the Jacobian must be employed when derivatives of the nuclear coordinates are involved.

The ONIOM method has been combines with the continuum PCM-IEF model (VREVEN, MENNUCCI, DA SILVA, MOROKUMA and TOMASI [2001]) we have described in Section 4 of Chapter II. Four versions of the method have been developed. These schemes differ mainly with respect to the level of coupling between the solute charge distribution and the continuum dielectric, which has important consequences for the computational efficiency. Any property that can be calculated by both ONIOM and PCM-IEF can also be calculated by the ONIOM-PCM method.

11. The effective fragment potential (EFP) method

The EFP method (DAY, JENSEN, GORDON, WEBB, STEVENS, KRAUSS, GARMER, BASCH and COHEN [1996]) is another QM/MM method. The EFP method has successfully been applied to the study of aqueous solvation effects, by using EFPs to represent solvent molecules while the solute molecules are treated with Hartree–Fock theory.

The basic idea behind the EFP method is to replace the chemically inert part of a system by EFPs, while performing a regular *ab initio* calculation on the chemically active part. Here "inert" means that no covalent bond breaking process occurs. This "spectator region" consists of one or more "fragments", which interact with the *ab initio* "active region" through nonbonded interactions, and so these EFP interactions affect the *ab initio* wavefunction. A simple example of an active region might be a solute molecule, with a surrounding spectator region of solvent molecules represented by fragments. Each discrete solvent molecule is represented by a single fragment potential, in marked contrast to continuum models for solvation.

The nonbonded interactions currently implemented are:

- Coulomb interaction: The charge distribution of the fragments is represented by an arbitrary number of charges, dipoles, quadrupoles, and octupoles, which interact with the *ab initio* Hamiltonian as well as with multipoles on other fragments.

It is possible to input a screening term that accounts for the charge penetration. Typically the multipole expansion points are located on atomic nuclei and at bond midpoints.

- Dipole polarizability: An arbitrary number of dipole polarizability tensors can be used to calculate the induced dipole on a fragment due to the electric field of the *ab initio* system as well as all the other fragments. These induced dipoles interact with the *ab initio* system as well as the other EFPs, in turn changing their electric fields. All induced dipoles are therefore iterated to self-consistency. Typically the polarizability tensors are located at the centroid of charge of each localized orbital of a fragment.

- Repulsive potential: Two different forms for the repulsive potentials are used: one for *ab initio*-EFP repulsion and one for EFP-EFP repulsion. The form of the potentials is empirical, and consists of distributed Gaussian or exponential functions, respectively. The primary contribution to the repulsion is the QM exchange repulsion, but the fitting technique used to develop this term also includes the effects of charge transfer. Typically these fitted potentials are located on atomic nuclei within the fragment.

The EFP method for treating discrete solvent effects begins with the *ab initio* Hamiltonian of the "solute", which may include a small number of solvent molecules. The remaining solvent molecules are then treated by adding their effect on the system as one-electron terms in the *ab initio* Hamiltonian:

$$H = H_{AR} + V,$$

where H is the Hamiltonian for the entire system, H_{AR} is the *ab initio* Hamiltonian of the "solute", or active region, and V represents the one-electron terms that describe the potential due to the fragment molecules. This potential includes *ab initio*–fragment, *ab initio*(nuclei)–fragment, and fragment–fragment interactions, each including the three terms mentioned above (except for the *ab initio*(nuclei)–fragment interaction; there are no exchange repulsion/charge transfer terms there).

Recently the EFP method has been combined with the PCM-IEF continuum model to give a new discrete/continuum solvation model (BANDYOPADHYAY, GORDON, MENNUCCI and TOMASI [2002]). The first applications have shown that EFP/PCM model gives results that are in close agreement with the much more expensive full *ab initio*/PCM-IEF calculation.

12. Car–Parrinello *ab initio* molecular dynamics: AIMD

The term *ab initio* MD is used to refer to a class of methods for studying the dynamical motion of atoms, where computational work is spent in solving, as exactly as it is required, the entire QM electronic structure problem. When the electronic wavefunctions are reliably known, it will be possible to derive the forces on the atomic nuclei using the Hellmann–Feynman theorem. The forces may then be used to move the atoms, as in standard MD.

The most widely used theory for studying the QM electronic structure problem of solids and large molecular systems is the density-functional theory of Hohenberg and Kohn in the local-density approximation (LDA; KOHN and VASHISHTA [1993]). The self-consistent Schrödinger equation (or more precisely, the Kohn–Sham equations) for single-electron states is solved for the solid-state or molecular system, usually in a finite basis-set of analytical functions. The electronic ground state and its total energy is thus obtained. One widely used basis set is "plane waves", or simply the Fourier components of the numerical wavefunction with a kinetic energy less than some cutoff value. Such basis sets can only be used reliably for atomic potentials whose bound states are not too localized, and hence plane waves are almost always used in conjunction with pseudo-potentials (BACHELET, HAMANN and SCHLÜTER [1982]) that effectively represent the atomic cores as relatively smooth static effective potentials in which the valence electrons are treated.

CAR and PARRINELLO [1985] developed a method which is based upon the LDA, and uses pseudopotentials and plane wave basis sets, but they added the concept of updating iteratively the electronic wavefunctions simultaneously with the motion of atomic nuclei (electron and nucleus dynamics are coupled). This is implemented in a standard MD paradigm, associating dynamical degrees of freedom with each electronic Fourier component (with a small but finite mass). The efficiency of this iteration scheme has allowed not only for the mentioned pseudopotential based MD studies, but also for static calculations for far larger systems than had previously been accessible. Part of this improvement is due to the fact that some terms of the Kohn–Sham Hamiltonian can be efficiently represented in real-space, other terms in Fourier space, and that Fast Fourier Transforms (FFT) can be used to quickly transform from one representation to the other. Since the original paper by Car and Parrinello, a number of modifications (TUCKERMAN, UNGAR, VON ROSENVINGE and KLEIN [1996], PAYNE, TETER, ALLAN, ARIAS and JOANNOPOULOS [1992]) have been presented that improve significantly the efficiency of the iterative solution of the Kohn–Sham equations. The modifications include the introduction of the conjugate gradients method and a direct minimization of the total energy.

Until recently first principles electronic calculations where based on techniques which required the computationally expensive matrix diagonalization methods. Car and Parrinello formulated a new more efficient method which can be expressed in the language of MD. The essential step was to treat the expansion coefficients of the wavefunction as dynamical variables. In a conventional MD simulation a Lagrangian can be written in terms of the dynamical variables which normally are atomic positions $\{\mathbf{R}_I\}$ and the unit cell dimensions $\{\mathbf{B}\}$. The Car–Parrinello Lagrangian can similarly be written, but also includes a term for the electronic wavefunction. Ignoring any constraints for the moment, it is

$$\mathcal{L}' = \sum_i (\mu \langle \dot{\varphi}_i | \dot{\varphi}_i \rangle - E[\{\varphi_i\}, \{\mathbf{R}_I\}, \{\mathbf{B}\}]),$$

where μ is a fictitious mass which is associated to the expansion coefficients of the Kohn–Sham electronic wavefunctions. E is the Kohn–Sham energy functional. This is analogous to the usual form of the Lagrangian where the kinetic energy term is replaced with the fictitious dynamics of the wavefunctions and the Kohn–Sham energy functional replaces the potential energy. The Kohn–Sham electronic orbitals are subject to the orthonormal constraints

$$\int \varphi_i^*(\mathbf{r})\varphi_j(\mathbf{r})d\mathbf{r} = \delta_{ij}.$$

These constraints can be simply incorporated into the Car–Parrinello Lagrangian as follows:

$$\mathcal{L} = \mathcal{L}' + \sum_{i,j} \Delta_{ij} \left(\int \varphi_i^*(\mathbf{r})\varphi_j(\mathbf{r})d\mathbf{r} - \delta_{ij} \right),$$

where $\Delta_{i,j}$ are the Lagrange multipliers ensuring that the wavefunctions remain orthonormal. In terms of MD, these can be thought of as additional forces on the wavefunctions which maintain orthonormality throughout the calculation. From $\mathcal{L}$, it follows that the Lagrange equations of motion

$$\frac{\mathrm{d}}{\mathrm{d}t}\left(\frac{\partial \mathcal{L}}{\partial \dot{\varphi}_i^*}\right) = \frac{\partial \mathcal{L}}{\partial \varphi_i^*}$$

give

$$\mu \ddot{\varphi}_i = -H\varphi_i + \sum_{i,j} \Delta_{ij}\varphi_j,$$

where H is the Kohn–Sham Hamiltonian and the force $-H\varphi_i$ is the gradient of the Kohn–Sham energy functional at the point in Hilbert space that corresponds to the wavefunction. This equation when coupled to the more standard equation of motion for the classical nuclear degrees of freedom, $\mathbb{R}_I$, fully describes the dynamics of the system.

CHAPTER V

Computing Properties

The formulation of theoretical models to describe liquids has its main application in the study of phenomena and properties involving molecular systems in solution. In this chapter we shall present results on this kind of application in two separate sections: one on approaches treating the liquid as a whole, and the other on models focussing on a subsystem of the liquid, the solute.

13. Liquids

In this section we will present some examples of how to determine properties of pure liquids by mean of the previously shown MC and MD computational procedures. Obviously the following discussion cannot be considered as a complete review of all the literature in the field, but it is rather intended to give the readers a few suggestions on how these methodologies can be used to obtain data about the most usual properties of pure liquids. From what we said in the previous section, it should be clear to the reader that continuum models cannot properly be used in order to determine properties of liquids: they are in fact concerned in the treatment of a solute in a solvent, and so they are not of direct use for the study of the solvent properties.

13.1. Calculation of equilibrium properties

Statistical mechanics concepts can supply many information from MD (or MC) runs, since they furnish prescriptions for how to calculate the macroscopic properties of the system (e.g., thermodynamic functions like free energy and entropy changes, heats of reaction, rate constant, etc.) from statistical averages of elementary steps on molecular trajectory (or of configurations). They also allow us to uncover a correlation between the motions of groups or fragments of the system as well as provide various distribution functions. Essentially, all expressions derived below apply to both MD and MC, with the only difference being that the "time step" in MD should be changed to a "new configuration" for MC. For the moment let us choose MD for illustration, since it is more often applied to chemical problems.

Three quantities form a foundation for these calculations: the partition function Z, the Boltzmann probability functional P, and the ensemble average, $\langle B \rangle$, of a quantity B:

$$Z = \iint e^{-H(\mathbf{p},\mathbf{q})/kT} \, d\mathbf{q} d\mathbf{p}, \tag{13.1}$$

$$P(\mathbf{p},\mathbf{q}) = \frac{1}{Z} e^{-H(\mathbf{p},\mathbf{q})/kT},$$

$$\langle B \rangle = \int B(\mathbf{p},\mathbf{q}) P(\mathbf{p},\mathbf{q}) \, d\mathbf{q} d\mathbf{p},$$

where $\mathbf{p}$ and $\mathbf{q}$ are all generalized momenta and coordinates of the system (e.g., angular and linear velocities and internal coordinates); $H = K + V$ is the Hamiltonian for the system, k denotes the Boltzmann constant, and T is the temperature. The integration extends over the whole range accessible to momenta and coordinates.

13.1.1. Free energy

For the isochoric-isothermic system of constant NTV, the Helmholtz free energy, $A = U - TS = G - PV$ (U, G, S and T are internal energy, Gibbs free energy, entropy and absolute temperature, respectively) for the system of unit volume is given as:

$$A = -kT \ln Z.$$

The above equation simplifies significantly when Cartesian coordinates and Cartesian momenta are used, since the kinetic energy factor of the total energy represented by the Hamiltonian in the expression for Z reduces to a constant depending upon the size, temperature and volume of the system:

$$A = -kT \ln \underbrace{\int e^{-V(\mathbf{X}^N)/kT} d\mathbf{X}^N}_{Z^c} + \text{const}(N, V, T),$$

where the potential energy function $V(\mathbf{X}^N)$ depends only on Cartesian coordinates of atoms and the integral Z^c is called a classical configurational partition function or a configurational integral. Hence, the free energy A can in principle be calculated from the integral of the potential energy function V. In practice, however, calculation of the absolute value of the free energy is not possible, except for very simple molecules or systems in which the motion of atoms is very restricted (e.g., crystals). Note that Cartesian coordinates span the range $(-\infty, +\infty)$ and for systems which are diffusional in nature (i.e., solutions, complexes, etc.) atoms can essentially occupy any position in space and contributions from the scattered locations to the value of the partition function are not negligible.

   The situation is more optimistic with differences in the free energy between two closely related states (STRAATSMA [1996]), since it can be assumed that contributions from points far apart in the configurational space will be similar, and hence will cancel each other in this case. The free energy difference between two states 0 and 1 is:

$$\Delta A_{0 \to 1} = A_1 - A_0 = -kT \ln \frac{Z_1^c}{Z_0^c},$$

$$\frac{Z_1^c}{Z_0^c} = \langle e^{-\Delta V_{01}(\mathbf{X}^N)} \rangle, \tag{13.2}$$

i.e., the ensemble average of the exponential function involving the difference between potential energy functions for state 0 and 1 calculated at coordinates obtained for the state 0. As was mentioned earlier, the difference between states 0 and 1 should be very small, i.e., a perturbation. For this reason the method is called a *free energy perturbation* (ZWANZIG [1954]). If the difference between states 0 and 1 is larger, the process has to be divided into smaller steps. In practice, the change from state 0 to 1 is represented as a function of a suitably chosen coupling parameter λ such that for state 0 $\lambda = 0$, and for state 1 $\lambda = 1$. This parameter is incorporated into the potential energy function to allow smooth transition between the two states.

Now, the path $\lambda = 0 \rightarrow \lambda = 1$ can be easily split into any number of desired substates spaced close enough to satisfy the small perturbation requirement. Assuming that the process was divided into n equal subprocesses we have: $\lambda_i = i\,\Delta\lambda$, where $\Delta\lambda = \lambda/n$. However, in actual calculations, the division is usually not uniform to account for the fact that the most rapid changes in free energy often occur at the beginning and at the end of the path. The simulation is run for each value of λ providing free energy changes between substates i and $i + 1$, where the coordinates calculated at the substate λ_i are used to calculate ΔV between states λ_i and λ_{i+1}. Then the total change of free energy is calculated as a sum of partial contributions. The most important conclusion from the above is that the conversion between two states does not have to be conducted along a physically meaningful path. Obviously, atoms cannot be converted stepwise to one another, since they are made of elementary particles. However, basic thermodynamics, namely, the Hess's law of constant heat summation, ensures that the difference between free energy depends only upon initial and final states. Any continuous path which starts at state 0 and ends at state 1, even if it is a completely fictitious one, can be used to calculate this value.

There are other methods than perturbation of free energy to calculate the difference in free energies between states. The most accepted one is the *thermodynamic integration method* (MITCHELL and McCAMMON [1991]) where the derivative of free energy A versus coupling parameter λ is calculated. It can be proven that the derivative of the free energy versus λ is equal to the ensemble average of the derivative of potential energy versus λ, and hence:

$$\Delta A_{0 \rightarrow 1} = A_1(\lambda = 1) - A(\lambda = 0) = \int_0^1 \left\langle \frac{\partial V(\mathbf{X}^N)}{\partial \lambda} \right\rangle \mathrm{d}\lambda.$$

In most cases, the calculations are run in both directions, to provide the value of hysteresis, i.e., the precision of the integration process. Large differences between $\Delta A_{0 \rightarrow 1}$ and $\Delta A_{1 \rightarrow 0}$ indicate that either the simulation was not run long enough for each of the substates, or that the number of substates is too small. While checking the hysteresis represents a proof of the precision of the calculations, it is not a proof of the accuracy of the results. Errors in estimating the free energy changes may result from an inadequate energy function and in most cases from inadequate probing of the energy surface of the system. MD, similar to molecular mechanics, can also suffer from the "local minimum syndrome". If atom positions corresponding to the minimum of energy in state 0 are substantially different from those for state 1, the practical limitations of the length of the computer simulation may not allow the system to explore the configuration space corresponding to the true minimum for state 1. For this reason the set of independent simulations, each starting at different atom positions should be performed to assess the adequacy in sampling of the configuration space for both states.

The methods above allow one to calculate in principle the free energies of solvation, binding, etc. In the case of free energies of solvation, for example, one solvent molecule is converted to a solute molecule in a stepwise fashion, by creating and annihilating atoms or changing their type which in effect is creating the solute molecule. Binding energies could in principle be calculated in a similar way by "creating" the ligand molecule in the enzyme cavity. In practice, these calculations are very difficult, since they involve a massive displacement of solvent, i.e., the initial and final state differ dramatically. However, in drug design for example, in most cases one is interested not in the actual free energies of solvation or binding, but in differences between these parameters for different molecules, i.e., in values of $\Delta\Delta A$ (or $\Delta\Delta G$ if an isothermic-isobaric system is being considered). The quantitative method of finding these parameters is often called a *thermodynamic cycle* approach and is becoming a routine procedure in finding the differences in free energy of binding for two different ligands or the influence of mutation in the macromolecular receptor or enzyme on binding.

Besides free energies other thermodynamic quantities can be computed applying statistical concepts to computer simulations. Here we present just two of them, namely pressure and distribution functions. Here, for simplicity's sake we shift to MC notation.

13.1.2. Distribution functions

The evaluation of the pair correlation function $g(\mathbf{r})$ (either center-of-mass or site–site) is a direct application of the formula:

$$\rho g(\mathbf{r}) = \frac{1}{N} \left\langle \sum_{i \neq j} \delta(\mathbf{r} - (\mathbf{r}_i - \mathbf{r}_j)) \right\rangle.$$

For an isotropic system such as a liquid or gas, for which there is no preferred direction in space, only the magnitude r ($|\mathbf{r}| = r$) is of relevance, and thus, the function $g(r)$ depends only on the distance r between two particles and not on the relative orientation.

It is easy to see that the quantity $\rho g(\mathbf{r}) \, d^3\mathbf{r}$ is proportional to the mean number of molecules which are at a distance $\mathbf{r}$ from one supposed to be in the center of the coordinate system (averaged over all the molecules in the sample) or, stated differently, the number of the pairs whose distance $\mathbf{r}$ is divided by the number of molecules. In the large $\mathbf{r}$ limit the position of the pairs are uncorrelated, so this number becomes $\rho \, d^3\mathbf{r}$ and we have the limit

$$\lim_{\tau \to \infty} g(\mathbf{r}) = 1$$

on the other hand, since the particles cannot occupy the same position, we have also the limit

$$\lim_{\tau \to 0} g(\mathbf{r}) = 0.$$

In general $g(\mathbf{r})$ has a large peak at short distances, and an oscillating behavior around the limiting value after the peak.

To compute $g(\mathbf{r})$ an histogram is made with slots of length dr and the number of pairs whose distance is in the range $(r; r + dr)$ is collected in analyzing the data. Since $\rho g(r)\, d^3 r$ is the mean number $\mathcal{N}(r)$ of pairs whose relative distance is r, we have the equality

$$\rho g(\mathbf{r})4\pi r^2 d\mathbf{r} = \frac{1}{M}\mathcal{N}(r)$$

after M configuration have been used; from which it is easily obtained that

$$g(\mathbf{r}) = \frac{1}{\rho 4\pi r^2 d\mathbf{r}}\mathcal{N}(r).$$

It is interesting to notice that, according to the definition, the pair (i, j) and (j, i) (with the obvious condition $i \neq j$) must both be counted. Usually the average is performed only on the pairs such that $i > j$, and every contribution counted twice.

The function $g(\mathbf{r})$, that can be measured in neutron and X-ray diffraction experiments, is important for many reasons. It tells us about the structure of complex, isotropic systems and it determines the thermodynamic quantities at the level of the pair potential approximation, such as, for example the pressure

$$P = \frac{NKT}{V} - \frac{N^2}{6V^2}\int r\frac{du}{dr}g(r)dr,$$

where $u(r)$ is the pair potential and V the volume.

13.2. Calculation of dynamical quantities

MD simulations give access to the time evolution of the system. The time correlation function formalism provides a systematic way to study this evolution. For a MD-simulation of an equilibrium system, the time correlation function C is defined:

$$C_{AB}(\tau) = \langle A(t)B(\tau)\rangle = \lim_{T\to\infty}\frac{1}{T-\tau}\int_0^{T-\tau} A(t)B(t+\tau)dt,$$

where A and B are dynamical variables. In the computational practice, C can be obtained as a straightforward application of the discrete time version of such definition,

$$C_{AB}(\tau) = \frac{1}{T-\tau}\sum_{t=0}^{T-\tau} A_i(t)B_i(t+\tau).$$

One has to take care of the fact that, since we have only a finite number of time-steps saved for the analysis, the correlation function at longer times have less statistics than the correlation at small times, and so the total number of steps must be large enough.

If $A = B$ the term autocorrelation function is used. An autocorrelation function commonly calculated in simulations of atomic liquids is the velocity autocorrelation function. It is of the form $\langle \mathbf{v}_i(t) \cdot \mathbf{v}_i(0) \rangle$. If all the atoms are of the same type, then the result should be independent of i, so we may as well calculate this for each of the atoms and average.

$$\frac{1}{N} \sum_i^N \langle \mathbf{v}_i(t) \cdot \mathbf{v}_i(0) \rangle.$$

Typical behavior of a velocity autocorrelation function
- The value at $t = 0$ is $\langle \mathbf{v}^2 \rangle = 3kT/m$.
- It goes to zero for large times.
- The decay might be monotonic, or go negative and then to zero, or oscillate slightly.

The reason for interest in the velocity autocorrelation function is that it is related to the self-diffusion coefficient

$$D = \frac{1}{3} \int_0^\infty dt \langle \mathbf{v}_i(t) \cdot \mathbf{v}_i(0) \rangle$$

and it is the simplest and easiest transport coefficient to calculate. Other things, like the shear viscosity coefficient, that are more commonly measured experimentally, are much harder to calculate from simulations. Thus, if any transport or time fluctuation properties are calculated, this is the one that is chosen. To calculate the self-diffusion coefficient using this formula, it is necessary to calculate the velocity autocorrelation function for all positive times out to the time where the function is essentially zero.

There is a somewhat easier way of calculating the self-diffusion coefficient in simulations, using the mean squared displacement (MSD). The MSD is defined as

$$\langle |\mathbf{r}_i(t) - \mathbf{r}_i(0)|^2 \rangle.$$

Since all particles have the same statistical properties in a one component system, we might as well calculate

$$\frac{1}{N} \sum_i^N \langle |\mathbf{r}_i(t) - \mathbf{r}_i(0)|^2 \rangle.$$

The relationship to the self-diffusion coefficient is that the MSD is equal to $6Dt$ at long times.

The Fourier transform of velocity autocorrelation function is called the spectral density of the particle motions. By making additional assumptions it can be related to Infrared-, Raman-, and neutron inelastic scattering intensities. More in general, time autocorrelation functions are related via Fourier transformation to spectral densities or simply spectra. For example the infrared spectrum is given by

$$I(\omega) \propto \int_0^T dt \exp(i\omega\tau) \langle \mu_i(t) \cdot \mu_i(0) \rangle,$$

where the quantity in brackets is the normalized time autocorrelation function of the system dipole moment vector.

The importance of correlation functions is large. Besides being used to compute dynamical properties as shown above, they are also used to see if simulation has progressed far enough so that calculated quantities are no longer dependent on initial conditions They can be used to look at molecular order in simulation, such as dipole moment, and angular velocity (molecular orientation).

14. Molecules in solution

In this section we shall consider an important application of solvation models: the study of the interactions between molecular systems and external electromagnetic fields in the presence of a solvent. The subject is so large and the related physical phenomena so numerous – suffice it to quote the various spectroscopies – that we shall necessarily limit our exposition to some specific aspects. In particular, we shall present properties that cannot be computed by computer simulation techniques (either MC or MD) as they require accurate QM calculations. In this context, continuum solvation methods represent the most effective approach and for this reason we shall focus on these methods only; in particular, we shall describe how a specific continuum method, the PCM-IEF (see Section 4 of Chapter II), has been generalized to evaluate response properties of molecules in solution to electric and magnetic fields, or to their combination. We recall that PCM-IEF is here used as it is the method developed by our group and thus the one we know better but other continuum models can be alternatively used to get response properties; it is however worth saying that PCM-IEF represents one of the most accurate continuum model at the QM level of calculation and certainly, the one with the largest applicability in the study of phenomena related to the interaction of solvated systems and external fields.

The generalization of continuum models to aspects of solvation going beyond the energetics has been made possible due the their specific characteristic to reduce solvent effects to a set of operators which can be cast in a physically and formally simple form. In this framework, the inclusion of the formalism of continuum models into the various approaches provided by the QM theory becomes almost straightforward. In addition, the use of an accurate representation of the solvent field through an apparent surface charge when joined to the definition of a realistic molecular cavity embedding the solute immersed in the dielectric (see previous chapters for more details), makes the continuum methods particularly suitable for this kind of studies. Finally, their recent extension to different environments, like anisotropic solvents, ionic solutions, immiscible solvents with a contact surface, etc., allows, for the first time, the analysis of important response properties also for molecular systems immersed in not standard external matrices, such as liquid crystalline phases or symmetric crystalline frames, charged solutions like those representing the natural neighborhood of proteins and other biological molecules and membranes, just to quote few examples.

Before passing to present the methodologies developed to compute electric and magnetic properties of solvated molecules, an introductory presentation of the basic aspects of the inclusion of solvation continuum models into the QM formalism is necessary; as said we exploit here the PCM formalism as representative example of QM solvation continuum models.

The generalization of continuum models to QM calculations implies to define an _Effective Hamiltonian_, i.e., an Hamiltonian to which solute–solvent interactions are added in terms of a solvent reaction potential. The basic hypothesis of this kind of approach is that one can always define a free energy functional $\mathcal{G}(\Psi)$ depending on the solute electronic wavefunction Ψ. This energy functional can be expressed in the following general form (AMOVILLI, BARONE, CAMMI, CANCÈS, COSSI, MENNUCCI, POMELLI and TOMASI [1998]):

$$\mathcal{G}(\Psi) = \langle \Psi | \hat{H}^0 | \Psi \rangle + \langle \Psi | \hat{\rho}_r | \Psi \rangle V_r^R + \frac{1}{2} \langle \Psi | \hat{\rho} | \Psi \rangle V_{rr'}^R \langle \Psi | \hat{\rho}_{r'} | \Psi \rangle. \tag{14.1}$$

In Eq. (14.1) the Born–Oppenheimer (BO) approximation is employed. This means a standard partition of the Hamiltonian into an electronic and a nuclear part, as well as the factorization of the wavefunction into an electronic and a nuclear component. In this approximation Eq. (14.1) refers to the electronic wavefunction with the electronic Hamiltonian dependent on the coordinates of the electrons, and, parametrically, on the coordinates of the nuclei.

The detailed description of the various terms comparing in the equation above will be given in the following, here it suffices to say that $\hat{H}^0$ is the Hamiltonian describing the isolated molecule, $\hat{\rho}_r$ represents the operator of the solute electronic charge density, V_r^R is the solvent permanent potential, and $\mathbf{V}_{rr'}^R$ describes the _response function of the reaction potential_ associated with the solvent. Here an extension of the Einstein convention on the sum has been exploited: the space variables r and r', appearing as repeated subscripts, imply an integration in the 3-dimensional space.

By applying the variational principle on this functional we can derive the nonlinear Schrödinger equation specific for the system under scrutiny:

$$\hat{H}_{\text{eff}} | \Psi \rangle = [\hat{H}^0 + \hat{\rho}_r V_r^R + \hat{\rho}_r \mathbf{V}_{rr'}^R \langle \Psi | \hat{\rho}_{r'} | \Psi \rangle)] | \Psi \rangle = E | \Psi \rangle \tag{14.2}$$

where E is the Lagrange multiplier introduced to fulfill the normalization condition on the electronic wavefunction. Eq. (14.2) defines the specific 'Effective Hamiltonian', $\hat{H}_{\text{eff}}$, giving the name to the whole procedure.

The first solvent term V_r^R does not lead to any difficulty, neither from the theoretical point of view, nor from the practical. Many examples are known in which an external potential is introduced in the molecular calculations. On the contrary, the treatment of the reaction potential operator $\hat{\rho}_r \mathbf{V}_{rr'}^R \langle \Psi | \hat{\rho}_{r'} | \Psi \rangle$ is rather delicate, as this term induces a nonlinear character to the solute Schrödinger equation.

By imposing that first-order variation of $\mathcal{G}$ with respect to an arbitrary variation of the solute wavefunction Ψ is zero, and following the standard scheme developed for the self consistent field theory in vacuo, we finally obtain the generalized Fock operator $\hat{F}$ and the corresponding equation from which the final wavefunction has to be derived. By introducing the common finite-basis approximation and a closed shell system, we can eliminate the spin dependence (the occupied spin-orbitals occur in pairs) and expand the molecular orbitals (MOs) as a linear combination of atomic orbitals (LCAO). On performing the spin integration in the equations used so far, we find, for the free energy:

$$\mathcal{G} = \text{tr}\mathbf{Ph} + \frac{1}{2}\text{tr}\mathbf{PG}(\mathbf{P}) + \text{tr}\mathbf{Ph}^R + \frac{1}{2}\text{tr}\mathbf{PX}^R(\mathbf{P}) \tag{14.3}$$

and for the generalized Fock matrix:

$$\mathbf{F}' = \mathbf{h} + \mathbf{G}(\mathbf{P}) + \mathbf{h}^R + \mathbf{X}^R(\mathbf{P}), \tag{14.4}$$

where $\mathbf{P}$ is the one-electron density matrix and $\mathbf{h}$ and $\mathbf{G}(\mathbf{P})$ are the matrices used in standard calculations *in vacuo* to collect one- and two-electron integrals, respectively. In Eqs. (14.3), (14.4) the solvent contributions, previously introduced as potential functions, have been translated into a form recalling the vacuum system, just to emphasize the parallelism between the two calculations; in fact, also solvent effects can be partitioned in one- and two-electron contributions, indicated as $\mathbf{h}^R$ and $\mathbf{X}^R(\mathbf{P})$, respectively. In general, $\mathbf{h}^R$ and $\mathbf{X}^R(\mathbf{P})$ will contain different terms related to all possible interactions (dispersive, repulsive and electrostatic) between solute and solvent. In particular for the electrostatic parts we have

$$h^{el} = -\sum_k V(s_k)q^N(s_k),$$
$$X^{el}(P) = -\sum_k V(s_k)q^e(s_k),$$

where q^N and q^e represent the apparent charges induced on the cavity surface by the solute nuclei and electrons, respectively, and $\mathbf{V}$ collects the AO potential integrals on the cavity; the sum runs on all the tessera forming the cavity surface (see Sections 4 and 6 of Chapter II).

In this framework the generalized Fock equation can be solved with the same iterative procedure of the problem *in vacuo*; the only difference introduced by the presence of the continuum dielectric is that, at each SCF cycle, one has to simultaneously solve the standard QM problem and the additional problem of the evaluation of the interaction matrices. In this scheme the apparent charges are obtained through a self-consistent technique which has to be nested in that determining the solute wavefunction; as a consequence, at the convergency, solute and solvent distribution charges are mutually equilibrated.

As final note, we observe that the nonlinear equation (14.2) is a direct consequence of the variational principle applied to $\mathcal{G}$. The (free) energy functional $\mathcal{G}$ has a privileged role in the theory, as the solution of the Schrödinger equation gives a minimum of this functional even though it is not the eigenvalue of the nonlinear Hamiltonian, here indicated as E. We stress that in these habitual linear Hamiltonians these two quantities, the Hamiltonian eigenvalue and the variational functional, coincide. The difference between E and $\mathcal{G}$ has, however, a clear physical meaning; it represents the polarization work which the solute does to create the charge density inside the solvent. It is worth remarking that this interpretation is equally valid for zero-temperature models and for those in which the thermal agitation is implicitly or explicitly taken into account.

To pass from the free energy functional $\mathcal{G}$ defined above to the thermodynamical analog G, further aspects have to be considered. As said in the introductory chapter, first we have to choose the reference state (in our case it is given by noninteracting nuclei and electrons of M, supplemented by the unperturbed, i.e., unpolarized, pure liquid S), and to take into account the contributions of the interactions not included in the reaction potential.

14.1. Electric dipole polarizabilities

In this section we are concerned with the calculation of electric properties which measure the response of a charge distribution (the solute molecule) to an external electric field. If the charge distribution is mobile, then it will redistribute itself until its energy in the external field is minimized, and this is the phenomenon of polarization. The electric moments will therefore change in the external field, and we can study the change by expanding the dipole moments as a Taylor series. The expansion terms are collected by order into the so-called polarizability tensors $\gamma^{(n)}$. We shall follow the usual convention which indicates as α, β, γ the tensors corresponding to the first three $\gamma^{(n)}$ sets of coefficients, namely the polarizability, the first and the second hyperpolarizabilities.

When the external field has an oscillatory behavior, all these quantities depend on the frequency of such oscillations; for a given $\gamma^{(n)}$ we have to consider the frequencies and phases of the various components of external fields that can be combined in all possible ways to give different electric molecular response. These elements constitute the essential part of the linear and nonlinear optics, a subject for which there is a remarkable interest to know the influence of solvation effects.

Nonlinear optics (NLO) deals with the interaction of electromagnetic fields (light) with matter to generate new electromagnetic fields, altered with respect to phase, frequency, amplitude or other propagation characteristics from the incident field. A major advantage of the use of photonics instead of electronics is the possibility to increase the speed of information processes such as photonic switching and optical computing. One of the most intensively studied nonlinear optical phenomena is second harmonic generation (SHG) or frequency doubling. By this process, near infrared laser light (frequency ω) can be converted by a nonlinear optical material to blue light (2ω). The resulting wavelength is half the incident wavelength and hence it is possible to store information with a higher density. It is obvious that the required properties of the materials depend on the application that they are used for. Traditionally, the materials used to measure second order nonlinear optical behavior were inorganic crystals. Organic materials, such as organic crystals and polymers, have been shown to offer better nonlinear optical and physical properties, such as ultrafast response times, lower dielectric constants, better processability characteristics and a remarkable resistance to optical damage, when compared to the inorganic materials. The ease of modification of organic molecular structures makes it possible to synthesize tailor-made molecules and to fine-tune the properties to the desired application. In the case of second-order nonlinear optical processes, the macroscopic nonlinearity of the material (bulk susceptibility) is derived from the microscopic molecular nonlinearity and the geometrical arrangement of the NLO-chromophores. So, optimizing a material's nonlinearity begins at the molecular structural level.

The molecular response to an electric field regards its whole charge distribution, electron and nuclei. We may introduce also for molecules in solution the usual partition of the theoretical chemistry into electronic and nuclear parts, without neglecting couplings.

14.1.1. Electronic contribution to dipole polarizabilities

The electronic contribution can be computed using two derivative schemes involving QM calculations of the energy or, alternatively, of the dipole moment followed by derivatives with respect to the perturbing external field, computed at zero intensity. At Hartree–Fock (HF) or Density Functional (DF) level both approaches lead to the use of the coupled HF or Kohn–Sham theory either in its time-independent (CHF or CKS) or time-dependent (TDHF or TDKS) version according to the case (McWEENY [1992]).

The solute Hamiltonian must be now supplemented by further terms describing the interaction with the external field; in particular, the corresponding one-electron contribution to the Fock matrix $\mathbf{F}'$ (14.4) becomes

$$\mathbf{h}' = \mathbf{h} + \mathbf{h}^R + \sum_a \mathbf{m}_a E_a,$$

where $\mathbf{m}_a$ is the matrix of the ath Cartesian component of the dipole moment operator, and E_a is the corresponding component of the electric field vector.

To obtain the various electric response functions we have to determine the density matrix of the unperturbed system $\mathbf{P}^0$ and its derivatives ($\mathbf{P}^a$, $\mathbf{P}^{ab}$, etc.) with respect to the components of the electric field, corresponding to the order of the electric response functions.

To obtain time-dependent properties, we have to shift from the basic static model to an extended version in which the solute is described thorough a time-dependent Schrödinger equation. In this extended version of the model we have also to introduce the time-dependence of the solvent polarization, which is expressed in terms of a Fourier expansion and requires the whole frequency spectrum of the dielectric permittivity $\varepsilon(\omega)$ of the solvent.

Applying the Frenkel variational principle at HF or DFT level, we arrive at the following time-dependent equation (McWEENY [1992])

$$\left(\mathbf{F}' - i\frac{\partial}{\partial t}\right)\mathbf{T} = \mathbf{T}\varepsilon, \tag{14.5}$$

where $\mathbf{T}$ is the matrix containing the expansion coefficients of the MOs on the atomic orbital basis set, and ε is the matrix collecting the Lagrangian multipliers. The solutions of he TDHF (or TDKS) equation can be obtained in a time-dependent coupled perturbation scheme. We first expand Eq. (14.5) in its Fourier components and then each component is expanded in terms of the components of the external field. The separation by orders leads to a set of coupled perturbed equations whose Fock matrices can be written in the following form:

$$\mathbf{F}' = \mathbf{h}' + \mathbf{G}(\mathbf{P}_0) + \mathbf{X}_0(\mathbf{P}_0),$$

$$\mathbf{F}'^a(\omega_\sigma; \omega) = \mathbf{m}_a + \mathbf{G}(\mathbf{P}^a(\omega_\sigma; \omega)) + \mathbf{X}_{\omega_\sigma}(\mathbf{P}^a(\omega_\sigma; \omega)),$$

$$\mathbf{F}'^{ab}(\omega_\sigma; \omega_1, \omega_2) = \mathbf{G}(\mathbf{P}^{ab}(\omega_\sigma; \omega_1, \omega_2)) + \mathbf{X}_{\omega_\sigma}(\mathbf{P}^{ab}(\omega_\sigma; \omega_1, \omega_2)),$$

$$\mathbf{F}'^{abc}(\omega_\sigma; \omega_1, \omega_2, \omega_3) = \mathbf{G}(\mathbf{P}^{abc}(\omega_\sigma; \omega_1, \omega_2, \omega_3)) + \mathbf{X}_{\omega_\sigma}(\mathbf{P}^{abc}(\omega_\sigma; \omega_1, \omega_2, \omega_3)),$$

$$(14.6)$$

where ω_σ is the frequency of the resulting wave (i.e., $\omega_\sigma = \sum_k \omega_k$, with ω_k including the sign), and matrices $\mathbf{X}_{\omega_\sigma}(\mathbf{P}^{ab\cdots}(\omega_\sigma))$ represent the interactions with the solvent which has been induced by the perturbed electron density $\mathbf{P}^{ab\cdots}(\omega_\sigma)$ oscillating at frequency ω_σ (CAMMI, COSSI, MENNUCCI and TOMASI [1996]). There is a number of equations of this type, corresponding to the different combinations of frequencies, each one related to a different phenomenon of the nonlinear optics.

Once the solution of the different TDHF (or TDKS) equations has been obtained, the dynamic (hyper)polarizabilities of interest can be expressed in the following forms:

$$\alpha_{ab}(\omega_\sigma; \omega) = -\mathrm{tr}[\mathbf{m}_a \mathbf{P}^b(\omega_\sigma; \omega)],$$

$$\beta_{abc}(\omega_\sigma; \omega_1, \omega_2) = -\mathrm{tr}[\mathbf{m}_a \mathbf{P}^{bc}(\omega_\sigma; \omega_1, \omega_2)], \qquad (14.7)$$

$$\gamma_{abcd}(\omega_\sigma; \omega_1, \omega_2, \omega_3) = -\mathrm{tr}[\mathbf{m}_a \mathbf{P}^{bcd}(\omega_\sigma; \omega_1, \omega_2, \omega_3)],$$

where tr indicates a trace operation on the product of the dipole integral matrix $\mathbf{m}_a$ and the derivatives of the electronic densities $\mathbf{P}^{abc}$.

Since the frequency-dependent response of the solvent is included in the kernel of the integrals collected into $\mathbf{X}_{\omega_\sigma}$, the resulting (hyper) polarizabilities will also depend on the frequency spectrum of the dielectric function $\varepsilon(\omega)$ of the solvent. When $\varepsilon(\omega)$ is described by the Debye formula (i.e., in terms of a single relaxation mode), the resulting dispersion curve depends on the frequency corresponding to the inverse of the Debye relaxation time.

We have remarked that there is a large number of coupled Fock equations to be solved in order to get electric response properties. However, as for the case of a molecule in vacuo, the computational effort can be reduced by resorting the exploiting to so called $(2n + 1)$ rule that permits to get the $(2n + 1)$th response using nth derivative of $\mathbf{P}$. This is a formal property of the perturbative scheme which is known since a long time and fulfilled by all good computational codes, when applied to molecules in vacuo: the same property also holds for molecules in solution.

14.1.2. Nuclear contribution to electric polarizabilities

The global effect of an applied external field on a molecule involves distortions both in the electronic charge distribution and in the nuclear charge distribution; the latter leads to the so-called vibrational, or nuclear, contribution to the (hyper)polarizabilities (BISHOP [1998]).

The analysis of the vibrational components reveals the presence of the distinct components, the "curvature" related to the effect of the field vibrational motion and including the zero point vibrational (ZPV) correction, and the "nuclear relaxation" (nr) originates from the shift of the equilibrium geometry induced by the field.

The nuclear relaxation is the dominant contribution and can be computed in two ways: by perturbation theory or by finite field approximation. We shall limit ourselves to the perturbation theory. In this scheme the polarizabilities which can be written in terms of a sum over vibrational states k of energy-weighted transition moments, for example:

$$\alpha_{ab} = \sum_{k \neq 0} \frac{\langle 0|\mu_a|k\rangle \langle k|\mu_b|0\rangle}{\hbar \nu_k}, \tag{14.8}$$

where $\hbar \nu_k$ is the energy of the vibrational state $|k\rangle$ relative to the ground state $|0\rangle$, which is excluded from the sum.

The vibrational transition moments can be obtained through an expansion of the electronic properties (dipole and polarizability functions) in terms of the nuclear coordinate of the molecule, analogous to the expansion of the vibrational potential within the Born–Oppenheimer approximation. For a molecule in solution the potential energy is given by $\mathcal{G}(\mathbf{R})$ as function of the nuclei configuration $\mathbf{R}$. If, we separate the components according their anharmonicity, at level of double harmonicity (electric and mechanical) only linear term are considered in the expansion of the properties and only quadratic terms are considered in the expansion of the vibrational potential. The corresponding expressions of the vibrational contribution to the electric (hyper) polarizabilities become:

$$\alpha_{ab}^v = \sum_{i}^{3N-6} \left(\frac{\partial \mu_a}{\partial Q_i}\right)_0 \left(\frac{\partial \mu_b}{\partial Q_i}\right)_0 \bigg/ (4\pi^2 \nu_i^2),$$

$$\beta_{abc}^v = \sum_{i}^{3N-6} \left[\left(\frac{\partial \mu_c}{\partial Q_i}\right)_0 \left(\frac{\partial \alpha_{ab}}{\partial Q_i}\right)_0 + \left(\frac{\partial \mu_b}{\partial Q_i}\right)_0 \left(\frac{\partial \alpha_{ac}}{\partial Q_i}\right)_0 + \left(\frac{\partial \mu_a}{\partial Q_i}\right)_0 \left(\frac{\alpha_{bc}}{\partial Q_i}\right)_0 \right] \bigg/ (4\pi^2 \nu_i^2),$$

where ν_i is the harmonic frequency obtained by the eigenvalues of the nuclear Hessian computed in presence of solvent effects; its elements are the second derivatives of the free energy $\mathcal{G}(\mathbf{R})$ with respect the mass weighted nuclear Cartesian components. Q_i denotes the normal mode associated to ν_i and each partial derivative is evaluated at the equilibrium geometry of the solvated system (CAMMI, MENNUCCI and TOMASI [1998b]).

14.2. Macroscopic susceptibilities: Local field effects and effective polarizabilities

The phenomenological description of the polarization of a macroscopic medium subjected to an electric field and of the connected phenomenon of linear and nonlinear response is given in terms of the macroscopic susceptibilities $\chi^{(n)}$. As of particular interest, here we focus on the optical phenomena connected with linear and nonlinear response properties of a medium subjected to an electric field given by the superposition of a static and an optical component:

$$\mathbf{E} = \mathbf{E}^0 + \mathbf{E}^\omega \cos \omega t. \tag{14.9}$$

In this case the polarization density $P(t)$ of the medium can be written in terms of Fourier components (BUTCHER and COTTER [1990]):

$$P(t) = \mathbf{P}^0 + \mathbf{P}^\omega \cos(\omega t) + \mathbf{P}^{2\omega} \cos(2\omega t) + \mathbf{P}^{3\omega} \cos(3\omega t),$$

where

$$\mathbf{P}^0 = \chi^{(0)} + \chi^{(1)}(0;0)\cdot\mathbf{E}^0 + \chi^{(2)}(0;0,0) : \mathbf{E}^0\mathbf{E}^0 + \frac{1}{2}\chi^{(2)}(0;-\omega,\omega) : \mathbf{E}^\omega\mathbf{E}^\omega$$
$$+ \chi^{(3)}(0;0,0,0) : \mathbf{E}^0\mathbf{E}^0\mathbf{E}^0 + \cdots,$$

$$\mathbf{P}^\omega = \chi^{(1)}(-\omega;\omega)\cdot\mathbf{E}^\omega + 2\chi^{(2)}(-\omega;\omega,0) : \mathbf{E}^\omega\mathbf{E}^0 + 3\chi^{(3)}(-\omega;\omega,0,0) : \mathbf{E}^\omega\mathbf{E}^0\mathbf{E}^0 + \cdots,$$

$$\mathbf{P}^{2\omega} = \frac{1}{2}\chi^{(2)}(-2\omega;\omega,\omega) : \mathbf{E}^\omega\mathbf{E}^\omega + \frac{3}{2}\chi^{(3)}(-2\omega;\omega,\omega,0) : \mathbf{E}^\omega\mathbf{E}^\omega\mathbf{E}^0 + \cdots. \tag{14.10}$$

In Eqs. (14.10) the argument of the susceptibilities tensors $\chi^{(n)}$ describes the nature of the frequency-dependence at the various order; in all cases the frequency of the resulting wave (which is indicated as ω_σ) is stated first, then the frequency of the incident interacting wave(s) (two in a first-order process, three in the second-order analog and four in the third-order case).

The various susceptibilities may be obtained through specific experiments in linear and nonlinear optics. The first-order static susceptibility $\chi^{(1)}$ (0; 0) is related to the dielectric constant at zero frequency, $\varepsilon(0)$, while $\chi^{(1)}(-\omega; \omega)$ is the linear optical susceptibility related to the refractive index n^ω at frequency ω. Passing to nonlinear effects it is worth recalling that $\chi^{(2)}(-2\omega; \omega, \omega)$ describes frequency doubling process which is usually indicated as SHG, and $\chi^{(3)}(-2\omega; \omega, \omega, 0)$ describes the influence of an external field on the SHG process, measured in the third-order process of electric-field induced second harmonic generation (EFISHG).

If we consider as macroscopic sample a liquid solution of different molecular components, each at a concentration c_J, the effects of the single components are assumed to be additive so that the global measured response becomes:

$$\chi^{(n)} = \sum_J \zeta_J^{(n)} c_J,$$

where $\zeta_J^{(n)}$ are the nth-order molar polarizabilities of the constituent J. The values of the single $\zeta_J^{(n)}$ can be extracted from measurements of $\chi^{(n)}$ at different concentrations.

The electric response properties of molecular materials, as a liquid solution, on which the macroscopic electric susceptibilities $\chi^{(n)}$ depend upon are the (hyper)polarizabilities of the constituting molecules. For the gas phase the connection between the molecular hyperpolarizabilities and the corresponding susceptibilities is given in terms of a Boltzmann thermal average process regarding the permanent and the induced molecular dipole moments. To establish the same connection for a condensed phase we have also to face the so called "Local Field Problem". The problem arises because the field acting locally on a molecule is different from the applied Maxwell field $\mathbf{E}(t)$ due to the presence of all the solvent molecules around it (WORTMANN and BISHOP [1998]).

Studies on this problem date to the beginning of the past century: 'almost' empirical recipes are based on the Onsager local field correction factor for polar liquids and the Lorentz factor for nonpolar liquids. Both factors apply to cavities of regular shape, spheres or ellipsoid (BÖTTCHER [1973], BÖTTCHER and BORDEWIJK [1978]).

In the solvation PCM-IEF model a more general description of the local field effect can be obtained (CAMMI, MENNUCCI and TOMASI [1998a], CAMMI, MENNUCCI and TOMASI [2000]). In this method the effect is described by supplementing the solute Hamiltonian by a further term describing the interaction of the solute with the apparent surface charges $\mathbf{q}^{ex}$ induced by the external field $E(\omega)$. The resulting one-electron contribution to the Fock matrix is then given by:

$$\mathbf{h}' = \mathbf{h} + \mathbf{h}^R + \sum_a \mathbf{m}_a \left[E_a^\omega (e^{i\omega t} + e^{-i\omega t}) + E_a^0 \right] + \sum_a \widetilde{\mathbf{m}}_a^\omega E_a^\omega (e^{i\omega t} + e^{-i\omega t})$$
$$+ \sum_a \widetilde{\mathbf{m}}_a^0 E_a^0,$$

$$(14.11)$$

where the elements of $\widetilde{\mathbf{m}}_a$ are defined in terms of the additional set of apparent charges, $\mathbf{q}^{ex}$, induced on the cavity surface by the external field, namely (CAMMI, MENNUCCI and TOMASI [1998a], CAMMI, MENNUCCI and TOMASI [2000]):

$$\widetilde{\mathbf{m}}_\alpha(\omega) = \sum_k \mathbf{V}(s_k) \frac{\partial \mathbf{q}^{ex}(\omega; s_k)}{\partial E_\alpha},$$

where we have indicated the dependence of the apparent charges and of the following $\widetilde{\mathbf{m}}$ matrix on the frequency ω of the applied field; such dependence is obtained by using a frequency-dependent permittivity $\varepsilon(\omega)$ to compute the apparent charges (CAMMI, MENNUCCI and TOMASI [2000]). The two last terms in the r.h.s. of Eq. (14.11) influence the evaluation of the derivatives of the density matrices required to get the (hyper)polarizabilities (see Eqs. (14.7)); they in fact modify the first-order derivative of the Fock matrix given in (14.6) as follows:

$$\mathbf{F}'^a(\omega_\sigma; \omega) = \mathbf{m}_a + \widetilde{\mathbf{m}}_a^\omega + \mathbf{G}(\widetilde{\mathbf{P}}^a(\omega_\sigma; \omega)) + \mathbf{X}_{\omega_\sigma}(\widetilde{\mathbf{P}}^a(\omega_\sigma; \omega)). \qquad (14.12)$$

The new set of density matrix derivatives, $\widetilde{\mathbf{P}}^{ab\cdots}(\omega_\sigma)$, obtained by applying the usual procedure to this new system, can be then exploited to compute the so called *effective (hyper)polarizabilities*: (see Eqs. (14.7)) which describe directly the electric response of the molecular solute to the applied Maxwell field $E(\omega)$.

The macroscopic susceptibilities can now be obtained as statistical average of the corresponding effective (hyper)polarizabilities. Expressions of the molar polarizabilities for linear processes are:

$$\zeta^{(1)}(0; 0) = N_A \left(\frac{\mu^* \cdot \widetilde{\mu}}{3kT} + \widetilde{\alpha}_{is}(0; 0) \right), \qquad (14.13)$$

$$\zeta^{(1)}(-\omega; \omega) = N_A \widetilde{\alpha}_{is}(-\omega; \omega) \qquad (14.14)$$

and as an example for higher-order processes, we quote here the EFISHG:

$$\zeta^{(3)}_{ZZZZ}(-2\omega;\omega,\omega,0) = N_A\left(\frac{\widetilde{\beta}_v(-2\omega;\omega,\omega)\cdot\mu^*}{15kT} + \widetilde{\gamma}_s(-2\omega;\omega,\omega,0)\right), \qquad (14.15)$$

where N_A is the Avogadro number and k is the Boltzmann constant. In Eqs. (14.13)–(14.14) $\widetilde{\alpha}_{is}$ is 1/3 of the trace of the effective polarizability tensor while μ^* of Eq. (14.15) is the *effective dipole moment* of the solute given by the first derivative of the free energy $\mathcal{G}$ with respect to the static component $\mathbf{E}^0$ of the field (14.9); $\widetilde{\beta}_v(-2\omega;\omega,\omega)$ and $\widetilde{\gamma}_s(-2\omega;\omega,\omega,0)$ of Eq. (14.15) are, respectively, the "vector part" of the second order polarizability and the "scalar part" of the third order polarizability. Parallel expressions for other NLO process can be easily formulated.
As the molar polarizabilities $\zeta^{(n)}_J$ represent an easily available 'experimental' set of data, the expressions above become important for the comparison between theoretical and experimental evaluation of bulk response properties, as we shall show in the next section.

14.3. A numerical example: Susceptibilities of nitroanilines

Organic molecules that exhibit large nonlinear optical properties usually consist of a frame with a delocalized π-system, end-capped with either a donor (D) or acceptor (A) substituent or both. This asymmetry results in a high degree of intramolecular charge-transfer (ICT) interaction from the donor to acceptor, which seems to be a prerequisite for a large nonlinearity. Extensively studied classes of NLO-chromophores of this type are 1,4-disubstituted benzenes from which p-nitroaniline (pNA) is a prototypical example. The nonlinear part of the induced molecular polarization is the result of the polarizability of the π-electron system. Because of the low mass of electrons compared to that of ions in inorganic crystals, organic molecules can respond to electromagnetic fields with much higher frequencies (up to 10^{14} Hz). This faster optical response is particularly interesting for applications that depend on the speed of information processing, such as optical switching. The nonlinearity can be enhanced by using stronger donor and acceptor substituents to increase the electronic asymmetry or by increasing the conjugation length between the substituents.

Here we present results of a study on molar polarizabilities for pNA in different liquid solutions. In this presentation we collect results extracted from a research we have published on Journal of Physical Chemistry A (CAMMI, MENNUCCI and TOMASI [2000]) and unpublished data. The attention will be focussed on two specific experimental processes from which data of first and third-order molar polarizabilities have been extracted: refractometric and EFISH measurements.

The results refer to HF and DFT calculations with a Dunning double-zeta valence (DZV) basis set to which $d(0.2)$ function for C, N and O and a $p(0.1)$ function on H have been added, (the numbers in parentheses are the exponents of the extra functions). For DFT calculations the hybrid functional which mixes the Lee, Yang and Parr functional for the correlation part and Becke's three-parameter functional for the exchange (B3LYP) has been used. The solvation method used is the PCM-IEF with a molecular cavity obtained in terms of interlocking spheres centered on the six carbons of the aromatic ring, and on all the nuclei of the external groups.

The geometry of the two solutes has been optimized at B3LYP/6–311G* level in the presence of the solvent. All the effective electronic properties (both static and dynamic) have been computed with the TDHF/TDKS procedures implemented in two standard computational packages Gaussian (FRISCH ET AL. [1998]) and GAMESS (SCHMIDT, BALDRIDGE, BOATZ, ELBERT, GORDON, JENSEN, KOSEKI, MATSUNAGA, NGUYEN, SU, WINDUS, DUPUIS and MONTGOMERY [1993]) and properly modified to take into account the solvent effects.

In Table 14.1 we report the refractometric molar polarizability of pNA in three different solutions: dioxane, acetone and acetonitrile. The specific expressions to be used for the property is reported in (14.14), where the exploited frequency is that corresponding to $\lambda = 589$ nm. Three sets of computed values are reported corresponding to static, dynamic and effective calculations, respectively. Here the terms static and dynamic refer to the model exploited to describe the solvent response: namely a medium described by its static dielectric constant $\varepsilon(0)$ or by its dynamical analog $\varepsilon(\omega)$. Finally, the effective values take into account both the dynamical response of the solvent and the "local field effects" described in the previous section.

The comparison with the experimental values clearly shows the increasing accuracy of the computed results going from the static, to the dynamic and finally to the effective model. It has to be noted that exploiting a static or a dynamic model makes significant differences in the two polar solvents (acetone and acetonitrile) but not in the apolar dioxane: in apolar solvents, in fact, the static and the dynamic permittivities are very similar as the dielectric response of the solvent molecules is almost exclusively described by the electronic motions (those inducing the dynamic $\varepsilon(\omega)$) while the small or null dipolar character does not significantly contribute.

The further important improvement obtained passing from dynamic to effective values indicates that the solvation model has to take into account not only the molecular aspects of the phenomenon (through an accurate description of the solute-solvent interactions) but also the macroscopic aspects involved in the experimental measurement. This is here achieved by explicitly introducing the effect of the oscillating external field on the solvent through a further operator to be considered in the TDHF/TDKS scheme (see Eq. (14.11)).

Let us pass now to the second application we have introduced at the beginning of the section, namely that of the evaluation of EFISH second order susceptibilities.

TABLE 14.1

Computed and experimental frequency dependent first order molar polarizabilities $\zeta^{(1)}(\omega)$ of pNA in different solutions. All molar polarizabilities are in SI units (10^{-16} cm^2 V^{-1} mol^{-1}). The frequency corresponds to $\lambda = 589$ m

	Static	Dynamic	Effective	exp[a]
Dioxane	13.2	13.2	15.1	15.7 ÷ 0.5
Acetone	19.3	14.5	16.4	16.26 ± 0.5
Acetonitrile	19.8	14.6	16.5	16.28 ± 0.3

[a]WOLFF and WORTMANN [1999].

The EFISH technique (PRASAD and WILLIAMS [1991]) is one of the most used for obtaining information on the molecular hyperpolarizability, β; here, once again, we do not report any details but just some basic notes. The operating frequency is that related to the fundamental beam of 1064 nm from a Q-switched, mode-locked Nd: YAG laser. A symmetry consideration shows that the third-order nonlinearity that is measured in the EFISH experiment for a medium which is isotropic in the absence of any external electric field has only two independent components. In order to determine these tensor elements two EFISH measurements are usually performed for two polarization conditions, the electric field vector of the fundamental being parallel ($\|$) or perpendicular ($\perp$) to the external electric field $\mathbf{E}^0$. The frequency doubled photons are detected with polarization parallel to $\mathbf{E}^0$ in both cases. The exact expressions for the corresponding two molar polarizabilities can be derived from the general equation (14.15), by taking into account also the symmetry of the molecules under examination, in our case C_{2v}.

To compute the frequency-dependent first and second hyperpolarizabilities necessary to get the third-order molar polarizabilities (14.15), HF calculations have been performed: geometries and basis set, as well as all the parameters of the PCM-IEF model are the same we have used to get first order $\zeta^{(1)}(\omega)$. In this case only a set of computed values are presented: they have obtained with the full effective model.

Once again the agreement between computed third-order molar polarizabilities and experimental EFISH data are well within the experimental error. Here, however the complex nature of the final property (given by a combination of dipole, first and second hyperpolarizabilities) is difficult to analyze in terms of clear solvent effects. Actually, the contribution given by the second hyperpolarizability $\widetilde{\gamma}$ can be neglected, being at least an order of magnitude smaller that the $\widetilde{\beta} \cdot \mu$ term (as usually observed by the experimentalists), and thus an attempt of explanation could be based on the different effects of solvation in the calculations of dipoles and higher-order properties (Table 14.2).

14.4. Magnetic response properties

All nuclei are surrounded by electrons. When a magnetic field is applied to an atom it induces a circulation of electrons round the nucleus. This movement of electrons produces a tiny, localized magnetic field which opposes the applied field. As a result, the nucleus experiences a reduced overall field and is described as shielded. The

TABLE 14.2
Computed (HF) effective EFISHG third-order molar polarizabilities of pNA in dioxane. Molar polarizabilities are in SI units (10^{-36} cm^4 V^{-3} mol^{-1}). The frequency corresponds to $\lambda = 1064$ nm

	calc	exp[a]
$\zeta^{(3)}(\|)$	110	120 ± 11
$\zeta^{(3)}(\perp)$	36	39 ± 4

[a]WOLFF and WORTMANN [1999].

extent of the shielding depends on the nature of the electron density in that region. Spectrometers are sensitive enough to show the amount of shielding a nucleus experiences and this can be seen on a spectrum as the chemical shift in the Nuclear-Magnetic-Resonance (NMR) spectroscopy.

If the charge distribution about a nucleus is spherically symmetric, the induced field at the nucleus $\mathbf{B}'$ opposes the applied magnetic field $\mathbf{B}^0$. For rapidly tumbling molecules in solution, $\mathbf{B}'$ is related to the external field by the nuclear shielding tensor σ:

$$\mathbf{B}' = -\sigma \cdot \mathbf{B}^0$$

which means the local field is defined by

$$\mathbf{B}_{\mathrm{loc}} = (\mathbf{1} - \sigma) \cdot \mathbf{B}^0.$$

The components of the nuclear shielding are expressed in dimensionless units (ppm) which are independent of the applied magnetic field, and are therefore molecular characteristics.

There are several sources contributing to the secondary magnetic fields. The nuclear shielding is generally partitioned into two major components: the diamagnetic and paramagnetic shieldings. Diamagnetic nuclear shielding arises from circulation of electrons in the s orbitals and filled shells surrounding the nucleus and it depends on the ground state of the molecule. The diamagnetic contribution to nuclear shielding is normally associated with increased shielding, with the local field at the nucleus anti-parallel to the external magnetic field. Paramagnetic shielding, which arises from the nonspherical orbitals, is associated with the orbital angular momentum of electrons, and it is therefore dependent on excited states of the molecule. Paramagnetic contributions normally result in deshielding, with local fields at the nuclei aligned parallel with $\mathbf{B}^0$.

The nuclear shielding is very sensitive to the molecular environment, and the NMR spectra, which are usually obtained for molecules in solution, strongly depend on the solvent. In general, the main effects on NMR spectra arise from intermolecular interactions between solute and solvent molecules.

In the past semiclassical models have been proposed to describe solvent effects on nuclear shieldings; of particular importance is the well-known analysis elaborated by Buckingham (BUCKINGHAM, SCHAEFER and SCHNEIDER [1960]). In this scheme the solvent effect on the solute shielding for nucleus λ may be partitioned as follows

$$\Delta\sigma(\lambda) = \sigma_E(\lambda) + \sigma_w(\lambda) + \sigma_a(\lambda) + \sigma_b(\lambda), \tag{14.16}$$

where σ_E is the "polar effect caused by the charge distribution in the neighboring solvent molecules, thereby perturbing its electronic structure and hence the nuclear screening constants", σ_w is "due to the Van der Waals forces between the solute and the solvent", σ_a "arises from anisotropy in the molecular susceptibility of the solvent molecules", and σ_b is the "contribution proportional to the bulk magnetic susceptibility of the medium". In the context of the subject treated in this chapter, the most important component is that due to the 'polar effect' σ_E, and in fact below we shall describe how this quantity can be evaluated within PCM-IEF solvation method. Here however we would also like to note that all terms appearing in (14.16) in principle could be analytically included into solvation models; in particular for $\sigma_E(\lambda)$, and $\sigma_w(\lambda)$ continuum approaches should be sufficient while σ_a could maybe require the inclusion of discrete solvent molecules around the target solute. A more complex analysis is that required for the bulk susceptibility effect σ_b; a possible approach could be to consider this effect as that already mentioned for other molecular response properties computed in a condensed medium. The external field acting on the molecule is modified by the presence the solvent molecules with an extra local modification which can be related to the shape and the dimension of the volume occupied by the molecule.

14.4.1. Nuclear magnetic shielding

The nuclear magnetic shielding for molecule *in vacuo* can be described in terms of the influence on the total energy of the molecule of the nuclear magnetic moment and of the applied uniform magnetic field. Translating this analysis to molecular solutes in the presence of solvent interactions, leads to define the components of the shielding tensor σ, as the following second derivatives of the free energy functional:

$$\sigma_{ab}^X = \frac{\partial^2 \mathcal{G}}{\partial B_a \partial \mu_b^X},$$

where B_a and $\mu_b^X(a, b = x, y, z)$ are the Cartesian components of the external magnetic field **B**, and of the nuclear magnetic moment μ^X (X refers to a given nucleus).

It is well known that the presence of the magnetic field introduces the problem of the definition of the origin of the corresponding vector potential. However, since σ is a molecular properties, it must be invariant with respect changes of the gauge origin. To obtain this gauge invariance in the *ab initio* calculation, two ways can be adopted. One is to employ a sufficiently complete basis set so that the consequences of the choice of the gauge origin on th calculated value of σ are minimal. The second method is to introduce gauge factors into either the atomic orbitals of the basis set or the MOs of a coupled Hartree–Fock calculation in such a manner that the results are independent on the gauge origin even though the calculation is approximate. Inclusion of gauge factors in the atomic orbitals may be accomplished by using gauge invariant atomic orbitals (GIAO) (DITCHFIELD [1974], WOLINSKI, HINTON and PULAY [1990]):

$$\chi_v(B) = \chi_v(0) \exp\left[-\frac{i}{2c}(B \times R_v) \cdot r\right], \tag{14.17}$$

where R_v is the position vector of the basis function, and $\chi_v(0)$ denotes the usual field-independent basis function.

The GIAO method is used in conjunction with analytical derivative theory; in this approach the magnetic field perturbation is treated in an analogous way to the perturbation produced by changes in the nuclear coordinates. In this framework, the components of the nuclear magnetic shielding tensor are obtained as:

$$\sigma_{ab} = \text{tr}[\mathbf{Ph}^{B_a \mu_b^X} + \mathbf{P}^{B_a} \mathbf{h}^{\mu_b^X}],$$

where $\mathbf{P}^{B_a}$ is the derivative of the density matrix with respect to the magnetic field. Matrices $\mathbf{h}^{\mu_b^X}$ and $\mathbf{h}^{B_a\mu_b^X}$ contain the first derivative of the standard one-electron Hamiltonian with respect to the nuclear magnetic moment and the second derivative with respect the magnetic field and the nuclear magnetic moment, respectively. Both terms do no contain explicit solvent-induced contributions as the latter do not depend on the nuclear magnetic moment of the solute and thus the corresponding derivatives are zero.

On the contrary, explicit solvent effects act on the first derivative of the density matrix $\mathbf{P}^{B_a}$ which can be obtained as solution of the corresponding first-order HF (or KS) equation characterized by the following derivative of the Fock matrix (CAMMI, MENNUCCI and TOMASI [1998], CAMMI, MENNUCCI and TOMASI [1999]):

$$\mathbf{F}'^{B_a} = \mathbf{h}^{B_a} + \mathbf{h}_R^{B_a} + \mathbf{G}^{B_a}(\mathbf{P}) + \mathbf{X}_R^{B_a}(\mathbf{P})$$

where $\mathbf{P}$ is the unperturbed density matrix and the solvent-induced terms, $\mathbf{h}_R^{B_a} + \mathbf{X}_R^{B_a}(\mathbf{P})$, appear due to the magnetic field dependence of the atomic orbitals.

14.5. Chiroptical solvated molecules: OR and VCD

Optical activity, and the related spectroscopy (generally called polarimetry), is one of the oldest research tools that is routinely practiced by chemists. Polarimetry is in fact a sensitive, nondestructive technique for measuring the optical activity exhibited by inorganic and organic compounds. A compound is considered to be optically active if linearly polarized light is rotated when passing through it. The amount of optical rotation is determined by the molecular structure and concentration of chiral molecules in the substance.

The polarimetric method is a simple and accurate means for determination and investigation of structure in macro, semi-micro and micro analysis of expensive and nonduplicable samples. Polarimetry is employed in quality control, process control and research in the pharmaceutical, chemical, essential oil, flavor and food industries. Research applications for polarimetry are found in industry, research institutes and universities as a means of: (i) Evaluating and characterizing optically active compounds by measuring their specific rotation and comparing this value with the theoretical values found in literature; (ii) Investigating kinetic reactions by measuring optical rotation as a function of time; (iii) Monitoring changes in concentration of an optically active component in a reaction mixture, as in enzymatic cleavage; (iv) Analyzing molecular structure by plotting optical rotatory dispersion curves over a wide range of wavelengths; (v) Distinguishing between optical isomers.

Optical rotatory dispersion (optical rotation, OR, versus wavelength) and circular dichroism (CD) are the two components of the optical activity that can be used to elucidate the absolute stereochemistry of chiral molecules. Vibrational circular dichroism (VCD), and vibrational Raman optical activity (VROA) are alternative properties that are being currently used for the same scope. The main aspect to stress here is that both sets of investigative tools are strictly related to the condensed phase, as experimental measurements are generally limited to this. Despite this, the extension of solvation models to general QM methods for predicting chiro-optical properties has been very limited. Only in this last few years some efforts (many unpublished as still in their phase of testing) have been appeared; below we shall try to give a preliminary summary of these new applications, once again with almost exclusive attention to PCM-IEF solvation scheme.

14.5.1. OR

As shown above chiral molecules exhibit optical rotation. With very few exceptions, optical rotation measurements are carried out in the condensed phase, most often in liquid solutions. Optical rotations of solutions of chiral molecules are solvent dependent. In the case of flexible molecules, which exhibit multiple conformations in solution, solvent effects can often be attributed predominantly to changes in conformational populations with solvent. However, in the case of rigid molecules, exhibiting a single conformation, optical rotations can still exhibit substantial solvent dependence.

According to the canonical treatment of the theory of optical rotatory power, the optical rotation at a frequency ω of an isotropic dilute solution of a chiral molecule is given by

$$\phi(\omega) = \frac{4\pi N \omega^2}{c^2} \gamma_{LF}(\omega) \beta(\omega),$$

where $\phi(\omega)$ is the rotation in radians/cm, c is the velocity of light and $\beta(\omega)$ is the frequency-dependent electric dipole-magnetic dipole polarizability of the chiral molecule. In the equation above, $\gamma_{LF}(\omega)$ is the "local field correction factor" (i.e., the ratio of the microscopic electric field acting on the chiral molecule to the macroscopic electric field of the light wave) of Lorentz, given by $\gamma_{LF}(\omega) = (n^2(\omega) + 2)/3$, where $n^2(\omega)$ is the refractive index of the solvent. A variety of ab initio methods have recently been applied to the calculation of optical rotation, but solvent effects on $\phi(\omega)$ have been generally ignored or included using the simplified Lorentz equation. None of these calculations satisfactorily take account of solvent effects. Here, we present a theory of solvent effects on optical rotations which allows to evaluate the total solvent effects on $\phi(\omega)$ by computing an "effective" electric dipole-magnetic dipole polarizability, i.e., without the need to add a scaling parameter as that represented by γ_{LF} (MENNUCCI, TOMASI, CAMMI, CHEESEMAN, FRISCH, DEVLIN, GABRIEL and STEPHENS [2002]). In this scheme, the optical rotations are calculated using ab initio Density Functional Theory (DFT) and solvent effects are incorporated using the PCM-IEF solvation model.

The ab initio theoretical treatment of the molecular optical rotation is based on the calculation of the electric dipole-magnetic dipole polarizability tensor, given by the expression (CONDON [1937]):

$$\beta_{ab} = \frac{4\pi c}{3h} \sum_{n \neq s} \frac{1}{\omega_{ns}^2 - \omega^2} \mathrm{Im}\{\langle \Psi_s | \mu_a^{el} | \Psi_n \rangle \langle \Psi_n | \mu_b^{mag} | \Psi_s \rangle\}, \tag{14.18}$$

where μ_a^{el} and μ_b^{mag} are, respectively, the electric and magnetic dipole operators, and Ψ_s and Ψ_n represent the ground and excited electronic states. The angular frequency

ω_{ns} and ω are those related to the $s \rightarrow n$ transition and the exciting radiation, respectively. The scalar β required to get the optical rotation $\phi(\omega)$ is obtained as $1/3$ of the trace of the tensor.

An explicit evaluation of the sum over excited states in Eq. (14.18) can be avoided by exploiting the expression based on the ground state function only:

$$\beta_{ab}(\omega) = \frac{hc}{3\pi} \text{Im} \left\langle \frac{\partial \Psi(\omega)}{\partial E_a} \bigg| \frac{\partial \Psi(\omega)}{\partial H_b} \right\rangle,$$

where E_a and H_b are electric and magnetic field directions, respectively.

The change in the ground state wavefunction with respect to (oscillating) applied electric or magnetic field perturbations are determined from the TDHF or TDKS procedure described in the paragraph entitled "Electric dipole polarizabilities". For the magnetic perturbation atomic basis functions χ_μ which are magnetic field dependent (GIAOs) (see Eq.(14.17)) are usually exploited.

Following what reported in the paragraph entitled "Macroscopic susceptibilities: local field effects and effective polarizabilities", in the presence of an PCM-IEF continuum dielectric, the frequency-dependent perturbation to be added to the one-electron operator of the unperturbed system can be rewritten as:

$$\mathbf{h}'(\omega) = \frac{1}{2} \mathbf{m}^{\text{elec}} \cdot \mathbf{E}(e^{-i\omega t} + e^{+i\omega t})$$
$$+ \frac{1}{2} \sum_s \mathbf{V}(s) \frac{\partial \mathbf{q}^{\text{ex}}(s)}{\partial \mathbf{E}} \mathbf{E}(e^{-i\omega t} + e^{+i\omega t}) \tag{14.19}$$

where we have assumed that the oscillating field is an electric field with strength $\mathbf{E}$ and $\mathbf{m}^{\text{elec}}$ is the electric dipole integrals matrix. In Eq. (14.19) $\mathbf{V}$ is the matrix collecting the potential integrals computed on the cavity tesserae and $\mathbf{q}^{\text{ex}}$ is the apparent charge induced on the cavity by the external oscillating field $\mathbf{E}$.

Expression (14.19) when summed to the Hamiltonian modified by the solvent terms described by $\mathbf{q}^N$ and $\mathbf{q}^e$ (see Section 14) allows one to take into account the complete reaction of the solvent to the combined action of the internal (due to the solute) and the external fields.

Approximate solutions of the time-dependent Schrödinger equation associated to the resulting effective Hamiltonian can be obtained by using the same procedures formulated for isolated systems but now the Fock operator includes explicit solvent terms. The inclusion of such additional solvent terms in the coupled perturbed equations will lead to different values of the derivatives of the ground state wavefunction for an electric perturbation.

The mixed nature of the electric dipole – magnetic dipole polarizability β requires an additional coupled perturbed procedure, this time containing a magnetic perturbation. Due to the imaginary nature of such perturbation solvent induced terms do not appear in the corresponding first-order expansion term of the Fock operator as explicit terms but only through the dependence of the atomic orbital basis set on the magnetic field (CAMMI, MENNUCCI and TOMASI [1998], CAMMI, MENNUCCI and TOMASI

[1999]); as a consequence also the derivatives of the wavefunction with respect to the magnetic field will be modified by the solvent (see details in the paragraphs entitled "Electric dipole polarizabilities" and "Nuclear magnetic shielding").

It is important to note that solvent contributions obtained in terms of solvent charges which are computed using the value of the dielectric constant at the frequency of the external field. In the present case such frequency is that corresponding to the sodium D line, and thus the value for $\varepsilon(\omega)$ coincides with the so-called optical ε_{opt} defined as the square of the refractive index. For polar solvents ε_{opt} is by far smaller that the static ε_0 analog and thus introducing a solvent response determined by it instead of ε_0 means to account for only a part of the response which would appear in the presence of a static field; such situation is usually defined as *nonequilibrium* solute-solvent regime while that corresponding to a full solvent response is indicated as *equilibrium* regime.

14.5.2. Vibrational Circular Dichroism

After recently celebrating twenty years of development since its early years of discovery, VCD has matured to a point where the phenomenon is well understood theoretically, can be measured and calculated routinely, and is being used to uncover exciting new information about the structure of optically active molecules (Stephens and Devlin [2000]). Beyond this, VCD has been shown to be a sensitive, noninvasive diagnostic probe of chiral purity or enantiomeric separation with potential use in the synthesis and manufacture of chiral drugs and pharmaceutical products. In descriptive terms, VCD is the coupling of optical activity to infrared vibrational spectroscopy. More specifically, VCD spectra are vibrational difference spectra with respect to left and right circularly polarized radiation. The essence of VCD is to combine the stereochemical sensitivity of natural optical activity with the rich structural content of vibrational spectroscopy. The result of a VCD measurement is two vibrational spectra of a sample, the VCD and its parent infrared spectrum. These can used together to deduce information about molecular structure. The principal area of application of VCD is structure elucidation of biologically significant molecules including peptides, proteins, nucleic acids, carbohydrates, natural products and pharmaceutical molecules; also, as mentioned above, it has growing potential as a chiral diagnostic probe. VCD complements the relatively slow time scale of NMR since molecular vibrations and conformational sensitivity occur in the subpicosecond time domain. VCD is also complementary to X-ray crystallography by virtue of its applicability to molecules in gas, liquid and solution phases. Here, we present a quantum-mechanical method to simulate VCD spectra of molecules in solution.

The differential response of a chiral sample to left and right circularly polarized light can be represented by the quantity $\Delta\varepsilon$, defined as:

$$\Delta\varepsilon = \varepsilon_L - \varepsilon_R,$$

where $\varepsilon_{L, R}$ are the molar absorption coefficients for left and right circularly polarized light, respectively. Ignoring for the moment solvent effects, the differential molar absorption coefficient at frequency v, $\Delta\varepsilon(v)$, is:

$$\Delta\varepsilon(v) = 4\gamma v \sum_i R_i f(v_i, v), \tag{14.20}$$

where v_i is the frequency of the ith transition, γ a numerical coefficient, $f(v_i, v)$ the normalized line-shape function and:

$$R_i = \text{Im}[\langle 0|\mu_{el}|1\rangle_i \cdot \langle 0|\mu_{mag}|1\rangle_i] \tag{14.21}$$

the rotational strength. In Eq. (14.21) μ_{el} and μ_{mag} are the electric and magnetic dipole moment operators, respectively, and $|0\rangle$, $|1\rangle$ are vibrational states. The computation of the transition moments in Eq. (14.21) can be avoided by resorting to a coupled perturbed HF, or KS, approach in which the derivatives of the ground state wavefunction with respect to nuclear displacement and magnetic field respectively are computed, as well as the vibrational frequencies v_i and the normal coordinates Q_i. v_i and Q_i are obtained simultaneously by diagonalization of the mass-weighted Cartesian force field (the Hessian).

The calculation of $\Delta\varepsilon$ in the presence of a solvent medium still relies on Eqs. (14.20), (14.21), but some refinements are needed. As for electric response properties (see the paragraph entitled "Macroscopic susceptibilities: local field effects and effective polarizabilities"), as well as for infrared intensities (CAMMI, CAPPELLI, CORNI and TOMASI [2000]), the μ_{el} operator in Eq. (14.21) has to be replaced by the sum of the dipole moment μ_{el} of the molecule and the dipole moment $\widetilde{\mu}_{el}$ arising from the polarization induced by the molecule on the solvent (here with the symbol $\widetilde{\mu}_{el}$ we indicate the operator from which the previous matrix $\widetilde{\mathbf{m}}$ is derived when we shift to the representation on the atomic basis set). As already observed, $\widetilde{\mu}_{el}$ takes into account effects due to the field generated from the solvent response to the probing field once the cavity has been created (i.e., the cavity field). In principle also μ_{mag} should be similarly reformulated. However, by assuming the response of the solvent to magnetic perturbations to be described only in terms of its magnetic permittivity (which is usually close to unity), it is reasonable to consider that the magnetic analogous of the electric "cavity field" gives minor contributions to R_i^{sol}. Thus, the final expression for R_i^{sol} to use is (CAPPELLI, CORNI, MENNUCCI, CAMMI and TOMASI [2002]):

$$R_i^{sol} = \text{Im}\langle 0|(\mu_{el} + \widetilde{\mu}_{el})|1\rangle_i \cdot \langle 0|\mu_{mag}|1\rangle_i$$

where the frequencies and the normal coordinates are obtained by diagonalizing the Hessian for the molecule in solution. The calculation of the modified $\mu_{el} + \widetilde{\mu}_{el}$ term is performed by defining an additional set of external field-induced charges $\mathbf{q}^{ex}$ spread on the cavity surface: details on this point can be found in the previous Section 14. The derivatives of the solvent-modified wavefunction with respect to nuclear displacement and magnetic field can be obtained with the methods presented in the previous sections. It is also possible to account for vibrational nonequilibrium solvent effects on such derivative and on the normal modes of the molecule (CAPPELLI, CORNI, CAMMI, MENNUCCI and TOMASI [2000]). We remark that nonequilibrium effects arise since the solvent cannot instantaneously equilibrate to the charge distribution of the vibrating molecule.

CHAPTER VI

Excited States

Solvation models, used to study solvated molecules in their ground state, can be extended to electronically excited states: this extension however is neither immediate nor straightforward. There are in fact many fundamental characteristics which are specific of the molecules in their excited states and which make them unique systems to be treated with completely new techniques.

This is also true for isolated systems, for which the computational tools devoted to the excited molecules are generally different from those used for ground state systems; as well-known example suffice here to quote the Hartree–Fock approach which gives very satisfactory results for properties and reactivity of molecules in their ground state but it completely fails in treating excited systems. This specificity of the electronically excited molecules acquires an even larger importance when the molecule can interact with an external medium. In this case in fact it becomes compulsory to introduce the concepts of time and time progress, concepts which can be safely neglected in treating molecules in their ground states. In these cases in fact, and also when introducing reaction processes, one can always reduce the analysis to a completely equilibrated solute-solvent system.

On the contrary, when the attention is shifted towards dynamical phenomena as those involved in electronically transitions (absorptions and/or emissions), or towards relaxation phenomena as those which describe the time evolution of the excited state, one has to introduce new models, in which solute and solvent have proper response times which have not to be coherent or at least not before very long times, of no interest in the present research. To explain better this fundamental issue, it is useful to summarize some aspects of the dynamical behavior of the medium which have been already used in the previous chapter but without a detailed description.

In general, the dynamics of solvent can be basically characterized by two main components. One, of more immediate comprehension, is represented by the internal molecular motions inside the solvent due to changes in the charge distribution, and eventually in the geometry, of the solute system. As already shown, the solute when immersed in the solvent produces an electric field inside the bulk of the medium which can modify its structure, for example inducing phenomena of alignment and/or preferential orientation of the solvent molecules around the cavity embedding the solute. These molecular motions are characterized by specific time scales of the order of the rotational and translational times proper of the condensed phases $(10^{-9}-10^{-11}$ s). In a analogous way, we can assume that the single solvent molecules are subjected to internal geometrical variations, i.e., vibrations, due to the changes in the solute field; once again these will be described by specific well-known time scales (10^{-14}–10^{-12} s). Both the translational, the rotational and/or the vibrational motions involve nuclear displacements and therefore, in the following, they will be collectively indicated as 'nuclear motions'. The other important component of the dynamical nature of the medium, complementary to the nuclear one, is that induced by motions of the electrons inside each solvent molecule; these motions are extremely fast (of the order of 10^{-16}–10^{-15} s) and they represent the electronic polarization of the solvent.

The dynamical phenomena we have described above, which are usually neglected in standard calculations on solvated systems. Solvation continuum models have been generalized to treat phenomena involving excited states since a long time (see the large literature in TOMASI and PERSICO [1994], CRAMER and TRUHLAR [1999]).

However, these examples are generally limited to some specific aspects of the whole problem, and/or applied to classical or low-quality QM computational levels. On the contrary, the aim is to treat the phenomenon of the excited states in solution in its totality, by using computational tools of the same quality as those, by now standard, for isolated systems.

To do that it will be necessary to introduce in the formalism of continuum models the specificities proper of the excited systems. As said, the main innovation is the dynamical nature of the phenomenon and the related time-dependent response of the solvent. The brief analysis reported above on the time scales characterizing the various motions inside the medium represents the physical bases on which to build up a theoretical model which is realistic and sensitive enough to allow distinctions among the various components of the solvent response. In fact, it is easy to see that two main components we have previously described as nuclear and electronic motions, due to their different dynamic behavior, will give rise to different effects.

In the same way, it is easy to accept that the electronic motions can be considered as instantaneous and thus the part of the solvent response they originate is always equilibrated to any change, even if fast, in the charge distribution of the solute (as in the case of vertical electronic transitions, in which no nuclear relaxations have time to happen inside the molecule which remains frozen in its initial geometry also when radical changes happen in its electronic state). On the contrary, solvent nuclear motions, by far slower than the electronic ones, can be delayed with respect to fast changes, and thus they can give origin to solute–solvent systems not completely equilibrated. This condition of nonequilibrium will successively evolve towards a more stable and completely equilibrated state in a time interval which will depend on the specific system under scrutiny.

15. Continuum solvation approaches

The PCM-like continuum models, in which the solvent response is described in terms of apparent charges on the cavity surface have been generalized to treat dynamical phenomena (AGUILAR, OLIVARES DEL VALLE and TOMASI [1993]); in particular the solvent charges can be made time dependent through the time dependence of the solute wavefunction from which they are directly originated, and also due to the time dependence of the solvent permittivity (or dielectric constant). The latter, in fact, is the numerical quantification, in terms of an electric response, of the solvent polarity. It is thus immediate to extract from its value various contributions related to different motions in the solvent. In particular, experimental measurements can give information on its purely electronic component; this coincides with the so-called optical permittivity. In general, the dielectric constant presents a time dependence which is determined by the intrinsic characteristics of the solvent and thus it will be different passing from one solvent to another, and finally not representable through a general analytic function. However, by applying theoretical models based on measurements of relaxation times (as that formulated by Debye) and by knowing by experiments the behavior of the permittivity with respect to the frequency of an external applied field (the so-called Maxwell field), it is possible to derive semiempirical formula of the dielectric constant as a function of time (BÖTTCHER [1973], BÖTTCHER and BORDEWIJK [1978]). These formula, in turn, will allow to obtain induced charges which take into account, through its dependence on the time-dependent dielectric constant, the dynamic behavior of the specific solvent they represent.

It is worth noting that in the limit case of vertical electronic transitions, in which we can assume a Franck–Condon-like response of the solvent, exactly as for the solute molecule, the nuclear motions inside and among the solvent molecules will not be able to immediately follow the fast changes in the solute charge distribution, which will instantaneously shift from that characterizing the ground state to that proper of the excited state (in an absorption process), or vice versa from that of an equilibrated excited state to that of the ground state (in an emission). During, and immediately after the transition, the solvent will thus respond to the variations only through its electronic component, while the nuclear part of the response (also indicated as inertial) will remain frozen in the state immediately previous the transition.

In the PCM framework, this will lead to a partition of the apparent charges in two separate sets (AGUILAR, OLIVARES DEL VALLE and TOMASI [1993], CAMMI and TOMASI [1995a], MENNUCCI, CAMMI and TOMASI [1998]), one related to the instantaneous electronic response, σ_{fast}, and the other to the slower response connected to the nuclear motions, σ_{slow} (their sum is just the already defined total apparent charge σ). According to what said before about the origin of the apparent charges, the electronic components will depend on the instantaneous wavefunction of the solute, on the optical dielectric constant and on the complementary set of slower apparent charges. The latter, on the contrary, will still depend on the solute wavefunction of the initial state.

From a physical point of view, we can always define the equations giving the total and the electronic surface charge by considering two parallel systems of Poisson-like equations in which both the operator and the related boundary conditions are defined in terms of ε or of ε_∞, respectively. On the contrary, the orientational surface charge has to be defined as difference of the two other charges, being the orientational contribution to the solvent polarization not related to a physically stated dielectric constant.

By applying this model to the problem of a solute electronic transition from the ground state to a given excited state, we have:

$$
\begin{cases}
-\Delta V = 4\pi\rho_M^{\text{fin}} & \text{within the cavity,} \\
-\varepsilon_\infty \Delta V = 0 & \text{out of the cavity,} \\
V_i - V_e = 0 & \text{on the cavity surface,} \\
(\partial V/\partial n)_i - \varepsilon_\infty (\partial V/\partial n)_e = 4\pi\sigma_{\text{slow}} & \text{on the cavity surface,}
\end{cases}
$$

where ρ_M^{fin} is the charge distribution of the solute in its final excited state. The second jump condition on the gradient of V has here a different form with respect to that of the equilibrium situation given before, due to the presence of the constant slow charge σ_{slow}.

As shown before for ground state equilibrated solutes, the system (15) can be solved by defining an electronic surface charge σ_{fast} which can be discretized into point charges, $\mathbf{q}_{\text{fast}}$, exploiting to the usual partition of the surface cavity into tesserae. However, in the nonequilibrium case the definition of the electrostatic free energy has to be changed; in particular the contribution of the slow polarization has to be seen an external fixed field which does not follow the variationally derived rules.

As recalled before, this is just a part of the more general process of creation and evolution of the excited states in solution. However, this aspect, even if partial, represents a valid and immediate check on the quality of the adopted model; frequencies and intensity of absorption and/or emission peaks of spectra of molecular systems in liquid solutions are in fact the most common experimental data one can compare to. Data with richer information about the dynamical aspects of the process can be derived from other experimental measurements of time-dependent fluorescence spectroscopy. In particular, the time analysis of the so-called 'time-dependent Stokes-shift', i.e., the time evolution of the difference between the maxima of emission and absorption, has a large importance for the comparison with theoretical results. Such quantity, experimentally measured, can in fact give a valid test for the models used to represent the solvent relaxation after an electronic excitation inside the solute, on the one hand, and to describe the way this relaxation interacts with that proper of the excited state, on the other hand. In fact, the time evolution of this spectroscopic quantity strongly depends on the solvent.

In the "dynamic Stokes shift" experiment, a short light pulse is used to electronically excite a probe solute and thereby change its electron distribution and hence its interactions with the solvent instantaneously (as far as solvent motions are concerned). Relaxation of the solvent is monitored by measuring the frequency shift of the solute's emission spectrum, which red-shifts in a way that directly reflects the time evolution of the solvation energy. Since solvation takes only a few picoseconds in most room-temperature solvents, emission spectra must be measured on a sub-picosecond time scale in order to see solvation taking place. To do so the technique of fluorescence upconversion is used; this affords effective time resolution of 30–50 fs. The solvation times that can be observed vary over wide ranges, with characteristic solvation times as fast as 50–100 fs in water an acetonitrile, to $\geqslant$ 500 ps in long-chain alcohols. This broad range of times, and the highly nonexponential nature of the solvation relaxation are captured with reasonable fidelity using dielectric continuum models of solvation (HSU, SONG and MARCUS [1997]). In particular, recent QM studies using PCM-IEF have shown that continuum representations of the solvent can be a fruitful way to reach a deeper understanding of solvation dynamics.

References

AGUILAR, M.A. and F.J. OLIVARES DEL VALLE (1986), A computation procedure for the dispersion component of the interaction energy in continuum solute–solvent models, *Chem. Phys.* **138**, 327–335.

AGUILAR, M.A., F.J. OLIVARES DEL VALLE and J. TOMASI (1993), Nonequilibrium solvation: An ab-initio quantum-mechanical method in the continuum cavity model approximation, *J. Chem. Phys.* **98**, 7375–7384.

AKKIRAJU, N. and H. EDELSBRUNNER (1996), Triangulating the surface of a molecule, *Discrete Appl. Math.* **71**, 5–22.

ALLEN, M.P. and D.J. TILDESLEY (1987), *Computer Simulation of Liquids* (Clarendon, Oxford).

AMOVILLI, C., V. BARONE, R. CAMMI, E. CANCÈS, M. COSSI, B. MENNUCCI, C.S. POMELLI and J. TOMASI (1998), Recent advances in the description of solvent effects with the polarizable continuum model, *Adv. Quant. Chem.* **32**, 227–261.

AMOVILLI, C, and B. MENNUCCI (1997), Self-consistent field calculation of Pauli repulsion and dispersion contributions to the solvation free energy in the polarizable continuum model, *J. Phys. Chem. B* **101**, 1051–1057.

ARRIGHINI, G.P. (1981), *Intermolecular Forces and their Evaluation by Perturbation Theory* (Springer, Berlin).

BACHELET, G.B., D.R. HAMANN and M. SCHLÜTER (1982), Pseudopotentials that work: from hydrogen to plutonium, *Phys. Rev. B* **26**, 4199–4206.

BAJAJ, C., H.Y. LEE, R. MERKERT and V. PASCUCCI (1997), *NURBS based B-rep models for macro-molecules and their properties* (Fourth Symposium on Solid Modeling and Applications, Atlanta).

BALESCU, R. (1991), *Equilibrium and Non-Equilibrium Statistical Mechanics* (Krieger, Melbourne, FL).

BANDYOPADHYAY, P., M.S. GORDON, B. MENNUCCI and J. TOMASI (2002), An integrated effective fragment-polarizable continuum approach to solvation: Theory and application to glycine, *J. Chem. Phys.* **116**, 5023–5032.

BEN-NAIM, A. (1974), *Water and Aqueous Solutions* (Plenum, New York).

BEN-NAIM, A. (1987), *Solvation Thermodynamics* (Plenum, New York).

BISHOP, D.M. (1998), *Adv. Chem. Phys.* **104**, 1.

BOROUCHAKI, H., P. LAUG and P.L. GEORGE (2000), Parametric surface meshing using a combined advancing-front – generalized-Delaunay approach, *Internat. J. Numer. Methods. Engrg.* **49**(1–2), 233–259.

BÖTTCHER, C.J.F. (1973), *Theory of Electric Polarization*, Vol. I (Elsevier, Amsterdam).

BÖTTCHER, C.J.F. and P. BORDEWIJK (1978), *Theory of Electric Polarization*, Vol. II (Elsevier, Amsterdam).

BUCKINGHAM, A.D., T. SCHAEFER and W.G. SCHNEIDER (1960), Solvent effects in nuclear magnetic resonance spectra, *J. Chem. Phys.* **32**, 1227–1233.

BUTCHER, P.N. and D. COTTER (1990), *The Elements of Nonlinear Optics* (Cambridge Univ. Press, Cambridge).

CAMMI, R., C. CAPPELLI, S. CORNI and J. TOMASI (2000), On the calculation of infrared intensities in solution within the polarizable continuum model, *J. Phys. Chem. A* **104**, 9874–9879.

CAMMI, R., M. COSSI, B. MENNUCCI and J. TOMASI (1996), Analytical Hartree–Fock calculations of the dynamical polarizabilities α, β and γ of molecules in solution, *J. Chem. Phys.* **105**, 10556–10564.

CAMMI, R., B. MENNUCCI and J. TOMASI (1998a), On the calculation of local field factors for microscopic static hyperpolarizabilities of molecules in solution with the aid of quantum-mechanical methods, *J. Phys. Chem. A* **102**, 870–875.

CAMMI, R., B. MENNUCCI and J. TOMASI (1998b), Solvent effects on linear and nonlinear optical properties of Donor–Acceptor polyenes: investigation of electronic and vibrational components in terms of structure and charge distribution changes, *J. Amer. Chem. Soc.* **34**, 8834–8847.

CAMMI, R., B. MENNUCCI and J. TOMASI (1999), Nuclear magnetic shieldings in solution: Gauge invariant atomic orbital calculation using the polarizable continuum model, *J. Chem. Phys.* **110**, 7627–7638.

CAMMI, R., B. MENNUCCI and J. TOMASI (2000), An attempt to bridge the gap between computation and experiment for nonlinear optical properties: macroscopic susceptibilities in solution, *J. Phys. Chem. A* **104**, 4690–4698.

CAMMI, R. and J. TOMASI (1995a), Nonequilibrium solvation theory for the polarizable continuum model: a new formulation at the SCF level with application to the case of the frequency-dependent linear electric response function, *Int. J. Quantum Chem.* **29**, 465–474.

CAMMI, R. and J. TOMASI (1995b), Remarks in the use of the apparent surface charges (ASC) methods in solvation problems: iterative versus matrix-inversion procedures and the renormalization of the apparent charges, *J. Comput. Chem.* **16**, 1449–1458.

CANCÈS, E. and B. MENNUCCI (1998a), Analytical derivatives for geometry optimization in solvation continuum models, I: Theory, *J. Chem. Phys.* **109**, 249–259.

CANCÈS, E. and B. MENNUCCI (1998b), New applications of integral equation methods for solvation continuum models: ionic solutions and liquid crystals, *J. Math. Chem.* **23**, 309–326.

CANCÈS, E., B. MENNUCCI and J. TOMASI (1997), A new integral equation formalism for the polarizable continuum model: theoretical background and applications to isotropic and anisotropic dielectrics, *J. Chem. Phys.* **107**, 3032–3041.

CANCÈS, E., B. MENNUCCI and J. TOMASI (1998), Analytical derivatives for geometry optimization in solvation continuum models, II: numerical applications, *J. Chem. Phys.* **109**, 260–266.

CAPPELLI, C., S. CORNI, R. CAMMI, B. MENNUCCI and J. TOMASI (2000), Nonequilibrium formulation of infrared frequencies and intensities in solution: Analytical evaluation within the polarizable continuum model, *J. Chem. Phys.* **113**, 11270–11279.

CAPPELLI, C., S. CORNI, B. MENNUCCI, R. CAMMI and J. TOMASI (2002), Vibrational circular dichroism within the polarizable continuum model: A theoretical evidence of conformation effects and hydrogen bonding for (s)-(-)-3-butyn-2-d in CCl_4 solution, *J. Phys. Chem. A* **106**, 12331–12339.

CAR, R. and M. PARRINELLO (1985), Unified approach for molecular dynamics and density functional theory, *Phys. Rev. Lett.* **55**, 2471–2474.

CHEN, S.W. and B. HONIG (1997), Monovalent and divalent salt effects on electrostatic free energies defined by the nonlinear Poisson–Boltzmann equation: Application to DNA reactions, *J. Phys. Chem. B* **101**, 9113–9118.

CONDON, E.U. (1937), Theories of optical rotatory power, *Rev. Mod. Phys.* **9**, 432–457.

CONNOLLY, M.L. (1996), *Molecular surfaces: a review*, http://www.netsci.org/Science/Compchem/feature14.html.

CRAMER, C.J. and D.G. TRUHLAR (1991), General parameterized SCF model for free energies of solvation in aqueous solution, *J. Amer. Chem. Soc.* **113**, 8305–8311; Molecular orbital theory calculations of aqueous solvation effects on chemical equilibria, *J. Amer. Chem. Soc.* **113**, 8552–8554.

CRAMER, C.J. and D.G. TRUHLAR (1999), Implicit solvation models: Equilibria, structure, spectra, and dynamics, *Chem. Rev.* **99**, 2161–2200.

DAY, P.N., J.H. JENSEN, M.S. GORDON, S.P. WEBB, W.J. STEVENS, M. KRAUSS, D. GARMER, H. BASCH and D. COHEN (1996), An effective fragment method for modeling solvent effects in quantum mechanical calculations, *J. Chem. Phys.* **105**, 1968–1986.

DITCHFIELD, R. (1974), Self-consistent perturbation theory of diamagnetism, I. A gauge-invariant LCAO method for N.M.R. chemical shifts, *Mol. Phys.* **27**, 789–792.

FIELD, M.J., P.A. BASH and M. KARPLUS (1990), A combined quantum mechanical and molecular mechanical potential for molecular dynamics simulations, *J. Comput. Chem.* **11**, 700–733.

FLORIS, F.M., J. TOMASI and J.L. PASCUAL-AHUIR (1991), Dispersion and repulsion contributions to the solvation energy: Refinements to a simple computational model in the continuum approximation, *J. Comput. Chem.* **12**, 784–791.

FRENKEL, D. and B. SMIT (1996), *Understanding Molecular Simulation* (Academic Press, San Diego).

FREY, P.J. and P.L. GEORGE (2000), *Mesh Generation – Application to Finite Elements* (Hermès Science Europe).

FRISCH, M.J. et al. (1998), *Gaussian 98* (Gaussian, Pittsburgh).

FROESE, R.D.J. and K. MOROKUMA (1998), in: P.v.R. Schleyer, N.L. Allinger, T. Clark, J. Gasteiger, P.A. Kollman, H.F. Schaefer and P.R. Schreiner, eds., *The Encyclopedia of Computational Chemistry* (John Wiley, New York) 1245–1300.

GAO, J. (1995), Methods and applications of combined quantum mechanical and molecular mechanical potentials, in: K.B. Lipkowitz and D.B. Boyd, eds., *Reviews in Computational Chemistry*, Vol. 7 (VCH, New York) 119–185.

GRAY, C.G. and K.E. GUBBINS (1984), *Theory of Molecular Fluids, Vol. 1, Fundamentals* (Clarendon, Oxford).

HACKBUSCH, W. (1995), *Integral Equations – Theory and Numerical Treatment* (Birkhäuser, Basel).

HANSEN, J.-P. and I.R. MCDONALD (1986), *Theory of Simple Liquids* (Academic Press, San Diego).

HELGAKER, H. et al. (2001), Dalton, an ab into electronic structure program, Release 1.2. See http://www.kjemi.uio.no/software/dalton/dalton.html.

HOBZA, P. and R. ZAHRADNIK (1988), *Intermolecular Complexes* (Elsevier, Amsterdam).

HSU, C.-P., X. SONG and R.A. MARCUS (1997), Time-dependent Stokes shift and its calculation from solvent dielectric dispersion data, *J. Phys. Chem. B* **101**, 2546–2551.

KIRKPATRICK, S., C.D. GELATT JR. and M.P. VECCHI (1983), Optimization by simulated annealing, *Science* **220**, 671–680.

KLAMT, A. and G. SCHÜÜRMANN (1993), COSMO: A new approach to dielectric screening in solvents with explicit expressions for the screening energy and its gradient, *J. Chem. Soc. Perkin Trans.* **2**, 799–805.

KOHN, W. and P. VASHISHTA (1993), General density functional theory, in: Lundqvist and March, eds., *Theory of the inhomogeneous Electron Gas* (Plenum, New York).

LAUG, P. and H. BOROUCHAKI (1999), *BLSURF* – Mesh generator for composite parametric surfaces – user's manual, INRIA Technical Report RT-0232.

LAUG, P. and H. BOROUCHAKI (2001), Molecular surface modeling and meshing, in: 10th International Meshing Roundtable, Newport Beach, California, USA, 31–41, to appear in: *Engineering with Computers*.

LAUG, P. and H. BOROUCHAKI (2002), BLMOL, molecular surface mesher, http://www-rocq.inria.fr/Patrick.Laug/blmol/index.html.

LEE, B. and F.M. RICHARDS (1971), The interpretation of protein structures: estimation of static accessibility, *J. Mol. Biol.* **55**, 379–400.

MARGENAU, H. and N.R. KESTNER (1971), *Theory of Intermolecular Forces* (Pergamon, Elmsford).

MCQUARRIE, D.A. (1976), *Statistical Mechanics* (Harper and Row, New York).

MCWEENY, R. (1992), *Methods of Molecular Quantum Mechanics*, 2nd edn. (Academic Press, San Diego).

MENNUCCI, B., R. CAMMI and J. TOMASI (1998), Excited states and solvatochromic shifts within a nonequilibrium solvation approach: a new formulation of the integral equation formalism method at the self-consistent field, configuration interaction, and multiconfiguration self-consistent field level, *J. Chem. Phys.* **109**, 2798–2807.

MENNUCCI, B., E. CANCÈS and J. TOMASI (1997), Evaluation of solvent effects in isotropic and anisotropic dielectrics, and in ionic solutions with a unified integral equation method: theoretical bases, computational implementation and numerical applications, *J. Phys. Chem. B* **101**, 10506–10517.

MENNUCCI, B., J. TOMASI, R. CAMMI, J.R. CHEESEMAN, M.J. FRISCH, F.J. DEVLIN, S. GABRIEL and P.J. STEPHENS (2002), Polarizable continuum model (PCM) calculations of solvent effects on optical rotations of chiral molecules, *J. Phys. Chem. A* **106**, 6102–6113.

METROPOLIS, M., A.W. ROSENBLUTH, M.N. ROSENBLUTH, A.N. TELLER and E. TELLER (1953), Equation of state calculations by fast computing machines, *J. Chem. Phys.* **21**, 1087–1092.

MIERTUŠ, S., E. SCROCCO and J. TOMASI (1981), Electrostatic interaction of a solute with a continuum, *Chem. Phys.* **55**, 117–129.

MIKKELSEN, K.V., A. CESAR, H. ÅGREN and H.J.A. JENSEN (1995), Multiconfigurational self-consistent reaction field theory for nonequilibrium solvation, *J. Chem. Phys.* **103**, 9010–9023, and references therein.

MITCHELL, M.J. and J.A. MCCAMMON (1991), Free energy difference calculations by thermodynamic integration: difficulties in obtaining a precise value, *J. Comp. Chem.* **12**, 271–275.

NÉDÉLEC, J.C. (1994), New trends in the use and analysis of integral equations, *Proc. Sympos. Appl. Math.* **48**, 151–176.

NÉDÉLEC, J.C. and J. PLANCHARD (1973), Une méthode variationelle d'éléments finis pour la résolution d'un problème extérieur dans $\mathbb{R}^3$, *RAIRO* **7**, 105.

NEW YORK UNIVERSITY, (2002), MathMol Library, http://www.nyu.edu/pages/mathmol/.

ONSAGER, L. (1936), Electric moments of molecules in liquids, *J. Amer. Chem. Soc.* **58**, 1486–1493.

PAYNE, M.C., M.P. TETER, D.C. ALLAN, T.A. ARIAS and J. JOANNOPOULOS (1992), Iterative minimization techniques for ab initio total energy calculations: Molecular dynamics and conjugate gradients, *Rev. Mod. Phys.* **64**, 1045–1097.

PIEROTTI, R.A. (1976), A scaled particle theory of aqueous and nonaqueous solutions, *Chem. Rev.* **76**, 717–726.

PRASAD, P.N. and D.J. WILLIAMS (1991), *Introduction to Nonlinear Optical Effects in Organic Molecules and Polymers* (Wiley, New York).

QUARTERONI, A. and A. VALLI (1997), *Numerical Approximation of Partial Differential Equations*, 2nd edn. (Springer, Berlin).

REICHL, L.E. (1980), *A Modern Course in Statistical Physics* (Edward, London).

RICHARDS, F.M. (1977), Areas, volumes, packing, and protein structure, *Ann. Rev. Biophys. Bioeng.* **6**, 151–176.

RICHMOND, T.J. and F.M. RICHARDS (1978), Packing of a-helices: Geometrical constraints and contact areas, *J. Mol. Biol.* **119**, 537–555.

RINALDI, D., B.J. COSTA-CABRAL and J.L. RIVAIL (1986), *Chem. Phys. Lett.* **125**, 495–500.

RIVAIL, J.-L. and D. RINALDI (1976), A quantum chemical approach to dielectric solvent effects in molecular liquids, *Chem. Phys.* **18**, 233–242.

RYCKAERT, J.P., G. CICCOTTI and H.J.C. BERENDSEN (1977), Numerical integration of the Cartesian equation of motion of a system with constraints: molecular dynamics of N-alkanes, *J. Comput. Phys.* **23**, 327–341.

SANNER, M.F., A.J. OLSON and J.C. SPEHNER (1996), Reduced surface: An efficient way to compute molecular surfaces, *Biopolymers* **38**, 305–320.

SCHMIDT, M.W., K.K. BALDRIDGE, J.A. BOATZ, S.T. ELBERT, M.S. GORDON, J.H. JENSEN, S. KOSEKI, N. MATSUNAGA, K.A. NGUYEN, S.J. SU, T.L. WINDUS, M. DUPUIS and J.A. MONTGOMERY (1993), General atomic and molecular electronic structure system, *J. Comput. Chem.* **14**, 1347–1363.

STEPHENS, P.J. and F.J. DEVLIN (2000), Determination of the structure of chiral molecules using ab initio vibrational circular dichroism spectroscopy, *Chirality* **12**, 172–179.

STRAATSMA, T.P. (1996), in: K.B. Lipkowitz and D.B. Boyd, eds., *Reviews in Computational Chemistry*, Vol. 9 (VCH, New York) 81–127.

TAPIA, O. and O. GOSCINSKI (1975), Self-consistent reaction field theory of solvent effects, *Mol. Phys.* **29**, 1653–1661.

TOMASI, J., B. MENNUCCI and C. CAPPELLI (2001), Interaction in solvents and solutions, in: G. Wypych, ed., *Handbook of Solvents* (Chemtech, Toronto) 387–472.

TOMASI, J. and M. PERSICO (1994), Molecular interactions in solution: An overview of methods based on continous distributions of the solvent, *Chem. Rev.* **94**, 2027–2094.

TRUONG, T.N. and E.V. STEFANOVICH (1995), A new method for incorporating solvent effect into the classical, ab initio molecular orbital and density functional theory frameworks for arbitrary shape solute, *Chem. Phys. Lett.* **240**, 253–260.

TUCKERMAN, M.E., P.J. UNGAR, T. VON ROSENVINGE and M.L. KLEIN (1996), Ab initio molecular dynamics simulations, *J. Phys. Chem.* **100**, 12878–12887.

VREVEN, T., B. MENNUCCI, C.O. DA SILVA, K. MOROKUMA and J. TOMASI (2001), The ONIOM–PCM method: Combining the hybrid molecular orbital method and the polarizable continuum model for solvation. Application to the geometry and properties of a merocyanine in solution, *J. Chem. Phys.* **115**, 62–72.

WARSHEL, A. and M. LEVITT (1976), Theoretical studies of enzymic reactions: Dielectric, electrostatic and steric stabilization of the carbonium ion in the reaction of lysozyme, *J. Mol. Biol.* **103**, 227–249.

WOLFF, J.J. and R. WORTMANN (1999), Organic materials for second-order non-linear optics, *Adv. Phys. Org. Chem.* **32**, 121–217.

WOLINSKI, K., J.F. HINTON and P. PULAY (1990), Efficient implementation of the gauge-independent atomic orbital method for NMR chemical shift calculations, *J. Am. Chem. Soc.* **112**, 8251–8260.

WORTMANN, R. and D.M. BISHOP (1998), Effective polarizabilities and local field corrections for nonlinear optical experiments in condensed media, *J. Chem. Phys.* **108**, 1001–1007.

ZHU, T., J. LI, G.D. HAWKINS, C.J. CRAMER and G.G. TRUHLAR (1998), Density functional solvation model based on CM2 atomic charges, *J. Chem. Phys.* **109**, 9117–9133; Errata (1999), **111**, 5624 and (2000), **113**, 3930.

ZWANZIG, R.W. (1954), High temperature equation of state by a perturbation method, I. Nonpolar gases, *J. Chem. Phys.* **22**, 1420–1426.

Computational Approaches
of Relativistic Models
in Quantum Chemistry

J.P. Desclaux

Sassenage, France.
E-mail: jean-paul.desclaux@wanadoo.fr

J. Dolbeault

CEREMADE, Unité Mixte de Recherche du CNRS no. 7534 et Université Paris
IX-Dauphine, Paris cedex, France.
E-mail: dolbeaul@ceremade.dauphine.fr

M.J. Esteban

CEREMADE, Unité Mixte de Recherche du CNRS no. 7534 et Université Paris
IX-Dauphine, Paris cedex, France.
E-mail: esteban@ceremade.dauphine.fr

P. Indelicato

Laboratoire Kastler-Brossel, Unité Mixte de Recherche du CNRS no. C8552, École
Normale Supérieure et Université Pierre et Marie Curie, Paris cedex, France.
E-mail: paul.indelicato@spectro.jussieu.fr

E. Séré

CEREMADE, Unité Mixte de Recherche du CNRS no. 7534 et Université Paris
IX-Dauphine, Paris cedex, France.
E-mail: sere@ceremade.dauphine.fr

Computational Chemistry
Special Volume (C. Le Bris, Guest Editor) of
HANDBOOK OF NUMERICAL ANALYSIS
P.G. Ciarlet (Editor)

1. Introduction

1.1. QED and relativistic models in Quantum Chemistry

It is now well known, following many experimental and theoretical results, that the use of *ab initio* relativistic calculations are mandatory if one is to obtain an accurate description of heavy atoms and ions. This is true whether one is considering highly charged ions, inner shells of neutral or quasi neutral atoms or outer shells of very heavy atoms.

From a physics point of view, the natural formalism to treat such a system is Quantum Electrodynamics (QED), the prototype of field theories. For recent reviews of different aspects of QED in few electron ions see, for example, EIDES, GROTCH and SHELYUTO [2001], MOHR, PLUNIEN and SOFF [1998], BEIER [2000]. Yet a direct calculation using only QED is impractical for atoms with more than one electron because of the complexity of the calculation. This is due to the slow rate of convergence of the so-called Ladder approximation (1/Z), that in nonrelativistic theory amounts to a perturbation expansion using the electron–electron interaction as a perturbation. The only known method to do an accurate calculation is to attempt to treat to all orders the electron–electron interaction, and reserve QED for radiative corrections (interaction of the electron with its own radiation field, creation of virtual electron–positron pairs). The use of a naive approach however, taking a non-relativistic Hamiltonian and replacing one-electron Schrödinger Hamiltonian by Dirac Hamiltonian fails. This approach does not take into account one of the two main features of relativity: the possibility of particle creation, and leads to severe problems as noted already in BROWN and RAVENHALL [1951] and studied in SUCHER [1980]. This theory, for example, does not preserve charge conservation in intermediate states and leads to divergence already in the second-order of perturbation expansion. The only way to derive a proper relativistic many-electron Hamiltonian is to start from QED. The Hamiltonian of an N electron system can be written formally

$$H = H_0[Ne^-, 0e^+] + H_1[(N+1)e^-, 1e^+] + H_2[(N+2)e^-, 2e^+] + \cdots. \quad (1.1.1)$$

Keeping only the first term, the so-called "no-pair" Hamiltonian reads

$$H^{\mathrm{np}} = \sum_{i=1}^{Ne^-} h_D(r_i) + \sum_{i<j} \mathcal{U}_{ij}, \quad (1.1.2)$$

where (in atomic units) $h_D(r_i) = c\,\boldsymbol{\alpha} \cdot \mathbf{p} + \beta mc^2 + V_N(\mathbf{r}_i)$ is a one-electron Dirac Hamiltonian in a suitable classical central potential V_N, that represents the interaction of the electron with the atomic nucleus. The speed of light is denoted by c, $\boldsymbol{\alpha}$, β are the Dirac matrices, with

$$\beta = \begin{pmatrix} 1 & 0 \\ 0 & -1 \end{pmatrix}, \quad \alpha_i = \begin{pmatrix} 0 & \sigma_1 \\ \sigma_1 & 0 \end{pmatrix}, \quad (1.1.3)$$

$$\sigma_1 = \begin{pmatrix} 0 & 1 \\ 1 & 0 \end{pmatrix}, \quad \sigma_2 = \begin{pmatrix} 0 & -i \\ i & 0 \end{pmatrix}, \quad \sigma_3 = \begin{pmatrix} 1 & 0 \\ 0 & -1 \end{pmatrix}, \tag{1.1.4}$$

$p = -i\nabla_i$ and

$$\mathcal{U} = \Lambda_i^+ \Lambda_j^+ V(|\mathbf{r}_i - \mathbf{r}_j|) \Lambda_i^+ \Lambda_j^+, \tag{1.1.5}$$

where Λ_i^+ is the positive spectral projection operator of a one-particle Hamiltonian similar to $h_D(r_i)$ [i.e., $\Lambda_i^+ \phi = \phi$ for all eigenfunctions ϕ of $h_D(r_i)$ corresponding to positive eigenvalues]. Usually the potential used in this Hamiltonian is the direct Dirac–Fock potential (see Section 1.3). Moreover,

$$V(|\mathbf{r}_i - \mathbf{r}_j|) = \frac{1}{r_{ij}} - \frac{\alpha_i \cdot \alpha_j}{r_{ij}} + \left(\frac{1}{r_{ij}} - \frac{\alpha_i \cdot \alpha_j}{r_{ij}} \right) (\cos(\omega_{ij} r_{ij}/c) - 1)$$
$$+ c^2 (\alpha_i \cdot \nabla_i)(\alpha_j \cdot \nabla_\varphi) \frac{\cos(\omega_{ij} r_{ij}/c) - 1}{\omega_{ij}^2 r_{ij}} \tag{1.1.6}$$

is the electron–electron interaction of order 1 in $\alpha = 1/c \approx 1/137$, the fine structure constant. This expression is in Coulomb gauge, and is derived directly from QED. Here $r_{ij} = |\mathbf{r}_i - \mathbf{r}_j|$ is the inter-electronic distance, ω_{ij} is the energy of the photon exchanged between the electron i and j, which usually reduces to $\varepsilon_i - \varepsilon_j$, where the ε_i are the one-electron energies in the problem under consideration (e.g., diagonal Lagrange multipliers in the case of Dirac–Fock). Note that in Eq. (1.1.6) gradient operators act only on the r_{ij} and not on the following wave functions. The presence of the ω_{ij} in this expression originates from the multitime nature of the relativistic problem due to the finiteness of the speed of light. From this interaction, one can deduce the Breit operator, that contains retardation only to second order in $1/c$, in which the ω_{ij} can be eliminated by use of commutation relations between r and the one-particle Dirac Hamiltonian. This operator can then be readily used in the evaluation of correlation, while the higher-order in $1/c$ in the interaction Eq. (1.1.6) can only be evaluated perturbatively.

Finding bound states of Eq. (1.1.2) is difficult and requires approximations. The different methods of solution are inspired from the nonrelativistic problem. The three main categories of methods are the Relativistic Many-Body perturbation theory (RMBPT, see, e.g., LINDGREN and MORRISON [1982] for the nonrelativistic case), the Relativistic Random Phase Approximation (RRPA, see, e.g., JOHNSON and LIN [1976]), which has been heavily used for evaluation of photoionization cross-sections, and Multiconfiguration Dirac–Fock (MCDF). The RMBPT method requires the use of basis sets to sum over intermediate states. The MCDF method is a variational method.

1.2. Relativistic RMBPT and RRPA

In its most general version, the RMBPT method starts from a multidimensional model space and uses Rayleigh–Schrödinger perturbation theory. The concept of model space is mandatory if there are several levels of quasi-degenerate energy as in the ground state of Be-like ions ($1s^2 2s^2$ $^1S_0$ and $1s^2 2p^2$ $^1S_0$ are very close in energy, leading to very strong intra-shell correlation). In that case one gets would get very bad convergence of the perturbation expansion, because of the near-zero energy denominators, if building the perturbation theory on a single level.

Following LINDGREN and MORRISON [1982] we separate the Hamiltonian in a sum

$$H_T = H_0 + V_0. \tag{1.2.1}$$

We assume that we know a set of N eigenfunctions Ψ_α^0 of eigenenergies E_α^0 which are all the solutions obtained by diagonalizing H_0 on a subspace $\mathcal{P}$ (these solutions can be obtained with the Dirac–Fock method in a suitable average potential). The unperturbed Hamiltonian is then chosen as

$$H_0^N = P_0 H_0 P_0 = \sum_{\alpha=1}^{N} E_\alpha^0 |\Psi_\alpha^0\rangle\langle\Psi_\alpha^0|, \tag{1.2.2}$$

where and P_0 is the projector on $\mathcal{P}$, defined by

$$P_0 = \sum_{\alpha=1}^{N} |\Psi_\alpha^0\rangle\langle\Psi_\alpha^0|. \tag{1.2.3}$$

We define the perturbation potential by

$$V = H_T - P_0 H_0 P_0 = H_T - H_0^N. \tag{1.2.4}$$

We also define $Q_0 = 1 - P_0$ as the projector on the orthogonal space $\mathcal{Q}$. We now define the wave operator, which build the exact solution of the Hamiltonian equation (1.2.1) from the Ψ_α^0

$$\Psi_\alpha = \Omega \Psi_\alpha^0, \tag{1.2.5}$$

so that

$$H_T \Psi_\alpha = E_\alpha \Psi_\alpha, \tag{1.2.6}$$

with the property

$$P_0 \Omega P_0 = P_0. \tag{1.2.7}$$

The *exact* eigenenergies can be obtained by the application of the Model-space wave functions on the effective Hamiltonian

$$H_{\text{eff}} = P_0 H_T \Omega P_0 = P_0 H_0 P_0 + P_0 V \Omega P_0 = H_0^N + P_0 V \Omega P_0, \tag{1.2.8}$$

using Eqs. (1.2.2) and (1.2.7), $P_0^2 = P_0$ and the fact that H_0^N and P_0 commute. This operator, acting on the unperturbed wave functions give the exact eigenenergies:

$$H_{\text{eff}} \Psi_\alpha^0 = E_\alpha \Psi_\alpha^0. \tag{1.2.9}$$

The wave operator obeys the generalized Bloch equation

$$[\Omega, H_0] P_0 = V \Omega P_0 - \Omega P_0 V \Omega P_0 \tag{1.2.10}$$

using Eq. (1.2.7). This can be expanded in a series

$$\Omega = 1 + \Omega^{(1)} + \Omega^{(2)} + \cdots. \tag{1.2.11}$$

Eqs. (1.2.10) and (1.2.11) leads to the sequence of equations

$$[\Omega^{(1)}, H_0]P_0 = Q_0 V P_0, \tag{1.2.12}$$

$$[\Omega^{(2)}, H_0]P_0 = Q_0 V \Omega^{(1)} P_0 - \Omega^{(1)} P_0 V P_0. \tag{1.2.13}$$

The RRPA method is based on the solution of the Hamiltonian Eq. (1.1.2) subjected to a time-dependent perturbation (like a classical electromagnetic radiation of known frequency). This time-dependent Dirac–Fock equation is solved over a set of solutions of the unperturbed problem, leading to a set of time-dependent mixing coefficients in the usual fashion of time-dependent perturbation theory. The phases of those coefficients are approximated (leading to the name "Random Phase"), leading to differential equations very similar to the Dirac–Fock ones. This method include to all orders some classes of correlation contribution that can be easily also evaluated in the framework of RMBPT. It is mostly used for the ground state of atoms and ions to study photoionization. It is more difficult to use for excited states.

This paper is mostly devoted to the MCDF method for atoms and molecules, and to preliminary results for the linear Dirac operator.

1.3. The MCDF wave function

We first start by describing shortly the formalism used to build the Dirac–Fock solutions for a spherically-symmetric system like an isolated atom.

If we define the angular momentum operators $L = r \wedge p$, $J = L + \frac{\sigma}{2}$, the parity Π as βP, then the total wave function is expressed in term of configuration state functions (CSF) as antisymmetric products of one-electron wave functions so that they are eigenvalues of the parity Π, the total angular momentum J and its projection M. The label v stands for all other values (angular momentum recoupling scheme, seniority numbers, ...) necessary to define unambiguously the CSF. For an N-electron system, a CSF is thus a linear combination of Slater determinants:

$$|v\Pi JM\rangle = \sum_{i=1} d_i^v \begin{vmatrix} \Phi_1^{i,v}(r_1) & \cdots & \Phi_N^{i,v}(r_1) \\ \vdots & \ddots & \vdots \\ \Phi_1^{i,v}(r_N) & \cdots & \Phi_N^{i,v}(r_N) \end{vmatrix}, \tag{1.3.1}$$

all of them with the same Π and M values while the d_i's are determined by the requirement that the CSF is an eigenstate of J^2.

The total MCDF wave function is constructed as a superposition of CSFs, tha is,

$$\Psi(\Pi JM) = \sum_{v=1}^{NCF} c_v |v\Pi JM\rangle, \tag{1.3.2}$$

where NCF is the number of configurations and the c_v are called the configurations mixing coefficients.

The MCDF method has two variants. In one variant, one uses numerical or analytic basis sets to construct the CSF. In the other one, direct numerical solution of the

MCDF equation is used. Both methods have been used in atomic and molecular physics. The numerical MCDF method is better suited for small systems, while analytic basis set techniques are better suited for cases with millions of determinants.

This chapter is organized as follows. In Section 2, different choices of basis sets for the Dirac equation are presented. In Section 3, the MCDF equations are presented, and numerical techniques adapted to the numerical MCDF method in atoms are described. In Section 4, we deal with techniques for the numerical MCDF method in molecules.

2. Linear Dirac equations

2.1. Properties of the linear Dirac operator

The unboundedness from below of the Dirac operator

$$H_0 = -ic\alpha\cdot\nabla + mc^2\beta \tag{2.1.1}$$

creates important difficulties when trying to find its eigenvalues. The so-called *variational collapse* is indeed related to this unboundedness property. On the other hand, finite-dimensional approximations to this problem may lead to finding *spurious solutions*: some eigenvalues of the finite-dimensional problem do not approach the eigenvalues of the Dirac operator and destroy the monotonicity of the approximated eigenvalues with respect to the basis dimension. These problems seem to be much more acute in molecular than in atomic computations, but they are already present in one-electron systems. In this section, we address this difficulty for one-electron systems by describing various methods used to deal with this problem. Well-behaved approximation methods should also provide good nonrelativistic limits, that is, variational problems whose eigenvalues and eigenfunctions converge well to those of the corresponding nonrelativistic Schrödinger Hamiltonian.

A way often used to find good numerical approximations of eigenvalues of an operator A consists in projecting the eigenvalue equation

$$Ax = \lambda x \tag{2.1.2}$$

over a well chosen finite-dimensional space X_N of dimension N, in order to find an approximation (λ_N, x_N) satisfying

$$A_N x_N = \lambda_N x_N, \tag{2.1.3}$$

such that (λ_N, x_N) converges to (λ, x) as $N \to \infty$. Then one looks for the eigenvalues of the $N\times N$ matrix A_N and these eigenvalues will converge either to eigenvalues of A or to points in the essential spectrum of A. As N increases, the limit set of the eigenvalues of A_N is the spectrum of A.

The difficulty with the Dirac operator is that for most physically interesting potentials V, the spectrum of $H_0 + V$ is made of its essential spectrum $(-\infty, -mc^2] \cup [mc^2, +\infty)$ and a discrete set of eigenvalues lying in the *gap* $(-mc^2, mc^2)$. Hence, the choice of the finite-dimensional space, or equivalently, of the finite basis set, is fundamental if we want to ensure that for some N large, the eigenvalues of $(H_0 + V)_N$, or at least some of them, will be approximations of the eigenvalues of $H_0 + V$ in the gap $(-mc^2,$ $mc^2)$. The question of how to choose a good basis set has been addressed in many papers, among which DRAKE and GOLDMAN [1981], DRAKE and GOLDMAN [1988], GRANT [1989], GRANT [1982], JOHNSON, BLUNDELL and SAPIRSTEIN [1988], KUTZEL-NIGG [1984], LEE [2001], that we will describe with further details in Sections 2.2 and 2.3 below. In particular, Section 2.3 is devoted to the description of numerical techniques based either on discretization or on B-splines, and shows that with appropriate boundary conditions one can avoid the variational collapse.

When the operator A is bounded from below, it is often possible to characterize its spectrum by variational methods, for instance by looking for critical values of the Rayleigh quotient

$$Q(x) := \frac{(Ax, x)}{(x, x)} \qquad (2.1.4)$$

over the domain of A. More concretely, when A is bounded from below, under appropriate assumptions, its ground state energy can be found by minimizing the above Rayleigh quotient. However, this cannot be done directly in the context of the Dirac operator, since it is unbounded from below (and also from above). A large number of works have been devoted to the variational resolution of this problem in view of the Dirac operator. Most of them use the approximation of an effective Hamiltonian which is bounded from below. The idea hidden behind this kind of techniques is that there is no explicit way of diagonalizing the Dirac Hamiltonian $H + V$, but this can be done at an abstract level. The diagonalized operator is then approximated via a finite expansion or an iterative procedure. These methods are therefore perturbative and contain an approximation at the operator level. They will be referred to as *perturbation theories and effective Hamiltonian methods* and will be described below (see DRAKE and GOLDMAN [1981], KUTZELNIGG [1984], GRANT [1982], JOHNSON, BLUNDELL and SAPIRSTEIN [1988] and Section 2.4 for more details).

Other variational techniques are based on a correspondence between the eigenvalues of A and those of $T(A)$, for some operator function T, like the inverse function $Tx = x^{-1}$ (see HILL and KRAUTHAUSER [1994]) or the function $Tx = x^2$ (see WALLMEIER and KUTZELNIGG [1981], BAYLISS and PEEL [1983]). Finally, some authors solve the variational problem in a subspace of the domain in which the operator is bounded from below and "avoids" the negative continuum. Section 2.5 will be devoted to these more *direct variational approaches*, based on either linear or nonlinear constraints.

Before going into the details of the computational methods, let us start with some notations and preliminary considerations. For any ψ with values in $\mathbb{C}^4$, if we write

$$\psi = \begin{pmatrix} \varphi \\ \chi \end{pmatrix},$$ with φ, χ taking values in $\mathbb{C}^2$, then the eigenvalue equation

$$H\psi = (H_0 + V)\psi = \lambda\psi \qquad (2.1.5)$$

is equivalent to the following system:

$$\begin{cases} R\chi = (\lambda - mc^2 - V)\varphi, \\ R\varphi = (\lambda + mc^2 - V)\chi, \end{cases} \quad (2.1.6)$$

with $R = ic\left(\vec{\sigma} \cdot \vec{\nabla}\right) = \sum_{j=1}^{3} ic\sigma_j \frac{\partial}{\partial x_j}$. Here σ_j, $j = 1, 2, 3$, are the Pauli matrices. As long as $\lambda + mc^2 - V \neq 0$, the system Eq. (2.1.6) can be written as

$$H^\mu \varphi := R\left(\frac{R\varphi}{g_\mu}\right) + V\varphi = \mu\varphi, \quad \chi = \frac{R\varphi}{g_\mu}, \quad (2.1.7)$$

where $g_\mu = \mu + 2mc^2 - V$ and $\mu = \lambda - mc^2$. Note that the Hamiltonian operator H^μ is eigenvalue dependent. Reducing the 4-component spinor ψ to an equation for the 2-spinor φ is often called *partitioning*. Let us immediately notice that at least formally, the partitioned equation (2.1.7) converges to its nonrelativistic counterpart

$$-\frac{1}{2m}\Delta\varphi + V\varphi = \mu\varphi \quad (2.1.8)$$

(see, for instance, WOOD, GRANT and WILSON [1985]). For this reason, but also because the principal part of the second order operator in Eq. (2.1.7) is semibounded for not too large potentials V, the partitioned equation has been extensively studied for finding eigenvalues of linear Dirac operators.

To end these preliminary considerations on linear Dirac equations, note that in the case of *rotationally invariant* potentials, the solutions can be put in the form

$$\psi = \frac{1}{r}\begin{pmatrix} P_\kappa(r)\chi_{\kappa m}(\theta, \varphi) \\ iQ_\kappa(r)\chi_{-\kappa m}(\theta, \varphi) \end{pmatrix}. \quad (2.1.9)$$

The dependence on the angular coordinates is contained in the 2-spinors $\chi_{\pm\kappa m}(\theta, \varphi)$, which are eigenfunctions of the angular momentum operators J, its third component J_z (with eigenvalues $j(j+1)$ and m, respectively) and of parity. On the other hand, the radial dependence is contained in the functions f and g which are called the upper and lower radial components of ψ.

In the ansatz defined in Eq. (2.1.9), for a given $\kappa = \pm(j + \frac{1}{2})$, with $j = \mp\frac{1}{2}$, $l = 0, 1$, ..., the eigenvalue equation (2.1.5) is equivalent to

$$(H_r^\kappa + V)\Phi = \lambda\Phi, \quad (2.1.10)$$

with

$$H_r^\kappa = \begin{pmatrix} mc^2 & c\left(-\frac{\mathrm{d}}{\mathrm{d}r} + \kappa_j\right) \\ c\left(\frac{\mathrm{d}}{\mathrm{d}r} + \kappa_j\right) & -mc^2 \end{pmatrix}, \quad (2.1.11)$$

$$\Phi = \begin{pmatrix} P_\kappa \\ Q_\kappa \end{pmatrix} \text{begin a } 2 - \text{vector with two scalar real components.}$$

2.2. Finite basis set approaches

The choice of *finite-dimensional spaces* is essential for the discretization of the operator and the approximation of its eigenvalues. The presence of the negative continuum makes this task difficult in the case of the Dirac operator. The basic criterium to decide whether a particular space, or a generating basis set, is good, is to check that the approximated eigenvalues found are either negative and lying in the negative continuum or positive. In this case, if they are below the positive continuum, they are approximations of the discrete exact (positive) eigenvalues. Many attempts to construct finite basis sets can be found in the literature.

DRAKE and GOLDMAN [1981] introduced the so-called *Slater type orbitals* (STO):

$$\Phi(r) = r^{\gamma-1} e^{-\nu r} \sum_{i=1}^{N} r^i \left[a_i \begin{pmatrix} 1 \\ 0 \end{pmatrix} + b_i \begin{pmatrix} 0 \\ 1 \end{pmatrix} \right], \tag{2.2.1}$$

with a particular choice of γ and ν which depends on κ and V. They showed numerical evidence that such a finite basis satisfies the above properties in the case of hydrogen-like atoms. Note that the STOs exhibit the same behavior near 0 and at ∞ as the exact eigenfunctions. The properties of STO basis sets are made more explicit in GRANT [1982], where STO basis sets are replaced by orthonormal sets of *Laguerre polynomials*. The main drawback of this approach is that some of the eigenvalues of the approximated matrix are spurious roots which do not approximate any of the exact eigenvalues.

Another way to construct *basis sets* with good properties consists in imposing the so-called *kinetic balance* condition relating the *upper* and *lower components* of the functions in the basis set. See, for instance, KUTZELNIGG [1984].

Other types of basis sets proposed in the literature include those generated by *B-splines* (see JOHNSON, BLUNDELL and SAPIRSTEIN [1988]), which have very good properties since, in this approach, the matrices are very sparse: only a finite number (depending on the degree of the splines) of diagonal lines are nonzero. This kind of basis sets has been widely used in atomic and molecular computations (see Section 2.3).

The choice of a good basis set can be quite effective in some computations, but as it appears clearly in the literature that we quote, there is very often a risk of finding spurious roots or of variational collapse. In the next subsection, we give some more precise examples of how to use particular basis sets in the context of Dirac operators.

2.3. Numerical basis sets

This section is devoted to the special case of basis sets whose elements are computed numerically.

Discretisation method

The Göteborg group has developed an efficient technique to obtain basis sets for the Dirac equation (SALOMONSON and ÖSTER [1989a]). The Dirac equation is discretized and solved on a grid. The atom is placed in a spherical box, large enough not to disturb the bound state wave function considered. The method provides a finite number of orbitals which is complete over the discretized space (SALOMONSON and ÖSTER [1989b]), and resemble lattice gauge field calculation (WILSON [1974]). The method enables to eliminate spurious states and preserves the Hermiticity of the discretized Hamiltonian. The appearance of spurious states in a discretized method, is traced back to the "fermion doubling", first encoutered in gauge-field lattice calculations (KOGUT [1983]). On a lattice of dimension $(D + 1)$ (D spatial and one time dimensions), an equation for a massless fermion will describe not one but 2^D ones if no precaution is taken (STACEY [1982]).

As an example, let us consider a one-dimensional Dirac equation for a free fermion

$$
\begin{pmatrix} mc^2 & -c\dfrac{d}{dx} \\ c\dfrac{d}{dx} & -mc^2 \end{pmatrix} \begin{pmatrix} f(x) \\ g(x) \end{pmatrix} = (\varepsilon + mc^2) \begin{pmatrix} f(x) \\ g(x) \end{pmatrix}.
\tag{2.3.1}
$$

The derivatives are approximated over the lattice points using

$$
f_i' = \frac{f_{i+1} - f_{i-1}}{2h},
\tag{2.3.2}
$$

where h is the space between adjacent lattice sites. Eliminating the large component in Eq. (2.3.1), one gets the following equation

$$
-\frac{1}{2m} \left(\frac{f_{i+2} - 2f_i + f_{i-2}}{4h^2} \right) = \varepsilon \left(1 + \frac{\varepsilon}{2mc^2} \right) f_i,
\tag{2.3.3}
$$

in which the left-hand side is the kinetic energy operator $p_x^2/2m$ acting on f at the lattice point i. Yet this second order derivative does not connect even and odd lattice sites. The highest energy solution over the lattice is the one changing sign at each site so that

$$
\ldots \approx f_{i-2} \approx -f_{i-1} \approx f_i \approx -f_{i+1} \approx \ldots.
\tag{2.3.4}
$$

Using the expression Eq. (2.3.3) acting on this solution gives the same results as if it had no nodes. A high-energy eigenvector thus appears as a spurious state in the low energy part of the spectrum. For low-order derivative two equivalent ways can be used (STACEY [1982], SALOMONSON and ÖSTER [1989a]). One is to use forward derivatives for f and backward derivatives for g,

$$
f_i' = \frac{f_{i+1} - f_i}{h}, \quad g_i' = \frac{g_i - g_{i-1}}{h}.
\tag{2.3.5}
$$

The other consists in defining the large and small components on alternating sites on the lattice.

$$
\ldots f_{i-3} g_{i-2} f_{i-1} g_i f_{i+1} g_{i+2} \ldots,
\tag{2.3.6}
$$

with h being the separation between g_{i-2} and g_i. In this case the derivative is expressed as

$$f_i' = \frac{f_{i+1} - f_i}{h}, \quad g_j' = \frac{g_j - g_{j-1}}{h},$$ (2.3.7)

with $i = 2n - 1$, $j = 2n$, $n = 1, 2, \ldots, N$. These methods reduce to the same second-order equation (STACEY [1982]).

Salomonson and Öster use a more accurate six-point formula

$$
\begin{aligned}
f'(x) = \frac{1}{1920h} &\left[-9f\left(x - \frac{5}{2}h\right) + 125f\left(x - \frac{3}{2}h\right) - 2250f\left(x - \frac{1}{2}h\right) \right. \\
&+ 2250f\left(x + \frac{1}{2}h\right) - 125f\left(x + \frac{3}{2}h\right) \\
&\left. + 9f\left(x + \frac{5}{2}h\right) \right] + \mathcal{O}(h^6).
\end{aligned}
$$ (2.3.8)

This six-point formula combined with Eq. (2.3.6) provides a spurious-state-free solution, while using the same lattice for f and g and a forward-backward derivative scheme does not work.

In the spherical case, one needs to use a logarithmic lattice to get a good description of the wave function. The Hermiticity of the Hamiltonian must be preserved by doing the variable change

$$y(r) \to \frac{1}{\sqrt{r}} y(x), \quad x = \log(r).$$ (2.3.9)

The corresponding Dirac equation is

$$
\begin{pmatrix}
V(r) & -c\left(\frac{1}{\sqrt{r}}\frac{d}{dx}\frac{1}{\sqrt{r}} - \frac{\kappa}{\sqrt{r}}\frac{1}{\sqrt{r}}\right) \\
c\left(\frac{1}{\sqrt{r}}\frac{d}{dx}\frac{1}{\sqrt{r}} + \frac{\kappa}{\sqrt{r}}\frac{1}{\sqrt{r}}\right) & V(r) - mc^2
\end{pmatrix}
\begin{pmatrix} f(x) \\ g(x) \end{pmatrix}
= \varepsilon \begin{pmatrix} f(x) \\ g(x) \end{pmatrix}.
$$ (2.3.10)

Since the large and small component are defined on different lattices, one needs interpolation formulas to express $f(x)/\sqrt{r}$ and $g(x)/\sqrt{r}$ in the κ term.

The discretization finally provides a $2N \times 2N$ symmetric eigenvalue problem

$$
\begin{pmatrix} A & {}^tD + {}^tK \\ D + K & B \end{pmatrix}
\begin{pmatrix} F \\ G \end{pmatrix}
= \varepsilon \begin{pmatrix} F \\ G \end{pmatrix},
$$ (2.3.11)

with $(F, G) = (f_1, f_3, \ldots, f_{2N-1}, g_2, g_4, \ldots, g_{2N})$. For a point nucleus, the submatrices are $A_{ii} = -Z/r_i$ and $B_{jj} = -2mc^2 - Z/r_j$, $i = 2n - 1$, $j = 2n$, $n = 1, 2, \ldots, N$. With the 6 points interpolation and derivation formulas used in SALOMONSON and ÖSTER [1989a], one obtains

$$
D = \frac{c}{1920h}
\begin{pmatrix}
-\dfrac{2250}{\sqrt{r_2 r_1}} & \dfrac{2250}{\sqrt{r_2 r_3}} & -\dfrac{125}{\sqrt{r_2 r_5}} & \dfrac{9}{\sqrt{r_2 r_7}} & 0 & \cdots & \cdots \\[2ex]
\dfrac{125}{\sqrt{r_4 r_1}} & -\dfrac{2250}{\sqrt{r_4 r_3}} & \dfrac{2250}{\sqrt{r_4 r_5}} & -\dfrac{125}{\sqrt{r_4 r_7}} & \dfrac{9}{\sqrt{r_4 r_9}} & 0 & \cdots \\[2ex]
-\dfrac{9}{\sqrt{r_6 r_1}} & \dfrac{125}{\sqrt{r_6 r_3}} & -\dfrac{2250}{\sqrt{r_6 r_5}} & \dfrac{2250}{\sqrt{r_6 r_7}} & -\dfrac{125}{\sqrt{r_6 r_9}} & \dfrac{9}{\sqrt{r_6 r_{11}}} & \cdots \\[2ex]
0 & -\dfrac{9}{\sqrt{r_8 r_3}} & \dfrac{125}{\sqrt{r_8 r_5}} & -\dfrac{2250}{\sqrt{r_8 r_7}} & \dfrac{2250}{\sqrt{r_8 r_9}} & -\dfrac{125}{\sqrt{r_8 r_{11}}} & \cdots \\[2ex]
\vdots & & \vdots & \vdots & \vdots & \vdots & \ddots
\end{pmatrix},
\tag{2.3.12}
$$

and

$$
K = \frac{\kappa}{256h}
\begin{pmatrix}
\dfrac{150}{\sqrt{r_2 r_1}} & \dfrac{150}{\sqrt{r_2 r_3}} & -\dfrac{25}{\sqrt{r_2 r_5}} & \dfrac{3}{\sqrt{r_2 r_7}} & 0 & \cdots & \cdots \\[2ex]
-\dfrac{25}{\sqrt{r_4 r_1}} & \dfrac{150}{\sqrt{r_4 r_3}} & \dfrac{150}{\sqrt{r_4 r_5}} & -\dfrac{25}{\sqrt{r_4 r_7}} & \dfrac{3}{\sqrt{r_4 r_9}} & 0 & \cdots \\[2ex]
\dfrac{3}{\sqrt{r_6 r_1}} & -\dfrac{25}{\sqrt{r_6 r_3}} & \dfrac{150}{\sqrt{r_6 r_5}} & \dfrac{150}{\sqrt{r_6 r_7}} & -\dfrac{25}{\sqrt{r_6 r_9}} & \dfrac{3}{\sqrt{r_6 r_{11}}} & \cdots \\[2ex]
0 & \dfrac{3}{\sqrt{r_8 r_3}} & -\dfrac{25}{\sqrt{r_8 r_5}} & \dfrac{150}{\sqrt{r_8 r_7}} & \dfrac{150}{\sqrt{r_8 r_9}} & -\dfrac{25}{\sqrt{r_8 r_{11}}} & \cdots \\[2ex]
\vdots & & \vdots & \vdots & \vdots & \vdots & \ddots
\end{pmatrix}.
\tag{2.3.13}
$$

Eq. (2.3.11) is symmetric even though K and D are not. In the upper left corner of D, use has been made of the approximation

$$
f(r) \sim r^{\gamma+1/2} + \frac{2[\gamma + \kappa - (Z\alpha)^2]}{Z\alpha^2(2\gamma+1)} r^{\gamma+3/2} + \varepsilon \frac{[\gamma + \kappa - 2(Z\alpha)^2]}{Z(2\gamma+1)} r^{\gamma+3/2},
\tag{2.3.14}
$$

with $\gamma = \sqrt{\kappa^2 - (Z\alpha)^2}$, and the equivalent expression for g. To avoid nonlinear terms in the eigensystem Eq. (2.3.11), only the contribution independent of ε has been kept. This is a good approximation for bound states for which $\varepsilon \ll mc^2$.

Numerical basis sets based on B-splines

B-splines have been used (JOHNSON, BLUNDELL and SAPIRSTEIN [1988]) to provide numerically efficient basis sets. A knot sequence t_i is used for the radial coordinate, on which B-spline of order k provide a complete basis for piecewise polynomials of order $k - 1$. This radial coordinate extends to a distance R from the origin. The solutions of the Dirac equation are expressed as linear combinations of B-splines. A *Galerkin method* is employed to obtain the solution. The Dirac equation is derived from an action principle $\delta S = 0$, with

$$
S = \frac{1}{2} \int_0^R \left\{ cP_\kappa(r) \left(\frac{d}{dr} - \frac{\kappa}{r} \right) Q_\kappa(r) - cQ_\kappa(r) \left(\frac{d}{dr} + \frac{\kappa}{r} \right) P_\kappa(r) \right.
$$

$$
\left. + V_N(r)[P_\kappa(r)^2 + Q_\kappa(r)^2] - 2mc^2 Q_\kappa(r)^2 \right\} dr - \frac{1}{2}\varepsilon \int_0^R [P_\kappa(r)^2 + Q_\kappa(r)^2]dr
$$

$$
\tag{2.3.15}
$$

using the notations of Eq. (2.1.9) (note that in this representation the gap lies between $-2mc^2$ and 0), to which suitable boundary conditions are added through

$$S' = \begin{cases} \frac{c}{4}[P_\kappa(R)^2 - Q_\kappa(R)^2] + \frac{c}{2}P_\kappa(0)^2 - \frac{c^2}{2}P_\kappa(0)Q_\kappa(0) & \text{for } \kappa < 0, \\ \frac{c}{4}[P_\kappa(R)^2 - Q_\kappa(R)^2] + c^2 P_\kappa(0)^2 - \frac{c}{2}P_\kappa(0)Q_\kappa(0) & \text{for } \kappa > 0. \end{cases}$$

(2.3.16)

From the point of view of the variational principle, ε is a Lagrange multiplier introduced to ensure that the solutions of the Dirac equation are normalized. The boundary constraint Eq. (2.3.16) is designed to avoid a hard boundary at the box radius R, following the idea behind the MIT bag model for quark confinement, and provides $P_\kappa(R) = Q_\kappa(R)$. Forcing $P_\kappa(R) = Q_\kappa(R) = 0$ would amount to introduce an infinite potential at the boundary and possibly leads to the Klein paradox. Other choices of boundary conditions are possible. This particular choice avoids the appearance of spurious solutions. Expanding the radial wave function as

$$P_\kappa(r) = \sum_{i=1}^n p_i B_{i,k}(r), \quad Q_\kappa(r) = \sum_{i=1}^n q_i B_{i,k}(r),$$

(2.3.17)

the variational principle reduces to

$$\frac{d(S+S')}{dp_i} = 0, \quad \frac{d(S+S')}{dq_i} = 0, \quad i = 1, 2, \ldots, n.$$

(2.3.18)

This leads to a $2n \times 2n$ symmetric, generalized eigenvalue equation

$$Av = \varepsilon Bv,$$

(2.3.19)

where $v = (p_1, p_2, \ldots, p_n, q_1, q_2, \ldots, q_n)$,

$$A = \begin{pmatrix} (V) & c\left[(D) - \left(\frac{\kappa}{r}\right)\right] \\ -c\left[(D) + \left(\frac{\kappa}{r}\right)\right] & (V) - 2mc^2(C) \end{pmatrix} + A'$$

(2.3.20)

and

$$B = \begin{pmatrix} (C) & 0 \\ 0 & (C) \end{pmatrix}.$$

(2.3.21)

The $2n \times 2n$ matrix A' comes from the boundary term. The $n \times n$ matrix (C) is the B-spline overlap matrix defined by

$$(C)_{ij} = \int B_{i,k}(r)B_{j,k}(r)dr,$$

(2.3.22)

(D) comes from the differential operator

$$(D)_{ij} = \int B_{i,k}(r) \frac{dB_{j,k}(r)}{dr} dr, \tag{2.3.23}$$

(V) is the potential term

$$(V)_{ij} = \int B_{i,k}(r) V_N(r) B_{j,k}(r) dr \quad \text{and} \quad \left(\frac{\kappa}{r}\right)_{ij} = \int B_{i,k}(r) \frac{\kappa}{r} B_{j,k}(r) dr. \tag{2.3.24}$$

Diagonalization of Eq. (2.3.20) provides $2n$ eigenvalues and eigenfunctions, n of which have energies below $-2mc^2$, a few correspond to bound states (typically 5–6 for $k = 7$–9) and the rest belongs to the positive energy continuum.

2.4. Perturbation theory and effective Hamiltonians

An alternative way to find the eigenvalues of the unbounded relativistic operator H consists in looking for a so-called *effective Hamiltonian* H^{eff}, which is semi-bounded, such that both Hamiltonians have *common eigenvalues* on an interval above the negative continuous spectrum. Such a Hamiltonian H^{eff} cannot usually be found in an explicit way, but can be viewed as the limit of an iterative procedure. This leads to families of Hamiltonians which approach the effective Hamiltonian and yield approximated eigenvalues for H.

One of the most popular procedure in this direction is due to FOLDY and WOUTHUYSEN [1950], whose main idea was to apply a unitary transformation Ω to $H_0 + V$ such that

$$\Omega^*(H_0 + V)\Omega = H^{FW} = \begin{pmatrix} H_+^{FW} & 0 \\ O & H_-^{FW} \end{pmatrix}, \tag{2.4.1}$$

so that electronic and positronic states are decoupled: electrons (resp. positrons) would be described by the eigenfunctions of H_+^{FW} (resp. H_-^{FW}). Moreover, the Hamiltonians $H_+^{FW} - mc^2$ (resp. $H_-^{FW} + mc^2$) are bounded from below (resp. above) and have correct nonrelativistic limits. Although this procedure looks very promising, the problem is that Ω is unknown in closed form, and so there is no way of diagonalizing $H_0 + V$ in an explicit way. However, approximations of Ω, and therefore of H^{FW}, can be constructed either by writing a formal series expansion for $H_\pm^{FW}$ in the perturbation parameter c^{-2}:

$$H_\pm^{FW} = \sum_{k=0}^{+\infty} c^{-2k} H_{2k}^\pm, \tag{2.4.2}$$

and cutting it at level $k \geq 0$, or by approaching it by an iterative procedure.

In general one identifies the effective Hamiltonian H^{eff} as a solution to a nonlinear equation $H^{eff} = f(H^{eff})$, which can be solved approximately in an iterative way. By instance, one can produce an equation like the above one by "eliminating" the lower component χ of the spinor as in Eq. (2.1.7), that is, by *partitioning*.

Many proposals of effective Hamiltonians for the Dirac operator can be found in the literature. Some are Hermitian, some are not, some act on 4 component spinors, others on 2-spinors. A good review about various approaches to this problem and the corresponding difficulties has been written by KUTZELNIGG [1999] (see also KUTZEL-NIGG [1990], RUTKOWSKI and SCHWARZ [1996], RUTKOWSKI [1999]). An important difficulty arising in this context is that most of the proposed effective Hamiltonians are quite nice when the potential V is regular, but in the case of the Coulomb potential they contain very singular terms, which are not even well defined near the nucleus. These serious singularities are avoided by a method used by CHANG, PÉLISSIER and DURAND [1986] (see also DURAND [1986], DURAND and MALRIEU [1987]), where it is proposed to use $(2mc^2 - V)^{-1}$ as an expansion parameter in the formal series defining H^{eff}, instead of c^{-2}. They obtain a 2-component Pauli-like Hamiltonian which is bounded from below, contains only well defined terms and approaches H. Similar ideas have been used by HEULLY, LINDGREN, LINDROTH, LUNDQVIST and MÅRTENSSON-PENDRILL [1986] and by VAN LENTHE, VAN LEEUWEN, BAERENDS and SNIJDERS [1994a], VAN LENTHE, VAN LEEUWEN, BAERENDS and SNIJDERS [1994b]. The latter have also made a systematic numerical analysis of this method in self-consistent calculations for the uranium atom.

2.5. *Direct variational approaches*

To begin with, let us mention two variational methods based on *nonlinear transformations of the Hamiltonian*. WALLMEIER and KUTZELNIGG [1981] look for eigenvalues of the squared Hamiltonian $(H_0 + V)^2$. The practical difficulty arises from the need to compute complicated matrix representations. HILL and KRAUTHAUSER [1994] use the Rayleigh–Ritz variational principle applied to the inverse of the Dirac Hamiltonian, $1/H$. A difficulty arises here in the computation of the matrix elements for the inverse operator. This is avoided by working in the special set of test functions defined by those which are in the image by H of a regular set of spinors. The use of these two methods can be useful in some cases, but not when the eigenvalues become close to 0.

As already noticed, the eigenvalues of the operator $H_0 + V$ are critical points of the Rayleigh quotient

$$Q_V(\psi) := \frac{((H_0 + V)\psi, \psi)}{(\psi, \psi)} \tag{2.5.1}$$

in the domain of $H_0 + V$. We are now going to describe other more sophisticated variational approaches yielding exact eigenvalues of $H_0 + V$. The particular structure of the spectrum of H_0 clearly shows that eigenvalues of $H_0 + V$ lying in the gap of the essential spectrum should be given by some kind of *min–max approach*. This had been mentioned in several papers dealing with numerical computations of Dirac eigenvalues, before it was proved in a series of papers: ESTEBAN and SÉRÉ [1997], GRIESEMER and SIEDENTOP [1999], DOLBEAULT, ESTEBAN and SÉRÉ [2000a], GRIESEMER, LEWIS and SIEDENTOP [1999], DOLBEAULT, ESTEBAN and SÉRÉ [2000b]). Basically, in all those papers, it was shown that under appropriate assumptions on the potential V, the eigenvalues are indeed characterized as a sequence of min–max values defined for Q_V on well chosen sets. A theorem in DOLBEAULT, ESTEBAN and SÉRÉ [2000b] proves that for a large class of potentials V, the ground state energy of $H_0 + V$ is given by the smallest λ in the gap $[-mc^2, mc^2]$ such that there exists ϕ satisfying

$$\lambda \int_{\mathbb{R}^3} |\varphi|^2 dx = \int_{\mathbb{R}^3} \left(\frac{|(\sigma \cdot \nabla)\varphi|^2}{1 - V + \lambda} + (1 + V)|\varphi|^2 \right) dx \tag{2.5.2}$$

and the corresponding eigenfunction is the spinor function

$$\psi = \begin{pmatrix} \varphi \\ -i\dfrac{(\sigma \cdot \nabla)\varphi}{1 - V + \lambda} \end{pmatrix}. \tag{2.5.3}$$

Note that the idea to build a semibounded energy functional had already been introduced by BAYLISS and PEEL [1983] in another context. It is closely related to previous works of DATTA and DEVIAH [1988] and TALMAN [1986], where a particular min–max procedure for the Rayleigh quotient Q_V is proposed without proof. We will not give here further details on these theoretical aspects (for tractable numerical applications, see below).

An alternative variational method has been proposed by DOLBEAULT, ESTEBAN and SÉRÉ [2000a]. It is based on rigorous results proving that for a very large class of potentials (including all those relevant in atomic models), the ground state of $H_0 + V$ can be found by a *minimization* problem posed in a class of functions defined by a *nonlinear constraint*. The main idea is to eliminate the lower component of the spinor and solve a minimization problem for the upper one. With the notations of the introduction, $\psi = \begin{pmatrix} \varphi \\ \chi \end{pmatrix}$ is an eigenfunction of $H_0 + V$ if and only if Eq. (2.1.7) takes place. The first equation in Eq. (2.1.7) is an elliptic second order equation for the upper component φ, while the second part of Eq. (2.1.7) gives the lower component χ as a function of φ and the eigenvalue λ. The dependence of H^μ on $\lambda = \mu + mc^2$ makes this problem nonlinear, since λ is still to be found, but the difficulty of finding the unitary transformation Ω in the Foldy–Wouthuysen approach is now replaced by a much simpler problem.

We may reformulate the question as follows. Let $A(\lambda)$ be the operator defined by the quadratic form acting on 2-spinors:

$$\varphi \mapsto \int_{\mathbb{R}^3} \left(\frac{|(\sigma \cdot \nabla)\varphi|^2}{1 - V + \lambda} + (1 - \lambda + V)|\varphi|^2 \right) dx =: (\varphi, A(\lambda)\varphi) \tag{2.5.4}$$

and consider its lowest eigenvalue, $\mu_1(\lambda)$. Because of the monotonicity with respect to λ, there exists at most one λ for which $\mu_1(\lambda) = 0$. This λ is the ground state level.

An algorithm to numerically solve the above problem has been proposed in DOLBEAULT, ESTEBAN, SÉRÉ and VANBREUGEL [2000]. The idea consists in discretizing Eq. (2.5.2) in a finite-dimensional space E_n of dimension n of 2-spinor functions. The discretized version of Eq. (2.5.4) is

$$A^n(\lambda)x_n \cdot x_n = 0, \tag{2.5.5}$$

where $x_n \in E_n$ and $A^n(\lambda)$ is a λ-dependent $n \times n$ matrix. If E_n is generated by a basis set $\{\varphi_i, \ldots \varphi_n\}$, the entries of the matrix $A^n(\lambda)$ are the numbers

$$\int_{\mathbb{R}^3} \left(\frac{((\sigma \cdot \nabla)\varphi_i, (\sigma \cdot \nabla)\varphi_j)}{1 - V + \lambda} + (1 - \lambda + V)(\varphi_i, \varphi_j) \right) dx. \tag{2.5.6}$$

The ground state energy will then be approached from above by the unique λ for which the first eigenvalue of $A^n(\lambda)$ is zero. This method has been tested on a basis of Hermite polynomials (see DOLBEAULT, ESTEBAN, SÉRÉ and VANBREUGEL [2000] for some numerical results). More efficient computations have been made recently on radially symmetric configurations with B-splines basis sets, involving very sparse matrices. Approximations from above of the excited levels can also be computed by requiring successively that the second, third, ... eigenvalues of $A^n(\lambda)$ are equal to zero.

3. The MCDF method for atoms

3.1. The MCDF method

The MCDF equations are obtained from Eq. (1.1.2) by a variational principle. The energy functional is written

$$E_{\text{tot}} = \frac{\langle v\Pi JM | H^{\text{np}} | v\Pi JM \rangle}{\langle v\Pi JM || v\Pi JM \rangle}. \tag{3.1.1}$$

A Hamiltonian matrix which provides the mixing coefficients by diagonalization is obtained from Eq. (3.1.1) with the help of

$$\frac{\partial}{\partial c_v} E_{\text{tot}} = 0, \tag{3.1.2}$$

and a set of integro-differential equations for the radial wave functions $P_\kappa(r)$ and $Q_\kappa(r)$ is obtained from the functional derivatives

$$\begin{cases} \dfrac{\delta}{\delta P_\kappa(r)} E_{\text{tot}} = 0, \\[2mm] \dfrac{\delta}{\delta Q_\kappa(r)} E_{\text{tot}} = 0. \end{cases} \tag{3.1.3}$$

One assumes the orthogonality condition (restricted Dirac–Fock)

$$\int_0^\infty [P_A(r)P_B(r) + Q_A(r)Q_B(r)]dr = \delta_{\kappa_A, \kappa_B} \delta_{n_A, n_B}, \tag{3.1.4}$$

in order to make the angular calculations possible. Eq. (3.1.3) then leads to the inhomogeneous Dirac equation for a given orbital A

$$
\begin{pmatrix}
\dfrac{d}{dr} + \dfrac{\kappa_A}{r} & -\dfrac{2}{\alpha} + \alpha V_A(r) \\[2ex]
-\alpha V_A(r) & \dfrac{d}{dr} - \dfrac{\kappa_A}{r}
\end{pmatrix}
\begin{pmatrix}
P_A(r) \\[1ex]
Q_A(r)
\end{pmatrix}
= \alpha \sum_B \varepsilon_{A,B}
\begin{pmatrix}
Q_B(r) \\[1ex]
-P_B(r)
\end{pmatrix}
+
\begin{pmatrix}
X_{Q_A}(r) \\[1ex]
-X_{P_A}(r)
\end{pmatrix},
$$

$$(3.1.5)$$

where V_A is the sum of the nuclear potential and the direct Coulomb potential, while the exchange terms X_{P_A} and X_{Q_A} include all the two-electron interactions except for the direct Coulomb instantaneous repulsion. The constants $\varepsilon_{A,B}$ are Lagrange parameters used to enforce the orthogonality constraints of Eq. (3.1.4) and thus the summation over B runs only for orbitals with $\kappa_B = \kappa_A$. The exchange terms can be very large if the orbital A has a small effective occupation (the exchange term is a sum of exchange potentials divided by the effective occupation of the orbitals). This effective occupation is the sum

$$
o_A = \sum_{i=1}^{NCF} c_v^2 q_v^{(A)},
\tag{3.1.6}
$$

where $q_v^{(A)}$ is the number of electrons in the orbital A in the vth configuration.

The numerical MCDF methods are based on a fixed-point method, or to be precise on an iteration scheme which provides a self-consistent field (SCF) state in a way very similar to the method which is used to solve the Hartree–Fock model. Initial wave functions must be chosen, for example, hydrogenic wave functions, wave functions in a Thomas–Fermi potential or wave functions already optimized with a smaller set of configurations. One then builds the Hamiltonian matrix Eq. (3.1.2) and obtains the mixing coefficients. Those coefficients and the initial wave functions enter the direct and exchange potential in Eq. (3.1.5), which become normal differential equations, and are solved numerically for each orbital. A new set of potential terms is then evaluated until all the wave functions are stable to a given accuracy ($\approx 10^{-2}$ in the first cycle of diagonalization to $\approx 10^{-6}$ at the last cycle, at the point where the largest variation occurs). A new Hamiltonian matrix is then built and new mixing coefficients are calculated. This process is repeated until convergence is reached. As it is a highly nonlinear process, this can be very tricky, and trial and error on the initial conditions is often required when many configuration and correlation orbitals (i.e., orbitals with very small effective occupations) are involved. All those calculations are done using direct numerical solutions of the MCDF differential equations (3.1.5), which has the advantage of providing very accurate results with relatively limited set of configurations, while MCDF methods using basis set require orders of magnitude more configurations to achieve similar accuracies.

Explicit expressions for V_A, X_{P_A}, and X_{Q_A} can be found in GRANT [1987], GRANT and QUINEY [1988], and DESCLAUX [1993]. All potentials can be expressed in term of the functions

$$
Z_{i,j}^k(x) = \frac{1}{x^k} \int_0^x dr \rho_{ij}(r) r^k,
\tag{3.1.7}
$$

$$Y_{i,j}^k(x) = \frac{1}{x^k} \int_0^x dr \rho_{ij}(r) r^k + x^{k+1} \int_x^\infty dr \frac{\rho_{ij}(r)}{r^{k+1}}, \tag{3.1.8}$$

where $\rho_{ij}(r) = P_i(r) P_j(r) + Q_i(r) Q_j(r)$ for the Coulomb part of the interaction, to which are added terms with $\rho_{ij}(r) = P_i(r) Q_j(r)$ or $\rho_{ij}(r) = Q_i(r) P_j(r)$ when Breit retardation is included in the SCF process. These potential terms can be obtained very efficiently numerically by solving a second-order differential equation (Poisson equation), as a set of two first-order differential equations, with the predictor-corrector method presented in Section 3.2.

3.2. Numerical solution of the inhomogeneous Dirac–Fock radial equations

In order to increase the numerical stability, the direct numerical computation of Eq. (3.1.5) is done by shooting techniques. First one chooses a change of variables to make the method more efficient because bound orbitals exhibit a rapid variation near the origin and exponential decay at large distances. One can choose either

$$t = r_0 \log(r) \quad \text{or} \quad t = r_0 \log(r) + br. \tag{3.2.1}$$

The first choice leads to a pure exponential grid, while the second leads to an exponential grid at short distances and to a linear grid at infinity, and is better suited to represent, for example, Rydberg states. One then takes a linear grid in the new variable t, $t_n = nh$ with h ranging from 0.02 to 0.05. In order to provide the few values needed to start the numerical integration at $r = 0$, and to have accurate integrals (for evaluation of the norm for example) the wave function is represented by its series expansion at the origin, which is of the form

$$\begin{cases} P_\kappa(r) = r^\lambda (p_0 + p_1 r + \ldots), \\ Q_\kappa(r) = r^\lambda (q_0 + q_1 r + \ldots), \end{cases} \tag{3.2.2}$$

where $\lambda = \sqrt{\kappa^2 - (Z\alpha)^2}$ if $V_N(r) = -Z/r$ is a pure Coulomb potential and $\lambda = |\kappa|$ if $V_N(r)$ represents the potential of a finite charge distribution. In this case if $\kappa > 0$, $p_0 = p_2 = \cdots = 0$ and $q_1 = q_3 = \cdots = 0$, and if $\kappa < 0$, $p_1 = p_3 = \cdots = 0$, $q_0 = q_0 = \cdots = 0$.

Predictor–corrector methods

In the case of the atomic problem, the use of fancy techniques like adaptative grids is not recommended, as it is much more efficient to tabulate all wave functions over the same grid, particularly if other properties like transition probabilities are calculated as well. One then uses well proven differential equation solving techniques like predictor–corrector methods and finite difference schemes. The expansion (3.2.2) is substituted into the differential equation (3.1.5) to obtain the coefficients p_i and q_i, for $i > 0$. These coefficients are used to generate values for the wave function at the few first n points of the grid, with an arbitrary value of p_0. Then the value of the function at the next grid point is obtained using the differential equation solver. At infinity the same procedure is used. An exponential approximation of the wave function is made, and the same differential equation solver is used downward to some matching point r_m, usually chosen close to the classical turning point in the potential $V_A(r)$. In the predictor–corrector technique, an approximate value of the function at the mesh point $n + 1$ is predicted from the known values at the preceding n points. This estimate is inserted in the differential equation to obtain the derivative that in turn is used to correct the first estimate, then the final value may be taken as a linear combination of the predicted and corrected values to increase the accuracy. As an example we consider the five points Adams' method that has been widely selected because of its stability properties (PRESS, FLANNERY, TEUKOLSKY and VETTERLING [1986]). The predicted, corrected and final values are given, respectively, by:

$$p_{n+1} = y_n + (1901y'_{n-1} + 2616y'_{n-2} - 1274y'_{n-3} + 251y'_{n-4})/720,$$
$$c_{n+1} = y_n + (251p'_{n+1} + 646y'_n - 264y'_{n-1} + 106y'_{n-2} - 19y'_{n-3})/720, \qquad (3.2.3)$$
$$y_{n+1} = (475c_{n+} + 27p_{n+1})/502,$$

where p' and y' stand for the derivatives with respect to the tabulation variable. The linear combination for the final value is defined as to cancel the term of order h^6, h being the constant interval step of the mesh. In the above equations, y represents either the large or small component of the radial wave function.

Since one starts with a somewhat arbitrary energy and slope at the origin, the components of the wave function obtained by the preceding method are not continuous. A strategy must be devised to obtain the real eigenenergy and slope at the origin from the numerical solution. In the case of a homogeneous equation, one can simply make the large component continuous by multiplying the wave function by the ratio of the inward and outward values of the large component at the matching point and then change the energy until the small component is continuous, using the default in the norm. To first order the correction to the eigenvalue is

$$\delta\varepsilon = \frac{cP(r_m)[Q(r_m^-) - Q(r_m^+)]}{\int_0^\infty [P^2(u) + Q^2(u)]du}, \qquad (3.2.4)$$

where $Q(r_m^\pm)$ are the solutions from each side of the matching point. One then checks that the solution is the desired one by verifying that it has the right number of nodes.

In the inhomogeneous case such a strategy cannot work. In order to obtain a solution which is continuous everywhere, it is possible to proceed in the following way. One uses the well-known fact that the solution of an inhomogeneous differential equation can be written as the sum of a particular solution of the inhomogeneous equation and of the solution of the associated homogeneous equation (in the present case the equation obtained by neglecting the exchange potentials). Thus if P^o and P^i are, respectively, the outward and inward solutions for the large component, one obtains, with the same labels for the small component:

$$[P_I^o + aP_H^o]_{r=r_m^-} = [P_I^i + bP_H^i]_{r=r_m^+},$$
$$[Q_I^o + aQ_H^o]_{r=r_m^-} = [Q_I^i + bQ_H^i]_{r=r_m^+}, \qquad (3.2.5)$$

where the subscripts I and H stand for the inhomogeneous and homogeneous solutions. The coefficients a and b can be obtained from the differential equation. Obviously this continuous solution will not be normalized for an arbitrary value of the diagonal parameter $\varepsilon_{A,A}$ of Eq. (3.1.5). The default in the norm is then used to modify $\varepsilon_{A,A}$ until the proper eigenvalue is found. This method is very accurate but not very efficient since it requires to solve both the inhomogeneous and the homogeneous equations to obtain a continuous solution.

Finite differences methods

As seen above, the predictor-corrector method has some disadvantages. In the nonrelativistic case the Numerov method associated with tail correction (FROESE FISCHER [1977]) provides directly a continuous approximation (the derivative remains discontinuous until the eigenvalue is found). We consider now alternative methods that easily allow to enforce the continuity of one of the two radial components. Let us define the solution at point $n + 1$ as:

$$y_{n+1} = y_n + h(y'_n + y'_{n+1}) + \Delta_n, \tag{3.2.6}$$

where Δ_n is a difference correction given, in terms of central differences, by:

$$\Delta_n = \frac{-1}{12} \delta^3 y_{n+\frac{1}{2}} + \frac{1}{120} \delta^5 y_{n+\frac{1}{2}}, \tag{3.2.7}$$

with:

$$\delta^3 y_{n+\frac{1}{2}} = y_{n+2} - 3y_{n+1} + 3y_n - y_{n-1},$$

$$\delta^5 y_{n+\frac{1}{2}} = y_{n+3} - 5y_{n+2} + 10y_{n+1} - 10y_n + 5y_{n-1} - y_{n-2}. \tag{3.2.8}$$

Accurate solutions are required only when self-consistency is reached. Consequently, the difference correction Δ_n can be obtained at each iteration from the wave functions of the previous iteration as it is done for the potential terms. One can then design computationally efficient schemes (DESCLAUX, MAYERS and O'BRIEN [1971]). We define

$$a_n = 1 + \frac{\kappa h \, r'_n}{2 \, r_n}, \qquad u_n = \Delta_n^P + \frac{h}{2}[r'_n X_n^Q + r'_{n+1} X_{n+1}^Q],$$

$$b_n = -1 + \frac{\kappa h \, r'_n}{2 \, r_n}, \qquad v_n = \Delta_n^P + \frac{h}{2}[r'_n X_n^P + r'_{n+1} X_{n+1}^P], \tag{3.2.9}$$

$$\varphi_n = \alpha \frac{h}{2}[\varepsilon_n - V_n] r'_n, \qquad \theta_n = \frac{h}{\alpha} r'_n + \theta_n,$$

where r' stands for dr/dt (to take into account the fact that the tabulation variable t is a function of r) and $X^{P(Q)} = X_{P_{A(Q_A)}} + \sum_{B \neq A} \varepsilon_{A,B} P_B(Q_B)$. All the functions of r are evaluated using wave functions obtained at the previous iteration. Then the system of algebraic equations:

$$a_{n+1} P_{n+1} - \theta_{n+1} Q_{n+1} + b_n P_n - \theta_n Q_n = u_n,$$

$$\varphi_{n+1} P_{n+1} - b_{n+1} Q_{n+1} + \varphi_n P_n - a_n Q_n = v_n, \tag{3.2.10}$$

determines P_{n+1} and Q_{n+1} if P_n and Q_n are known. For the outward integration, this system is solved step by step from near the origin to the matching point after getting the solution at the first point by series expansion. For the inward integration, an elimination process is used by expressing the solution in the matrix form $[M](PQ) = (uv)$ with the matrix M given by:

$$
M = \begin{bmatrix}
-a_m & \varphi_{m+1} & -b_{m+1} & & & & & & & & \\
-\theta_m & a_{m+1} & -\theta_{m+1} & & & & & & & & \\
& \varphi_{m+1} & -a_{m+1} & \varphi_{m+2} & -b_{m+2} & \cdot & & & & & \\
& b_{m+1} & -\theta_{m+1} & a_{m+2} & -\theta_{m+2} & \cdot & & & & & \\
& \cdot & \cdot & & & \cdot & \cdot & \cdot & \cdot & \cdot & \cdot \\
& & & & & \cdot & b_{N-2} & -\theta_{N-2} & a_{N-1} & \theta_{N-1} & \\
& & & & & & & \varphi_{N-1} & -a_{N-1} & \varphi_N \\
& & & & & & & b_{N-1} & -\theta_{N-1} & a_N
\end{bmatrix}
$$

$$(3.2.11)$$

and the two column vectors (PQ) and (uv) defined as:

$$
(PQ) = \begin{bmatrix}
Q_m \\
P_{m+1} \\
Q_{m+1} \\
P_{m+2} \\
\cdot \\
P_{N-1} \\
Q_{N-1} \\
P_N
\end{bmatrix}, \quad
(uv) = \begin{bmatrix}
v_m - \varphi_m P_m \\
u_m - b_m P_m \\
v_{m+1} \\
u_{m+1} \\
\cdot \\
u_{N-2} \\
v_{N-1} + b_N Q_N \\
u_{N-1} + \theta_N Q_N
\end{bmatrix}.
$$

$$(3.2.12)$$

As displayed in Eq. (3.2.11) each row of the matrix M has at most four nonzero elements. To solve this system of equations, the matrix M is decomposed into the product of two triangular matrices $M = LT$ in which L is a lower matrix with only three nonzero elements on each row and T an upper matrix with the same property. Introducing an intermediate vector (pq) it is possible to solve $L(pq) = (uv)$ for m, $m+1, \ldots, N$ and then $T(PQ) = (pq)$ for $N, N-1, \ldots, m$. The last point of tabulation N is determined by the requirement that P_N should be lower than a specified small value when assuming $Q_N = 0$. Thus the number of tabulation points of each orbital is determined automatically during the self-consistency process. This elimination process produces, as written here, a large component P that is continuous everywhere. The discontinuity of the small component at the matching point r_m can then be used to adjust the eigenvalue $\varepsilon_{A,A}$. In practice this method works very well for occupied orbitals (i.e., orbitals with effective occupations at the Dirac–Fock level $q_o^{(A)} \equiv n, n$ integer larger or equal to 1). Yet it is not sufficiently accurate for correlation orbitals and leads to convergence instability. A good strategy DESCLAUX [1993] is thus to use the accurate predictor-corrector method for the outward integration and the finite differences method with the tail correction for the inward integration. However, the accuracy of the inward integration is increased by computing directly the difference correction Eq. (3.2.7) from the wave function being computed rather than from the one from the previous iteration.

Diagonal Lagrange multipliers

One can use differential techniques, when the gaining of the eigenenergy ε_{AA} is difficult. Their evaluation proceeds as follows. One can obtain the first order variation of
the large component P with respect to a change $\Delta\varepsilon_{AA}$ of one of the off-diagonal
Lagrange multipliers by substituting the development

$$P(\varepsilon_{AA}^0 + \Delta\varepsilon_{AA}) = P(\varepsilon_{AA}^0) + \Delta\varepsilon_{AA} \frac{\partial P}{\partial \varepsilon_{AA}}\bigg|_{\varepsilon_{AA}=\varepsilon_{AA}^0} \tag{3.2.13}$$

(and the equivalent one for the small component Q) into the differential equation
(3.1.5). Defining

$$p_{AA} = \frac{\partial P}{\partial \varepsilon_{AA}}, \quad q_{AA} = \frac{\partial Q}{\partial \varepsilon_{AA}}, \tag{3.2.14}$$

leads to the new set of differential equations

$$\begin{pmatrix} \frac{d}{dr} + \frac{k_A}{r} & -\frac{2}{\alpha} + \alpha V_A(r) \\ -\alpha V_A(r) & \frac{d}{dr} - \frac{k_A}{r} \end{pmatrix} \begin{pmatrix} P_{AA}(r) \\ q_{AA}(r) \end{pmatrix} = \alpha \varepsilon_{A,A} \begin{pmatrix} q_{AA}(r) \\ -p_{AA}(r) \end{pmatrix} + \alpha \begin{pmatrix} Q_B(r) \\ -P_B(r) \end{pmatrix} \tag{3.2.15}$$

which is very similar to Eq. (3.1.5), with the replacement of $X_{P_A}(r)$ (resp. X_{Q_A}) by
$P_B(r)$ (resp. $Q_B(r)$). This system can be solved in $p_{AA}(r)$ and $q_{AA}(r)$ by the above techniques. With this solutions $\Delta\varepsilon_{AA}$ can be calculated in first order from

$$\Delta\varepsilon_{AA} = \frac{1 - \int_0^\infty [P_A(r)P_B(r) + Q_A(r)Q_B(r)]dr}{2\int_0^\infty [p_{AA}(r)P_B(r) + q_{AA}(r)Q_B(r)]dr}. \tag{3.2.16}$$

Note that such relations could be established to provide the change in the nondiagonal
Lagrange multipliers ε_{AB} as well, if one were to solve for several orbitals of identical
symmetry simultaneously.

Off-diagonal Lagrange multipliers

The self-consistent process outlined in Section 1.3 requires the evaluation of the off-
diagonal Lagrange parameters to satisfy the ortho-normality constraint Eq. (3.1.4). As
in the nonrelativistic case, the off-diagonal Lagrange multiplier between closed[1]
shells can be set to zero, which only amounts to perform a unitary transformation in
the subspace of the closed shells. If the generalized occupation numbers o_A and o_B
of two orbitals are different, one can use the symmetry relation

$$\varepsilon_{AB}o_A = \varepsilon_{BA}o_B \tag{3.2.17}$$

and Eq. (3.1.5) to obtain

---

[1]Closed shells are the shell filled with the maximum number of electrons as allowed by the Pauli princi-
ple, that is, $2|\kappa|$.

$$\frac{\varepsilon_{AB}(o_B - o_A)}{o_B} = \int_0^\infty [V_A(r) - V_B(r)][P_A(r)P_B(r) + Q_A(r)Q_B(r)]dr$$

$$-\frac{1}{\alpha}\int_0^\infty [X_{Q_A}(r)Q_B(r) - X_{Q_B}(r)Q_A(r) + X_{P_A}(r)P_B(r) - X_{P_B}(r)P_A(r)]dr.$$

$$(3.2.18)$$

This equation shows that many terms will cancel out in the determination of the Lagrange multipliers (e.g., the closed shell contribution to $V_A(r)$ and $V_B(r)$) and thus provides an accurate method to calculate them provided one retains only the nonzero contributions. If $(o_B - o_A) \ll 1$, however, one must use Eqs. (3.2.15) and (3.2.16) to evaluate the Lagrange multipliers.

3.3. Solution of the inhomogeneous Dirac–Fock equation over a basis set

It has been found however (INDELICATO and DESCLAUX [1993], INDELICATO [1995]) that even the enhanced numerical techniques presented in Section 3.2 would not work for correlation orbitals with very small effective occupation, particularly when the contribution of the Breit interaction is used in Eq. (3.1.5). This leads to point out that in the numerical MCDF calculations, the projection operators which should be used according to Eq. (1.1.5) are absent, as they have no explicit expression. A new method has been proposed that retains the advantages of the numerical MCDF. The idea is to expand P_A, Q_A, X_{P_A} and X_{Q_A} over a finite basis set, for example, the one based on the B-spline calculated following the method of Section 2.3, using the full MCDF direct potential $V_A(r)$. Let us thus assume that one has a complete set of solutions $\{\varphi_1^{(A)}, \ldots, \varphi_{2n}^{(A)}\}$, with eigenvalues $\{\varepsilon_1^{(A)}, \ldots, \varepsilon_{2n}^{(A)}\}$ of the homogeneous equation associated to Eq. (3.1.5). One then writes

$$\begin{pmatrix} P_A(r) \\ Q_A(r) \end{pmatrix} = \sum_{i=1}^{2n} s_i^{(A)} \phi_i^{(A)}(r) \quad \text{and} \quad \begin{pmatrix} X_{P_A}(r) \\ X_{Q_A}(r) \end{pmatrix} = \sum_{i=1}^{2n} x_i^{(A)} \phi_i^{(A)}(r). \quad (3.3.1)$$

Substituting back into Eq. (3.1.5) and using the orthonormality of the basis set functions, one easily obtains

$$s_i^{(A)} = \frac{x_i^{(A)} + \sum_{B \neq A} \varepsilon_{AB} s_i^{(B)}}{\alpha(\varepsilon_1^{(A)} - \varepsilon_{AA})}. \quad (3.3.2)$$

The square of the norm of the solution of Eq. (3.1.5) is then easily obtained as

$$N(\varepsilon_{AA}) = \sum_{i=1}^{2n} (s_i^{(A)})^2 = \sum_{i=1}^{2n} \left(\frac{x_i^{(A)} + \sum_{B \neq A} \varepsilon_{AB} s_i^{(B)}}{\alpha(\varepsilon_1^{(A)} - \varepsilon_{AA})} \right)^2. \quad (3.3.3)$$

One then can calculate the normalized solution of Eq. (3.1.5) if the off-diagonal Lagrange parameters are known, by solving $N(\varepsilon_{AA}) = 1$ for ε_{AA}. One can notice the interesting feature of Eq. (3.3.3) that the norm of the solution of the inhomogeneous equation (3.1.5) has a pole for each eigenenergy of the homogeneous equation.

This method has the advantage over purely numerical techniques that by restricting the sums in Eq. (3.3.1) to positive energy eigenstates, one can explicitly implement projection operators, thus solving readily the "no-pair" Hamiltonian Eq. (1.1.2), rather than an ill-defined equation. More details on this method and on the evaluation of the off-diagonal Lagrange multipliers can be found in INDELICATO [1995].

4. Numerical relativistic methods for molecules

Most of molecular methods that include relativistic corrections are based on the expansion of the molecular orbitals in terms of basis sets (most of the time taken to be Gaussian functions). We shall not review these methods here but refer the interested reader to a book to be published soon (SCHWERDTFEGER [2002]). Let us just point out that the sometimes observed lack of convergence to upper bounds in the total energy (the so-called variational collapse) is not unambiguously related to the Dirac negative energy continuum. Indeed this attractive explanation is unfortunately unable to explain the appearance of spurious solutions. Both the existence of spurious solutions and the lack of convergence to expected levels can be traced back to originate from poor basis sets and bad finite matrix representations of the operators (in particular the kinetic energy). For an extensive discussion see DYALL, GRANT and WILSON [1984]. Numerical methods successfully used are briefly sketched in the next two paragraphs.

4.1. Fully numerical two-dimensional method

For diatomic molecules, the one-electron Dirac wave functions may be written as

$$
\Phi = \begin{pmatrix} e^{i(m-1/2)\varphi}\phi_1^L(\xi,\eta) \\ e^{i(m+1/2)\varphi}\phi_2^L(\xi,\eta) \\ ie^{i(m-1/2)\varphi}\phi_3^S(\xi,\eta) \\ ie^{i(m+1/2)\varphi}\phi_4^S(\xi,\eta) \end{pmatrix}
\tag{4.1.1}
$$

where $L(S)$ stands for the large (small) component and elliptical coordinates (ξ, η, φ) are used with:

$$
\xi = (r_1 + r_2)/R, \quad \eta = (r_1 - r_2)/R,
\tag{4.1.2}
$$

where r_1 and r_2 are the distances between the electron and each of the nucleus, R is the inter-nuclear distance. The third variable φ is the azimuthal angle around the axis through the nuclei.

As usual for molecular calculations, the variational collapse is avoided by defining the small component in terms of the large one (KUTZELNIGG [1984]). Starting from the Dirac equation in a local potential V one possibility is to use:

$$
\phi^S = c\sigma.p\phi^L/[2c^2 + E - V].
\tag{4.1.3}
$$

After this substitution, the large component is given as solution of a second order differential equation that can be solved using well-known relaxation methods (VARGA [1963]).

For efficiency, the distribution of integration points must be chosen as to accumulate points where the functions are rapidly varying. It was found that the transformation,

$$\mu = \text{arccosh}(\xi), \quad \nu = \text{arccosh}(\eta), \tag{4.1.4}$$

which yields a quadratic distribution of points near the nuclei, is some kind of optimum to reduce the number of points needed to achieve a given accuracy. Then the derivatives of the Laplace operator are approximated by n-point finite differences. In so doing, the differential equations are replaced by a set of linear equations that can be written in a matrix form as

$$(A - ES)X = B, \tag{4.1.5}$$

where the matrix A, that represents the direct part of the Fock operator, is diagonal dominant but has nondiagonal elements arising from the discretization of the Laplace operator. Here E is the energy eigenvalue, S is the overlap diagonal matrix and B a vector due to the exchange part of the Fock operator whose values change during iterations. Then the relaxation method can be viewed as an iterative method to find the x_i component of X such that

$$(A - ES)x_i = b_i, \tag{4.1.6}$$

each iteration n being associated with a linear combination of the initial and final estimate of x_i at iteration $n - 1$, that is,

$$x_i^{\text{initial}_{n+1}} = (1 - \omega)x_i^{\text{initial}_n} + \omega x_i^{\text{final}_n}. \tag{4.1.7}$$

It was found that with overrelaxation (i.e., $\omega > 1$), the method may be slow in convergence but it is quite stable. Applications of the method outlined above may be found in SUNDHOLM, PYYKKÖ and LAAKSONEN [1987] and in references therein.

4.2. Numerical integrations with linear combinations of atomic orbitals

A widely used approximation in molecular calculations is to expand the molecular orbitals as a linear combination of atomic orbitals. If these atomic orbitals are chosen as the numerical solutions of some kind of Dirac–Fock atomic calculations, then small basis sets are sufficient to achieve good accuracy. The main disadvantage of this choice is that all multi-dimensional integrals have to be calculated numerically. This is compensated by two advantages: first the kinetic energy contribution can be computed by a single integral using the atomic Dirac equations (thus avoiding numerical differentiation), second, by including only positive energy atomic wave functions, no "variational collapse" will occur.

In this method, the molecular wave functions ψ are expanded in terms of symmetry molecular orbitals χ as:

$$\psi_\lambda = \sum_\nu c_\nu^\lambda \chi^\nu, \tag{4.2.1}$$

while the symmetry molecular orbitals χ are taken to be linear combinations of atomic orbitals φ:

$$\chi^v = \sum_i d_i^v \varphi^i. \tag{4.2.2}$$

The coefficients d_i^v are given by the symmetry of the molecular orbital and are obtained from the irreducible representations of the double point groups. Computing all necessary integrals (overlaps, matrix elements of the Dirac operator, the Coulomb interaction, etc., ...) the Dirac–Fock equations are reduced to a generalized matrix eigenvalue problem that determines both the eigenvalues and the c_v^λ coefficients of Eq. (4.2.1).

To compute the various matrix elements in the case of diatomic molecules, SEPP, KOLB, SENGLER, HARTUNG and FRICKE [1986] used Gauss–Laguerre and Gauss–Legendre integration schemes on a grid of points defined by the same variables as those of Eq. (4.1.4). Unfortunately this approach is not easy to extend beyond diatomic molecules and other methods have to be implemented. It has been shown, see for example ROSEN and ELLIS [1975], that the adaptation to molecules of the so-called Discrete Variational Method (DVM) developed for solid state calculations (ELLIS and PAINTER [1970]) may be both efficient and accurate. The DVM may be viewed as performing a multidimensional integral via a weighted sum of sampling points, that is, to compute a matrix element $\langle f \rangle$ by:

$$\langle f \rangle = \sum_{n=1}^{N} \omega(r_i) f(r_i), \tag{4.2.3}$$

where the weight function $\omega(r_i)$ can be considered as an integration weight corresponding to a local volume per point. This function is also constrained to force the error momenta to vanish on the grid points following the work of HASELGROVE [1961]. Furthermore the set of the sampling points $[r_i]$ must be chosen to preserve the symmetries of the system under configuration (this is accomplished by taking a set of sampling points that includes all points Rr_i, R standing for operations of the symmetry group). A full description of the DVM can be found in the references given above.

References

BAYLISS, W.E. and S.J. PEEL (1983), Stable variational calculations with the Dirac Hamiltonian, *Phys. Rev. A* **28** (4), 2552–2554.

BEIER, T. (2000), The g_j factor of a bound electron and the hyperfine structure splitting in hydrogenlike ions, *Phys. Rep.* **339** (2–3), 79–213.

BROWN, G.E. and D.E. RAVENHALL (1951), On the interaction of two electrons, *Proc. R. Soc. London, Ser. A* **208** (1951), 552–559.

CHANG, C.H., J.P. PÉLISSIER, and P.H. DURAND (1986), Regular two-component Pauli-like effective Hamiltonians in Dirac theory, *Phys. Scripta* **34**, 394–404.

DATTA, S.N. and G. DEVIAH (1988), The minimax technique in relativistic Hartree–Fock calculations, *Pramana* **30** (5), 393–416.

DESCLAUX, J.P. (1993), A relativistic multiconfiguration Dirac–Fock package, In: Clementi, E. (ed.), *Methods and Techniques in Computational Chemistry: METECC-94, Vol. A: Small Systems* (STEF, Cogliari).

DESCLAUX, J.P., D.F. MAYERS and F. O'BRIEN (1971), Relativistic atomic wavefunction, *J. Phys. B At. Mol. Opt. Phys.* **4**, 631–642.

DOLBEAULT, J., M.J. ESTEBAN and E. SÉRÉ (2000a), Variational characterization for eigenvalues of Dirac operators, *Calc. Var. Partial. Differ. Equ.* **10**, 321–347.

DOLBEAULT, J., M.J. ESTEBAN and E. SÉRÉ (2000b), On the eigenvalues of operators with gaps. Application to Dirac operators, *J. Funct. Anal.* **174**, 208–226.

DOLBEAULT, J., M.J. ESTEBAN, E. SÉRÉ and M. VANBREUGEL (2000), Minimization methods for the one-particle Dirac equation, *Phys. Rev. Lett.* **85** (19), 4020–4023.

DRAKE, G.W.F. and S.P. GOLDMAN (1981), Application of discrete-basis-set methods to the Dirac equation, *Phys. Rev. A* **23**, 2093–2098.

DRAKE, G.W.F. and S.P. GOLDMAN (1988), Relativistic Sturmian and finite basis set methods in atomic physics, *Adv. At. Mol. Phys.* **23**, 23–29.

DURAND, P.H. (1986), Transformation du Hamiltonien de Dirac en Hamiltoniens variationnels de type Pauli, Application à des atomes hydrogenoï des. *C. R. Acad. Sci. Paris, sér. II* **303** (2), 119–124.

DURAND, P.H. and J.P. MALRIEU (1987), Effective Hamiltonians and pseudo-potentials as tools for rigorous modelling, In: Lawley, K.P. (ed.), *Ab initio Methods in Quantum Chemistry I* (Wiley).

DYALL, K.G., I.P. GRANT and S. WILSON (1984), Matrix representation of operator products, *J. Phys. B At. Mol. Phys.* **17**, 493–503.

EIDES, M.I., H. GROTCH and V.A. SHELYUTO (2001), Theory of light hydrogenlike atoms, *Phy. Rep.* **342** (2–3), 63–261.

ELLIS, D.E. and G.S. PAINTER (1970), Discrete variational method for the energy-band problem with general crystal potentials, *Phys. Rev. B* **2** (8), 2887–2898.

ESTEBAN, M. J. and E. SÉRÉ (1997), Existence and multiplicity of solutions for linear and nonlinear Dirac problems. In: Greiner, P.C., Ivrii, V., Seco, L.A., Sulem, C. (eds.), *Partial Differential Equations and Their Applications*, CRM Proceedings and Lecture Notes, vol. 12 (AMS).

FOLDY, L.L. and S.A. WOUTHUYSEN (1950), On the Dirac theory of spin-1/2 particles and its nonrelativistic limit, *Phys. Rev.* **78**, 29–36.

FROESE FISCHER, C. (1977), *The Hartree–Fock Method for Atoms* (Wiley).

GRANT, I.P. (1982), Conditions for convergence of variational solutions of Dirac's equation in a finite basis, *Phys. Rev. A* **25** (2), 1230–1232.

GRANT, I.P. (1987), Relativistic atomic structure calculations, *Meth. Comp. Chem.* **2**, 132.

GRANT, I. P. (1989), Notes on basis sets for relativistic atomic structure and QED. In: Mohr, P.J., Johnson, W.R., Sucher, J. (eds.), *A.I.P. Conf. Proc.*, vol. 189, 235–253.

GRANT, I.P. and H.M. QUINEY (1988), Foundation of the relativistic theory of atomic and molecular structure, *Adv. At. Mol. Phys.* **23**, 37–86.

GRIESEMER, M., R.T. LEWIS and H. SIEDENTOP (1999), A minimax principle in spectral gaps: Dirac operators with Coulomb potentials, *Doc. Math.* **4**, 275–283 (electronic).

GRIESEMER, M. and H. SIEDENTOP (1999), A minimax principle for the eigenvalues in spectral gaps, *J. London Math. Soc.* **60** (2), 490–500.

HASELGROVE, C.B. (1961), A method for numerical integration, *Math. Comput.* **15**, 323–337.

HEULLY, J.L., I. LINDGREN, E. LINDROTH, S. LUNDQVIST and A.M. MÅRTENSSON-PENDRILL (1986), Diagonalisation of the Dirac Hamiltonian as a basis for a relativistic many-body procedure, *J. Phys. B At. Mol. Phys.* **19**, 2799–2815.

HILL, R.N. and C. KRAUTHAUSER (1994), A solution to the problem of variational collapse for the one-particle Dirac equation, *Phys. Rev. Lett.* **72** (14), 2151–2154.

INDELICATO, P. (1995), Projection operators in multiconfiguration Dirac–Fock calculations, Application to the ground state of Heliumlike ions, *Phys. Rev. A* **51** (2), 1132–1145.

INDELICATO, P. and J.P. DESCLAUX (1993), Projection operators in the multiconfiguration Dirac–Fock method, *Phys. Scr.* **T 46**, 110–114.

JOHNSON, W.R., S. BLUNDELL and J. SAPIRSTEIN (1988), Finite basis sets for Dirac equation constructed from B splines, *Phys. Rev. A* **37** (2), 307–315.

JOHNSON, W.R. and C.D. LIN (1976), Relativistic random phase approximation applied to atoms of the He isoelectronic sequence, *Phys. Rev. A* **14** (2), 565–575.

KOGUT, J.B. (1983), Lattice gauge theory approach to quantum chromodynamics, *Rev. Mod. Phys.* **55** (3), 775–836.

KUTZELNIGG, W. (1984), Basis set expansion of the Dirac operator without variational collapse, *Int. J. Quant. Chem.* **25**, 107–129.

KUTZELNIGG, W. (1990), Perturbation theory of relativistic corrections 2. Analysis and classification of known and other possible methods, *Z. Phys. D* **15**, 27–50.

KUTZELNIGG, W. (1999), Effective Hamiltonians for degenerate and quasidegenerate direct perturbation theory of relativistic effects, *J. Chem. Phys.* **110** (17), 8283–8294.

LEE, S.H. (2001), A new basis set for the radial Dirac equation. (Preprint).

VAN LENTHE, E., R. VAN LEEUWEN, E.J. BAERENDS and J.G. SNIJDERS (1994a), Exact solutions of regular approximate relativistic wave equations for hydrogen-like atoms, *J. Chem. Phys.* **101** (2), 1272–1281.

VAN LENTHE, E., xR. VAN LEEUWEN, E.J. BAERENDS and J.G. SNIJDERS (1994b), Relativistic regular two-component Hamiltonians, In: Broek, R. et al. (eds.), *New Challenges in Computational Quantum Chemistry* (Publications Dept. Chem. Phys. and Material sciences, University of Groningen).

LINDGREN, I. and J. MORRISON (1982), *Atomic Many-Body Theory* (Springer).

MOHR, P.J., G. PLUNIEN and G. SOFF (1998), QED corrections in heavy atoms, *Phys. Rep.* **293** (5 & 6), 227–372.

PRESS, W.H., B.P. FLANNERY, S.A. TEUKOLSKY and W.T. VETTERLING (1986), *Numerical Recipes* (Cambridge University Press).

ROSEN, A. and D.E. ELLIS (1975), Relativistic molecular calculations in the Dirac–Slater model, *J. Chem. Phys.* **62**, 3039–3049.

RUTKOWSKI, A. (1999), Iterative solution of the one-electron Dirac equation based on the Bloch equation of the 'direct perturbation theory, *Chem. Phys. Lett.* **307**, 259–264.

RUTKOWSKI, A. and W.H.E. SCHWARZ (1996), Effective Hamiltonian for near-degenerate states in direct relativistic perturbation theory. I. Formalism, *J. Chem. Phys.* **104** (21), 8546–8552.

SALOMONSON, S. and P. ÖSTER (1989a), Relativistic all-order pair functions from a discretized single-particle Dirac Hamiltonian, *Phys. Rev. A* **40** (10), 5548–5558.

SALOMONSON, S. and P. ÖSTER (1989b), Solution of the pair equation using a finite discrete spectrum, *Phys. Rev. A* **40** (10), 5559–5567.

SCHWERDTFEGER, P. (ed.) (2002). *Relativistic Electronic Structure Theory. Part 1: Fundamental Aspects* (Elsevier).

SEPP, W.D., D. KOLB, W. SENGLER, H. HARTUNG and B. FRICKE (1986), Relativistic Dirac–Fock–Slater program to calculate potential-energy curves for diatomic molecules, *Phys. Rev. A* **33**, 3679–3687.

STACEY, R. (1982), Eliminating lattice fermion doubling, *Phys. Rev. D* **26** (2), 468–472.

SUCHER, J. (1980), Foundation of the relativistic theory of many-electron atoms, *Phys. Rev. A* **22** (2), 348–362.

SUNDHOLM, D., P. PYYKKÖ and L. LAAKSONEN (1987), Two-dimensional fully numerical solutions of second-order Dirac equations for diatomic molecules. Part 3, *Phys. Scr.* **36**, 400–402.

TALMAN, J.D. (1986), Minimax principle for the Dirac equation, *Phys. Rev. Lett.* **57** (9), 1091–1094.

VARGA, R.S. (1963), *Matrix Iterative Analysis* (Prentice-Hall, Englewoods Cliffs, NJ).

WALLMEIER, H. and W. KUTZELNIGG (1981), Use of the squared Dirac operator in variational relativistic calculations. *Chem. Phys. Lett.* **78** (2), 341–346.

WILSON, K.G. (1974), Confinement of quarks. *Phys. Rev. D* **10** (8), 2445–2459.

WOOD, J., I.P. GRANT and S. WILSON (1985), The Dirac equation in the algebraic approximation: IV. Application of the partitioning technique, *J. Phys. B At. Mol. Phys.* **18**, 3027–3041.

Quantum Monte Carlo Methods for the Solution of the Schrödinger Equation for Molecular Systems

Alán Aspuru-Guzik, William A. Lester, Jr.

Department of Chemistry, University of California, Berkeley, California, USA

Preface

The solution of the time independent Schrödinger equation for molecular systems requires the use of modern computers, because analytic solutions are not available.

This review deals with some of the methods known under the umbrella term quantum Monte Carlo (QMC), specifically those that have been most commonly used for electronic structure. Other applications of QMC are widespread to rotational and vibrational states of molecules, such as the work of BENOIT and CLARY [2000], KWON, HUANG, PATEL, BLUME and WHALEY [2000], CLARY [2001], VIEL and WHALEY [2001], condensed matter physics (CEPERLEY and ALDER [1980], FOULKES, MITAS, NEEDS and RAJAGOPAL [2001]), and nuclear physics (WIRINGA, PIEPER, CARLSON and PANDHARIPANDE [2000], PIEPER and WIRINGA [2001]).

QMC methods have several advantages:

- Computer time scales with system size roughly as N^3, where N is the number of particles of the system. Recent developments have made possible the approach to linear scaling in certain cases.
- Computer memory requirements are small and grow modestly with system size.
- QMC computer codes are significantly smaller and more easily adapted to parallel computers than basis set molecular quantum mechanics codes.
- Basis set truncation errors are absent in the QMC formalism.
- Monte Carlo numerical efficiency can be arbitrarily increased. QMC calculations have an accuracy dependence of $\sqrt{T}$, where T is the computer time. This enables

Computational Chemistry
Special Volume (C. Le Bris, Guest Editor) of
HANDBOOK OF NUMERICAL ANALYSIS
P.G. Ciarlet (Editor)

one to choose an accuracy range and readily estimate the computer time needed for performing a calculation of an observable with an acceptable error bar.

The purpose of the present work is to present a description of the commonly used algorithms of QMC for electronic structure and to report some recent developments in the field.

The chapter is organized as follows. In Part I, we provide a short introduction to the topic, as well as enumerate some properties of wave functions that are useful for QMC applications. In Part II we describe commonly used QMC algorithms. In Part III, we briefly introduce some special topics that remain fertile research areas.

Other sources that complement and enrich the topics presented in this chapter are our previous monograph, HAMMOND, LESTER and REYNOLDS [1994] and the reviews of SCHMIDT [1986], LESTER, JR. and HAMMOND [1990], CEPERLEY and MITAS [1996], ACIOLI [1997], BRESSANINI and REYNOLDS [1998], MITAS [1998], ANDERSON [1999], LÜCHOW and ANDERSON [2000], FOULKES, MITAS, NEEDS and RAJAGOPAL [2001]. There are also chapters on QMC contained in selected computational physics texts (KOONIN and MEREDITH [1995], GOULD and TOBOCHNIK [1996], THIJSSEN [1999]). Selected applications of the method are contained in GREEFF, HAMMOND and LESTER, JR. [1996], SOKOLOVA [2000], GREEFF and LESTER, JR. [1997], FLAD, SAVIN and SCHULTHEISS [1994], GROSSMAN and MITAS [1995], FLAD and DOLG [1997]. FILIPPI and UMRIGAR [1996], BARNETT, SUN and LESTER, JR. [2001], FLAD, CAFFAREL and SAVIN [1997], GROSSMAN, LESTER, JR. and LOUIE [2000], OVCHARENKO, LESTER, JR., XIAO and HAGELBERG [2001].

QMC methods that are not covered in this review are the auxiliary field QMC method (CHARUTZ and NEUHAUSER [1995], ROM, CHARUTZ and NEUHAUSER [1997], BAER, HEAD-GORDON and NEUHAUSER [1998], BAER and NEUHAUSER [2000]) and path integral methods (CEPERLEY [1995], SARSA, SCHMIDT and MAGRO [2000]).

Atomic units are used throughout, the charge of the electron e and Planck's normalized constant are set to unity. In this metric system, the unit distance is the Bohr radius a_0.

Part I. Introduction

The goal of the quantum Monte Carlo (QMC) method is to solve the Schrödinger equation, which in the time independent representation is given by

$$\hat{H}\Psi_n(\mathbf{R}) = E_n\Psi_n(\mathbf{R}). \tag{0.1}$$

Here, $\hat{H}$ is the Hamiltonian operator of the system in state n, with wave function $\Psi_n(\mathbf{R})$ and energy E_n; $\mathbf{R}$ is a vector that denotes the $3N$ coordinates of the system of N particles (electrons and nuclei), $\mathbf{R} \equiv \{\mathbf{r}_1,\ldots,\mathbf{r}_n\}$. For molecular systems, in the absence of electric or magnetic fields, the Hamiltonian has the form $\hat{H} \equiv \hat{T} + \hat{V}$, where $\hat{T}$ is the kinetic energy operator, $\hat{T} \equiv -\frac{1}{2}\nabla_\mathbf{R}^2 \equiv -\frac{1}{2}\sum_i\nabla_i^2$, and $\hat{V}$ is the potential energy operator. For atomic and molecular systems $\hat{V}$ is the Coulomb potential between particles of charge q_i, $\hat{V} \equiv \sum_{ij}\frac{q_{ij}}{\mathbf{r}_{ij}}$.

The first suggestion of a Monte Carlo solution of the Schrödinger equation dates back to Enrico Fermi, based on METROPOLIS and ULAM [1949]. He indicated that a solution to the stationary state equation

$$-\frac{1}{2}\nabla_{\mathbf{R}}^2\Psi(\mathbf{R}) = E\Psi(\mathbf{R}) - V(\mathbf{R})\Psi(\mathbf{R}) \tag{0.2}$$

could be obtained by introducing a wave function of the form $\Psi(\mathbf{R}, \tau) = \Psi(\mathbf{R})\, e^{-E\tau}$. This yields the equation

$$\frac{\partial\Psi(\mathbf{R}, \tau)}{\partial\tau} = \frac{1}{2}\nabla^2\Psi(\mathbf{R}, \tau) - V(\mathbf{R})\Psi(\mathbf{R}, \tau). \tag{0.3}$$

Taking the limit $\tau \to \infty$, Eq. (0.2) is recovered. If the second term on the right hand side of Eq. (0.3) is ignored, the equation is isomorphic with a diffusion equation, which can be simulated by a random walk (EINSTEIN [1926], COURANT, FRIEDRICHS and LEWY [1928]), where random walkers diffuse in a $\mathbf{R}$-dimensional space. If the first term is ignored, the equation is a first-order kinetics equation with a position dependent rate constant, $V(\mathbf{R})$, which can also be interpreted as a stochastic survival probability. A numerical simulation in which random walkers diffuse through $\mathbf{R}$-space, reproduce in regions of low potential, and die in regions of high potential leads to a stationary distribution proportional to $\Psi(\mathbf{R})$, from which expectation values can be obtained.

1. Numerical solution of the Schrödinger equation

Most efforts to solve the Schrödinger equation are wave function methods. These approaches rely exclusively on linear combinations of Slater determinants, and include configuration interaction (CI) and the multiconfiguration self-consistent field (MCSCF). There are perturbation approaches including the Möller–Plesset series (MP2, MP4), and coupled cluster (CC) theory, which are presently popular computational procedures. Wave function methods suffer from scaling deficiencies. An exact calculation with a given basis set expansion requires $N!$ computer operations, where N is the number of basis functions. Competitive methods such as CC with singles and doubles, and triples perturbation treatment, CCSD(T), scale as N^7.[1]

A term that we will use later is correlation energy (CE). It is defined as the difference between the exact nonrelativistic energy, and the energy of a mean field solution of the Schrödinger equation, the Hartree–Fock method, in the limit of an infinite basis set (LÖWDIN [1959], SENATORE and MARCH [1994])

$$E_{\text{corr}} = E_{\text{exact}} - E_{\text{HF}}. \tag{1.1}$$

---

[1]For a more detailed analysis of the scaling of wave function based methods see, for example, HEAD-GORDON [1996], and RAGHAVACHARI and ANDERSON [1996]. For a general overview of these methods, the reader is referred to Chapter I of this book (CANCÈS, DEFRANCESCHI, KUTZELNIGG, LE BRIS and MADAY [2003]).

The CI, MCSCF, MP(N), and CC methods are all directed at generating energies that approach E_{exact}.

Other methods that have been developed include dimensional expansions (WATSON, DUNN, GERMANN, HERSCHBACH and GOODSON [1996]), and the contracted Schrödinger equation method (MAZZIOTTI [1998]). For an overview of quantum chemistry methods, see SCHLEYER, ALLINGER, CLARK, GASTEIGER, KOLLMAN, SCHAEFER III and SCHREINER [1998].

Since the pioneering work of the late forties to early sixties (METROPOLIS and ULAM [1949], DONSKER and KAC [1950], KALOS [1962]), the MC and related methods have grown in interest. QMC methods have an advantage with system size scaling, in the simplicity of algorithms and in trial wave function forms that can be used.

2. Properties of the exact wave function

The exact time independent wave function solves the eigenvalue Eq. (0.1). Some analytic properties of this function are very helpful in the construction of trial functions for QMC methods.

For the present discussion, we are interested in the discrete spectrum of the $\hat{H}$ operator. In most applications the total Schrödinger Eq. (0.1) can be represented into an "electronic" Schrödinger equation and a "nuclear" Schrödinger equation based on the large mass difference between electrons and nuclei. This is the Born–Oppenheimer (BO) approximation. Such a representation need not be introduced in QMC but here is the practical benefit of it that the nuclei can be held fixed for electronic motion results in the simplest form of the electronic Schrödinger equation.

The wave function also must satisfy the virial, hypervirial Hellman–Feynman and generalized Hellman–Feynman theorems (HELLMAN [1937], FEYNMANN [1939], WEISSBLUTH [1978]).[2] The local energy (FROST, KELLOG and CURTIS [1960])

$$E_L(\mathbf{R}) \equiv \frac{\hat{H}\Psi(\mathbf{R})}{\Psi(\mathbf{R})} \tag{2.1}$$

is a constant for the exact wave function.

When charged particles meet, there is a singularity in the Coulomb potential. This singularity must be compensated by a singularity in the kinetic energy, which results in a discontinuity in the first derivative, that is, a *cusp*, in the wave function when two or more particles meet (KATO [1957], MEYERS, UMRIGAR, SETHNA and MORGAN [1991]). For one electron coalescing at a nucleus, if we focus in a one electron function or orbital $\phi(\mathbf{r}) = \chi(r)Y_l^m(\theta, \phi)$, where $\chi(r)$ is a radial function, and $Y_l^m(\theta, \phi)$ is a spherical harmonic with angular and magnetic quantum numbers l and m, the electron–nucleus cusp condition is

$$\frac{1}{\eta(r)} \frac{d\eta(r)}{dr}\bigg|_{r=0} = -\frac{Z}{l+1}, \tag{2.2}$$

---

[2] The Hellman–Feynman theorem is discussed in Section 8.2.3.

where $\eta(r)$ is the radial wave function with the leading r dependence factored out, $\eta(r) \equiv \chi(r)/r^m$, and Z is the atomic number of the nucleus.
For electron–electron interactions, the cusp condition takes the form

$$\frac{1}{\eta_{ij}(r)} \frac{d\eta_{ij}(r)}{dr}\bigg|_{r_{ij}=0} = \frac{1}{2(l+1)}, \tag{2.3}$$

where $\eta_{ij}(r)$ is the r^m factored function for the electron–electron radial distribution function.

Furthermore, $\bar{\rho}(\mathbf{r})$, the spherical average of the electron density, $\rho(\mathbf{r})$,[3] must satisfy another cusp condition, namely,

$$\frac{\partial}{\partial r} \bar{\rho}(r)\bigg|_{r=0} = -2Z\bar{\rho}(r) \tag{2.4}$$

at any nucleus. Another condition on $\rho(\mathbf{r})$ is that asymptotically it decays exponentially:

$$\rho(r \to \infty) \approx e^{-2\sqrt{2I_0}\,r}, \tag{2.5}$$

where I_0 is the first ionization potential. This relation can be derived from consideration of a single electron at large distance. Details on these requirements can be found in MORELL, PARR and LEVY [1975] and DAVIDSON [1976].

We discuss how to impose properties of the exact wave function on QMC trial functions in Section 3.2.

2.1. Approximate wave functions

JAMES and COOLIDGE [1937] proposed three accuracy tests of a trial wave function, Ψ_T: the root mean square error in Ψ_T

$$\delta_\Psi = \left[\int (\Psi_T - \Psi_0)^2 d\mathbf{R}\right]^{\frac{1}{2}}, \tag{2.6}$$

the energy error

$$\delta_E = E(\Psi_T) - E_0 \tag{2.7}$$

and the root mean square local energy deviation

$$\delta_{E_L} = \left[\int |(\hat{H} - E_0)\Psi|^2 d\mathbf{R}\right]^{\frac{1}{2}} \tag{2.8}$$

where the local energy is defined as in Eq. (2.1).

The calculation of δ_Ψ by QMC requires sampling the exact wave function, a procedure that will be described in Section 8.1.

---

[3]If N is the number of electrons, then $\rho(r)$ is defined by $\rho(r) = N \int |\Psi(\mathbf{R})|^2 d\mathbf{R}$.

Several stochastic optimization schemes have been proposed for minimizing expressions (2.6)–(2.8). Most researchers have focused on (2.8), that is, minimizing δ_{E_L}; see, for example, McDOWELL [1981]. In Section 4 we turn to stochastic wave function optimization procedures.

Part II. Algorithms

In this part, we describe the computational procedures of QMC methods. All of these methods use MC techniques widely used in other fields, such as operations research, applied statistics, and classical statistical mechanics simulations. Techniques such as importance sampling, correlated sampling and MC optimization are similar in spirit to those described in MC treatises by BAUER [1958], HAMMERSLEY and HANDSCOMB [1964], HALTON [1970], McDOWELL [1981], KALOS and WHITLOCK [1986], WOOD and ERPENBECK [1976], FISHMAN [1996], SOBOL [1994], MANNO [1999], DOUCET, DE FREITAS, GORDON and SMITH [2001], LIU [2001]. The reader is referred to the former for more details on the techniques described in this part.

We present the simple, yet powerful variational Monte Carlo (VMC) method, in which the Metropolis MC[4] method is used to sample a known trial function Ψ_T. We follow with the projector Monte Carlo (PMC) methods that sample the unknown ground state wave function.

3. Variational Monte Carlo

3.1. Formalism

Variational methods involve the calculation of the expectation value of the Hamiltonian operator using a trial wave function Ψ_T. This function is dependent on a set of parameters, Λ, that are varied to minimize the expectation value, that is,

$$\langle \hat{H} \rangle = \frac{\langle \Psi_T | \hat{H} | \Psi_T \rangle}{\langle \Psi_T | \Psi_T \rangle} \equiv E[\Lambda] \geqslant E_0. \tag{3.1}$$

The expectation value (3.1) can be sampled from a probability distribution proportional to Ψ_T^2, and evaluated from the expression

$$\frac{\int d\mathbf{R} \left[\frac{\hat{H} \Psi_T(\mathbf{R})}{\Psi_T(\mathbf{R})} \right] \Psi_T^2(\mathbf{R})}{\int d\mathbf{R} \Psi_T^2(\mathbf{R})} = \frac{\int d\mathbf{R} E_L \Psi_T^2(\mathbf{R})}{\int d\mathbf{R} \Psi_T^2(\mathbf{R})} \geqslant E_0, \tag{3.2}$$

where E_L is the local energy defined in Section 2.0.1. The procedure involves sampling random points in $\mathbf{R}$-space from

---

[4]This algorithm is also known as the M(RT)^2, due to the full list of the authors that contributed to its development, Metropolis, Rosenbluth, Rosenbluth, Teller and Teller, see METROPOLIS, ROSENBLUTH, ROSENBLUTH, TELLER and TELLER [1953].

$$P(\mathbf{R}) \equiv \frac{\Psi_T^2(\mathbf{R})}{\int d\mathbf{R}\Psi_T^2(\mathbf{R})}. \tag{3.3}$$

The advantage of using (3.3) as the probability density function is that one need not perform the averaging of the numerator and denominator of Eq. (3.2). The calculation of the ratio of two integrals with the MC method is biased by definition: the average of a quotient is not equal to the quotient of the averages, so this choice of $P(\mathbf{R})$ avoids this problem.

In general, sampling is done using the Metropolis method (METROPOLIS, ROSENBLUTH, ROSENBLUTH, TELLER and TELLER [1953]), that is well described in Chapter 3 of KALOS and WHITLOCK [1986], and briefly summarized here in Section 3.1.1.

Expectation values can be obtained using the VMC method from the following general expressions (BRESSANINI and REYNOLDS [1998]):

$$\langle\hat{O}\rangle \equiv \frac{\int d\mathbf{R}\Psi_T(\mathbf{R})^2\hat{O}(\mathbf{R})}{\int d\mathbf{R}\Psi_T(\mathbf{R})^2} \cong \frac{1}{N}\sum_{i=1}^{N}\hat{O}(\mathbf{R}_i), \tag{3.4}$$

$$\langle\hat{O}_d\rangle \equiv \frac{\int d\mathbf{R}\left[\frac{\hat{O}_d\Psi_T(\mathbf{R})}{\Psi_T(\mathbf{R})}\right]\Psi_T(\mathbf{R})^2}{\int d\mathbf{R}\Psi_T(\mathbf{R})^2} \cong \frac{1}{N}\sum_{i=1}^{N}\frac{\hat{O}_d\Psi_T(\mathbf{R}_i)}{\Psi_T(\mathbf{R}_i)}. \tag{3.5}$$

Equation (3.4) is for a coordinate operator, $\hat{O}$, and (3.5) is preferred for a differential operator, $\hat{O}_d$.

3.1.1. The generalized Metropolis algorithm

The main idea of the Metropolis algorithm is to sample the electronic density, given hereby, $\Psi_T^2(\mathbf{R})$ using fictitious kinetics that in the limit of large simulation time yields the density at equilibrium. A coordinate move is proposed, $\mathbf{R} \rightarrow \mathbf{R}'$, which has the probability of being accepted given by

$$P(\mathbf{R} \rightarrow \mathbf{R}') = \min\left(1, \frac{T(\mathbf{R}' \rightarrow \mathbf{R})\Psi_T^2(\mathbf{R}')}{T(\mathbf{R} \rightarrow \mathbf{R}')\Psi_T^2(\mathbf{R})}\right), \tag{3.6}$$

where $T(\mathbf{R} \rightarrow \mathbf{R}')$ denotes the transition probability for a coordinate move from $\mathbf{R}$ to $\mathbf{R}'$. In the original Metropolis procedure, T was taken to be a uniform random distribution over a coordinate interval $\Delta\mathbf{R}$. Condition (3.6) is necessary to satisfy the detailed balance condition

$$T(\mathbf{R}' \rightarrow \mathbf{R})\Psi_T^2(\mathbf{R}') = T(\mathbf{R} \rightarrow \mathbf{R}')\Psi_T^2(\mathbf{R}) \tag{3.7}$$

which is necessary for $\Psi_T^2(\mathbf{R})$ to be the equilibrium distribution of the sampling process.

Several improvements to the Metropolis method have been pursued both in classical and in QMC simulations. These improvements involve new transition probability functions and other sampling procedures. See, for example, KALOS, LEVESQUE and VERLET [1974], CEPERLEY, CHESTER and KALOS [1977], RAO and BERNE [1979],

PANGALI, RAO and BERNE [1979], CEPERLEY and ALDER [1980], ANDERSON [1980], BRESSANINI and REYNOLDS [1998], DEWING [2000].

A common approach for improving $T(\mathbf{R} \rightarrow \mathbf{R}')$ in VMC, is to use the quantum force,

$$\mathbf{F_q} \equiv \nabla \ln|\Psi_T(\mathbf{R})^2| \tag{3.8}$$

as a component of the transition probability. The quantum force can be incorporated by expanding $f(\mathbf{R}, \tau) = |\Psi_T(\mathbf{R})^2| = e^{-\ln|\Psi_T^2(\mathbf{R})|}$ in a Taylor series in $\ln|\Psi_T^2(\mathbf{R})|$ and truncating at first order

$$T(\mathbf{R} \rightarrow \mathbf{R}') \approx \frac{1}{N} e^{\lambda \mathbf{F_q}(\mathbf{R}) \cdot (\mathbf{R}' - \mathbf{R})}, \tag{3.9}$$

where N is a normalization factor, and λ is a parameter fixed for the simulation or optimized in some fashion, for example, see STEDMAN, FOWLKES and NEKOVEE [1998]. A usual improvement is to introduce a cutoff in $\Delta\mathbf{R} = (\mathbf{R}' - \mathbf{R})$, so that if the proposed displacement is larger than a predetermined measure, the move is rejected.

A good transition probability should also contain random displacements, so that all of phase space can be sampled. The combination of the desired drift arising from the quantum force of Eq. (3.9) with a Gaussian random move, gives rise to Langevin fictitious dynamics, namely,

$$\mathbf{R}' \rightarrow \mathbf{R} + \frac{1}{2}\mathbf{F_q}(\mathbf{R}) + \mathcal{G}_{\delta\tau}, \tag{3.10}$$

where $\mathcal{G}_{\delta\tau}$ is a number sampled from a Gaussian distribution with standard deviation $\delta\tau$. The propagator or transition probability for Eq. (3.10) is

$$T_L(\mathbf{R} \rightarrow \mathbf{R}') = \frac{1}{\sqrt{4\pi D\delta\tau}^{3N}} e^{-(\mathbf{R}' - \mathbf{R} - \frac{1}{2}\mathbf{F_q}(\mathbf{R})\delta\tau)^2/2\delta\tau} \tag{3.11}$$

which is a drifting Gaussian, spreading in $\delta\tau$. Using Eq. (3.10) is equivalent to finding the solution of the Fokker–Planck equation (COURANT, FRIEDRICHS and LEWY [1928])

$$\frac{\partial f(\mathbf{R}, \tau)}{\partial \tau} = \frac{1}{2}\nabla \cdot (\nabla - \mathbf{F_q})f(\mathbf{R}, \tau). \tag{3.12}$$

Equation (3.11) has proved to be a simple and effective choice for a VMC transition probability. More refined choices can be made, usually with the goal of increasing acceptance probabilities in regions of rapid change in $|\Psi_T(\mathbf{R})^2|$, such as close to nuclei. For a more detailed discussion of this formalism, the reader is directed to Chapter 2 of HAMMOND, LESTER and REYNOLDS [1994]. More elaborate transition rules can be found in UMRIGAR [1993], SUN, SOTO and LESTER, JR. [1994], STEDMAN, FOWLKES and NEKOVEE [1998], BRESSANINI and REYNOLDS [1999].

3.1.2. Statistics

Usually, VMC calculations are performed using an ensemble of N_W random walkers $\mathcal{W} \equiv \{\mathbf{R}_1, \mathbf{R}_2, \ldots, \mathbf{R}_n\}$ that are propagated following $T(\mathbf{R} \rightarrow \mathbf{R}')$ using the probability

$P(\mathbf{R} \rightarrow \mathbf{R}')$ to accept or reject proposed moves for ensemble members. Statistical averaging has to take into account autocorrelation between moves that arises if the mean square displacement for the ensemble, $\Delta(\mathbf{R} \rightarrow \mathbf{R}')^2/N_W$ is sufficiently large. In such cases, observables measured at the points $\mathbf{R}'$ will be statistically correlated with those evaluated at $\mathbf{R}$. The variance for an observable, $\hat{O}$, measured over N_s MC steps of a random walk is

$$\sigma_{\hat{O}} \equiv \frac{1}{N_s N_W}(O_i - \langle O \rangle), \tag{3.13}$$

where $\langle O \rangle$ is the average of the observations, O_i, over the sample. A simple approach to remove autocorrelation between samples is to define a number of blocks, N_b, where each block is an average of N_s steps, with variance

$$\sigma_B \equiv \frac{1}{N_B N_W}(O_b - \langle O \rangle), \tag{3.14}$$

where O_b is the average number of observations N_t in block b. If N_t is sufficiently large, σ_B is a good estimator of the variance of the observable over the random walk. The autocorrelation time is a good measure of computational efficiency, and is given by

$$T_{\text{corr}} = \lim_{N_s \to \infty} N_s \left(\frac{\sigma_B^2}{\sigma_{\hat{O}}^2}\right). \tag{3.15}$$

The efficiency of a method depends on time step (ROTHSTEIN and VBRIK [1988]). Serial correlation between sample points should vanish for an accurate estimator of the variance. For an observable $\langle O \rangle$, the serial correlation coefficient is defined as

$$\xi_k \equiv \frac{1}{(\langle O^2 \rangle - \langle O \rangle^2)(N - k)} \sum_{i=1}^{N-k}(O_i - \langle O \rangle)(O_{i+k} - \langle O \rangle), \tag{3.16}$$

where k is the number of MC steps between the points O_i and O_{i+k}. The function (3.16) decays exponentially with k. The correlation length, L, is defined as the number of steps necessary for ξ_k to decay essentially to zero. For an accurate variance estimator, blocks should be at least L steps long.

The efficiency of a simulation is inversely proportional to ξ_k. The ξ_k dependence on time step is usually strong (HAMMOND, LESTER and REYNOLDS [1994]); the larger the time step, the fewer steps/block L necessary, and the more points available for calculating the global average $\langle O \rangle$. A rule of thumb is to use an $N_t \approx 10$ times larger than the autocorrelation time to insure statistical independence of block averages, and therefore a reliable variance estimate.

The VMC method shares some of the strengths and weaknesses of traditional variational methods: the energy is an upper bound to the true ground state energy. If reasonable trial functions are used, often reliable estimates of properties can be obtained. For quantum MC applications, VMC can be used to obtain valuable results. In chemical applications, VMC is typically used to analyze and generate trial wave functions for PMC.

3.2. Trial wave functions

In contrast to wave function methods, where the wave function is constructed from linear combinations of determinants of orbitals, QMC methods can use arbitrary functional forms for the wave function subject to the requirements in Section 2. Because QMC trial wave functions are not restricted to expansions in one-electron functions (orbitals), more compact representations are routinely used. In this section, we review the forms most commonly used for QMC calculations.

Fermion wave functions must be antisymmetric with respect to the exchange of an arbitrary pair of particle coordinates. If they are constructed as the product of N functions of the coordinates, $\phi(r_1, r_2, \ldots, r_N)$, the most general wave function can be constructed enforcing explicit permutation:

$$\Psi(\mathbf{R}, \Sigma) = \frac{1}{\sqrt{(N \cdot M)!}} \sum_{n,m} (-1)^n \hat{S}_m \hat{P}_n \phi(r_1, \sigma_1 r_2, \sigma_2, \ldots, r_N, \sigma_N), \tag{3.17}$$

where $\hat{P}_n$ is the nth coordinate permutation operator, $\hat{P}_n \phi(r_1, r_2, \ldots, r_i, r_j, \ldots, r_N) = \phi(r_1, r_2, \ldots, r_j, r_i, \ldots, r_N)$, and $\hat{S}_m \phi(\sigma_1, \sigma_2, \sigma_i, \sigma_j, \ldots, \sigma_N) = \phi(\sigma_1, \sigma_2, \sigma_j, \sigma_i, \ldots, \sigma_N)$ is the mth spin coordinate permutation operator.

If the functions ϕ_i depend only on single-particle coordinates, their antisymmetrized product can be expressed as a Slater determinant

$$D(\mathbf{R}, \Sigma) = \frac{1}{\sqrt{N!}} \det|\phi_1, \ldots, \phi_i(r_j, \sigma_j), \ldots, \phi_n|. \tag{3.18}$$

Trial wave functions constructed from orbitals scale computationally as N^3, where N is the system size, compared to $N!$ for the fully antisymmetrized form.[5] The number of evaluations can be reduced by determining which permutations contribute to a particular spin state.

For QMC evaluation of properties that do not depend on spin coordinates, Σ, for a given spin state, the $M!$ configurations that arise from relabeling electrons, need not be evaluated. The reason is that the Hamiltonian of Eq. (0.1), contains no magnetic or spin operators and spin degrees of freedom remain unchanged. In this case, and for the remainder of this paper, $\sigma_\uparrow$ electrons do not permute with $\sigma_\downarrow$ electrons, so that the full Slater determinant(s) can be factored into a product of spin-up, $D^\uparrow$, and spin-down, $D^\downarrow$, determinants. The number of allowed permutations is reduced from $(N_\uparrow + N_\downarrow)!$ to $N_\uparrow! N_\downarrow!$ (CAFFAREL and CLAVERIE [1988], HUANG, SUN and LESTER, JR. [1990]).

The use of various wave function forms in QMC has been explored by ALEXANDER and COLDWELL [1997], as well as BERTINI, BRESSANINI, MELLA and MOROSI [1999]. Fully antisymmetric descriptions of the wave function are more flexible and require

---

[5]The evaluation of a determinant of size N requires N^2 computer operations. If the one-electron functions scale with system size as well, the scaling becomes N^3. In contrast, a fully antisymmetrized form requires the explicit evaluation of the $N!$ permutations, making the evaluation of this kind of wave functions in QMC prohibitive for systems of large N.

fewer parameters than determinants, but their evaluation is inefficient due to the $N!$ scaling.

A good compromise is to use a product wave function of a determinant or linear combination of determinants,for example, HF, MCSCF, CASSCF, CI, multiplied by a correlation function that is symmetric with respect to particle exchange,

$$\Psi_T = \mathcal{D}\mathcal{F}. \tag{3.19}$$

Here $\mathcal{D}$ denotes the antisymmetric wave function factor and $\mathcal{F}$ is the symmetric factor. We now describe some of the forms used for $\mathcal{D}$ and then we describe forms for $\mathcal{F}$. Such products are also known as the correlated molecular orbital (CMO) wave function.

In the CMO wave functions, the antisymmetric part of the wave function is constructed as a linear combination of determinants of independent particle functions, ϕ_i (see Eq. (3.18)). The ϕ_i are usually formed as a linear combination of basis functions centered on atomic centers, $\phi_i = \sum_j c_j \chi_j$. The most commonly used basis functions in traditional *ab initio* quantum chemistry are Gaussian functions, which owe their popularity to ease of integration of molecular integrals. Gaussian basis functions take the form

$$\chi_G \equiv x^a y^b z^c e^{-\xi r^2}. \tag{3.20}$$

For QMC applications, it is better to use the Slater-type basis functions

$$\chi_S \equiv x^a y^b z^c e^{-\xi r}, \tag{3.21}$$

because they rigorously satisfy the electron–nuclear cusp condition (see Eq. (2.2)), and the asymptotic property of Eq. (2.5). Nevertheless, in most studies, Gaussian basis functions have been used, and corrections for enforcing the cusp conditions can be made to improve local behavior close to a nucleus. For example, in one approach (MANTEN and LÜCHOW [2001]), the region close to a nucleus is described by a Slater-type function, and a polynomial fit is used to connect the Gaussian region to the exponential. This procedure strongly reduces fluctuations of the kinetic energy of these functions, a desirable property for guided VMC and Green's function methods.

The symmetric part of the wave function is usually built as a product of terms explicitly dependent on interparticle distance, $\mathbf{r}_{ij} = |\mathbf{r}_i - \mathbf{r}_j|$. These functions are usually constructed to reproduce the form of the wave function at electron–electron and electron–nucleus cusps. A now familiar form is that proposed by BIJL [1940], DINGLE [1949], JASTROW [1955], and known as the Jastrow *ansatz*:

$$\mathcal{F} \equiv e^{U(\mathbf{r}_{ij})} \equiv e^{\prod_{i<j} g_{ij}}, \tag{3.22}$$

where the correlation function g_{ij} is

$$g_{ij} \equiv \frac{a_{ij}\mathbf{r}_{ij}}{1 + b_{ij}\mathbf{r}_{ij}} \tag{3.23}$$

with constants specified to satisfy the cusp conditions

$$a_{ij} \equiv \begin{cases} \dfrac{1}{4} & \text{if } ij \text{ are like spins,} \\[2ex] \dfrac{1}{2} & \text{if } ij \text{ are unlike spins,} \\[2ex] -\mathcal{Z} & \text{if } ij \text{ are electron/nucleus pairs.} \end{cases} \qquad (3.24)$$

Electron correlation for parallel spins is taken into account by the Slater determinants.

This simple Slater–Jastrow *ansatz* has a number of desirable properties. First, as stated above, scaling with system size for the evaluation of the trial function is N^3, where N is the number of particles in the system, the correct cusp conditions are satisfied at 2-body coalescence points and the correlation function g_{ij} approaches a constant at large distances, which is the correct behavior as $r_{ij} \to \infty$.

In general, the inclusion of 3-body and 4-body correlation terms has been shown to improve wave function quality. The work of HUANG, UMRIGAR and NIGHTINGALE [1997] shows that if the determinant parameters λ_D are optimized along with the correlation function parameters, λ_C, one finds that the nodal structure of the wave function does not improve noticeably in going from 3- to 4-body correlation terms, which suggests that increasing the number of determinants, N_D is more important than adding fourth- and higher order correlation terms.

The use of Feynman–Cohen backflow correlations (FEYNMANN and COHEN [1956]), which has been suggested (SCHMIDT and MOSKOWITZ [1990]) for the inclusion of three body correlations in U, has been used in trial functions for homogeneous systems such as the electron gas (KWON, CEPERLEY and MARTIN [1993], KWON, CEPERLEY and MARTIN [1998]) and liquid helium (SCHMIDT, KALOS, LEE and CHESTER [1980], CASULLERAS and BORONAT [2000]). Feynman (FEYNMANN and COHEN [1956]) suggested replacing the orbitals by functions that include hydrodynamic *backflow* effects. His idea was based on the conservation of particle current and the variational principle. The procedure involves replacing mean field orbitals by *backflow*-corrected orbitals of the form

$$\phi_n(\mathbf{r}_i) \to \phi_n\left(\mathbf{r}_i + \sum_{j \neq i} \mathbf{r}_{ij} v(\mathbf{r}_{ij})\right), \qquad (3.25)$$

where $v(\mathbf{r}_{ij})$ is the backflow function. Others (PANDHARIPANDE and ITOH [1973]) proposed that $v(\mathbf{r}_{ij})$ should consist of the difference between the $l = 0$ and $l = 1$ states of an effective two-particle Schrödinger equation. Furthermore, they proposed (PANDHARIPANDE, PIEPER and WIRINGA [1986]) the inclusion of a $1/r^3$ tail, as originally suggested by Feynman and Cohen,

$$v(\mathbf{r}) = \lambda_v e^{-\left[\frac{r_i - r_j}{\omega_v}\right]^2} + \frac{\lambda_{v'}}{r^3}, \qquad (3.26)$$

where, λ_v, $\lambda_{v'}$, and ω_v are variational parameters. As recently noted by KWON, CEPERLEY and MARTIN [1993], the incorporation of the full backflow trial function into wave functions involves a power of N increase in computational expense, but yields a better DMC energy for the electron gas.[6]

Recently, one has seen the practice of taking orbitals from a mean field calculation and the inclusion of averaged backflow terms in the correlation function $\mathcal{F}$. The advantage of this approach is that orbitals are unperturbed and readily obtainable from mean field computer codes.

The correlation function form used by SCHMIDT and MOSKOWITZ [1990] is a selection of certain terms of the general form originally proposed in connection with the transcorrelated method (BOYS and HANDY [1969]):

$$\mathcal{F} = e^{\sum_{I, i<j} U_{Iij}}, \tag{3.27}$$

where

$$U_{Iij} = \sum_{v}^{N(I)} \Delta(m_{kI} n_{kI}) c_{kI} (g_{iI}^{m_{kI}} g_{jI}^{n_{kI}} + g_{jI}^{m_{kI}} g_{iI}^{n_{kI}}) g_{ij}^{o_{kI}}. \tag{3.28}$$

The sum in (3.27) goes over I nuclei, ij electron pairs, and the sum in Eq. (3.28) is over the $N(I)$ terms of the correlation function for each nucleus. The parameters m, n and o are integers. The function $\Delta(m, n)$ takes the value 1 when $m \neq n$ and $\frac{1}{2}$ otherwise. The functions g_{ij} are specified by Eq. (3.23).

This correlation function (3.27), (3.28) can be shown to have contributions to averaged backflow effects from the presence of electron–electron–nucleus correlations that correspond to values of m, n and o in Eq. (3.28) of 2, 2, 0 and 2, 0, 2. These contributions recover $\approx 25\%$ or more of the total CE of atomic and molecular systems above that from the simple Jastrow term (SCHMIDT and MOSKOWITZ [1990]).

3.3. The variational Monte Carlo algorithm

The VMC algorithm is an application of the generalized Metropolis MC method. As in most applications of the method, one needs to insure that the ensemble has achieved equilibrium in the simulation sense. Equilibrium is reached when the ensemble $\mathcal{W}$ is distributed according to $\mathcal{P}(\mathbf{R})$. This is usually achieved by performing a Metropolis random walk and monitoring the trace of the observables of interest. When the trace fluctuates around a mean, it is generally safe to start averaging in order to obtain desired properties.

An implementation of the VMC algorithm follows:
(1) Equilibration stage
  (a) Generate an initial set of random walker positions, $\mathcal{W}_0$; it can be read in from a previous random walk, or generated at random.
  (b) Perform a loop over N_s steps,
    (i) For each r_i of the N_p number of particles,

---

[6]As discussed in Section 5.6, an improved fixed-node energy is a consequence of better nodes of the trial wave function, a critically important characteristic for importance sampling functions in QMC methods.

(A) Propose a move from $\Psi(\mathbf{R}) \equiv \Psi(\mathbf{r}_1, \mathbf{r}_2, \ldots, \mathbf{r}_i, \ldots, \mathbf{r}_{N_p})$ to $\Psi(\mathbf{R}')$
$\equiv \Psi(\mathbf{r}_1, \mathbf{r}_2, \ldots, \mathbf{r}'_i, \ldots, \mathbf{r}_{N_p})$. Move from $\mathbf{r}$ to $\mathbf{r}'$ according to

$$\mathbf{r}' \leftarrow \mathbf{r} + \mathcal{G}_{\delta\tau} + \frac{1}{2}\mathbf{F}_\mathbf{q}\delta\tau, \tag{3.29}$$

where $\mathcal{G}_{\delta\tau}$ is a Gaussian random number with standard deviation $\delta\tau$, which is a proposed step size, and $\mathbf{F}_\mathbf{q}$ is the quantum force (see Eq. (3.9)). This is the Langevin dynamics of Eq. (3.10).

(B) Compute the Metropolis acceptance/rejection probability

$$P(\mathbf{R} \to \mathbf{R}') = \min\left(1, \frac{T_L(\mathbf{R}' \to \mathbf{R})\Psi_T^2(\mathbf{R}')}{T_L(\mathbf{R} \to \mathbf{R}')\Psi_T^2(\mathbf{R})}\right), \tag{3.30}$$

where T_L is given by Eq. (3.11).

(C) Compare $P(\mathbf{R} \to \mathbf{R}')$ with an uniform random number between 0 and 1, $\mathcal{U}_{[0,1]}$. If $P > \mathcal{U}_{[0,1]}$, accept the move, otherwise, reject it.

(D) Calculate the contribution to the averages $\frac{\hat{O}_d\Psi_T(\mathbf{R}')}{\Psi_T(\mathbf{R}')}$, and perform blocking statistics as described in Section 3.1.2.

(ii) Continue the loop until the desired accuracy is achieved.

4. Wave function optimization

Trial wave functions $\Psi_T(\mathbf{R}, \Lambda)$ for QMC are dependent on variational parameters $\Lambda = \{\lambda_1, \ldots, \lambda_n\}$. Optimization of Λ is a key element for obtaining accurate trial functions. Importance sampling using an optimized trial function increases the efficiency of DMC simulations. There is a direct relationship between trial-function accuracy and the computer time required to calculate accurate expectation values. Some of the parameters λ_i may be fixed by imposing appropriate wave function properties, such as cusp conditions (see Section 2).

It is useful to divide Λ into groups distinguished by whether the optimization changes the nodes of the wave function. The Slater determinant parameters, $\lambda_{\mathcal{D}\uparrow\downarrow}$ and the Slater determinant weights, λ_{k_i} change wave function nodal structure (UMRIGAR, WILSON and WILKINS [1988], SCHMIDT and MOSKOWITZ [1990], SUN, BARNETT and LESTER, JR. [1992], FLAD, CAFFAREL and SAVIN [1997], BARNETT, SUN and LESTER, JR. [1997], BARNETT, SUN and LESTER, JR. [2001]). The correlation function parameters, $\lambda_{\mathcal{F}}$ do not change the nodal structure of the overall wave function, and therefore the DMC energy. For some systems, the optimization of $\lambda_{\mathcal{F}}$ is sufficient for building reliable trial functions for PMC methods, because $\mathcal{F}$ is designed in part to satisfy cusp conditions (KATO [1957], MEYERS, UMRIGAR, SETHNA and MORGAN [1991], FILIPPI and UMRIGAR [1996]).

There have been several optimization methods proposed previously. Some involve the use of analytical derivatives (HUANG, SUN and LESTER, JR. [1990], BUECKERT, ROTHSTEIN and VRBIK [1992], LÜCHOW and ANDERSON [1996], HUANG and CAO [1996], HUANG, XIE, CAO, LI, YUE and MING [1999], and others focus on the use of a fixed sample for variance minimization (CONROY [1964]), and more recently others (UMRIGAR, WILSON and WILKINS [1988], SUN, HUANG, BARNETT and LESTER, JR. [1990], NIGHTINGALE and UMRIGAR [1997]). Yet another direction is the use of histogram analysis for optimizing the energy, variance, and molecular geometry for small systems (SNAJDR, DWYER and ROTHSTEIN [1999]). In the present study, we concentrate on fixed sample optimization to eliminate stochastic uncertainty during the random walk (SUN, HUANG, BARNETT and LESTER, JR. [1990]).

Other authors optimize the trial wave function using information obtained from a DMC random walk (FILIPPI and FAHY [2000]). This approach shows promise, because usually the orbitals obtained from a mean field theory, such as HF or local density approximation (LDA), are frozen and used in the DMC calculation without reoptimization specifically for correlation effects within the DMC framework.

The common variance functional (*VF*) (UMRIGAR, WILSON and WILKINS [1988]) is given by

$$
VF = \frac{\sum_{i=1}^{N} \left[\frac{\hat{H}\Psi(\mathbf{R}_i, \Lambda)}{\Psi(\mathbf{R}_i, \Lambda)} - E_T \right]^2 w_i}{\sum_{i=1}^{N} w_i},
\tag{4.1}
$$

where E_T is a trial energy, w_i is a weighting factor defined by

$$
w_i(\Lambda) = \frac{\Psi_i^2(\mathbf{R}_i, \Lambda)}{\Psi_i^2(\mathbf{R}_i, \Lambda_0)},
\tag{4.2}
$$

and Λ_0 is an initial set of parameters. The sum in Eq. (4.1) is over fixed sample configurations.

4.1. Trial wave function quality

The overlap of Ψ_T with the ground state wave function, $\langle \Psi_T | \Psi_0 \rangle$, by DMC methods (HORNIK, SNAJDR and ROTHSTEIN [2000]) is a very efficient way of assessing wave function quality. There is also a trend that correlates the variational energy of the wave function with associated variance in a linear relationship (KWON, CEPERLEY and MARTIN [1993], KWON, CEPERLEY and MARTIN [1998]). This correlation is expected because both properties, δ_E, and δ_{E_L}, approach limits $-E_0$ and zero respectively – as wave function quality improves. Observing this correlations is a good method of validating the optimization method, as well as assessing wave function quality.

5. Projector methods

QMC methods such as DMC and Green's function Monte Carlo (GFMC) are usefully called PMC methods.[7] The general idea is to project out a state of the Hamiltonian by

---

[7]The introductory section of this chapter follows the work of HETHERINGTON [1984], CHIN [1990], CERF and MARTIN [1995].

iteration of a projection operator, $\hat{P}$. For simplicity, we assume that the desired state is the ground state, Ψ_0, but projectors can be constructed for any state,

$$\lim_{i \to \infty} \hat{P}^i |\Psi_T\rangle \approx |\Psi_0\rangle. \tag{5.1}$$

After sufficient iterations i, the contribution of all excited states $|\Psi_i\rangle$, will be filtered out, and only the ground state is recovered.

If $|\Psi_T\rangle$ is a vector and $\hat{P}$ is a matrix, then the procedure implied by (5.1) is the algebraic power method: If a matrix is applied iteratively to an initial arbitrary vector for a sufficient number of times, only the dominant eigenvector, $|\Psi_0\rangle$, will survive. One can see for large i,

$$\hat{P}^i |\Psi_T\rangle = \lambda_0^i \langle \Psi_0 | \Psi_T \rangle |\Psi_0\rangle + \mathcal{O}(\lambda_1^i), \tag{5.2}$$

where λ_0 is the leading eigenvalue, and λ_1 is the largest subleading eigenvalue.

For this approach, it is possible to obtain an estimator of the eigenvalue, as described in HAMMERSLEY and HANDSCOMB [1964], given by

$$\lambda_0 = \lim_{i \to \infty} \left(\frac{\langle \phi | \hat{P}^{i+j} | \Psi_T \rangle}{\langle \phi | \hat{P}^i | \Psi_T \rangle} \right)^{\frac{1}{m}}. \tag{5.3}$$

5.2. *Markov processes and stochastic projection*

For high-dimensional vectors, such as those encountered in molecular electronic structure, the algebraic power method described previously needs to be generalized with stochastic implementation. For this to occur, the projection operator must be symmetric, so that all eigenvalues are real. This is the case for QMC methods, because $\hat{P}$ is a function of the Hamiltonian operator, $\hat{H}$, which is Hermitian by construction.

A stochastic matrix is a normalized nonnegative matrix. By normalization, we mean that the stochastic matrix columns add to one, $\sum_i M_{ij} = 1$. An $\mathbf{R}$-space representation would be a stochastic propagator $M(\mathbf{R}, \mathbf{R}')$ that satisfies the condition

$$\int M(\mathbf{R}, \mathbf{R}') d\mathbf{R}' = 1. \tag{5.4}$$

A Markov chain is a sequence of states obtained from subsequent transitions from state i to j with a probability related to the stochastic matrix element M_{ij}, in which the move only depends on the current state, i. For example, in $\mathbf{R}$-space, this is equivalent to the following process

$$\pi(\mathbf{R}') = \int M(\mathbf{R}', \mathbf{R}'') \pi(\mathbf{R}'') d\mathbf{R}'',$$
$$\pi(\mathbf{R}) = \int M(\mathbf{R}, \mathbf{R}') \pi(\mathbf{R}') d\mathbf{R}', \tag{5.5}$$
$$\cdots$$

The sequence of states $S = \{\pi(\mathbf{R}''), \pi(\mathbf{R}'), \pi(\mathbf{R}), \ldots\}$ is the Markov chain.

The propagators of QMC for electronic structure are not generally normalized, therefore they are not stochastic matrices, but we can represent them in terms of the latter by factoring,

$$\hat{P}_{ij} = M_{ij}w_j, \tag{5.6}$$

where the weights, w_j, are defined by $w_j = \sum_i \hat{P}_{ij}$. This definition unambiguously defines both the associated stochastic matrix M and the weight vector w.

A MC sampling scheme of $\hat{P}_{ij}|\Psi_T\rangle$ can be generated by first performing a random walk, and keep a weight vector $W(\mathbf{R})$ for the random walkers,

$$\begin{aligned}
\Psi(\mathbf{R}') &\equiv \pi(\mathbf{R}')W(\mathbf{R}') = \int P(\mathbf{R}', \mathbf{R}'')\Psi(\mathbf{R}'')d\mathbf{R}'' \\
&= \int M(\mathbf{R}', \mathbf{R}'')B(\mathbf{R}')\Psi(\mathbf{R}'')d\mathbf{R}'', \\
\Psi(\mathbf{R}) &\equiv \pi(\mathbf{R})W(\mathbf{R}) = \int P(\mathbf{R}, \mathbf{R}')\Psi(\mathbf{R}')d\mathbf{R}' \\
&= \int M(\mathbf{R}, \mathbf{R}')B(\mathbf{R})\Psi(\mathbf{R}')d\mathbf{R}',
\end{aligned} \tag{5.7}$$

$$\dots$$

Here, $B(\mathbf{R})$ is the function that determines the weight of the configurations at each state of the random chain. This leads to a generalized stochastic projection algorithm for unnormalized transition probabilities that forms the basis for population Monte Carlo (PopMC) algorithms, which are not only used for QMC, but they are also used for statistical information processing and robotic vision (IBA [2000]). A generalized PopMC stochastic projection algorithm, represented in $\mathbf{R}$-space follows:

(1) *Initialize*

Generate a set of **n** random walkers, located at different spatial positions, $\mathcal{W} \equiv \{\mathbf{R}_1, \mathbf{R}_2, \dots, \mathbf{R}_n\}$, where $\mathbf{R}_i$ denotes a Dirac delta function at that point in space, $\delta(\mathbf{R} - \mathbf{R}_i)$. These points are intended to sample a probability density function $\Phi(\mathbf{R})$.

(2) *Move*

(a) Each walker j is moved independently from $\mathbf{R}$ to a new position $\mathbf{R}'$, according to the transition probability

$$T(\mathbf{R} \rightarrow \mathbf{R}') \equiv M(\mathbf{R}, \mathbf{R}'). \tag{5.8}$$

(b) Ensure detailed balance if $T(\mathbf{R} \rightarrow \mathbf{R}') \neq T(\mathbf{R}' \rightarrow \mathbf{R})$ by using a Metropolis acceptance/rejection step as in Eq. (3.30).

(3) *Weight*

(a) Calculate a weight vector using a weighting function $B(\mathbf{R}_i)$,

$$w_i^* = B(\mathbf{R}_i). \tag{5.9}$$

The ideal weight function preserves normalization of $\hat{P}(\mathbf{R}, \mathbf{R}')$ and maintains individual weights w_i close to unity.

(b) Update the weight of the walker, multiplying the weight of the previous iteration by the weight of the new iteration,

$$w'_i = w_i^* w_i. \tag{5.10}$$

(4) *Reconfiguration*
  (a) Split walkers with large weights into multiple walkers with weights that add up to the original weight.
  (b) Remove walkers with small weight.

Step 4 is necessary to avoid statistical fluctuations in the weights. It is a form of importance sampling in the sense that makes the calculation stable over time. Some algorithms omit this step; see, for example, efforts by CAFFAREL and CLAVERIE [1988], but it has been proved that such calculations eventually diverge (ASSARAF, CAFFAREL and KHELIF [2000]). There is a slight bias associated with the introduction of step 4 together with population control methods, that will be discussed in Section 5.3. When step 4 is used, $B(\mathbf{R})$ is also referred in the literature as a *branching factor*.

It is important to recall that PopMC algorithms are not canonical Markov Chain Monte Carlo (MCMC) algorithms (BYRON and FULLER [1992], DOUCET, DE FREITAS, GORDON and SMITH [2001]), in the sense that the propagator used is not normalized, and therefore factoring the propagator into a normalized transition probability and a weighting function is required.

5.3. Projection operators or Green's functions

Different projection operators lead to different QMC methods. If the resolvent operator,

$$\hat{P}(\hat{H}) \equiv \frac{1}{1 + \delta\tau(\hat{H} - E_R)}, \tag{5.11}$$

is used, one obtains GFMC (KALOS [1962], CEPERLEY and KALOS [1986]). This algorithm will be described in Section 5.5.

If the imaginary time evolution operator is used, that is,

$$\hat{P}(\hat{H}) \equiv e^{-\delta\tau(\hat{H} - E_R)}, \tag{5.12}$$

one has the DMC method (ANDERSON [1976], REYNOLDS, CEPERLEY, ALDER and LESTER, JR. [1982]), which is discussed in Section 5.1.2.

For finite $\delta\tau$, and for molecular systems, the exact projector is not known analytically. In GFMC, the resolvent of Eq. (5.11) is sampled by iteration of a simpler resolvent, whereas for DMC, the resolvent is known exactly at $\tau \to 0$, so an extrapolation to $\delta\tau \to 0$ is done.

Note that any decreasing function of $\hat{H}$ can serve as a projector. Therefore, new QMC methods still await to be explored.

5.1. Imaginary propagator

If one transforms the time-dependent Schrödinger equation (Eq. (0.2)) to imaginary time, that is,

$$it \to \tau, \tag{5.13}$$

then one obtains

$$\frac{\partial}{\partial \tau} \Psi(\mathbf{R}, \tau) = (\hat{H} - E_R)\Psi(\mathbf{R}, \tau). \tag{5.14}$$

Here E_R is an energy offset, called the reference energy. For real $\Psi(\mathbf{R}, \tau)$, this equation has the advantage of being an equation in $\mathcal{R}^N$, whereas Eq. (0.2) has in general, complex solutions.

Equation (5.14) can be cast into integral form,

$$\Psi(\mathbf{R}, \tau + \delta\tau) = \lambda_\tau \int G(\mathbf{R}, \mathbf{R}', \delta\tau)\Psi(\mathbf{R}', \tau)d\mathbf{R}'. \tag{5.15}$$

The Green's function, $G(\mathbf{R}', \mathbf{R}, \delta\tau)$, satisfies the same boundary conditions as Eq. (5.14):

$$\frac{\partial}{\partial \tau} G(\mathbf{R}, \mathbf{R}', \delta\tau) = (\hat{H} - E_T)G(\mathbf{R}, \mathbf{R}', \delta\tau) \tag{5.16}$$

with the initial conditions associated with the propagation of a Dirac delta function, namely,

$$G(\mathbf{R}, \mathbf{R}', 0) = \delta(\mathbf{R} - \mathbf{R}'). \tag{5.17}$$

The form of the Green's function that satisfies Eq. (5.16), subject to (5.17), is

$$G(\mathbf{R}, \mathbf{R}', \delta\tau) = \langle \mathbf{R}|e^{-\tau(\hat{H} - E_R)}|\mathbf{R}\rangle. \tag{5.18}$$

This operator can be expanded in eigenfunctions, Ψ_α, and eigenvalues E_α of the system

$$G(\mathbf{R}, \mathbf{R}', \delta\tau) = \sum_\alpha e^{-\tau(E_\alpha - E_R)}\Psi_\alpha^*(\mathbf{R}')\Psi_\alpha(\mathbf{R}). \tag{5.19}$$

For an arbitrary initial trial function, $\Psi(\mathbf{R})$, in the long term limit, $\tau \to \infty$, one has

$$\begin{aligned}
\lim_{\tau \to \infty} e^{-\tau(\hat{H} - E_T)}\Psi &= \lim_{\tau \to \infty} \int G(\mathbf{R}', \mathbf{R}, \tau)\Psi(\mathbf{R}')d\mathbf{R}' \\
&= \lim_{\tau \to \infty} \langle \Psi|\Psi_0\rangle e^{-\tau(E_0 - E_R)}\phi_0,
\end{aligned} \tag{5.20}$$

and only the ground state wave function Ψ_0 is obtained from any initial wave function. Therefore, the imaginary time evolution operator can be used as a projection operator as mentioned at the beginning of this chapter.

5.1.1. Diffusion Monte Carlo stochastic projection

Due to the high dimensionality of molecular systems, a MC projection procedure is used for obtaining expectation values. In this approach, the wave function is represented as an ensemble of delta functions, also known as configurations, walkers, or *psips* (psi-particles):

$$\Phi(R) \leftrightarrow \sum_k \delta(\mathbf{R} - \mathbf{R_k}). \tag{5.21}$$

The wave function is propagated in imaginary time using the Green's function. In the continuous case, one can construct a Neumann series

$$\Psi^{(2)}(\mathbf{R}, \tau) = \lambda_1 \int G\left(\mathbf{R}', \mathbf{R}, \tau_2 - \tau_1\right) \Psi^{(1)}\left(\mathbf{R}'\right) d\mathbf{R},$$

$$\Psi^{(3)}(\mathbf{R}, \tau) = \lambda_2 \int G\left(\mathbf{R}', \mathbf{R}, \tau_3 - \tau_2\right) \Psi^{(2)}\left(\mathbf{R}'\right) d\mathbf{R},$$

$$\vdots \tag{5.22}$$

This Neumann series is a specific case of the PopMC propagation of Section 5.0.2. The discrete Neumann series can be constructed in a similar way:

$$\Phi^{(n+1)}(\mathbf{R}, \tau + \delta\tau) \leftrightarrow \lambda_k \sum_k G^{(n)}(\mathbf{R}, \mathbf{R}', \delta\tau). \tag{5.23}$$

Therefore, a stochastic vector of configurations $\mathcal{W} \equiv \{\mathbf{R}_1, \ldots, \mathbf{R}_n\}$ is used to represent $\Psi(\mathbf{R})$ and is iterated using $G^{(n)}(\mathbf{R}, \mathbf{R}', \delta\tau)$.

5.1.2. The form of the propagator

Sampling Eq. (5.18) can not be done exactly, because the argument of the exponential is an operator composed of two terms that do not commute with each other.

In the short-time approximation (STA), the propagator $G(\mathbf{R}, \mathbf{R}_k, d\tau)$ is approximated as if the kinetic and potential energy operators commuted with each other:

$$e^{(T+V)\delta\tau} \approx e^{T\delta\tau} \cdot e^{V\delta\tau} + \mathcal{O}((\delta\tau)^2) \equiv G_{ST} \equiv G_D \cdot G_B. \tag{5.24}$$

The Green's function then becomes the product of a diffusion factor G_D and a branching factor G_B. Both propagators are known:

$$G_D = (2\pi\tau)^{-3N/2} e^{-\frac{(\mathbf{R}-\mathbf{R}')^2}{2\tau}} \tag{5.25}$$

and

$$G_B = e^{-\delta\tau(V(\mathbf{R})-2E_T)}. \tag{5.26}$$

G_D is a fundamental solution of the Fourier equation (that describes a diffusion process in wave function space) and G_B is the fundamental solution of a first-order kinetic birth–death process.

The Campbell–Baker–Hausdorff (CBH) formula,

$$e^A e^B = e^{A+B+\frac{1}{2}[A,B]+\frac{1}{12}[(A-B),[A,B]]+\cdots} \tag{5.27}$$

can help in constructing more accurate decompositions, such as an expansion with a cubic error $\mathcal{O}((\delta\tau)^3)$,

$$e^{\delta\tau(T+V)} = e^{\delta\tau(V/2)} e^{\delta\tau T} e^{\delta\tau(V/2)} + \mathcal{O}((\delta\tau)^3). \tag{5.28}$$

There are more sophisticated second order (CHIN [1990]) and fourth order (CHIN [1997], FORBERT and CHIN [2001]) expansions that reduce the error considerably and make more exact DMC algorithms at the expense of a more complex propagator.

The most common implementation using G_D as a stochastic transition probability

$T(\mathbf{R} \rightarrow \mathbf{R}')$, and G_B as a weighting or branching factor, $B(\mathbf{R})$. Sampling of Eq. (5.25) can be achieved by obtaining random variates from a Gaussian distribution of standard deviation $\delta_\tau, \mathcal{G}_{\delta\tau}$.

5.2. Importance sampling

Direct application of the algorithm of the previous section to systems governed by the Coulomb potential leads to large population fluctuations. These arise because the potential $\hat{V}(\mathbf{R})$ becomes unbounded and induces large fluctuations in the random walker population. A remedy, importance sampling, was first used for GFMC by KALOS [1962] and extended to the DMC method by CEPERLEY and ALDER [1980].

In importance sampling, the goal is to reduce fluctuations, by multiplying the probability distribution by a known trial function, $\Psi_T(\mathbf{R})$, that is expected to be a good approximation for the wave function of the system. Rather than $\Psi(\mathbf{R}, \tau)$, one samples the product

$$f(\mathbf{R}, \tau) = \Psi_T(\mathbf{R})\Psi(\mathbf{R}, \tau). \tag{5.29}$$

Multiplying Eq. (5.15) by $\Psi_T(\mathbf{R})$, one obtains

$$f(\mathbf{R}, \tau + d\tau) = \int K(\mathbf{R}, \mathbf{R}, \delta\tau) f(\mathbf{R}', \tau) d\mathbf{R}', \tag{5.30}$$

where $K(\mathbf{R}, \mathbf{R}', \delta\tau) \equiv e^{-\tau(\hat{H}-E_T)} \frac{\Psi_T(\mathbf{R})}{\Psi_T(\mathbf{R}')}$. Expanding K in a Taylor series, at the $\delta\tau \rightarrow 0$ limit, one obtains the expression

$$K = Ne^{-(\mathbf{R}_2 - \mathbf{R}_1 + \frac{1}{2}\nabla \ln\Psi_T(\mathbf{R}_1)\delta\tau)^2/(2\delta\tau)} \times e^{-\left(\frac{\hat{H}\Psi_T(\mathbf{R}_1)}{\Psi_T(\mathbf{R}_1)} - E_T\right)\delta\tau} \equiv K_D \times K_B. \tag{5.31}$$

Equation (5.31) is closely associated to the product of the kernel of the Smoluchowski equation, which describes a diffusion process with drift, multiplied by a first order rate process. Here the rate process is dominated by the local energy, instead of the potential. The random walk is modified by appearance of a drift term that moves configurations to regions of high values of the wave function. This drift is the quantum force of Eq. (3.8).

The excess local energy $(E_T - E_L(\mathbf{R}))$ replaces the excess potential energy in the branching term exponent, see (5.26). The local energy has kinetic and potential energy contributions that tend to cancel each other, giving a smoother function: If $\Psi_T(\mathbf{R})$ is a reasonable function, the excess local energy will be nearly a constant. The regions where charged particles meet have to be taken care of by enforcing the cusp conditions on $\Psi_T(\mathbf{R})$ (see Section 2).

The local energy is the estimator of the energy with a lower statistical variance, so it is preferred over other possible choices for an estimator. A simple average of the local energy will yield the estimator of the energy of the quantum system,[8]

---

[8]For other energy estimators, refer to the discussion in CEPERLEY and KALOS [1986] and HAMMOND, LESTER and REYNOLDS [1994].

$$\langle E_L \rangle = \int f(\mathbf{R}, \tau \to \infty) E_L(\mathbf{R}) d\mathbf{R} / \int f(\mathbf{R}) d\mathbf{R}$$

$$= \int \Psi(\mathbf{R}) \Psi_T(\mathbf{R}) \left[\frac{\hat{H} \Psi_T(\mathbf{R})}{\Psi_T(\mathbf{R})} \right] d\mathbf{R} / \int \Psi(\mathbf{R}) \Psi_T(\mathbf{R}) \qquad (5.32)$$

$$= \int \Psi(\mathbf{R}) \hat{H} \Psi(\mathbf{R}) d\mathbf{R} / \int \Psi(\mathbf{R}) \Psi_T(\mathbf{R}) d\mathbf{R}$$

$$= E_0.$$

Therefore, a simple averaging of the local energy will yield the DMC energy estimator:

$$\langle E_L \rangle = \lim_{N_s \to \infty} \frac{1}{N_s} \sum_i^{N_w} E_L(\mathbf{R}_i). \qquad (5.33)$$

Because the importance sampled propagator, $K(\mathbf{R}, \mathbf{R}', \delta\tau)$, is only exact to a certain order, for obtaining an exact estimator is necessary to extrapolate to $\delta\tau = 0$ for several values of $\langle E_L \rangle$.

Importance sampling with appropriate trial functions, such as those used for accurate VMC calculations, can increase the efficiency of the random walk by several orders of magnitude. In the limit required to obtain the exact trial function, only a single evaluation of the local energy is required to obtain the exact answer. Importance sampling has made molecular and atomic calculations feasible. Note that the quantum force present in Eq. (5.31) also moves random walkers away from the nodal regions into regions of large values of the trial wave function, reducing the number of attempted node crossings by several orders of magnitude.

5.3. Population control

If left uncontrolled, the population of random walkers will eventually vanish or fill all computer memory. Therefore, some form of population control is needed to stabilize the number of random walkers. Control is usually achieved by slowly changing E_T as the simulation progresses. As more walkers are produced in the procedure, one needs to lower the trial energy, E_T, or if the population starts to decrease, then one needs to raise E_T. This can be achieved by periodically changing the trial energy. One version of the adjustment is to use

$$E_T = \langle E_0 \rangle + \alpha \ln \frac{N_w^0}{N_w}, \qquad (5.34)$$

where $\langle E_0 \rangle$ is the best approximation to the eigenvalue of the problem to this point, α is a parameter that should be as small as possible while still having a population control effect, N_w^0 is the number of desired random walkers, and N_w is the current number of random walkers.

This simple population control procedure has a slight bias if the population control parameter α is large, or if the population is small. The bias observed goes as $1/N_w$, and, formally a $N_w \to \infty$ extrapolation is required. The bias is absent in the limit of an infinite population.

A recently resurrected population control strategy, stochastic reconfiguration (SORELLA [1998], BUONAURA and SORELLA [1998], SORELLA and CAPRIOTTI [2000], ASSARAF, CAFFAREL and KHELIF [2000]) originally came from a the work of HETHERINGTON [1984]. In this algorithm, walkers carry a weight, but the weight is redetermined at each step to keep the population constant. The idea behind this method is to control the global weight $\bar{w}$ of the population,

$$\bar{w} = \frac{1}{N_w} \sum_{i=1}^{N_w} w_i. \tag{5.35}$$

The idea is to introduce a renormalized individual walker weight, ω_i, defined as

$$\omega_i \equiv \frac{w_i}{\bar{w}}. \tag{5.36}$$

Another stochastic reconfiguration scheme proposes setting the number of copies of walker i for the next step, proportional to the renormalized walker weight ω_i. This algorithm has shown to have less bias than the scheme of Eq. (5.34), and also has the advantage of having the same number of walkers at each step, simplifying implementations of the algorithm in parallel computers.

5.4. Diffusion Monte Carlo algorithm

There are several versions of the DMC algorithm. The approach presented here focuses on simplicity. For the latest developments, the reader is referred to REYNOLDS, CEPERLEY, ALDER and LESTER, JR. [1982], DEPASQUALE, ROTHSTEIN and VRBIK [1988], UMRIGAR, NIGHTINGALE and RUNGE [1993].

(1) Initialize an ensemble $\mathcal{W}$ of $N_{\mathcal{W}}$ configurations, distributed according to $P(\mathbf{R})$ for $\Psi_T(\mathbf{R})$; for example, use the random walkers obtained from a previous VMC run.

(2) For every configuration in $\mathcal{W}$:

(a) Propose an electron move from $\Psi(\mathbf{R}) \equiv \Psi(\mathbf{r}_1, \mathbf{r}_2, \ldots, \mathbf{r}_i, \ldots, \mathbf{r}_{N_p})$ to $\Psi(\mathbf{R}') \equiv \Psi(\mathbf{r}_1, \mathbf{r}_2, \ldots, \mathbf{r}_i', \ldots, \mathbf{r}_{N_p})$. The STA propagator, $K(\mathbf{R}, \mathbf{R}'; \delta\tau)$, has an associated stochastic move

$$\mathbf{R}' \rightarrow \mathbf{R} + \mathbf{F}_\mathbf{q}(\mathbf{R})\delta\tau + \mathcal{G}_{\delta\tau}. \tag{5.37}$$

(b) Enforce the fixed node constraint: if a random walker crosses a node, that is, sign $(\Psi_T(\mathbf{R})) \neq \text{sign}(\Psi_T(\mathbf{R}'))$, then reject the move for the current electron and proceed to treat the next electron.

(c) Compute the Metropolis acceptance/rejection probability

$$P(\mathbf{R} \rightarrow \mathbf{R}') = \min\left(1, \frac{K_D(\mathbf{R}, \mathbf{R}'; \delta\tau)\Psi_T^2(\mathbf{R}')}{K_D(\mathbf{R}', \mathbf{R}; \delta\tau)\Psi_T^2(\mathbf{R})}\right), \tag{5.38}$$

where K_D is the diffusion and drift transition probability given by Eq. (5.31).

(d) Compare $P(\mathbf{R} \rightarrow \mathbf{R}')$ with an uniform random number between 0 and 1, $\mathcal{U}_{[0,1]}$, if $P > \mathcal{U}_{[0,1]}$, accept the move, otherwise, reject it.

(3) Calculate the branching factor G_B for the current configuration

$$B(\mathbf{R}, \mathbf{R}') = e^{\left(E_R - \frac{1}{2}\left(\frac{\hat{H}\Psi_T(\mathbf{R}')}{\Psi_T(\mathbf{R}')} + \frac{\hat{H}\Psi_T(\mathbf{R})}{\Psi_T(\mathbf{R})}\right)\right)\delta\tau}. \tag{5.39}$$

(4) Accumulate all observables, such as the energy. All contributions, O_i, are weighted by the branching factor, that is,

$$O_T^{(n+1)} = O_T^{(n)} + B(\mathbf{R}, \mathbf{R}')O_i(\mathbf{R}), \tag{5.40}$$

where $O_T^{(n)}$ is the cumulative sum of the observable at step n.

(5) Generate a new generation of random walkers, reproducing the existing population, creating an average $B(\mathbf{R}, \mathbf{R}')$ new walkers out of a walker at $\mathbf{R}$. The simplest procedure for achieving this goal is to generate n new copies of $\mathbf{R}$ where $n = \int(B(\mathbf{R}, \mathbf{R}') + \mathcal{U}_{[0,1]})$.

(6) Perform blocking statistics (see Section 3.1.2), and apply population control (see Section 5.3)

   (a) One choice is to update the reference energy E_R at the end of each accumulation block,

$$E_R \leftarrow E_R + E_R^\omega * E_B, \tag{5.41}$$

where E_R^ω is a reweighting parameter, usually chosen to be ≈ 0.5, and E_B is the average energy for block B, $E_B = E^{\text{sum}}/N_B$.

   (b) Discard a relaxation time of steps, N_{Rel}, which is of the order of a tenth of a block, because moving the reference energy induces the most bias in about one relaxation time.

(7) Continue the loop until the desired accuracy is achieved.

UMRIGAR, NIGHTINGALE and RUNGE [1993] proposed several modifications to the above algorithm to reduce time-step error. These modifications concentrate on improving the propagator in regions where the STA performs poorly; namely, near wave function nodes and Coulomb singularities. These propagator errors are expected, because the STA propagator assumes a constant potential over the move interval, which is a poor approximation in Ψ regions where the Coulomb interaction diverges.

5.5. Green's function Monte Carlo

The GFMC method is a QMC approach that has the advantage of having no time-step error. It has been shown to require more computer time than DMC, and therefore, has been applied less frequently than DMC to atomic and molecular systems. Good descriptions of the method can be found in KALOS [1962], CEPERLEY [1983], SKINNER, MOSKOWITZ, LEE, SCHMIDT and WHITLOCK [1985], KALOS and WHITLOCK [1986], SCHMIDT and MOSKOWITZ [1986]. The GFMC approach is a PopMC method for which the projector for obtaining the ground state Green's function is the standard resolvent for the Schrödinger equation (see Eq. (5.11)). The integral equation for this case, takes the simple form

$$\Psi^{(n+1)} = \left[\frac{E_T + E_C}{\hat{H} + E_C}\right] \Psi^{(n)} \tag{5.42}$$

where the constant E_C is positive and fulfills the condition that $|E_C| > |E_0|$, and E_T is a trial energy. The resolvent of Eq. (5.42) is related to the DMC propagator by the one-sided Laplace transform

$$\frac{1}{\hat{H} + E_C} = \int_0^\infty e^{-(\hat{H} + E_C)\tau} d\tau. \tag{5.43}$$

This integral is evaluated by MC. After equilibration, the sampled times have a Poisson distribution with a mean of $\frac{N_s}{E_0 + E_C}$ after N_s steps. E_C is a parameter that controls the average time step.

The Green's function is not known in close form, so it has to be sampled by MC. This can be done by rewriting the resolvent in the form

$$\frac{1}{\hat{H} + E_C} = \frac{1}{\hat{H}_U + E_C} + \frac{1}{\hat{H}_U + E_C}(\hat{H}_U - \hat{H})\frac{1}{\hat{H} + E_C}. \tag{5.44}$$

The Hamiltonian $\hat{H}_U$ represents a family of solvable Hamiltonians. To sample the Green's function, one samples the sum of terms on the right-hand side of Eq. (5.44). The Green's functions associated with $\hat{H}$ and $\hat{H}_U$ satisfy the relations

$$(\hat{H} + E_C)G(\mathbf{R}, \mathbf{R}') = \delta(\mathbf{R} - \mathbf{R}'), \tag{5.45}$$

$$(\hat{H}_U + E_C)G_U(\mathbf{R}, \mathbf{R}') = \delta(\mathbf{R} - \mathbf{R}'). \tag{5.46}$$

The most commonly used form of $\hat{H}_U$ is

$$\hat{H}_U = \frac{1}{2}\nabla_R^2 + U, \tag{5.47}$$

where U is a potential that is independent of $\mathbf{R}$. It is convenient to have $G_U(\mathbf{R}, \mathbf{R}')$ vanish at the domain boundary. $\hat{H}_U$ should be a good approximation to $\hat{H}$ in the domain to achieve good convergence.

The $\mathbf{R}$-space representation of Eq. (5.44) is

$$G(\mathbf{R}, \mathbf{R}') = G_U(\mathbf{R}, \mathbf{R}') - \int_S d\mathbf{R}'' G(\mathbf{R}, \mathbf{R}'')\left[-\hat{n}\cdot\nabla G_U(\mathbf{R}'', \mathbf{R})\right]$$
$$+ \int_V d\mathbf{R}'' G(\mathbf{R}, \mathbf{R}'')[U - V(\mathbf{R}'')]G_U(\mathbf{R}'', \mathbf{R}'). \tag{5.48}$$

5.6. Fixed-node approximation

We have not discussed the implications of the fermion character of $\Psi(\mathbf{R})$. It is an excited state in a manifold containing all the fermionic and bosonic states. A fermion wave function has positive and negative regions that are difficult to sample with the DMC algorithm as described in Section 5.4. Considering real wave functions, $\Psi(\mathbf{R})$ contains positive and negative regions, $\Psi^+(\mathbf{R})$, and $\Psi^-(\mathbf{R})$ that, in principle, could be represented as probabilities. The sign of the wave function could be used as an extra weight for the random walk. In practice, this is a very slowly convergent method.

Returning to the importance sampled algorithm, recall that the initial distribution, $|\Psi(\mathbf{R})|^2$, is positive. Nevertheless, the Green's function, $K(\mathbf{R}, \mathbf{R}')$, can become negative, if a random walker crosses a node of the trial wave function. Again, the sign of $K(\mathbf{R}, \mathbf{R}')$ could be used as a weight for sampling $|K(\mathbf{R}, \mathbf{R}')|$. The problem is that the statistics of this process lead to exponential growth of the variance of any observable.

The simplest approach to avoid exponential growth is to forbid moves in which the product wave function, $\Psi(\mathbf{R})\Psi_T(\mathbf{R})$, changes sign. This boundary condition on permitted moves is the defining characteristic of the fixed-node approximation (FNA). The nodes of the sampled wave function are *fixed* to be the nodes of the trial wave function. The FNA is an inherent feature of the DMC method, which is, by far, the most commonly used method for atomic and molecular MC applications (ANDERSON [1976], REYNOLDS, CEPERLEY, ALDER and LESTER, JR. [1982]).

The fixed-node energy is an upper bound to the exact energy of the system. In fact, it is the best solution for that fixed set of nodes. The DMC method has much higher accuracy than the VMC method. For atomic and molecular systems, it is common to recover 95–100% of the CE, cf. Section 1, whereas the CE recovered with the VMC approach is typically less than 80% of the total.

5.7. Exact methods

Probably the most important algorithmic challenge that still remains to be explored is the "node problem". Although progress has been made in systems that contain up to a dozen of electrons (ARNOW, KALOS, LEE and SCHMIDT [1982], CEPERLEY and ALDER [1984], ZHANG and KALOS [1991], ANDERSON, TRAYNOR and BOGHOSIAN [1991], LIU, ZHANG and KALOS [1994]), a stable algorithm that can sample the exact wave function without resorting to the FNA remains to be determined. In this section, we discuss a family of methods that avoid the FNA. These approaches yield exact answers, usually associated with a large increase in computational time.

The Pauli antisymmetry principle imposes a boundary condition on the wave function. It is the requirement that the exchange of like-spin electrons changes the sign of the wave function. This condition is a global condition that has to be enforced within an algorithm that only considers individual random walkers, that is, a local algorithm. The FNA is the most commonly imposed boundary condition. It satisfies the variational principle, that is, FN solutions approach the exact energy from above. This is an useful property, but one that does not assist the search for exact results, because there is not an easy way to parametrize the nodal surface and vary it to obtain the exact solution. We now describe methods that impose no additional boundary conditions on the wave function.

5.7.1. The release node method
The evolution operator, $e^{-\tau(\hat{H}-E_T)}$, is symmetric and has the same form for both fermions and bosons. Straightforward application of it to an arbitrary initial wave function, $|\Psi_0\rangle$, leads to collapse to the ground state bosonic wave function, as can be seen from Eq. (5.20).

An arbitrary fermion wave function, $\Psi(\mathbf{R})$, can be separated into two functions $\Psi^+(\mathbf{R})$ and $\Psi^-(\mathbf{R})$ as follows,

$$\Psi^\pm(\mathbf{R}, \tau) \equiv \frac{1}{2}[|\Psi(\mathbf{R}, \tau)| \pm \Psi(\mathbf{R}, \tau)]. \tag{5.49}$$

Note that the original trial wave function is recovered as

$$\Psi(\mathbf{R}, \tau) = \Psi^+(\mathbf{R}, \tau) - \Psi^+(\mathbf{R}, \tau). \tag{5.50}$$

The released node (RN) algorithm involves two independent DMC calculations, using Ψ^+ and Ψ^- as the wave functions to evolve

$$\Psi(\mathbf{R}, \tau) = \int G\left(\mathbf{R}, \mathbf{R}', \delta\tau\right)\Psi\left(\mathbf{R}', 0\right)d\mathbf{R}'$$

$$= \int G\left(\mathbf{R}, \mathbf{R}', \tau\right)\Psi^+\left(\mathbf{R}', 0\right)d\mathbf{R}' - \int G\left(\mathbf{R}, \mathbf{R}', \tau\right)\Psi^-\left(\mathbf{R}', 0\right)d\mathbf{R}'$$

$$= \Psi^+(\mathbf{R}, \tau) - \Psi^-(\mathbf{R}, \tau). \tag{5.51}$$

The time evolution of the system can be followed from the difference of separate simulations for $\Psi^\pm(\mathbf{R})$. Note that both distributions are always positive during the simulation, and that they decay to the ground state bosonic wave function. This decay is problematic because the "signal-to-noise" ratio in this method depends on the difference between these two distributions. The decay of the difference $\Psi^+(\mathbf{R}, \tau) - \Psi^-(\mathbf{R}, \tau)$ goes roughly as $e^{\tau(E_F - E_B)}$, where E_F is the lowest fermion state energy and E_B is the bosonic ground state energy.

For this method to be practical, one needs to start with the distribution of a good fermion trial wave function. The distribution will evolve from this starting point to the bosonic ground state at large imaginary time τ. In an intermediate "transient" regime one can collect information on the exact fermion wave function.

The energy can be estimated from the expression

$$E_{\mathrm{RN}}(\tau) = \frac{\int \Psi(\mathbf{R}, \tau)\hat{H}\Psi_T(\mathbf{R})d\mathbf{R}}{\int \Psi(\mathbf{R}, \tau)\Psi_T d\mathbf{R}}$$

$$= \frac{\int \Psi^+(\mathbf{R}, \tau)\hat{H}\Psi_T(\mathbf{R})d\mathbf{R}}{\int[\Psi^+(\mathbf{R}, \tau) - \Psi^-(\mathbf{R}, \tau)]\Psi_T(\mathbf{R})d\mathbf{R}} \tag{5.52}$$

$$- \frac{\int \Psi^-(\mathbf{R}, \tau)\hat{H}\Psi_T(\mathbf{R})d\mathbf{R}}{\int[\Psi^+(\mathbf{R}, \tau) - \Psi^-(\mathbf{R}, \tau)]\Psi_T(\mathbf{R})d\mathbf{R}}$$

$$= E_F.$$

In the release node method (CEPERLEY and ALDER [1984]), a fixed-node distribution is propagated as usual, but now two sets of random walkers are retained: $\mathcal{W}_{\mathrm{FN}}$, the fixed node ensemble, and $\mathcal{W}_{\mathrm{RN}}$, the RN ensemble. Walkers are allowed to cross nodes, and when they do, they are transferred from $\mathcal{W}_{\mathrm{FN}}$ to $\mathcal{W}_{\mathrm{RN}}$. Also a account is made of the number of iterations that a walker has survived, $\mathcal{S}_{\mathrm{RN}} = \{s_1, \ldots, s_{N_w}\}$.

This index is used to bin the walkers by age. Each time a walker crosses a node, a summation weight associated with it, $\Omega_{RN} = \{\omega_1, \ldots, \omega_{N_w}\}$ changes sign. These weights determine the sign of the walker contribution to global averages.

The RN energy can be calculated using the estimator,

$$E_{RN} = \frac{\sum_{i=1}^{N_w} \omega_i \frac{\Psi_T(\mathbf{R}_i)}{\Psi(\mathbf{R}_i)} E_L(\mathbf{R}_i)}{\sum_{i=1}^{N_w} \omega_i \frac{\Psi_T(\mathbf{R}_i)}{\Psi(\mathbf{R}_i)}}. \tag{5.53}$$

5.7.2. Fermion Monte Carlo

From the previous section one can infer that if a method in which the distribution does not go to the bosonic ground state, but stays in an intermediate regime, will not have the deficiency of exponential growth of "signal to noise". This leads to the fermion Monte Carlo (FMC) method. The approach by KALOS and SCHMIDT [1997], KALOS and PEDERIVA [1998], PEDERIVA and KALOS [1999], KALOS and PEDERIVA [2000a], KALOS and PEDERIVA [2000b] involves correlated random walks that achieve a constant "signal to noise".

The expectation value of Eq. (5.53) for an arbitrary distribution of signed walkers can be rewritten as

$$\langle E_{FMC} \rangle = \frac{\sum_{i=1}^{N_w} \left[\frac{\hat{H}\psi_T(\mathbf{R}_i^+)}{\Psi_G^+(R_i^+)} - \frac{\hat{H}\psi_T(\mathbf{R}_i^-)}{\Psi_G^-(R_i^-)} \right]}{\sum_{i=1}^{N_w} \left[\frac{\psi_T(R_i^+)}{\psi_G^+(R_i^+)} \frac{\psi_T(R_i^-)}{\Psi_G^-(R_i^-)} \right]}, \tag{5.54}$$

where $\Psi_G^\pm(\mathbf{R}^\pm)$ are the guiding functions for a pair of random walkers $P_i = \{\mathbf{R}_i^+, \mathbf{R}_i^-\}$. Note that the variance of the energy estimator of Eq. (5.54) goes to infinity as the difference between the two populations goes to zero, that is, the denominator,

$$\mathcal{D} \equiv \sum_{i=1}^{N_w} \left[\frac{\psi_T(R_i^+)}{\Psi_G^+(R_i^+)} \frac{\psi_T(R_i^-)}{\Psi_G^-(R_i^-)} \right], \tag{5.55}$$

goes to zero as the simulation approaches to the bosonic ground state. A procedure that would not change $\langle E_{FMC} \rangle$ would be to cancel positive and negative random walkers whenever they meet (ARNOW, KALOS, LEE and SCHMIDT [1982]). Although random walks are guaranteed to meet in one dimension, they need not meet in several dimensions, due to the exponentially decaying walker density in $\mathbf{R}$-space. Besides, cancellation has to be combined with other procedures to insure a stable algorithm.

Cancellation can be increased by introducing correlation between the random walkers. Recall the diffusion step in DMC, in which walkers diffuse from $\mathbf{R}$ to $\mathbf{R}'$ following G_D of Eq. (5.25). In the DMC algorithm, this is implemented stochastically by updating the coordinates of the random walkers with a random displacement taken from a Gaussian distribution with a variance of $\delta\tau$,

$$\mathbf{R}'^+ \rightarrow \mathbf{R}^+ + \mathcal{G}_{\delta\tau}^+ \text{ and } \mathbf{R}^- \rightarrow \mathbf{R}^- + \mathcal{G}_{\delta\tau}^-. \tag{5.56}$$

If we introduce correlation between the Gaussian vectors, $\mathcal{G}_{\delta\tau}^+$ and $\mathcal{G}_{\delta\tau}^-$, the expectation value of Eq. (5.54) is not affected, because it is linear in the density of random walkers.

An efficient cancellation scheme can be achieved if the Gaussian vectors are correlated as follows:

$$\mathcal{G}_{\delta\tau}^- = \mathcal{G}_{\delta\tau}^+ U^- - 2\left(\mathcal{G}_{\delta\tau}^+ \cdot \frac{(\mathbf{R}^+ - \mathbf{R}^-)}{|\mathbf{R}^+ - \mathbf{R}^-|^2}\right) \cdot (\mathbf{R}^+ - \mathbf{R}^-). \tag{5.57}$$

Equation (5.57) accounts for reflection along the perpendicular bisector of the vector that connects the pair, $\mathbf{R}^+ - \mathbf{R}^-$. This cancellation scheme generates a correlated random walk in one dimension along the vector $\mathbf{R}^+ - \mathbf{R}^-$. This one-dimensional random walk is independent of the number of dimensions of the physical system, and therefore, overcomes the cancellation difficulties mentioned above. Walkers are guaranteed to meet under these conditions.

The modifications to the DMC algorithm mentioned to this point are necessary, but not sufficient for achieving a stable algorithm. If one were to interchange the random walker populations, $\left\{\mathbf{R}_1^+, \ldots, \mathbf{R}_{N_w}^+\right\} \leftrightarrow \left\{\mathbf{R}_1^-, \ldots, \mathbf{R}_{N_w}^-\right\}$, the fictitious dynamics would not be able to distinguish between the two populations, leading to a random walk with two degenerate ground states. Namely, a ground state in which all the positive walkers $\mathbf{R}^+$ are marginally on the positive region of the wave function, and vice versa,

$\{\Psi^+ (\mathbf{R}^+), \Psi^-(\mathbf{R}^-)\}$ and $\{\Psi^+ \{\mathbf{R}^-\}, \Psi^-(\mathbf{R}^+)\}$. This *plus–minus* symmetry can be broken by using two distinct guiding functions. For example, the guiding function

$$\Psi_G^\pm = \sqrt{\Psi_S^2(\mathbf{R}) + c^2\Psi_A^2(\mathbf{R})} \pm c\Psi_A(\mathbf{R}), \tag{5.58}$$

where $\Psi_S(\mathbf{R})$ is a symmetric function under permutation of electron labels; $\Psi_A(\mathbf{R})$ is an antisymmetric function, and c is a small adjustable parameter. The guiding functions of Eq. (5.58) are almost equal, which provides nearly identical branching factors for the walker pair. It is positive everywhere, a requirement for the DMC algorithm, and it is symmetric under permutation of the coordinates, $\Psi_G^+(\hat{P}R) = \Psi_G^-(\mathbf{R})$.

The use of different guiding functions is the last required ingredient for a stable algorithm. It breaks the *plus–minus* symmetry effectively, because the drift dynamics is different because the quantum force of Eq. (3.8) is distinct for each population.

For a complete description of the FMC algorithm, the reader is referred to KALOS and PEDERIVA [2000a].

The denominator of Eq. (5.55) is an indicator of stability of the algorithm. It is a measure of the antisymmetric component of the wave function. FMC calculations have shown stable denominators for thousands of relaxation times, indicating the stability of the fermion algorithm.

Early versions of the method (ARNOW, KALOS, LEE and SCHMIDT [1982]) do not scale well with system size, due to the use of uncorrelated cancellation schemes. Nevertheless, researchers have been applied successfully to several small molecular systems obtaining solutions to the Schrödinger equation with no systematic error (ZHANG and KALOS [1991], BHATTACHARYA and ANDERSON [1994a], BHATTACHARYA and ANDERSON [1994b], ANDERSON [2001]). This version of the FMC algorithm, with

GFMC propagation and without correlated dynamics, is known as exact quantum Monte Carlo (EQMC).

5.8. Zero variance principle

An increase in computational efficiency can be achieved by improving the observables $\hat{O}$ by renormalizing them to observables that have the same expectation value, but lower variance. Recent work by ASSARAF and CAFFAREL [1999], ASSARAF and CAFFAREL [2000] has shown that estimators for the energy and energy derivatives with respect to nuclear coordinates can be constructed.

One can propose a trial operator $\hat{H}_V$ and auxiliary trial function Ψ_V such that the evaluation of a renormalized observable $\bar{O}$ will have a variance that is smaller than that of the original observable $\hat{O}$, and in principle can even be suppressed.

To develop this concept, let us construct a trial operator $\hat{H}_V$ such that,

$$\int \hat{H}_V(\mathbf{R}, \mathbf{R}')\sqrt{\pi(\mathbf{R}')}d\mathbf{R}' = 0, \tag{5.59}$$

where $\pi(\mathbf{R}')$ is the MC distribution. For example, in VMC the MC distribution is the wave function squared, $\Psi_T(\mathbf{R})^2$, and in DMC it is the mixed distribution of Eq. (5.29). Next, propose a renormalized observable $\bar{O}(\mathbf{R})$ related to the observable $\hat{O}(\mathbf{R})$ given by

$$\bar{O}(\mathbf{R}) = \hat{O}(\mathbf{R}) + \frac{\int \hat{H}_V(\mathbf{R}, \mathbf{R}')\psi_V(\mathbf{R}')d\mathbf{R}'}{\sqrt{\pi(\mathbf{R}')}}. \tag{5.60}$$

The mean of the rescaled operator is formally

$$\langle \bar{O} \rangle = \frac{\int \hat{O}(\mathbf{R})\pi(\mathbf{R})d\mathbf{R} + \dfrac{\int\int \pi(\mathbf{R})\hat{H}_V(\mathbf{R}, \mathbf{R}')\Psi_V(\mathbf{R}')d\mathbf{R}d\mathbf{R}'}{\sqrt{\pi(\mathbf{R})}}}{\int \pi(\mathbf{R})d\mathbf{R}} \tag{5.61}$$

which, by property (5.59), is the same as the mean for the unnormalized operator:

$$\langle \bar{O} \rangle = \langle \hat{O} \rangle. \tag{5.62}$$

Operator $\bar{O}$ can be used as an unbiased estimator, even though statistical errors for $\bar{O}$ and $\hat{O}$ can be quite different. The goal of this kind of importance sampling is to reduce the fluctuations by construction of such an operator.

The implementation of the procedure requires the optimization of a set of parameters for the auxiliary trial wave function, $\Psi_V(\mathbf{R}, \Lambda_V)$, using the minimization functional

$$\int \hat{H}_V(\mathbf{R}, \mathbf{R}')\Psi_V(\mathbf{R}, \Lambda_V)d\mathbf{R}' = -[\bar{O}(x) - \langle \bar{O} \rangle]\sqrt{\pi(\mathbf{R})}. \tag{5.63}$$

After the parameters Λ_V are optimized, one can run a simulation to average $\bar{O}$, instead of $\hat{O}$. The choice of auxiliary Hamiltonian suggested by recent work of ASSARAF and CAFFAREL [1999] is

$$H_V(x,y) = -\frac{1}{2}\nabla_{\mathbf{R}}^2 + \frac{1}{2\sqrt{\pi(\mathbf{R})}}\nabla_{\mathbf{R}}^2\sqrt{\pi(\mathbf{R})}. \tag{5.64}$$

Note that when Eq. (5.64) is applied to $\sqrt{\pi(\mathbf{R})}$, the $\mathbf{R}'$ integration vanishes by construction. The choice of auxiliary wave function is open, and an interesting observation is that minimization of the normalization factor of $\Psi_V(\mathbf{R})$, for any choice of auxiliary trial wave function, will reduce fluctuations in the auxiliary observable

$$\sigma(\bar{O})^2 = \sigma(\hat{O})^2 - \frac{\left\langle \dfrac{\hat{O}(\mathbf{R})\int \hat{H}(\mathbf{R},\mathbf{R}')\Psi_V(\mathbf{R}')\mathrm{d}\mathbf{R}'}{\sqrt{\pi(\mathbf{R})}} \right\rangle^2}{\left\langle \left[\dfrac{\int \hat{H}(\mathbf{R},\mathbf{R}')\Psi_V(\mathbf{R}')\mathrm{d}\mathbf{R}'}{\sqrt{\pi(\mathbf{R})}} \right]^2 \right\rangle}, \tag{5.65}$$

because the second term on the right hand side of Eq. (5.65) always has a negative sign.

This variance reduction technique, applied to VMC and GFMC simulations, has achieved an order of magnitude reduction in computational effort (ASSARAF and CAFFAREL [1999]). It can also be used to calculate energy derivatives (ASSARAF and CAFFAREL [2000]).

Part III Special topics

6. Fermion nodes

As discussed briefly in Section 5, the simulation of a quantum system without approximation, obtaining exact results other than the numerical integration scheme's associated error bar is still an open research topic. Several solutions have been proposed, but the challenge is to have a general method that scales favorably with system size.

For the ground state of a bosonic system, for which the wave function has the same sign everywhere, QMC provides an exact solution in a polynomial amount of computer time, that is, is already a solved problem. The research in this field attempts to obtain an algorithm that has the same properties but that can treat wave functions that have both positive and negative regions, and therefore nodes, with the same favorable scaling.

Investigation of nodes has been pursued by CAFFAREL, CLAVERIE, MIJOULE, ANDZELM and SALAHUB [1989], CEPERLEY [1991], GLAUSER, BROWN, LESTER, JR., BRESSANINI, HAMMOND and KOSYKOWSKI [1992], CAFFAREL, KROKIDIS and MIJOULE [1992], BRESSANINI, CEPERLEY and REYNOLDS [2002] to understand the properties of the nodes of fermion wave functions.

The full nodal hyper-surfaces of a wave function, $\Psi(\mathbf{R})$, where $\mathbf{R}$ is a $3N$-dimensional vector and N is the number of fermions in the system is a $3(N-1)$-dimensional function, $\eta(\mathbf{R})$. Of that function, symmetry requirements determine a $(3N-3)$-dimensional surface, the *symmetry* subsurface, $\sigma(\mathbf{R})$. This is unfortunate, because even though that $\sigma(\mathbf{R}) \subset \eta(\mathbf{R})$, the remainder of the nodal surface, the *peculiar* nodal surface, $\mathcal{W}(\mathbf{R})$ which is a function of the specific form of the nuclear and interelectronic potential, is difficult to be known *a priori* for an arbitrary system. Note that $\sigma(\mathbf{R}) \cap \mathcal{W}(\mathbf{R}) = \eta(\mathbf{R})$.

Understanding nodal properties is important for further development of QMC methods: these shall be exploited for bypassing the node problem.

CEPERLEY [1991] discusses general properties of wave function nodes. General properties of nodes follow.

(1) The *coincidence planes* $\pi(\mathbf{r}_i = \mathbf{r}_j)$, are located at nodes when the two electrons have the same spin, that is, $\delta_{\sigma_{ij}} = 1$. In more than one dimension, $\pi(\mathbf{R})$ is a scaffolding where the complex nodal surface passes through. Note that $\pi(\mathbf{R}) \subset \sigma(\mathbf{R})$.

(2) The nodes possess all the symmetries of the ground state wave function.

(3) The nodes of the many-body wave function are distinct from orbital nodes $\phi_i(\mathbf{r})$, see Section 3.2.

(4) For degenerate wave functions, the node positions are arbitrary. For a p-fold degenerate energy level, one can pick $p - 1$ points in $\mathbf{R}$ and find a linear transformation for which the transformed wave functions vanish at all but one of these points.

(5) A *nodal cell* $\Omega(\mathbf{R})$ around a point $\mathbf{R}$ is defined as the set of points that can be reached from $\mathbf{R}$ without crossing a node. For potentials of our present interest, the ground state nodal cells have the *tiling-property*: any point $\mathbf{R}'$ not on the node is related by symmetry to a point in $\Omega(\mathbf{R})$. This implies that there is only one type of nodal cell: all other cells are copies that can be accessed by relabeling the particles. This property is the generalization to fermions of the theorem that the bosonic ground state is nodewave functionless.

CEPERLEY [1991] suggests that for DMC simulations benefit from the tiling property: one only needs to sample one nodal cell, because all the other cells are copies of the first. Any trial function resulting from a strictly mean field theory will satisfy the tiling property. Such as the LDA wave functions.

GLAUSER, BROWN, LESTER, JR., BRESSANINI, HAMMOND and KOSYKOWSKI [1992] showed that simple HF wave functions for the first-row atoms were shown to have four nodal regions (two nodal surfaces intersecting) instead of two. This is attributed to factorizing the wave function into two distinct Slater determinants, $D^{\uparrow}$ and $D^{\downarrow}$, each composed of two surfaces, one for the $\uparrow$ and one for the $\downarrow$ electron, as discussed in Section 3.2.

Recently, after analysis of the wave functions for He, Li and Be, it was conjectured by BRESSANINI, CEPERLEY and REYNOLDS [2002], that the wave function can be factored as follows:

$$\Psi(\mathbf{R}) = N(\mathbf{R})e^{f(\mathbf{R})}, \tag{6.1}$$

where $N(\mathbf{R})$ is antisymmetric polynomial of finite order, and $f(\mathbf{R})$ is a positive definite function. A weaker conjecture is that N may not be a polynomial, but can be closely approximated by a lower order antisymmetric polynomial. The variables in which N should be expanded are the interparticle coordinates.

For example, for all ^3S states of two-electron atoms, the nodal factor $N(\mathbf{R})$ in Eq. (6.1) is

$$N(\mathbf{r}_1, \mathbf{r}_2) = \mathbf{r}_1 - \mathbf{r}_2, \tag{6.2}$$

where $\mathbf{r}_1$ and $\mathbf{r}_2$ are the coordinates of the two electrons.

7. Treatment of heavy elements

So far, we have not discussed the applicability of QMC to systems with large atomic number. There is an steep computational scaling of QMC methods, with respect to the atomic number Z. The computational cost of QMC methods has been estimated to scale as $Z^{5.5-6.5}$ (HAMMOND, REYNOLDS and LESTER, JR. [1987], CEPERLEY [1986]). This has motivated the replacement of the core electrons by ECPs. With this modification, the scaling with respect to atomic number is improved to $Z^{3.4}$ (HAMMOND, REYNOLDS and LESTER, JR. [1987]). Other approaches involve the use of core-valence separation schemes (STAROVEROV, LANGFELDER, and ROTHSTEIN [1998]) and the use of model potentials (YOSHIDA and IGUCHI [1988]).

7.1. Effective core potentials

In the ECP method (SZASZ [1985], KRAUSS and STEVENS [1984], CHRISTIANSEN, ERMLER and PITZER [1984], BALASUBRAMANIAN and PITZER [1987], DOLG [2000]), the effect of the core electrons is simulated by an effective potential acting on the valence electrons. The effective Hamiltonian for these electrons is:

$$\mathcal{H}_{\text{val}} = \sum_i \frac{-Z_{\text{eff}}}{r_i} + \sum_{i<j} \frac{1}{r_{ij}} + \sum_i \mathcal{W}_i(\mathbf{r}), \tag{7.1}$$

where i and j designate the valence electrons, Z_{eff} is the effective nuclear charge in the absence of core electrons, and $\mathcal{W}$ is the pseudopotential operator. The latter can be written,

$$\mathcal{W}(\mathbf{r}) = \sum_{l=0}^{\infty} \mathcal{W}_l(\mathbf{r}) \sum_m |lm\rangle\langle lm|, \tag{7.2}$$

where l and m are the angular momentum and magnetic quantum numbers. The projection operator $\sum_m |lm\rangle\langle lm|$, connects the pseudopotential with the one-electron valence functions. A common approximation to this equation is to assume that the angular momentum components of the pseudopotential, $w_l(\mathbf{r})$ do not depend on l when $l > L$, the angular momentum of the core. This approximation leads to the expression

$$\mathcal{W}(\mathbf{r}) = \mathcal{W}_{L+1}(\mathbf{r}) + \sum_{l=0}^{L} (\mathcal{W}_l(\mathbf{r}) - \mathcal{W}_{L+1}(\mathbf{r})) \sum_m |lm\rangle\langle lm|. \tag{7.3}$$

The operator (7.3) can be applied to a valence orbital, that is, pseudo-orbital, $\phi_l(\mathbf{r})$. This function is usually represented by a polynomial expansion for distances less than a cutoff radius, $r < r_c$, and by a fit to the all-electron orbital for $r > r_c$.

Rapid fluctuations in the potential terms can cause the first order approximation of Eq. (5.28) to break down, therefore, seeking a slowly varying form of ECP is relevant to QMC simulations. GREEFF and LESTER, JR. [1998] proposed the use of normconserving soft ECPs for QMC. Soft ECPs derive their name from the property of being finite at the nucleus, this leads to a pseudo-orbital with no singularities at the origin in the kinetic energy. The associated effective potential has no discontinuities or divergences.

7.2. *Embedding methods*

A commonly used approach in wave function based methods, is to use embedding schemes, in which a region of high interest of a large system is treated by an accurate procedure, and the remainder is described by a less accurate method. Recent work by FLAD, SCHAUTZ, WANG and DOLG [2002] has extended the methodology to QMC methods. In this approach, a mean field calculation, for example HF, is performed for the whole system. An electron localization procedure is performed, the orbitals to be correlated are chosen and separated from the remaining core orbitals. An effective Coulomb and exchange potential is constructed, $\hat{V}_E$, which is added to the standard Hamiltonian of Eq. (0.1) to construct an effective Hamiltonian, $\hat{H}_E$, that then is used in QMC calculations. Localization procedures similar to those required for ECPs are needed for representing the effect of nonlocal terms.

The effective Hamiltonian, $\hat{H}_E$, takes the form

$$\hat{H}_E = \hat{H}_{int} + \hat{V}_{ext} + \hat{J}_{ext} + \hat{K}_{ext} + \hat{S}_{ext} \tag{7.4}$$

where $\hat{H}_{int}$ is the Hamiltonian for the QMC active region, $\hat{V}_{ext}$ is the Coulomb potential exerted by the external nuclei, $\hat{J}_{ext}$ represents the Coulomb repulsions, The term $\hat{K}_{ext}$ represents the exchange interactions, and $\hat{S}_{ext}$ is a shift operator that prevents the wave function to be expanded into core orbitals, ϕ_c, by shifting their energy to infinity, and is given by

$$\hat{S}_{ext} = \lim_{\lambda \to \infty} \lambda \sum_{\alpha}^{int} \sum_{\beta}^{ext} |\phi_p(\mathbf{r})\rangle\langle\phi_p(\mathbf{r})| d\mathbf{r}. \tag{7.5}$$

Here λ is an effective orbital coupling constant that is derived from considering single and double excitations into core and virtual orbitals of the system. The Coulomb term, $\hat{J}_{ext}$, and the external Coulomb potential, $\hat{V}_{ext}$, are local potentials, and can be evaluated within QMC without further approximation. The remaining terms require localization approximations that are discussed in detail in the original work.

8. Other properties

8.1. *Exact wave function and quantities that do not commute with the Hamiltonian*

Properties related to a trial wave function $\Psi_T(\mathbf{R})$, are readily available in VMC calculations ALEXANDER, COLDWELL, AISSING and THAKKAR [1992]. In this case, expectation values are calculated from directly from Eqs. (3.4) and (3.5). The accuracy of the results obtained with VMC depend on the quality of $\Psi_T(\mathbf{R})$.

For obtaining expectation values of operators that do not commute with the Hamiltonian in an importance sampled PMC calculation, one needs to extract the exact distribution $\Psi^2(\mathbf{R})$ from the mixed distribution $f(\mathbf{R}) = \Psi(\mathbf{R}) \Psi_T(\mathbf{R})$. The expectation values for an operator $\hat{O}$, $\langle\Psi(\mathbf{R})|\hat{O}|\Psi_T(\mathbf{R})\rangle$ and $\langle\Psi_T(\mathbf{R})|\hat{O}|\Psi(\mathbf{R})\rangle$, are different to each other. MC sampling requires knowledge of the exact ground state distribution: a mixed distribution does not suffice to obtain the exact answer.

If the operator $\hat{O}$ is a multiplicative operator, then the algorithms described in this section will be pertinent. Nonmultiplicative operators, which are exemplified by forces in this chapter, are described in Section 8.2.

8.1.1. Extrapolation method

An approximate procedure for estimating the ground state distribution can be extrapolated from the mixed and VMC distributions. This procedure is valuable because no extra changes are needed to the canonical VMC and PMC algorithms. Being an approximate method, it can fail even in very simple cases (SARSA, SCHMIDT and MAGRO [2000]), but also has provided very accurate results in more favorable cases (LIU, KALOS and CHESTER [1974]). The mixed estimator of a coordinate operator $\hat{O}$ is

$$\langle\hat{O}\rangle_m = \frac{\int \Psi(\mathbf{R})\hat{O}\Psi_T(\mathbf{R})d\mathbf{R}}{\int \Psi(\mathbf{R})\Psi_T(\mathbf{R})d\mathbf{R}}, \tag{8.1}$$

to be distinguished from the pure estimator

$$\langle\hat{O}\rangle_p = \frac{\int \Psi(\mathbf{R})\hat{O}\Psi(\mathbf{R})d\mathbf{R}}{\int \Psi(\mathbf{R})\Psi(\mathbf{R})d\mathbf{R}}. \tag{8.2}$$

We will label $\langle\hat{O}\rangle_v$ the VMC estimator of Eq. (3.4). $\langle\hat{O}\rangle_m$ can be rewritten in a Taylor series in the difference between the exact and approximate wave functions, $\delta\Psi \equiv \Psi(\mathbf{R}) - \Psi_T(\mathbf{R})$,

$$\langle\hat{O}\rangle_m = \langle\hat{O}\rangle_p + \int \Psi(\langle\hat{O}\rangle_p - \hat{O}(\mathbf{R}))\delta\Psi d\mathbf{R} + \mathcal{O}((\delta\Psi)^2). \tag{8.3}$$

A similar expansion can be constructed for $\langle\hat{O}\rangle_v$,

$$\langle\hat{O}\rangle_v = \langle\hat{O}\rangle_p + 2\int \Psi(\langle\hat{O}\rangle_p - \hat{O}(\mathbf{R}))\delta\Psi d\mathbf{R} + \mathcal{O}((\delta\Psi)^2). \tag{8.4}$$

Combining Eqs. (8.3) and (8.4), we can arrive to an expression with a second order error,

$$\langle\hat{O}\rangle_e = 2\langle\hat{O}\rangle_m - \langle\hat{O}\rangle_v = \langle\hat{O}\rangle_p + \mathcal{O}((\delta\Psi)^2), \tag{8.5}$$

where $\langle\hat{O}\rangle_e$ is an extrapolation estimate readily available from VMC and PMC calculations.

8.1.2. Future walking

The future walking method can be combined with any importance sampled PMC method that leads to a mixed distribution. If one multiplies both sides of Eq. (8.1) by the ratio

$\Psi(\mathbf{R})/\Psi_T(\mathbf{R})$, one can recover Eq. (8.2). The ratio is obtained from the asymptotic population of descendants of a single walker (BARNETT, REYNOLDS and LESTER, JR. [1991]).

A walker in **R**-space can be represented as a sum of eigenfunctions of $\hat{H}$:

$$\delta(\mathbf{R}' - \mathbf{R}) = \Psi(\mathbf{R}') \sum_{n=0}^{\infty} c_i(\mathbf{R}) \Psi_n(\mathbf{R}). \tag{8.6}$$

The coefficients $c_i(\mathbf{R})$ can be obtained by multiplying Eq. (8.6) by $\Psi(\mathbf{R}')/\Psi_T(\mathbf{R}')$ and integrating over $\mathbf{R}'$:

$$c_n(\mathbf{R}) = \int \delta(\mathbf{R}' - \mathbf{R}) \frac{\Psi(\mathbf{R}')}{\Psi_T(\mathbf{R}')} d\mathbf{R}' = \frac{\Psi(\mathbf{R})}{\Psi_T(\mathbf{R})}. \tag{8.7}$$

Clearly, we want to know the contribution to the ground state wave function, $c_0(\mathbf{R})$ of the walker at **R**. If propagated for sufficiently long time, all coefficients $c_i(\mathbf{R}) \neq c_0(\mathbf{R})$ for the random walker will vanish. This can be seen from the decay in τ of Eqs. (5.19) and (5.20).

If we define $P_\infty(\mathbf{R})$ to be the asymptotic population of walkers descended from a random walker at **R**, we find,

$$\begin{aligned} P_\alpha(R) &= \int c_0(R) e^{-(E_0 - E_T)\tau} \Psi(R') \Psi_T(R') dR' \\ &= \frac{\Psi(R)}{\Psi_T(R)} e^{-(E_0 - E_T)\tau} \langle \Psi(R) | \Psi_T(R) \rangle \end{aligned} \tag{8.8}$$

For obtaining $P_\infty(\mathbf{R})$ in a PMC algorithm, one needs to keep a list of all the descendants of each walker $\mathbf{R}_i$ at each time step τ_j. The number of steps for which one requires to keep track of the descendants, N_d, is a critical parameter. The statistical error of the asymptotic walker population grows in the limit $N_d \to \infty$, and if only few steps are used, a bias is encountered by nonvanishing contributions from excited states $c_i(\mathbf{R}) \neq c_0(\mathbf{R})$. Efficient algorithms for keeping track of the number of descendants can be found in the literature (LIU, KALOS and CHESTER [1974], REYNOLDS, BARNETT, HAMMOND, GRIMES and LESTER, JR. [1986], EAST, ROTHSTEIN and VRBIK [1988], CAFFAREL and CLAVERIE [1988], BARNETT, REYNOLDS and LESTER, JR. [1991], HAMMOND, LESTER and REYNOLDS [1994], LANGFELDER, ROTHSTEIN and VRBIK [1997]).

The wave function overlap with the ground state can also be obtained with these methods, as shown by HORNIK, SNAJDR and ROTHSTEIN [2000]. These methods have been applied for obtaining dipole moments (SCHAUTZ and FLAD [1999]), transition dipole moments (BARNETT, REYNOLDS and LESTER, JR. [1992a]) and oscillator strengths (BARNETT, REYNOLDS and LESTER, JR. [1992b]), among other applications.

Other methods for obtaining the exact distribution that are not discussed here for reasons of space are bilinear methods (ZHANG and KALOS [1993]), and time correlation methods (CEPERLEY and BERNU [1988]).

8.2. Force calculation

Most QMC applications have been within the BO (BORN and OPPENHEIMER [1927]) approximation. In this approximation, the nuclear coordinates $\mathcal{R}$ are fixed at a certain position during the calculation.[9] Therefore, the wave function and energy depends parametrically on the nuclear coordinates $E(\mathcal{R})$ and $\Psi(\mathbf{R}, \mathcal{R})$. We will omit this parametric dependence for the remainder of the discussion, and simplify the symbols to E and $\Psi(\mathbf{R})$, when appropriate.

Forces are derivatives of the energy with respect to nuclear displacements:

$$F(\mathcal{R}) = \nabla_{\mathcal{R}} E(\mathcal{R}). \tag{8.9}$$

Because the stochastic nature of the algorithm, obtaining forces in QMC is a difficult task. Generally, QMC calculations at critical points, for example, equilibrium and reaction barrier geometries have been carried out, in which the geometries are obtained with a different quantum chemical method such as density functional theory (DFT) (PARR and YANG [1989]), or wave function methods. Whereas DFT and wave function methods use the Hellman–Feynman theorem for the calculation of forces, a straightforward application of the theorem in QMC leads to estimators with very large variance.

8.2.1. Correlated sampling
An efficient approach for force calculation is the use of correlated sampling, which is a MC method that uses correlation between similar observations to reduce the statistical error of the sampling. If one were to represent Eq. (8.9) in a finite difference scheme for evaluating a derivative along $\mathbf{r}_d$,

$$\frac{\partial E}{\partial \mathbf{r}_d} \approx \frac{E(\mathcal{R} + \mathbf{r}_d) - E(\mathcal{R})}{\mathbf{r}_d}, \tag{8.10}$$

then one obtains an approximate energy derivative along the $\mathbf{r}_d$. If two separate calculations are carried out, with a statistical error of the energies of σ_E, the statistical error for the difference σ_d is approximately

$$\sigma_d \approx \frac{\sigma_E}{\mathbf{r}_d}. \tag{8.11}$$

One can see that because $\mathbf{r}_d$ is a vector of a small perturbation, of ≈ 0.01 a.u., that the statistical error of the difference will be several times higher than the statistical error of the energies. If $\mathbf{r}_d$ is sufficiently small, a single random walk can be performed, while evaluating the energy at the original and perturbed geometries, $E[\Psi(\mathcal{R})]$ and $E[\Psi(\mathcal{R} + \mathbf{r}_d)]$. In this case, both the primary ($\mathcal{R}$) and secondary ($\mathcal{R}$) walks will be correlated, and therefore present lower variance than uncorrelated random walks.

---

[9]The QMC method can be used for calculations without the BO approximation, but so far the applications have been to nodeless systems, such as H_2 (TRAYNOR, ANDERSON and BOGHOSIAN [1994]).

If correlated sampling is used for forces in a PMC algorithm, it was recently proposed by FILIPPI and UMRIGAR [2002] to use expressions including branching factors $B(\mathbf{R})$, reoptimizing the parameters of the wave function Λ for each perturbed geometry, and perform additional coordinate transformations. Practical implementations of the correlated sampling method for derivatives are described in detail in SUN, LESTER, JR. and HAMMOND [1992] and FILIPPI and UMRIGAR [2002].

8.2.2. Analytic derivative methods

The calculation of analytic derivative estimators is a costly process, both for wave function based methods, and for QMC methods. Fortunately, in QMC one needs not to evaluate derivatives at each step, but rather sample points more sporadically, both for reducing computer time and serial correlation.

The local energy estimator for a DMC mixed distribution is

$$E_0 = \langle E_L \rangle = \frac{\int \Psi_0(\mathbf{R}) E_L(\mathbf{R}) \Psi_T(\mathbf{R}) d\mathbf{R}}{\int \Psi_0(\mathbf{R}) \Psi_T(\mathbf{R}) d\mathbf{R}}. \tag{8.12}$$

The gradient of expression (8.12) involves derivatives of the unknown exact wave function $\Psi_0(\mathbf{R})$, and the trial wave function $\Psi_T(\mathbf{R})$. Derivatives of $\Psi_0(\mathbf{R})$ have to be obtained with a method devised for sampling operators that do not commute with the Hamiltonian, which are described in Section 8.1. This can lead to an exact estimator for the derivative, but with the added computational complexity of those methods. Therefore, a simple approximation can be used, replacing the derivatives of $\Psi_0(\mathbf{R})$ with those of $\Psi_T(\mathbf{R})$ to obtain

$$\nabla_\mathcal{R} E_0 \approx \langle \nabla_\mathcal{R} E_L(\mathbf{R}) \rangle + 2 \left\langle E_L \frac{\nabla_\mathcal{R} \Psi_T(\mathbf{R})}{\Psi_T(\mathbf{R})} \right\rangle - 2 E_0 \left\langle \frac{\nabla_\mathcal{R} \Psi_T(\mathbf{R})}{\Psi_T(\mathbf{R})} \right\rangle. \tag{8.13}$$

The derivatives of $\Psi_T(\mathbf{R})$ are readily obtainable from the known analytic expression of $\Psi_T(\mathbf{R})$.

The expression for the exact derivative involves the cumulative weight of configuration $\mathbf{R}_i$ at time step s, $\mathbf{R}_i^s$,

$$\bar{B}_i = \prod_{s_0}^{s} B(\mathbf{R}_i^s), \tag{8.14}$$

where $s - s_0$ is the number of generations for the accumulation of the cumulative weight, and $B(\mathbf{R}_i^s)$ is the PMC branching factor of Eqs. (5.9) and (5.39). The energy expression using cumulative weights is

$$E_0 = \frac{\int \Psi_T(\mathbf{R})^2 \bar{B}(\mathbf{R}) E_L(\mathbf{R}) d\mathbf{R}}{\int \Psi_T(\mathbf{R})^2 \bar{B}(\mathbf{R}) d\mathbf{R}}. \tag{8.15}$$

The energy derivative of expression (8.15) leads to the following derivative expression,

$$\nabla_{\mathcal{R}}E_0 = \langle\nabla_{\mathcal{R}}E_L(\mathbf{R})\rangle + 2\left\langle E_L(\mathbf{R})\frac{\nabla_{\mathcal{R}}\Psi_T(\mathbf{R})}{\Psi_T(\mathbf{R})}\right\rangle - 2E_0\left\langle\frac{\nabla_{\mathcal{R}}\Psi_T(\mathbf{R})}{\Psi_T(\mathbf{R})}\right\rangle$$
$$+ \left\langle E_L(\mathbf{R})\frac{\nabla_{\mathcal{R}}\bar{B}(\mathbf{R})}{\bar{B}(\mathbf{R})}\right\rangle - E_0\left\langle\frac{\nabla_{\mathcal{R}}\bar{B}(\mathbf{R})}{\bar{B}(\mathbf{R})}\right\rangle. \tag{8.16}$$

Analytic energy derivatives have been applied to H_2 (REYNOLDS, BARNETT, HAMMOND, GRIMES and LESTER, JR. [1986]), LiH and CuH (VRBIK, LAGARE and ROTHSTEIN [1990]). Higher order derivatives can be obtained as well. Details on the former can be found in VRBIK and ROTHSTEIN [1992] and BELOHOREC, ROTHSTEIN and VRBIK [1993].

8.2.3. Hellman–Feynman derivatives and the zero variance theorem

The Hellman–Feynman theorem states that the forces can be obtained by looking at the value of the gradient of the potential

$$\langle\nabla_{\mathcal{R}}E_0\rangle = -\frac{\int\Psi^2(\mathbf{R})\nabla_{\mathcal{R}}V(\mathbf{R})d\mathbf{R}}{\int\Psi^2(\mathbf{R})d\mathbf{R}}, \tag{8.17}$$

where $V(\mathbf{R})$ is the Coulomb potential for the system. A QMC estimator of the Hellman–Feynman forces, $\mathbf{F}_{HF} \equiv -\nabla_{\mathcal{R}}V(\mathbf{R})$, can be constructed, but it has infinite variance. This comes from the fact that at short electron–nucleus distances, r_{iN}, the force behaves as $\mathbf{F}_{HF} \approx \frac{1}{r_{iN}^2}$, therefore the variance associated with $\mathbf{F}_{HF}$ depends on $\langle\mathbf{F}_{HF}^2\rangle$, which is infinite. Furthermore, the Hellman–Feynman theorem only holds for exact wave functions, and basis set errors need to be accounted for (UMRIGAR [1989]). Also, the FNA introduces an extra requirement on the nodal surface. The former has to be independent of the position of the nuclei, or it has to be the exact one. An elaborate discussion of this issue can be found in HUANG, NEEDS and RAJAGOPAL [2000] and SCHAUTZ and Flad [2000].

A proposed solution to the infinite variance problem is to evaluate the forces at a cutoff distance close to the origin, and then extrapolation to a cutoff distance of zero (VRBIK and ROTHSTEIN [1992]). This has the problem that the extrapolation procedure is difficult due to the increase of variance as the cutoff values decrease.

As discussed in Section 5.8, renormalized operators can be obtained, in such a way that they have the same expectation value, but lower variance. Recently ASSARAF and CAFFAREL [2000], a renormalized operator was introduced

$$\bar{\mathbf{F}}_{HF} = \mathbf{F}_{HF} + \left[\frac{\hat{H}_V\Psi_V(\mathbf{R})}{\Psi_V(\mathbf{R})} - \frac{\hat{H}_V\Psi_T(\mathbf{R})}{\Psi_T(\mathbf{R})}\right]\frac{\Psi_V(\mathbf{R})}{\Psi_T(\mathbf{R})}. \tag{8.18}$$

Here, $\Psi_V(\mathbf{R})$ are an auxiliary wave function and $\hat{H}_V$ is an auxiliary Hamiltonian. The variance of operator (8.18) can be shown to be finite, and therefore smaller than $\mathbf{F}_{HF}$. The form of $\Psi_V(\mathbf{R})$ proposed by ASSARAF and CAFFAREL [2000] is a simple form that cancels the singularities of the force in the case of a diatomic molecule. Nevertheless, general forms of the auxiliary wave function can be constructed.

8.2.4. *Variational Monte Carlo dynamics*

Correlated sampling can be combined with a fictitious Lagrangian technique, similar to that developed by CAR and PARRINELLO [1985] in a way first proposed by TANAKA [1994] for geometry optimization. In this approach, the expectation value of the Hamiltonian is treated as a functional of the nuclear positions and the correlation parameters:

$$\langle \hat{H} \rangle = \frac{\langle \Psi | \hat{H} | \Psi \rangle}{\langle \Psi | \Psi \rangle} = E[\{\Lambda\}, \{\mathcal{R}\}]. \tag{8.19}$$

With the previous functional, a fictitious Lagrangian can be constructed of the form

$$L = \sum_\alpha \frac{1}{2} \mu_\alpha \lambda_\alpha^{'2} + \sum_I \frac{1}{2} M_I \mathcal{R}_I^{'2} - E[\{\Lambda\}, \{\mathcal{R}\}], \tag{8.20}$$

where M_I are the nuclear masses and μ_a are the fictitious masses for the variational parameters, λ_α. The modified Euler–Lagrange equations can be used for generating dynamics for the sets of parameters, $\{\mathcal{R}\}$ and $\{\Lambda\}$,

$$M_I \mathcal{R}_I'' = \nabla_{R_I} E, \tag{8.21}$$

$$\mu_\alpha \lambda_\alpha'' = \frac{\partial E}{\partial \lambda_\alpha}. \tag{8.22}$$

A dissipative transformation of Eqs. (8.21) and (8.22), where the masses M_I and μ_α are replaced by damped masses $\tilde{M}_I$ and $\tilde{\mu}_\alpha$ can be used for geometry optimization. A more elaborate approach that attempts to include quantum effects in the dynamics is described in TANAKA [2002].

To conclude, a method that is not described here due to reasons of space and that needs to be explored further, is the generalized reweighting method (BORONAT and CASULLERAS [1999]).

List of symbols

We denote:

Z_j Atomic nuclear charge,

$\mathcal{R}$ Set of coordinates of clamped particles (under the BO approximation): $\mathcal{R} \equiv \{\mathcal{R}_1, \ldots, \mathcal{R}_m\}$,

r_i Electronic coordinate in a Cartesian frame $r_i = \{x_i \ldots x_d\}$, where the number of dimensions, d is 3 for most chemical applications,

R Set of coordinates of all n particles treated by QMC, $\mathbf{R} \equiv \{r_1, \ldots, r_n\}$,

σ_i Spin coordinate for an electron, $\sigma_\uparrow$ is spin-up and $\sigma_\downarrow$ for spin-down particles,

Σ Set of spin coordinates for particles, $\Sigma \equiv \{\sigma_1, \ldots, \sigma_N\}$,

Ψ_0 Exact ground state wave function,

Ψ_i ith exact wave function,

$\Psi_{T,i}$ Approximate trial wave function for state i,

ϕ_k Single particle molecular orbital (MO),

$D(\phi_k)$ Slater determinant of k MOs: $D = \frac{1}{\sqrt{N!}} \det|\phi_1, \ldots, \phi_N|$,

$D^\uparrow(\phi_k^\uparrow), D^\downarrow(\phi_k^\downarrow)$, Spin factored Slater determinants for spin up ($\uparrow$) and spin down ($\downarrow$) electrons,

$\hat{H}$ Hamiltonian operator: $\hat{H} \equiv \hat{T} + \hat{V}$,

$\hat{T}$ Kinetic energy operator $-\frac{1}{2}\nabla_R^2 \equiv -\frac{1}{2}\sum_i \nabla_i^2$,

$\hat{V}$ Potential energy operator, for atomic and molecular systems: $\hat{V} = -\sum_{ij}\frac{Z_j}{r_{ij}} + \sum_{i<j}\frac{1}{r_{ij}} + \sum_{i<j}\frac{Z_iZ_j}{r_{ij}}$,

τ Imaginary time, $\tau = it$,

$\rho(\mathbf{r})$ Electronic density,

E_L Local Energy, $\hat{H}\Psi_T(\mathbf{R})/\Psi_T(\mathbf{R})$,

$\mathcal{U}_{[a,b]}$ Uniform random variate in the interval $[a, b]$,

$\mathcal{G}_\sigma$ Gaussian random variate of variance σ,

$\sigma\hat{O}$ Monte Carlo variance for observable $\hat{O}$,

$\mathcal{W}$ Ensemble of random walkers, $\mathcal{W} \equiv \{\mathbf{R}_1, \mathbf{R}_2, \ldots, \mathbf{R}_n\}$,

$\mathcal{P}(\mathbf{R})$ Monte Carlo probability density function,

$P(\mathbf{R} \to \mathbf{R}')$ Monte Carlo transition probability,

$B(\mathbf{R}, \mathbf{R}')$ Branching factor for Population Monte Carlo algorithms,

$G(\mathbf{R}, \mathbf{R}'; \tau - \tau')$ Time dependent Green's function,

$G_{ST}(\mathbf{R}, \mathbf{R}'; \delta\tau)$ Time dependent short time Green's function for the Schrödinger equation,

$\mathbf{F_q}$ Quantum force, $\mathbf{F_q} \equiv \nabla \ln |\Psi_T(\mathbf{R})^2|$,

$\mathcal{D}$ Denominator in FMC, $\mathcal{D} \equiv \sum_{i=1}^{Nw}\left[\frac{\psi_T(R_i^+)}{\Psi_G^+(R_i^+)}\frac{\psi_T(R_i^-)}{\Psi_G^-(R_i^-)}\right]$,

M_I Ionic mass,

μ_α Fictitious parameter mass.

Acknowledgments

A.A.-G. was a holder of a Gates Millenium Fellowship during the preparation of this chapter. W.A.L. was supported by the Director, Office of Science, Office of Basic Energy Sciences, Chemical Sciences Division of the U.S. Department of Energy under Contract No. DE-AC03-76SF00098, and by the National Science Foundation through the CREST Program (HRD-9805465).

References

ACIOLI, P.H. (1997), Review of quantum Monte Carlo methods and their applications, *J. Mol. Struct. (Theochem)* **394**, 75.

ALEXANDER, S.A. and R.L. COLDWELL (1997), Atomic wave function forms, *Int. J. Quant. Chem.* **63**, 1001.

ALEXANDER, S.A., R.L. COLDWELL, G. AISSING and A.J. THAKKAR (1992), Calculation of atomic and molecular properties using variational Monte Carlo methods, *Int. J. Quant. Chem.* **26**, 213–227.

ANDERSON, J.B. (1976), Quantum chemistry by random walk, *J. Chem. Phys.* **65** (10), 4121–4127.

ANDERSON, J.B. (1980), Quantum chemistry by random walk: Higher accuracy, *J. Chem. Phys.* **73** (8), 3897–3899.

ANDERSON, J.B. (1999), Quantum Monte Carlo: Atoms, molecules, clusters, liquids and solids, in: K.B. Lipkowitz and D.B. Boyd, eds., *Reviews in Computational Chemistry*, Vol. 13 (John Wiley & Sons, New York) 133.

ANDERSON, J.B. (2001), An exact quantum Monte Carlo calculation of the helium–helium intermolecular potential. II, *J. Chem. Phys.* **115**, 4546.

ANDERSON, J.B., C.A. TRAYNOR and B.M. BOGHOSIAN (1991), Quantum chemistry by random walk: Exact treatment of many-electron systems, *J. Chem. Phys.* **95**, 7418.

ARNOW, D.M., M.H. KALOS, M.A. LEE and K.E. SCHMIDT (1982), Green's function Monte Carlo for few fermion problems, *J. Chem. Phys.* **77**, 5562.

ASSARAF, R. and M. CAFFAREL (1999), Zero-variance principle for Monte Carlo algorithms, *Phys. Rev. Lett.* **83**, 4682.

ASSARAF, R. and M. CAFFAREL (2000), Computing forces with quantum Monte Carlo, *J. Chem. Phys.* **113**, 4028.

ASSARAF, R., M. CAFFAREL and A. KHELIF (2000), Diffusion Monte Carlo methods with a fixed number of walkers, *Phys. Rev. E* **61**, 4566.

BAER, R., M. HEAD-GORDON and D. NEUHAUSER (1998), Shifted-contour auxiliary field Monte Carlo for ab initio electronic structure: Straddling the sign problem, *J. Chem. Phys.* **109**, 6219.

BAER, R. and D. NEUHAUSER (2000), Molecular electronic structure using auxiliary field Monte Carlo, plane waves and pseudopotentials, *J. Chem. Phys.* **112**, 1679.

BALASUBRAMANIAN, K. and K.S. PITZER (1987), Relativistic quantum chemistry, *Adv. Chem. Phys.* **77**, 287.

BARNETT, R.N., P.J. REYNOLDS and W.A. LESTER, Jr. (1991), Monte Carlo algorithms for expectation values of coordinate operators, *J. Comp. Phys.* **96**, 258.

BARNETT, R.N., P.J. REYNOLDS and W.A. LESTER, Jr. (1992a), Computation of transition dipole moments by Monte Carlo, *J. Chem. Phys.* **96**, 2141.

BARNETT, R.N., P.J. REYNOLDS and W.A. LESTER, Jr. (1992b), Monte Carlo determination of the oscillator strength and excited state lifetime for the Li 2s to 2p transition, *Int. J. Quant. Chem.* **42**, 837.

BARNETT, R.N., Z. SUN and W.A. LESTER, Jr. (1997), Fixed sample optimization in quantum Monte Carlo using a probability density function, *Chem. Phys. Lett.* **273**, 321.

BARNETT, R.N., Z. SUN and W.A. LESTER, Jr. (2001), Improved trial functions in quantum Monte Carlo: Application to acetylene and its dissociation fragments, *J. Chem. Phys.* **114**, 2013.

BAUER, W.F. (1958), The Monte Carlo method, *J. Soc. Indust. Appl. Math.* **6** (4), 438–451. A very well described introduction to MC methods by an applied mathematician. Cites the Courant and Levy 1928 paper.

BELOHOREC, P., S.M. ROTHSTEIN and J. VRBIK (1993), Infinitesimal differential diffusion quantum Monte Carlo study of CuH spectroscopic constants, *J. Chem. Phys.* **98**, 6401.

BENOIT, D.M. and D.C. CLARY (2000), Quaternion formulation of diffusion quantum Monte Carlo for the rotation of rigid molecules in clusters, *J. Chem. Phys.* **113**, 5193.

BERTINI, L., D. BRESSANINI, M. MELLA and G. MOROSI (1999), Linear expansions of correlated functions: A variational Monte Carlo case study, *Int. J. Quant. Chem.* **74**, 23.

BHATTACHARYA, A. and J.B. ANDERSON (1994a), Exact quantum Monte Carlo calculation of the H–He interaction potential, *Phys. Rev. A* **49**, 2441.

BHATTACHARYA, A. and J.B. ANDERSON (1994b), The interaction potential of a symmetric helium trimer, *J. Chem. Phys.* **100**, 8999.

BIJL, A. (1940), *Physica* **7**, 869.

BORN, M. and J.R. OPPENHEIMER (1927), Zur Quantentheorie der Molekeln, *Ann. Phys.* **84**, 457.

BORONAT, J. and J. CASULLERAS (1999), Sampling differences in quantum Monte Carlo: A generalized reweighting method, *Comp. Phys. Comm.* **122**, 466.

BOYS, S.F. and N.C. HANDY (1969), *Proc. Roy. Soc. London Ser. A* **310**, 63.

BRESSANINI, D., D.M. CEPERLEY and P.J. REYNOLDS (2002), What do we know about wave function nodes? in: S.M. Rothstein, W.A. Lester and S. Tanaka, eds., *Recent Advances in Quantum Monte Carlo Methods, Part II* (World Scientific, Singapore).

BRESSANINI, D. and P.J. REYNOLDS (1998), Between classical and quantum Monte Carlo methods: "Variational" QMC, *Adv. Chem. Phys.* **105**, 37.

BRESSANINI, D. and P.J. REYNOLDS (1999), Spatial-partitioning-based acceleration for variational Monte Carlo, *J. Chem. Phys.* **111**, 6180.

BUECKERT, H., S.M. ROTHSTEIN and J. VRBIK (1992), Optimization of quantum Monte Carlo wavefunctions using analytical derivatives, *Canad. J. Chem.* **70**, 366.

BUONAURA, M.C. and S. SORELLA (1998), Numerical study of the two-dimensional Heisenberg model using a Green function Monte Carlo technique with a fixed number of walkers, *Phys. Rev. B* **61**, 2559.

BYRON, F.W. and R. FULLER (1992), Mathematics of Classical and Quantum Physics *(Dover)* .

CAFFAREL, M. and P. CLAVERIE (1988), Development of a pure diffusion quantum Monte Carlo method using a full generalized Feynman–Kac Formula, I. Formalism, *J. Chem. Phys.* **88**, 1088–1099.

CAFFAREL, M., P. CLAVERIE, C. MIJOULE, J. ANDZELM and D.R. SALAHUB (1989), Quantum Monte-Carlo method for some model and realistic coupled anharmonic-oscillators, *J. Chem. Phys.* **90**, 990.

CAFFAREL, M., X. KROKIDIS and C. MIJOULE (1992), On the nonconservation of the number of nodal cells of eigenfunctions, *Europhys. Lett.* **7**, 581.

CANCÈS, E., M. DEFRANCESCHI, W. KUTZELNIGG, C. LE BRIS and Y. MADAY (2003), Computational quantum chemistry: A primer, in: P.G. Ciarlet and C. Le Bris, eds., *Computational Chemistry, Handbook of Numerical Analysis*, Vol. X (Elsevier, Amsterdam) 3–270.

CAR, R. and M. PARRINELLO (1985), Unified approach for molecular dynamics and density-functional theory, *Phys. Rev. Lett.* **55**, 2471.

CASULLERAS, J. and J. BORONAT (2000), Progress in Monte Carlo calculations of Fermi systems: Normal liquid He-3, *Phys. Rev. Lett.* **84**, 3121.

CEPERLEY, D.M. (1983), The simulation of quantum systems with random walks: A new algorithm for charged systems, *J. Comp. Phys.* **51**, 404.

CEPERLEY, D.M. (1986), The statistical error of Green's function Monte Carlo, *J. Stat. Phys.* **43**, 815.

CEPERLEY, D.M. (1991), Fermion nodes, *J. Stat. Phys.* **63**, 1237–1267.

CEPERLEY, D.M. (1995), Path integral Monte Carlo in the theory of condensed matter helium, *Rev. Modern. Phys.* **67** (2), 279–356.

CEPERLEY, D.M. and B.J. ALDER (1980), Ground state of the electron gas by a stochastic method, *Phys. Rev. Lett.* **45**, 566.

CEPERLEY, D.M. and B.J. ALDER (1984), Quantum Monte Carlo for molecules: Green's function and nodal release, *J. Chem. Phys.* **81**, 5833.

CEPERLEY, D.M. and B. BERNU (1988), The calculation of excited state properties with quantum Monte Carlo, *J. Chem. Phys.* **89**, 6316.

CEPERLEY, D., G.V. CHESTER and M.H. KALOS (1977), Monte Carlo simulation of a many fermion study, *Phys. Rev. B.* **16**, 3081.

CEPERLEY, D.M. and M.H. KALOS (1986), Quantum many-body problems, in: K. Binder, ed., *Monte Carlo Methods in Statistical Physics,* (Springer-Verlag, New York) 2nd edn.

CEPERLEY, D.M. and L. MITAS (1996), Quantum Monte Carlo methods in chemistry, in: I. Prigogine and S.A. Rice, eds., *New Methods in Computational Quantum Mechanics,* Adv. Chem. Phys., Vol. XCIII (John Wiley & Sons, New York).

CERF, N.J. and O.C. MARTIN (1995), Projection Monte Carlo methods: An algorithmic analysis, *Int. J. Mod. Phys.* **6**, 693.

CHARUTZ, D.M. and D. NEUHAUSER (1995), Electronic structure via the auxiliary-field Monte-Carlo algorithm, *J. Chem. Phys.* **102**, 4495.

CHIN, S.A. (1990), Quadratic diffusion Monte Carlo algorithms for solving atomic many-body problems, *Phys. Rev. A* **42**, 6991.

CHIN, S.A. (1997), Symplectic integrators from composite operator factorizations, *Phys. Lett. A* **226**, 344.

CHRISTIANSEN, P.A., W.C. ERMLER and K.S. PITZER (1984), Relativistic effects in chemical systems, *Annu. Rev. Phys. Chem.* **35**, 357.

CLARY, D.C. (2001), Torsional diffusion Monte Carlo: A method for quantum simulations of proteins, *J. Chem. Phys.* **114**, 9725.

CONROY, H. (1964), Molecular Schrödinger equation, 2. Monte Carlo evaluation of integrals, *J. Chem. Phys.* **41**, 1331.

COURANT, R., K.O. FRIEDRICHS and H. LEWY (1928), On the partial difference equations of mathematical physics, *Math. Ann.* **100**, 32.

DAVIDSON, E.R. (1976), *Reduced Density Matrices in Quantum Chemistry* (Academic Press, New York).

DEPASQUALE, M.F., S.M. ROTHSTEIN and J. VRBIK (1988), Reliable diffusion quantum Monte Carlo, *J. Chem. Phys.* **89**, 3629.

DEWING, M. (2000), Improved efficiency with variational Monte Carlo using two level sampling, *J. Chem. Phys.* **113**, 5123.

DINGLE, R.B. (1949), *Phil. Mag.* **40**, 573.

DOLG, M. (2000), Effective core potentials, in: J. Grotendorst, ed., *Modern Methods and Algorithms of Quantum Chemistry*, Vol. 1 ((John von Neumann Institute for Computing)) 479–508.

DONSKER, M.D. and M. KAC (1950), A sampling method for determining the lowest eigenvalue and the principal eigenfunction of Schrödinger's equation, *J. Res. Nat. Bur. Standards* **44** (50), 551–557.

DOUCET, A., N. DE FREITAS, N. GORDON and A. SMITH, eds. (2001), *Sequential Monte Carlo Methods in Practice* (Springer–Verlag, New York).

A.Doucet, A., N.de Freitas, N., N.Gordon, N., & A.Smith, A. eds., (2001). *Sequential Monte Carlo Methods in Practice* (Springer–Verlag, New York).

EAST, A.L.L., S.M. ROTHSTEIN and A. VRBIK (1988), Sampling the exact electron distribution by diffusion quantum Monte Carlo, *J. Chem. Phys.* **89**, 4880.

EINSTEIN, A. (1926), *Investigations in the theory of Brownian Motion* (Metheun & Co. Ltd), English translation of Einstein's original paper.

FEYNMANN, R.P. (1939), Forces in molecules, *Phys. Rev.* **56**, 340.

FEYNMANN, R.P. and M. COHEN (1956), Energy spectrum of the excitations in liquid helium, *Phys. Rev.* **102**, 1189.

FILIPPI, C. and S. FAHY (2000), Optimal orbitals from energy fluctuations in correlated wave functions, *J. Chem. Phys.* **112**, 3532.

FILIPPI, C. and C.J. UMRIGAR (1996), Multiconfiguration wave functions for quantum Monte Carlo calculations of first-row diatomic molecules, *J. Chem. Phys.* **105**, 213.

FILIPPI, C. and C.J. UMRIGAR (2002), Interatomic forces and correlated sampling in quantum Monte Carlo, in: S.M. Rothstein, W.A. Lester Jr. and S. Tanaka, eds., *Recent Advances in Quantum Monte Carlo Methods, Part II* (World Scientific, Singapore).

FISHMAN, G.S. (1996), *Monte Carlo: Concepts, Algorithms and Applications,* (Springer-Verlag, Berlin)1st edn.

FLAD, H.-J., M. CAFFAREL and A. SAVIN (1997), Quantum Monte Carlo calculations with multi-reference trial wave functions, in: W.A. Lester, Jr. ed., *Recent Advances in Quantum Monte Carlo Methods* (World Scientific, Singapore), Chapter 5, 73–98.

FLAD, H.-J. and M. DOLG (1997), Probing the accuracy of pseudopotentials for transition metals in quantum Monte Carlo calculations, *J. Chem. Phys.* **107**, 7951–7959.

FLAD, H.-J., A. SAVIN and M. SCHULTHEISS (1994), A systematic study of fixed-node and localization error in quantum Monte Carlo calculations with pseudopotentials for group III elements, *Chem. Phys. Lett.* **222**, 274.

FLAD, H.-J., F. SCHAUTZ, Y. WANG and M. DOLG (2002), Quantum Monte Carlo study of mercury clusters, in: S.M. Rothstein, W.A. Lester, Jr. and S. Tanaka eds., *Recent Advances in Quantum Monte Carlo Methods, Part II* (World Scientific, Singapore).

FORBERT, H.A. and S.A. CHIN (2001), Fourth order diffusion Monte Carlo algorithms for solving quantum many-body problems, *Int. J. Mod. Phys. B* **15**, 1752.

FOULKES, M., L. MITAS, R. NEEDS and G. RAJAGOPAL (2001), Quantum Monte Carlo for solids, *Rev. Mod. Phys.* **73**, 33.

FROST, A.A., R.E. KELLOG and E.A. CURTIS (1960), Local energy method for electronic energy calculations, *Rev. Mod. Phys.* **32**, 313–317.

GLAUSER, W.A., W.R. BROWN, W.A. LESTER, JR. D. BRESSANINI, B.L. HAMMOND and M.L. KOSYKOWSKI (1992), Random-walk approach to mapping nodal regions of *N*-body wave functions: Ground-state Hartree–Fock wave functions for Li–C, *J. Chem. Phys.* **97**, 9200.

GOULD, H. and J. TOBOCHNIK (1996), *An Introduction to Computer Simulation Methods: Applications to Physical Systems*, (Addison–Wesley, Reading, MA), 2nd edn.

GREEFF, C.W., B.L. HAMMOND and W.A. LESTER, JR. (1996), Electronic states of Al and Al_2 using quantum Monte Carlo with an effective core potential, *J. Chem. Phys.* **104**, 1973–1978.

GREEFF, C.W. and W.A. LESTER, JR. (1997), Quantum Monte Carlo binding energies for silicon hydrides, *J. Chem. Phys.* **106**, 6412–6417.

GREEFF, C.W. and W.A. LESTER, JR. (1998), A soft Hartree–Fock pseudopotential for carbon with application to quantum Monte Carlo, *J. Chem. Phys.* **109**, 1607–1612.

GROSSMAN, J.C., W.A. LESTER, JR. and S.G. LOUIE (2000), Quantum Monte Carlo and density functional theory characterization of 2-cyclopentenone and 3-cyclopentenone formation from O(3P) + cyclopentadiene, *J. Amer. Chem. Soc.* **122**, 703.

GROSSMAN, J.C. and L. MITAS (1995), Quantum Monte Carlo determination of electronic and structural properties of Si(*n*) clusters ($n < 20$), *Phys. Rev. Lett.* **74**, 1323.

HALTON, J.H. (1970), A retrospective and prospective survey of the Monte Carlo method, *SIAM Review* **12** (1), 1–53.

HAMMERSLEY, J.M. and D.C. HANDSCOMB (1964), *Monte Carlo Methods* (Methuen, London).

HAMMOND, B.L., W.A. LESTER, JR and P.J. REYNOLDS (1994), *Monte Carlo Methods in Ab Initio Quantum Chemistry* (World Scientific, Singapore).

HAMMOND, B.L., P.J. REYNOLDS and W.A. LESTER, JR. (1987), Valence quantum Monte Carlo with ab initio effective core potentials, *J. Chem. Phys.* **87**, 1130–1136.

HEAD-GORDON, M. (1996), Quantum chemistry and molecular processes, *J. Phys. Chem.* **100**, 13213.

HELLMAN, H. (1937), *Einführung in die Quanten Theorie* (Deuticke).

HETHERINGTON, J.H. (1984), Observations on the statistical iteration of matrices, *Phys. Rev. A.* **30** 2713.

HORNIK, M., M. SNAJDR and S.M. ROTHSTEIN (2000), Estimating the overlap of an approximate with the exact wave function by quantum Monte Carlo methods, *J. Chem. Phys.* **113**, 3496.

HUANG, C., C.J. UMRIGAR and M.P. NIGHTINGALE (1997), Accuracy of electronic wave functions in quantum Monte Carlo: The effect of high-order correlations, *J. Chem. Phys.* **107**, 3007.

HUANG, H. and Z. CAO (1996), A novel method for optimizing quantum Monte Carlo wave functions, *J. Chem. Phys.* **104**, 1.

HUANG, H., Q. XIE, Z. CAO, Z. LI, Z. YUE and L. MING (1999), A novel quantum Monte Carlo strategy: Surplus function approach, *J. Chem. Phys.* **110**, 3703.

HUANG, K.C., R.J. NEEDS and G. RAJAGOPAL (2000), Comment on "Quantum Monte Carlo study of the dipole moment of CO" [J. Chem. Phys. 110, 11700 (1999)]*J. Chem. Phys.* **112**, 4419.

HUANG, S.-Y., Z. SUN and W.A. LESTER, JR. (1990), Optimized trial functions for quantum Monte Carlo, *J. Chem. Phys.* **92**, 597.

IBA, Y. (2000), Population Monte Carlo algorithms, *Trans. Japanese Soc. Artif. Intell.* **16** (2), 279–286.

JAMES, H.H. and A.S. COOLIDGE (1937), Criteria for goodness for approximate wave functions, *Phys. Rev.* **51**, 860–863.

JASTROW, R. (1955), Many-body problem with strong forces, *Phys. Rev.* **98**, 1479.

KALOS, M.H. (1962), Monte Carlo calculations of the ground state of three- and four-body nuclei, *Phys. Rev.* **68** (4).

KALOS, M.H., D. LEVESQUE and L. VERLET (1974), Helium at zero temperature with hard-sphere and other forces, *Phys. Rev. A* **9**, 2178.

KALOS, M.H. and F. PEDERIVA (1998), Fermion Monte Carlo, in: M.P. Nightingale and C.J. Umrigar, eds., *Quantum Monte Carlo Methods in Physics and Chemistry*, Vol. 525 (Kluwer Academic, Dordrecht) 263–286.

KALOS, M.H. and F. PEDERIVA (2000a), Exact Monte Carlo method for continuum fermion systems, *Phys. Rev. Lett.* **85**, 3547.

KALOS, M.H. and F. PEDERIVA (2000b), Fermion Monte Carlo for continuum systems, *Physica A* **279**, 236.

KALOS, M.H. and K.E. SCHMIDT (1997), Model fermion Monte Carlo with correlated pairs II, *J. Stat. Phys.* **89**, 425.

KALOS, M.H. and P.A. WHITLOCK (1986), *Monte Carlo Methods, Vol. 1: Basics* (Wiley, New York).

KATO, T. (1957), On the eigenfunctions of many-particle systems in quantum mechanics, *Comm. Pure. and Appl. Math.* **10**, 151.

KOONIN, S.E. and MEREDITH (1995), *Computational Physics, FORTRAN Version*, (Addison–Wesley) 3rd edn.

KRAUSS, M. and W.J. STEVENS (1984), Effective potentials in molecular quantum chemistry, *Annu. Rev. Phys. Chem.* **35**, 357.

KWON, Y., D.M. CEPERLEY and R.M. MARTIN (1993), Effects of three-body and backflow correlations in the two-dimensional electron gas, *Phys. Rev. B.* **48**, 12037.

KWON, Y., D.M. CEPERLEY and R.M. MARTIN (1998), Effects of backflow correlation in the three-dimensional electron gas: Quantum Monte Carlo study, *Phys. Rev. B.* **58**, 6800.

KWON, Y., P. HUANG, M.V. PATEL, D. BLUME and K.B. WHALEY (2000), Quantum solvation and molecular rotations in superfluid helium clusters, *J. Chem. Phys.* **113**, 6469.

LANGFELDER, P., S.M. ROTHSTEIN and J. VRBIK (1997), Diffusion quantum Monte Carlo calculation of non-differential properties for atomic ground states, *J. Chem. Phys.* **107**, 8525.

LESTER, W.A., JR. and B.L. HAMMOND (1990), Quantum Monte Carlo for the electronic structure of atoms and molecules, *Annu. Rev. Phys. Chem.* **41**, 283–311.

LIU, J.S. (2001), *Monte Carlo Strategies in Scientific Computing* (Springer–Verlag, New York).

LIU, K.S., M.H. KALOS and G.V. CHESTER (1974), Quantum hard spheres in a channel, *Phys. Rev. A* **10**, 303.

LIU, Z., S. ZHANG and M.H. KALOS (1994), Model fermion Monte Carlo method with antithetical pairs, *Phys. Rev. E* **50**, 3220.

LÖWDIN, P.-O. (1959), Correlation problem in many-electron quantum mechanics, *Adv. Chem. Phys.* **2**, 207.

LÜCHOW, A. and J.B. ANDERSON (2000), Monte Carlo methods in electronic structures for large systems, *Ann. Rev. Phys. Chem.* **51**, 501.

LÜCHOW, L. and J.B. ANDERSON (1996), First-row hydrides: Dissociation and ground state energies using quantum Monte Carlo, *J. Chem. Phys.* **105**, 7573.

MANNO, I. (1999), *Introduction to the Monte-Carlo Method* (Akademiai Kiado, Budapest).

MANTEN, S. and A. LÜCHOW (2001), On the accuracy of the fixed-node diffusion quantum Monte Carlo method, *J. Chem. Phys.* **115**, 5362.

MAZZIOTTI, D.A. (1998), Contracted Schrödinger equation: Determining quantum energies and two-particle density matrices without wavefunctions, *Phys. Rev. A* **57**, 4219.

McDOWELL, K. (1981), Assessing the quality of a wavefunction using quantum Monte Carlo, *Int. J. Quant. Chem: Quant. Chem. Symp.* **15**, 177–181.

METROPOLIS, N., A.W. ROSENBLUTH, M.N. ROSENBLUTH, N.M. TELLER and E. TELLER (1953), Equation of state calculations by fast computing machines, *J. Chem. Phys.* **21**, 1087–1091.

METROPOLIS, N. and S. ULAM (1949), The Monte Carlo method, *J. Amer. Statist. Assoc.* **44**, 335.

MEYERS, C.R., C.J. UMRIGAR, J.P. SETHNA and J.D. MORGAN (1991), The Fock expansion, Kato's cusp conditions, and the exponential Ansatz, *Phys. Rev. A* **44**, 5537.

MITAS, L. (1998), Diffusion Monte Carlo, in: M.P. Nightingale and C.J. Umrigar, eds., *Quantum Monte Carlo Methods in Physics and Chemistry*, Vol. 525 (Kluwer Academic, Dordrecht) 247.

MORELL, M.M., R.G. PARR and M. LEVY (1975), Calculation of ionization-potentials from density matrices and natural functions, and long-range behavior of natural orbitals and electron density, *J. Chem. Phys.* **62**, 549.

NIGHTINGALE, M.P. and C.J. UMRIGAR (1997), Monte Carlo optimization of trial wave functions in quantum mechanics and statistical mechanics, in: W.A. Lester Jr., ed., *Recent Advances in Quantum Monte Carlo Methods* (World Scientific) Chapter 12, 201–227.

OVCHARENKO, I.V., W.A. LESTER, JR., C. XIAO and F. HAGELBERG (2001), Quantum Monte Carlo characterization of small Cu-doped silicon clusters: CuSi4 and CuSi6, *J. Chem. Phys.* **114**, 9028.

PANDHARIPANDE, V.R. and N. ITOH (1973), Effective mass of 3He in liquid 4He+, *Phys. Rev. A* **8**, 2564.

PANDHARIPANDE, V.R., S.C. PIEPER and R.B. WIRINGA (1986), Variational Monte Carlo calculations of ground states of liquid 4He and 3He drops, *Phys. Rev. B* **34**, 4571.

PANGALI, C., M. RAO and B.J. BERNE (1979), On the force bias Monte Carlo simulation of water: Methodology, optimization and comparison with molecular dynamics, *Mol. Phys.* **37**, 1773.

PARR, R.G. and W. YANG (1989), *Density Functional Theory of Atoms and Molecules* (Oxford Univ. Press).

PEDERIVA, F. and M.H. KALOS (1999), Fermion Monte Carlo, *Comp. Phys. Comm.* **122**, 445.

PIEPER, S.C. and R.B. WIRINGA (2001), Quantum Monte Carlo calculations of light nuclei, *Ann. Rev. Nucl. Part. Sci.* **51**, 53.

RAGHAVACHARI, K. and J.B. ANDERSON (1996), Electron correlation effects in molecules, *J. Phys. Chem.* **100**, 12960.

RAO, M. and B.J. BERNE (1979), On the force bias Monte Carlo simulation of simple liquids, *J. Chem. Phys.* **71**, 129.

REYNOLDS, P.J., R.N. BARNETT, B.L. HAMMOND, R.M. GRIMES and W.A. LESTER, JR. (1986), Quantum chemistry by quantum Monte Carlo: Beyond ground-state energy calculations, *Int. J. Quant. Chem.* **29**, 589.

REYNOLDS, P.J., D.M. CEPERLEY, B. ALDER and W.A. LESTER, JR. (1982), Fixed-node quantum Monte Carlo for molecules, *J. Chem. Phys.* **77**, 5593–5603.

ROM, N., D.M. CHARUTZ and D. NEUHAUSER (1997), Shifted-contour auxiliary-field Monte-Carlo: Circumventing the sign difficulty for electronic structure calculations, *Chem. Phys. Lett.* **270**, 382.

ROTHSTEIN, S.M. and J. VRBIK (1988), Statistical error of diffusion Monte Carlo, *J. Comp. Phys.* **74**, 127.

SARSA, A., K.E. SCHMIDT and W.R. MAGRO (2000), A path integral ground state method, *J. Chem. Phys.* **113**, 1366.

SCHAUTZ, F. and H.-J. FLAD (1999), Quantum Monte Carlo study of the dipole moment of CO, *J. Chem. Phys.* **110**, 11700.

SCHAUTZ, F. and H.-J. FLAD (2000), Response to "Comment on 'Quantum Monte Carlo study of the dipole moment of CO'" [J. Chem. Phys. 112, 4419 (2000)], *J. Chem. Phys.* **112**, 4421.

SCHLEYER, P.V.R., N.L. ALLINGER, T. CLARK, J. GASTEIGER, P.A. KOLLMAN, H.F. SCHAEFER III and P.R. SCHREINER, eds. (1998), *The Encyclopedia of Computational Chemistry* (John Wiley & Sons, Chichester).

SCHMIDT, K., M.H. KALOS, ML. A. LEE and G.V. CHESTER (1980), Variational Monte Carlo calculations of liquid 4He with three-body correlations, *Phys. Rev. Lett.* **45**, 573.

SCHMIDT, K.E. (1986), Variational and Green's function Monte Carlo calculations of few body systems, in: *Conference on Models and Methods in Few Body Physics*, (Lisbon).

SCHMIDT, K.E. and J.W. MOSKOWITZ (1986), Monte Carlo calculations of atoms and molecules, *J. Stat. Phys.* **43**, 1027.

SCHMIDT, K.E. and J.W. MOSKOWITZ (1990), Correlated Monte Carlo wave functions for the atoms He through Ne, *J. Chem. Phys.* **93** (4172).

SENATORE, G. and N.H. MARCH (1994), Recent progress in the field of electron correlation, *Rev. Mod. Phys.* **66**, 445.

SKINNER, D.W., J.W. MOSKOWITZ, M.A. LEE, K.E. SCHMIDT and P.A. WHITLOCK (1985), The solution of the Schrödinger equation in imaginary time by Green's function Monte Carlo, the rigorous sampling of the attractive Coulomb singularity, *J. Chem. Phys.* **83**, 4668.

SNAJDR, M., J.R. DWYER and S.M. ROTHSTEIN (1999), Histogram filtering: A technique to optimize wave functions for use in Monte Carlo simulations, *J. Chem. Phys.* **111**, 9971.

SOBOL, I.M. (1994), *A Primer for the Monte Carlo. Method* (CRC Press).

SOKOLOVA, S.A. (2000), An ab initio study of TiC with the diffusion quantum Monte Carlo method, *Chem. Phys. Lett.* **320**, 421–424.

SORELLA, S. (1998), Green's function Monte Carlo with stochastic reconfiguration, *Phys. Rev. Lett.* **80**, 4558.

SORELLA, S. and L. CAPRIOTTI (2000), Green function Monte Carlo with stochastic reconfiguration: An effective remedy for the sign problem, *Phys. Rev. B* **61**, 2599.

STAROVEROV, V.N., P. LANGFELDER and S.M. ROTHSTEIN (1998), Monte Carlo study of core-valence separation schemes, *J. Chem. Phys.* **108**, 2873.

STEDMAN, M.L., W.M.C. FOWLKES and M. NEKOVEE (1998), An accelerated Metropolis method, *J. Chem. Phys.* **109**, 2630.

SUN, Z., R.N. BARNETT and W.A. LESTER JR. (1992), Quantum and variational Monte Carlo interaction potentials for Li2 ($X^1 \Sigma g^+$)), *Chem. Phys. Lett.* **195**, 365.

SUN, Z., S.-Y. HUANG, R.N. BARNETT and W.A. LESTER, JR. (1990), Wave function optimization with a fixed sample in quantum Monte Carlo, *J. Chem. Phys.* **93** 5.

SUN, Z., W.A. LESTER, JR. and B.L. HAMMOND (1992), Correlated sampling of Monte Carlo derivatives with iterated-fixed sampling, *J. Chem. Phys.* **97**, 7585.

SUN, Z., M.M. SOTO and W.A. LESTER, JR. (1994), Characteristics of electron movement in variational Monte Carlo simulations, *J. Chem. Phys.* **100**, 1278.

SZASZ, L. (1985), *Pseudopotential Theory of Atoms and Molecules* (Wiley, New York).

TANAKA, S. (1994), Structural optimization in variational quantum Monte Carlo, *J. Chem. Phys.* **100**, 7416.

TANAKA, S. (2002), Ab initio approach to vibrational properties and quantum dynamics of molecules, in: S. M. Rothstein, W.A. Lester, Jr., S. Tanaka, eds., *Recent Advances in Quantum Monte Carlo Methods, Part II* (World Scientific, Singapore).

THIJSSEN, J.M. (1999), *Computational Physics* (Cambridge Univ. Press).

TRAYNOR, C.A., J.B. ANDERSON and B.M. BOGHOSIAN (1994), A quantum Monte Carlo calculation of the ground state energy of the hydrogen molecule, *J. Chem. Phys.* **94**, 3657.

UMRIGAR, C.J. (1989), Two aspects of quantum Monte Carlo: Determination of accurate wavefunctions and determination of potential energy surfaces of molecules, *Int. J. Quant. Chem.* **23**, 217.

UMRIGAR, C.J. (1993), Accelerated Metropolis method, *Phys. Rev. Lett.* **71**, 408.

UMRIGAR, C.J., M.P. NIGHTINGALE and K.J. RUNGE (1993), A diffusion Monte Carlo algorithm with very small time-step error, *J. Chem. Phys.* **99**, 2865.

UMRIGAR, C.J., K.G. WILSON and J.W. WILKINS (1988), Optimized trial wave functions for quantum Monte Carlo calculations, *Phys. Rev. Lett.* **60**, 1719–1722.

VIEL, A. and K.B. WHALEY (2001), Quantum structure and rotational dynamics of HCN in helium clusters, *J. Chem. Phys.* **115**, 10186.

VRBIK, J., D.A. LAGARE and S.M. ROTHSTEIN (1990), Infinitesimal differential diffusion quantum Monte Carlo: Diatomic molecular properties, *J. Chem. Phys.* **92**, 1221.

VRBIK, J. and S.M. ROTHSTEIN (1992), Infinitesimal differential diffusion quantum Monte Carlo study of diatomic vibrational frequencies, *J. Chem. Phys.* **96**, 2071.

WATSON, D., M. DUNN, T.C. GERMANN, D.R. HERSCHBACH and D.Z. GOODSON (1996), Dimensional expansions for atomic systems, in: C.A. Tsipis, V.S. Popov, D.R. Herschbach and J. Avery, eds., *New Methods in Quantum Theory* (Kluwer Academic, Dordrecht) 83.

WEISSBLUTH, M. (1978), *Atoms and Molecules* (Academic Press, New York) 570–572.

WIRINGA, R.B., S.C. PIEPER, J. CARLSON and V.R. PANDHARIPANDE (2000), Quantum Monte Carlo calculations of $A = 8$ nuclei, *Phys. Rev. C* **62**, 14001.

WOOD, W.W. and J.J. ERPENBECK (1976), Molecular dynamics and Monte Carlo calculations in statistical mechanics, *Ann. Rev. Phys. Chem.* **27**, 319.

YOSHIDA, T. and K. IGUCHI (1988), Quantum Monte Carlo method with the model potential, *J. Chem. Phys.* **88**, 1032.

ZHANG, S. and M.H. KALOS (1991), Exact Monte Carlo calculation for few-electron systems, *Phys. Rev. Lett.* **67**, 3074.

ZHANG, S.W. and M.H. KALOS (1993), Bilinear quantum Monte Carlo – expectation values and energy differences, *J. Stat. Phys.* **70**, 515.

Finite Difference Methods for *Ab Initio* Electronic Structure and Quantum Transport Calculations of Nanostructures

Jean-Luc Fattebert

Center for Applied Scientific Computing (CASC), Lawrence Livermore National Laboratory, P.O. Box 808, L-561, Livermore, CA, 94551, USA
E-mail: fattebert1@llnl.gov

Marco Buongiorno Nardelli

Department of Physics, North Carolina State University, Raleigh, NC, and Center for Computational Sciences (CCS) and Computational Science and Mathematics Division, Oak Ridge National Laboratory, Oak Ridge, TN, 37830, USA

1. Introduction

Among the numerical discretization methods used to solve the equations of density functional theory (DFT), the most widely used are linear combination of atomic orbitals (LCAO) – usually Gaussian-type orbitals (GTO) –, plane waves (PW) and finite differences (FD). Of these three methods, FD is the most recent and less common. Generally, fully 3D grid-based electronic structure representation using FD as approximate numerical schemes for partial differential operators have started being widely used in the last 10 years only. However, real-space FD approaches have already shown to be an efficient tool in a substantial number of large scale electronic structure calculations. Among its various applications, we can cite optical properties of surfaces (SCHMIDT, BECHSTEDT and BERNHOLC [2001]), surface reconstruction

Computational Chemistry
Special Volume (C. Le Bris, Guest Editor) of
HANDBOOK OF NUMERICAL ANALYSIS
P.G. Ciarlet (Editor)

(RAMAMOORTHY, BRIGGS and BERNHOLC [1998]), properties of GaN surfaces (BUN-GARO, RAPCEWICZ and BERNHOLC [1999]), excitation energies and photoabsorption spectra of atoms and clusters (VASILIEV, OGUT and CHELIKOWSKY [1999]), diffusion of oxygen ions in SiO_2 (JIN and CHANG [2001]), first-principles molecular dynamics of carbon nanotubes (BUONGIORNO NARDELLI, YAKOBSON and BERNHOLC [1998]) and processes in solution (TAKAHASHI, HORI, HASHIMOTO and NITTA [2001], FATTEBERT and GYGI [2002]). In this chapter, we review the numerical aspects of this approach for total-energy pseudopotential calculations and show the reasons why it is becoming a method of choice. We also illustrate the method with an application to calculations of electronic structure and conductance of carbon nanotube on a metallic contact. We will limit the discussion to *ab initio* DFT models, where no parameters are fitted to experimental data.

Traditionally, chemists have mostly used LCAO methods. Atomic orbitals expressed as linear combinations of Gaussian functions provide an efficient limited basis set that allows for a good description of the electronic structure of localized finite molecules. The wide spread commercial code GAUSSIAN (Gaussian, Inc.) is based on such an approach. On the other hand, PW (e.g., PAYNE, TETER, ALLAN, ARIAS and JOANNOPOULOS [1992], GALLI and PASQUARELLO [1993], PARRINELLO [1997]) are very efficient to describe periodic systems. Since a PW represents a free electron, this approach has been very successful at describing systems with almost free electrons, like metals. By their nature, they have been mostly used and developed by solid state physicists. In this method, known as pseudospectral method in the mathematics community, the numerical basis set is completely independent of the positions of the atoms present in the simulation. It can be made as accurate as desired by systematically increasing the number of basis functions included in the basis set.

Like PW, the FD method is an alternative to the LCAO when highly accurate electronic wave functions are required. Both approaches try to estimate the electronic structure of a physical system without any assumption on where the atoms and electrons are located. While PW discretizations benefit from the long and extensive experience of many groups of solid state physicists, there are only a few groups around the world that have developed fully functional codes that allow first-principles molecular dynamics simulations on real-space grids.

In recent years, large parallel supercomputers have become an essential tool in first-principles molecular dynamics simulations. They allow not only calculations that would take months or years on single processor machines, but also calculations that would just not fit into the memory of a workstation or high-end PC. In a parallel environment, the electronic wave functions described in a PW approach can be distributed between the processors, for example. Such an approach allows an efficient local application of the FFT algorithm to transform wave functions between real-space and reciprocal space (where the Laplacian is computed). However, every time a matrix element between two wave functions is required – in an orthogonalization process, for example – a huge traffic of data through the whole machine is necessary. In a real-space approach, all the expensive operations can be done locally, thanks to the real-space nature of the DFT Hamiltonian operator and the wave functions. To compute matrix elements between wave functions, local contributions are computed on every processing element (PE) before being summed up at the end over all the PEs. This is one of the main advantages of FD over PW for nowadays simulations.

Another key element in the development of efficient grid-based FD approaches in large scale electronic structure calculations is the multigrid method (BRANDT [1977]). Indeed, real-space large-scale *ab initio* calculations involve large sparse matrices, and multigrid methods, either as solvers or as preconditioners, allow to design very efficient scalable algorithms (BRIGGS, SULLIVAN and BERNHOLC [1995], BRIGGS, SULLIVAN and BERNHOLC [1996], FATTEBERT [1996], FATTEBERT [1999], ANCILOTTO, BLANDIN and TOIGO [1999], JIN, JEONG and CHANG [1999], WANG and BECK [2000], FATTEBERT and BERNHOLC [2000], HEISKANEN, TORSTI, PUSKA and NIEMINEN [2001]).

More recently, in the context of the search for linear scaling algorithms (see, e.g., GOEDECKER [2003], this Handbook), real-space methods have appeared to be appropriate for imposing natural localization constraints on the orbitals (HERNANDEZ and GILLAN [1995], HOSHI and FUJIWARA [1997], FATTEBERT and BERNHOLC [2000]). Such an approach leads to a dramatic reduction in computer time and memory requirements for very large systems. However, since these algorithms are very recent and useful only for systems of sizes close to the limit of computing resources available today, these methods are still in development and many open questions remain.

Other advantages of FD over standard PW approaches include the possibility of introducing local mesh refinements (GYGI and GALLI [1995], MODINE, ZUMBACH and KAXIRAS [1996], FATTEBERT [1999]) and Dirichlet boundary conditions for non-periodic systems in a very natural fashion. Some of these aspects are discussed, for example, in a recent review paper by BECK [2000]. If local mesh refinement is a requirement for all electrons calculations, many pseudopotential calculations can be carried out with a perfectly regular mesh. Also, local mesh refinements involve many complications such as load balancing for implementation on parallel computers, or Pulay forces. To our knowledge, no large scale first-principles molecular dynamics with local mesh refinements have been carried out so far and this aspect of FD methods will not be further discussed here.

The computational advantages of real-space approaches have also motivated some research in 3D finite elements (FE) methods for electronic structure calculations (WHITE, WILKINS and TETER [1989], MURAKAMI, SONNAD and CLEMENTI [1992], TSUCHIDA and TSUKADA [1995], KOHN, WEARE, ONG and BADEN [1997], PASK, KLEIN, FONG and STERNE [1999]). However, since electronic structure calculations in general require computation domains of very simple shapes like rectangular cells, so far FE methods have not demonstrated real advantages over FD methods for regular grids like those currently used with pseudopotentials. On the other hand, FE methods can be considerably more expensive.

Advanced numerical methods like those presented in this chapter are useful only if they allow to treat real problems in solid state physics or physical chemistry. Besides the foreseen improvement in performance for calculations of large scale systems, localized orbital adapted to their chemical environment offer the possibility of novel computational applications. In particular, the possibility of obtaining accurate localized orbital basis would be, in principles, a useful starting point for formal developments such as the semiclassical theory of electron dynamics or the theory of magnetic interactions in solids, in a natural extension of the Wannier representation of crystal wave functions (ASHCROFT and MERMIN [1976], Chapter 10, p. 187 and ff.). Among the possible applications of localized orbital methods, one of increasing technological importance is the calculation of the quantum transport properties of nanostructures.

The current limits of semiconductor electronics and the challenges for future developments involve the continuous shrinking of the physical dimensions of the devices and the attainment of higher speeds. The drive to produce smaller devices has lead the current research toward a new form of electronics in which nanoscale objects, such as clusters or molecules, replace the transistors of today's silicon technology. However, the production and integration of nanoscale individual components into easily reproducible device structures presents many challenges, both experimentally and theoretically. From the theoretical point of view, the design of such devices requires explicit modelling of quantum propagation of electrons in nanoscale systems. The quantity to be calculated is the quantum conductance, that is the measure of the ease with which electrons will transmit through a conductor, or alternatively of the resistance that electrons will encounter in their flow. As we will see in the following sections, the evaluation of conductance in nanostructures requires an electronic structure calculation of the system under consideration, the computation of its Green's function, and an accurate treatment of the coupling to and scattering at the contacts.

Recent years have witnessed a great amount of research in the field of quantum conductance in nanostructures (BEENAKKER and VAN HOUTEN [1991]). These have become the systems of choice for investigations of electrical conduction at mesoscopic scale. The improvements in nanostructured material production have stimulated developments in both experiment and theory. In particular, the formal relation between conduction and transmission, the Landauer formula (LANDAUER [1970]), has enhanced the understanding of electronic transport in extended systems and has proven to be very useful in interpreting experiments involving the conductance of nanostructures.

The problem of understanding the transport behavior of nanoscale structures cannot be effectively solved without a fully *ab initio* methodology. Only the latter is able to accurately describe the behavior of the electrons in the highly inhomogeneous environment of the nanoscale device, as well as account for the charge transfer and the interactions within the nano-system. Most of the existing methods to compute conductance from *ab initio* methods are based on the solution of the quantum scattering problem for the electronic wave functions through the conductor using a number of related techniques. Lippman–Schwinger and perturbative Green's function methods have been used to study conductance in metallic nanowires and recently in small molecular nanocontacts (LANG [1995], DI VENTRA, PANTELIDES and LANG [2000]). Conduction in nanowires, junctions and nanotube systems has been addressed using nonlocal pseudopotentials methods (CHOI and IHM [1999], CHOI, IHM, LOUIE and COHEN [2000]) and through the solution of the coupled channels equations in a scattering-theoretic approach (HIROSE and TSUKADA [1995], KOBAYASHI, BRANDBYGE and TSUKADA [2000], LANDMAN, BARNETT, SCHERBAKOV and AVOURIS [2000]). The above methods compute *ab initio* transport using a PW representation of the electronic wave functions. This imposes severe restrictions on the size of the system because of the large number of basis functions necessary for an accurate description of the electron transmission process. Only recently real-space approaches been considered for a more efficient solution of the electronic transport problem. They are based on the use of LCAO (YOON, MAZZONI, CHOI, IHM and LOUIE [2001], TAYLOR, GUO and WANG [2000]) or Gaussian orbital bases (YALIRAKI, ROITBERG, GONZALEZ, MUJICA and RATNER [1999]). These are combined with either a scattering state solution for the transmission (YOON, MAZZONI, CHOI, IHM and LOUIE [2001]) or Green's function-based techniques (YALIRAKI, ROITBERG, GONZALEZ, MUJICA and RATNER [1999], TAYLOR, GUO and WANG [2000]).

In this chapter we do not intend to cover the wide variety of techniques that have been developed to compute quantum conductance, both from *ab initio* or from more phenomenological approaches. For such an exhaustive task, we refer the interested reader to the excellent monographs by DATTA [1995] and FERRY and GOODNICK [1997] and to the references at the end of this chapter. On the contrary, we have chosen to outline the main steps of the approach derived by us in the context of localized orbital methods (BUONGIORNO NARDELLI, FATTEBERT and BERNHOLC [2001]). In this chapter we will limit the discussion to the linear response regime and thus to zero bias across the conductor-lead junctions.

This chapter is organized as follows. In Section 2, we review the FD method in the context of the Kohn–Sham (KS) equations and pseudopotentials approach, also describing some specific features of iterative algorithms used to solve the KS equations in real-space. We also review the computation of the forces to optimize geometries and carry out first principles molecular dynamics. We finish Section 2 with some more advanced features designed to reduce the computational cost of the method in a localized orbitals representation. In Section 3, we review a numerical method to compute quantum conductance through nanostructures. This method is based on a description of the electronic structure in a basis of localized orbitals. In Section 4, we conclude this chapter by an illustration that brings together the localized grid-based orbitals and the quantum conductance algorithm to compute *ab initio* quantum conductance of a carbon nanotube on an aluminium surface.

2. Electronic structure calculation by finite differences

2.1. Kohn–Sham equations

KS theory (KOHN and SHAM [1965]) is a widely used model for first-principles calculations (see, e.g., CANCÈS, DEFRANCESCHI, KUTZELNIGG, LE BRIS and MADAY [2003], this volume). It states that the electronic ground state of a physical system can be described by a system of orthogonal one-particle electronic wave functions ψ_j, $j = 1, \ldots, N$, that minimizes the KS total energy functional E_{KS}. To simplify the discussion we neglect here the spin of the electrons by allowing double occupations of the orbitals, so that the electronic density is defined as

$$\rho_e(\vec{r}) = \sum_{i=1}^{N} f_i |\psi_i(\vec{r})|^2, \tag{2.1}$$

where $0 \leqslant f_i \leqslant 2$, $i = 1, \ldots, N$, are the occupation numbers. We also assume that the system is neutral, that is, the total charge of the electrons neutralizes exactly the nuclei charges.

For a molecule composed of N_a atoms located at positions $\{\vec{R}_a\}_{a=1}^{N_a}$ in a computation domain Ω, the KS energy functional is given by (in atomic units)

$$
\begin{aligned}
E_{KS}\left[\{\psi_i\}_{i=1}^{N}, \{\vec{R}_a\}_{a=1}^{N_a}\right] = \sum_{i=1}^{N} f_i \int_\Omega \psi_i^*(\vec{r}) \left(-\frac{1}{2}\nabla^2\right) \psi_i(\vec{r}) \mathrm{d}\,\vec{r} \\
+ \frac{1}{2} \int_\Omega \int_\Omega \frac{\rho_e(\vec{r}_1)\rho_e(\vec{r}_2)}{|\vec{r}_1 - \vec{r}_2|} \mathrm{d}\vec{r}_1 \mathrm{d}\vec{r}_2 \\
+ E_{xc}[\rho_e] + \int_\Omega \psi_i^*(\vec{r})(V_{ext}\psi_i(\vec{r})\mathrm{d}\vec{r} .
\end{aligned}
\tag{2.2}
$$

The first term represents the kinetic energy of the electrons. The second represents the electrostatic energy of interaction between electrons that we will note E_{es}. E_{xc} models the exchange and correlation between electrons. In this chapter, we will use the local density approximation (LDA), or the first-principles exchange-correlation functional proposed by Perdew–Burke–Ernzerhof (PBE) (PERDEW, BURKE and ERNZERHOF [1996]) which often provides results in better agreement with experiments and is appropriate for a grid-based implementation. In the last term of Eq. (2.2), the potential V_{ext} represents the total potential produced by the atomic nuclei at positions $\{\vec{R}_a\}_{a=1}^{N_a}$.

The ground state of a physical system is represented by orbitals that minimize the energy functional (2.2) under the constraints that the ψ_j are orthonormal. This minima can be found by solving the associated Euler–Lagrange equations – KS equations (KOHN and SHAM [1965]) –

$$H\psi_j = \left[-\frac{1}{2}\nabla^2 + v_H(\rho_e) + \mu_{xc}(\rho_e) + V_{ext}\right]\psi_j = \varepsilon_j \psi_j, \tag{2.3}$$

which must be solved self-consistently for the N lowest eigenvalues ε_j, while imposing the orthonormality constraints $\langle \psi_i \mid \psi_j \rangle = \delta_{ij}$. We use the usual quantum mechanics notation $\langle \cdot | \cdot \rangle$ for the L^2 scalar product. The Hartree potential v_H represents the Coulomb potential due to the electronic charge density ρ_e, and $\mu_{xc} = \delta E_{xc}[\rho_e]/\delta \rho_e$ is the exchange and correlation potential.

2.2. Finite differences approach

In order to discretize the KS equations, we introduce a real-space rectangular grid Ω_h of mesh spacing h_x, h_y, h_z in the directions x, y, z that covers the computation domain Ω. Let M be the number of grid points. The wave functions, potentials and the electronic density are represented by their values at the grid points $\vec{r}_{ijk} = (x_i, y_i, z_k)$. Integrals over Ω are performed using the discrete summation rule

$$\int_\Omega u(\vec{r})\,d\vec{r} \approx h_x h_y h_z \sum_{i,j,k\in\Omega_h} u(\vec{r}_{ijk}).$$

Given the values of a function $u(\vec{r})$ on a set of nodes $\vec{r}_{i,j,k}$ the traditional FD approximation $w_{i,j,k}$ to the Laplacian of the function at a given node is expressed as a linear combination of values of the function at the neighboring nodes

$$w_{i,j,k} = \sum_{n=-p}^{p} c_n\big(u(x_i + nh_x, y_j, z_k) + u(x_i, y_j + nh_y, z_k) + u(x_i, y_j, z_k + nh_z)\big) \qquad (2.4)$$

where the coefficients $\{c_n\}$ can be computed from the Taylor expansion of u near $\vec{r}_{i,j,k}$. Such an approximation has an order of accuracy $2p$, that is for a sufficiently smooth function u, $w_{i,j,k}$ will converge at the rate $O(h^{2p})$ as the mesh spacing $h \to 0$. For the second order approximation, for example ($p = 1$), we have $c_0 = 2/h^2$ and $c_1 = c_{-1} = -1/h^2$. High order versions of this scheme were first used in electronic structure calculations by CHELIKOWSKY, TROUILLER and SAAD [1994].

As an alternative, one can also use a compact FD scheme (also called *Mehrstellenverfahren* in COLLATZ [1966]). For example, a fourth order FD scheme for the Laplacian on a cubic grid is based on the relation

$$
\begin{aligned}
&\frac{1}{6h^2}\left\{24u(\vec{r}_0) - 2 \sum_{\substack{\vec{r}\in\Omega_h, \\ \|\vec{r}-\vec{r}_0\|=h}} u(\vec{r}) - \sum_{\substack{\vec{r}\in\Omega_h, \\ \|\vec{r}-\vec{r}_0\|=\sqrt{2}h}} u(\vec{r})\right\} \\
&= \frac{1}{72}\left\{48(-\nabla^2 u)(\vec{r}_0) + 2 \sum_{\substack{\vec{r}\in\Omega_h, \\ \|\vec{r}-\vec{r}_0\|=h}} (-\nabla^2 u)(\vec{r}) + \sum_{\substack{\vec{r}\in\Omega_h, \\ \|\vec{r}-\vec{r}_0\|=\sqrt{2}h}} (-\nabla^2 u)(\vec{r})\right\} \\
&+ O(h^4),
\end{aligned}
\qquad (2.5)
$$

valid for a sufficiently differentiable function $u(\vec{r})$. For simplicity, we have assumed here that $h_x = h_y = h_z$, but this expression is easy to extend to the general case. This FD scheme requires only values at grid points within a sphere of radius $\sqrt{2}h$. Beside its good numerical properties, the compactness of this scheme reduces the amount of communications in a domain-decomposition based parallel implementation. While increasing the order of the FD scheme improves the accuracy for very fine grids, it may not be the case for a given computational grid. In practice, this compact fourth order scheme consistently improves the accuracy compared to a standard fourth order scheme, as illustrated in Fig. 2.1.

REMARK 2.1. The FD method is not variational, and by refining the mesh the total energy generally increases toward convergence.

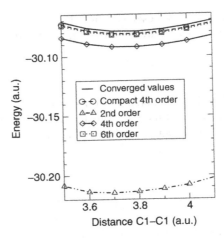

FIG. 2.1 Total energy for a Cl_2 molecule as a function of the distance Cl–Cl for several finite difference schemes, using the PBE exchange and correlation functional and pseudopotentials by HAMANN [1989]. The grid spacing used ($h = 0.34$) is sufficient to accurately compute the equilibrium bond length and binding energy of the molecule using the schemes of order 4 and 6, while it is clearly too coarse for the second order scheme. The top line shows a fully converged result.

It is easy to see that a compact FD scheme like (2.5) leads to an eigenvalue problem of the form

$$(L_h + B_h V_h)\vec{\psi}_i = \varepsilon_i B_h \vec{\psi}_i, \tag{2.6}$$

where L_h represents the FD scheme on the left-hand side of Eq. (2.5) and $B_h \in \mathcal{M}_M$ is a sparse well conditioned matrix that represents the FD-like scheme on the right-hand side of Eq. (2.5). V_h represents the potential on the grid. Let $H_h = L_h + B_h V_h$. One can show that $B_h^{-1} H_h$ is symmetric (B_h and L_h commute) so that the eigenvalues ε_i are real and the eigenvectors ψ_i can be chosen orthogonals.

This type of compact FD scheme was simultaneously introduced in electronic structure calculations by BRIGGS, SULLIVAN and BERNHOLC [1995] and FATTEBERT [1996]. The simulations presented in this chapter are based on this scheme. However, to simplify the notations, we will drop the matrix B_h in the rest of this chapter. In general, from the point of view of computer time and memory requirements, FD schemes of order larger than 2 are clearly worthwhile since they allow to work with much coarser grids. This reduces the cost of operations like scalar products between trial eigenfunctions or linear combinations of trial eigenfunctions in the iterative solver (see Section 2.5).

2.3. General formulation in nonorthogonal orbitals

Most of the time, we are not interested in the individual eigenfunctions solutions of the KS equations, but only in the subspace spanned by these functions. It means that one can represent the subspace of the electronic orbitals by a more general basis of nonorthogonal functions, $\{\phi_1, \ldots, \phi_N\}$. We write these functions as vectors, columns of a matrix Φ,

$$\Phi = (\vec{\phi}_1, \ldots, \vec{\phi}_N).$$

An orthonormal basis of approximate eigenfunctions (Ritz functions) can be obtained by a diagonalization in this subspace of dimension N (Ritz procedure). We denote by $C \in \mathcal{M}_N$ the matrix that transforms Φ into the basis Ψ of orthonormal Ritz functions,

$$\Psi = (\vec{\psi}_1, \ldots, \vec{\psi}_N) = \Phi C. \tag{2.7}$$

The matrix C satisfies

$$CC^T = S^{-1},$$

where $S = \Phi^T \Phi$ is the overlap matrix.

In the following, for an operator A we will use the notation

$$A^{(\Phi)} = \Phi^T A \Phi.$$

We then have the relation

$$A^{(\Psi)} = C^T A^{(\Phi)} C.$$

Equation (2.7) defines a transformation to a Ritz basis only if C is a solution of the generalized symmetric eigenvalue problem

$$H^{(\Phi)} C = SC\Lambda, \tag{2.8}$$

where $\Lambda \in \mathcal{M}_N$ is a diagonal matrix that satisfies $\Lambda = \Psi^T H \Psi$. The matrix C can actually be decomposed as a product $C = L^{-T} U$, where L is the Cholesky factorization of S,

$$S = LL^T,$$

and U is an orthogonal matrix. Knowing L, the generalized eigenvalue problem (2.8) is reduced to a standard symmetric eigenvalue problem

$$L^{-1} H^{(\Phi)} L^{-T} U = U\Lambda. \tag{2.9}$$

For a chemical potential μ, let us define $\Upsilon \in \mathcal{M}_N$ by its matrix elements

$$\Upsilon_{ij} = \delta_{ij} f[(\varepsilon_i - \mu)/k_B T],$$

where f is a Fermi–Dirac distribution at temperature T and k_B is the Boltzmann constant. The density operator $\hat{\rho}$ is then defined as

$$\hat{\rho} = \Psi \Upsilon \Psi^T = \Phi C \Upsilon C^T \Phi^T.$$

For $T = 0$, $\hat{\rho}$ is a projector onto the states of eigenvalues lower than μ. The dimension of this density matrix is given by the number of degrees of freedom, that is, the number of grid points in a grid-based approach. This number is in general so large that it is impossible to apply numerical methods that require $\rho(\vec{r}, \vec{r}')$ (matrix of size $M \times M$). However, it is useful to represent $\hat{\rho}$ in the basis Φ

$$\rho^{(\Phi)} = \Phi^T \hat{\rho} \Phi = C^{-T} \Upsilon C^{-1}.$$

Even more useful is the matrix $\bar{\rho}^{(\Phi)}$

$$\bar{\rho}^{(\Phi)} = S^{-1} \rho^{(\Phi)} S^{-1} = C \Upsilon C^T. \tag{2.10}$$

This matrix appears naturally in the expression used to compute the expectation value $\bar{A}$ of an operator A represented in the basis Φ,

$$\bar{A} = 2\mathrm{tr}\left(\Upsilon A^{(\Psi)} \right) = 2\mathrm{tr}(\bar{\rho}^{(\Phi)} A^{(\Phi)}).$$

In particular, the total number of electrons in the system is given by

$$N_e = 2\mathrm{tr}(\bar{\rho}^{(\Phi)} S).$$

Also, the electronic density in a nonorthogonal orbitals formulation is simply given by

$$\rho_e(\vec{r}) = 2 \sum_{j,k=1}^{N} (\bar{\rho}^{(\Phi)})_{jk} \phi_j(\vec{r}) \phi_k(\vec{r}). \tag{2.11}$$

REMARK 2.2. If all the computed states are fully occupied, we have $\rho^{(\Phi)} = S$ and $\bar{\rho}^{(\Phi)} = S^{-1}$.

2.4. Pseudopotentials

On regular grids, FD methods, like PW, are not very efficient to describe singular atomic potentials accurately. In particular for microcanonical molecular dynamics, it is difficult to guarantee a good conservation of the total energy of the system. These singularities can however be removed by replacing the atomic nuclei and core electrons – which can be approximated as frozen in their atomic state – by pseudopotentials. In the pseudopotential approach, the electronic structure calculation problem is reduced to the computation of a cloud of valence electrons living in a background of positive ions represented by smooth nonsingular pseudopotentials.

Accurate calculations can be performed by representing each atomic core by a nonlocal separable pseudopotential in its Kleinman–Bylander form (KLEINMAN and BYLANDER [1982])

$$V_{\mathrm{ps}} = V_{\mathrm{local}} + V_{\mathrm{nl}} = v_{\mathrm{ps}}^{\mathrm{local}}(\vec{r}) + \sum_{\ell=0}^{\ell_{\mathrm{max}}} \sum_{m=-\ell}^{\ell} |v_\ell^m\rangle E_\ell^{\mathrm{KB}} \langle v_\ell^m|, \tag{2.12}$$

where E_ℓ^{KB} are normalization coefficients. The function $v_{\mathrm{ps}}^{\mathrm{local}}$ contains the long range effects and is equal to $-Z/r$ outside of the core. The functions $v_\ell^m(\vec{r})$ are the product of a spherical harmonics Y_ℓ^m by a radial function $v_\ell(r)$ – centered on an atom – which vanishes beyond some critical radius. Being separable means that the matrix elements $\langle \psi_i | V_{\mathrm{nl}} | \psi_j \rangle$ can be computed efficiently according to

$$\langle \phi_j | V_{\mathrm{nl}} | \phi_k \rangle = \sum_{\ell=0}^{\ell_{\mathrm{max}}} \sum_{m=-\ell}^{\ell} E_\ell^{\mathrm{KB}} f_{\ell m}(\phi_j) f_{\ell m}(\phi_k), \tag{2.13}$$

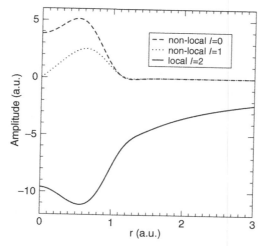

FIG. 2.2 Example of atomic pseudopotential: Chlorine for the PBE functional. The radial functions plotted are the radial components of the projector functions, v_ℓ for $\ell = 0, 1$ and the local pseudopotential $v_{\text{ps}}^{\text{local}}(\ell = 2)$.

where the quantities $f_{\ell m}(\phi_n) = \langle \phi_n | v_\ell^m \rangle$ can be computed independently of each other. Since the functions v_ℓ^m are localized in real-space, the evaluation of all the matrix elements $\langle \psi_j | V_{\text{nl}} | \psi_k \rangle$ for a system of N atoms scales like $O(N^2)$.

In the applications presented in this chapter, we use the pseudopotentials proposed by HAMANN [1989]. An example is represented in Fig. 2.2 (Chlorine).

REMARK 2.3. All the atoms of the periodic table cannot be represented by pseudopotentials with the same degree of smoothness. To be represented accurately, atomic species like Oxygen or Nitrogen require finer discretization grids than Silicon for example. This makes the calculations of the electronic structure of a crystal of 64 Silicon atoms 5–10 times cheaper than the simulation of a cell of liquid water with 32 molecules, even if the number of valence electrons to compute is the same in both cases.

Using periodic boundary conditions, the total energy of a system is in principle invariant under spatial translations. Unlike in a PW approach, a real-space finite grid representation breaks this invariance (BRIGGS, SULLIVAN and BERNHOLC [1995]). However, a discretization grid fine enough ensures that the total energy is conserved under translation within the minimal accuracy required in the calculation. Also, in order to avoid problems related to energy variations under spatial translations, the pseudopotentials can be filtered (BRIGGS, SULLIVAN and BERNHOLC [1996]). The pseudopotentials (local part and projectors) are first transformed to Fourier space by the Fourier transform

$$V_{\ell m}(\vec{k}) = \frac{1}{(2\pi)^{3/2}} \int_{\mathbb{R}^3} v_\ell^m(\vec{r}) e^{-i\vec{k}\cdot\vec{r}} d\vec{r} = C_{\vec{k},\ell,m} \int_0^\infty v_\ell(r) j_\ell(|\vec{k}|r) r^2 dr. \qquad (2.14)$$

The index (ℓ, m) denotes the symmetry of the functions $v_\ell^m(\vec{r}) = v_\ell(r) Y_\ell^m(\theta, \phi).j_\ell(r)$ is a spherical Bessel function of order ℓ and $C_{\vec{k},\ell,m}$ is a complex factor depending on $\vec{k}, \ell, m$. For $|\vec{k}|$ larger than a cutoff k_{cut}, the coefficients $V_{\ell m}(\vec{k})$ are filtered by a Gaussian function $\mathrm{e}^{-\beta(|\vec{k}|/k_{\text{cut}}-1)^2}$ before applying an inverse Fourier transform. One can use, for example, $k_{\text{cut}} = 2\pi/3h$. In practice, since the filtering depends only on $|\vec{k}|$, the coefficients $C_{\vec{k},\ell,m}$ do not need to be computed and only the 1D radial integral is required for an appropriate range of $|\vec{k}|$. The pseudopotentials lose in general their localization properties in real-space after being filtered in Fourier space and a second filtering in real-space is required to ensure that the nonlocal projectors remain confined within a limited radius. This second filtering can also be done by a Gaussian function to avoid reintroducing too many high frequency components.

Of course this filtering procedure modifies the pseudopotentials to a degree set by the grid spacing. These filtered functions will however converge toward the true pseudopotentials – together with the wave functions – as one decreases the mesh spacing. For every atomic species, one should then carefully check what grid spacing is required to ensure that the physical quantities of interest are well converged.

2.5. *Solving the eigenvalue problem*

The KS equations discretized by FD result in a huge 3D eigenvalue problem. Fortunately, the matrices involved are very sparse and efficient iterative methods can be used to solve this problem. In this chapter, we are going to restrict the discussion to the particularities of eigenvalue solvers for FD discretizations. After introducing some general features of minimization processes, we describe an appropriate multigrid preconditioner for real-space discretizations.

The multigrid full-approximation scheme (FAS) originally proposed by BRANDT [1977] is an efficient solver for nonlinear problems on a grid. In this algorithm, the entire problem has to be represented on all the grids, from the coarsest to the finest, in order to treat on equal footing all the length scales of the solution. Such an approach is not obvious to apply to the KS eigenvalue problem. Indeed, the numerous eigenfunctions of interest loose their meaning on very coarse grids, and one may be limited in the number of usable coarse grids. Only a few successful applications of the FAS algorithm for electronic structure calculations have been reported so far. They have been limited to purely academic problems (COSTINER and TAASAN [1995]) and static all electrons calculations of atoms and diatomic molecules (WANG and BECK [2000]). Recently, HEISKANEN, TORSTI, PUSKA and NIEMINEN [2001] proposed to use the Rayleigh quotient multigrid (RQMG) method (MANDEL and MC CORMICK [1989]) that avoids the coarse grid representation problem. So far it was applied only for static electronic structure calculations.

Various other methods are based on minimization schemes that make use of the steepest descent (SD) direction along which the energy functional decreases at the fastest rate. This direction is given by the gradient of the energy functional with the opposite sign. In the basis Ψ, this direction – in a space of dimension $M \times N$ – is easy to compute since it is given by the negative residual of the KS equations (2.3) and can be expressed as an $N \times M$ matrix

$$D^{(\Psi)} = \Psi\Lambda - H\Psi. \tag{2.15}$$

One verifies that this gradient satisfies the relation $\Psi^T D^{(\Psi)} = 0$.

For an optimum convergence rate, it is important to use the *true* SD direction in algorithms expressed in nonorthogonal orbitals formulation. This direction can differ substantially from the derivative with respect to Φ if the basis Φ is highly nonorthogonal. This SD direction is easy to compute for the eigenfunctions Ψ and a simple way to obtain it in the basis Φ is to use the matrix C (from Eq. (2.7)) to derive

$$D^{(\Phi)} = D^{(\Psi)}C^{-1} = (\Psi\Lambda - H\Psi)C^{-1} = \Phi\Theta - H\Phi, \tag{2.16}$$

where $\Theta = S^{-1}H^{(\Phi)}$. In the following, we consider an SD algorithm with a linear preconditioning operator K. The basis Φ is updated according to

$$\Phi^{new} = \Phi + \eta K(\Phi\Theta - H\Phi), \tag{2.17}$$

where η is a pseudo-time step. In this algorithm all the trial wave functions are updated simultaneously. In the particular case $K = Identity$, (2.17) is equivalent to the method proposed in GALLI and PARRINELLO [1992]. Since by definition

$$\Phi^{new} = (\Psi + \eta K D^{(\Psi)})C^{-1} = \Psi^{new}C^{-1},$$

the same subspace is generated at each iteration, independently of the choice of the basis Φ. Note also that Eq. (2.17) does not depend on C and therefore does not require the solution of the eigenvalue problem (2.8).

REMARK 2.4. Alternatively, the above SD directions can be used in a conjugate gradient (CG) method (EDELMAN, ARIAS and SMITH [1998]). In the applications presented here, the convergence rate of the preconditioned steepest descent (PSD) algorithm with the preconditioning presented later in this Section is fast enough to make the implementation of the CG approach unnecessary.

In actual calculations, the basis functions Φ are corrected at each iteration using the PSD directions as in (2.17). A new electronic density ρ_e is then computed as well as new Hartree and exchange-correlation potentials. To avoid large oscillations of the charge distribution from step to step, these potentials can be mixed linearly with those used at the previous step. The basis Φ is refined by iterative updates until self-consistency (SC) is achieved and, at convergence accurately describe the true KS ground state of the system.

REMARK 2.5. For iterative algorithms based on the Ritz vectors, the transformation (2.7) is one of the most expensive part of the calculation for large scale simulations. In a nonorthogonal representation, this operation is not required anymore. The cost is however transferred to the computation of the SD directions (Eq. (2.16)).

In the particular case of a linear Hamiltonian operator and if all the orbitals are fully occupied, the KS functional expressed in nonorthogonal orbitals can be written as a functional without constraints

$$E_{KS} = 2\text{tr}(S^{-1}\Phi^T H\Phi). \tag{2.18}$$

To motivate the introduction of a preconditioner, we first estimate the expected rate of convergence of iterative algorithms based on SD directions to minimize Eq. (2.18). The rate of convergence of the SD method is determined by the condition number $\chi(\mathcal{H})$ of the Hessian matrix $\mathcal{H}$ associated to the problem. To estimate $\chi(\mathcal{H})$ we compute the eigenvalues of $\mathcal{H}$. As in PFROMMER, DEMMEL and SIMON [1999], we consider electronic states $\{\phi_i\}_{i=1}^N$ expressed as a perturbation of the ground state eigenfunctions $\{\psi_i\}_{i=1}^N$

$$\phi_i = \psi_i + \sum_{l=1}^{M} c_l^{(i)} \psi_l. \tag{2.19}$$

Inserting (2.19) into (2.18), we obtain to the second order in the coefficients $c_l^{(i)}$

$$E_{KS} - E_0 = 2 \sum_{i=1}^{N} \sum_{k=N+1}^{M} (\varepsilon_k - \varepsilon_i)(c_k^{(i)})^2. \tag{2.20}$$

The coefficients $c_l^{(i)}$ for $l \leq N$ correspond to directions in the parameter space along which the objective function is constant. In the complementary parameter space, we obtain that the condition number of the Hessian matrix is given by

$$\chi(\mathcal{H}) = \frac{\varepsilon_M - \varepsilon_1}{\varepsilon_{N+1} - \varepsilon_N}. \tag{2.21}$$

REMARK 2.6. In general, as the number of atoms in a physical system increases, the spectrum of the Hamiltonian becomes denser, while the extreme eigenvalues (ε_M and ε_1) remain about the same.

As in PW calculations, real-space representations of the electronic wave functions require a very large number of degrees of freedom. In particular in the presence of atoms represented by very hard pseudopotentials. A very fine grid implies a quite large value for ε_M in Eq. (2.21) that negatively affects the condition number of the Hessian matrix. If Ψ is corrected at each step according to an SD algorithm without preconditioning

$$\Psi^{\text{new}} = \Psi + \eta D^{(\Psi)},$$

the parameter η has to be very small – for numerical stability reasons – and the convergence can be very slow.

REMARK 2.7. Equation (2.21) also points out at problems one can observe if the *band gap*
($\varepsilon_{N+1} - \varepsilon_N$) is very small. In practice, one can overcome this limitation by including more eigenstates than needed in the search subspace Ψ, the highest eigenstates being empty or fractionally occupied.

To introduce an appropriate preconditioner, we start by discussing the Rayleigh Quotient Iteration (RQI) method and some of its variants use in real-space electronic structure calculations. RQI is a very fast algorithm to compute one single eigenvalue of a matrix (e.g., PARLETT [1998], Chapter 4). Here we describe this method in a different form that we find more suitable for matrices of size too large to make use of direct linear solvers. Starting from an approximate eigenpair $(\varepsilon^{(k)}, \vec{\psi}^{(k)})$ of a discretized Hamiltonian matrix H, $\varepsilon^{(k)}$ given by the Rayleigh Quotient of $\vec{\psi}^{(k)}$ at step k, we look for an improved approximation $\vec{\psi}^{(k+1)}$. Specifically, we write

$$\vec{\psi}^{(k+1)} = \frac{1}{\xi}(\vec{\psi}^{(k)} + \delta\vec{\psi}^{(k)}),$$

where the correction $\delta\vec{\psi}^{(k)}$ is chosen orthogonal to $\vec{\psi}^{(k)}$. ξ is a normalization factor. Improving $\vec{\psi}^{(k)}$ by RQI requires then to find $\delta\vec{\psi}^{(k)} \perp \vec{\psi}^{(k)}$ and ξ such that

$$(H - \varepsilon^{(k)})\frac{1}{\xi}(\vec{\psi}^{(k)} + \delta\vec{\psi}^{(k)}) = \vec{\psi}^{(k)}. \tag{2.22}$$

We eliminate ξ by projecting the whole equation onto $(\vec{\psi}^{(k)})^\perp$ and then rewrite it as

$$(I - \vec{\psi}^{(k)}(\vec{\psi}^{(k)})^T)(H - \varepsilon^{(k)})\delta\vec{\psi}^{(k)} = -(H - \varepsilon^{(k)})\vec{\psi}^{(k)}. \tag{2.23}$$

By the properties of the Rayleigh Quotient, the projector on the right hand side of the equation has been omitted.

REMARK 2.8. Equation (2.23) is the same equation used to define the iterative corrections in the Jacobi–Davidson method (SLEIJPEN and VAN DER VORST [1996]).

This algorithm can be generalized to the simultaneous search of N eigenfunctions, with possible degeneracy of some eigenvalues (DESCLOUX, FATTEBERT and GYGI [1998], FATTEBERT [1998]). The idea is to replace the Rayleigh quotient by the Rayleigh–Ritz (RR) procedure (e.g., PARLETT [1998], Chapter 11) and to look for corrections orthogonal to the whole subspace $\Psi^{(k)}$ of trial eigenfunctions. Denoting $\Psi^{(k)} = (\vec{\psi}_1^{(k)}, \ldots, \vec{\psi}_N^{(k)})$ as a matrix made of vector columns $\vec{\psi}_j^{(k)}$, we can write the following iterative algorithm:

ALGORITHM 2.1
   (1). *Let $\Psi^{(0)}$ be a trial subspace of dimension N.*
   (2). *For $k = 0,1,2, \ldots$, do:*
      (a) *Rayleigh–Ritz for H in the subspace $\Psi^{(k)} \rightarrow (\varepsilon_j^{(k)}, \vec{\psi}_j^{(k)}), j = 1, \ldots, N,$*
      (b) *For $j = 1, \ldots, N$, compute $\delta\vec{\psi}_j^{(k)} \perp \Psi^{(k)}$ solution of*

$$(I - \Psi^{(k)}\Psi^{(k)T})(H - \varepsilon_j^{(k)})\delta\vec{\psi}_j^{(k)} = -(H - \varepsilon_j^{(k)})\vec{\psi}_j^{(k)}, \tag{2.24}$$

      (c) *Define $\Psi^{(k+1)} = (\vec{\psi}_1^{(k)} + \delta\vec{\psi}_1^{(k)}, \ldots, \vec{\psi}_N^{(k)} + \delta\vec{\psi}_N^{(k)}).$*

An exact solution of Eq. (2.24) would lead to a locally quadratic convergence rate close to the solution of the eigenvalue problem for a nonself-consistent (i.e., *linear*) Hamiltonian as proved in FATTEBERT [1998]. Such an algorithm has actually been applied in FATTEBERT [1996], FATTEBERT [1999], where the linear systems are solved by multigrid. Slightly different versions have been proposed by JIN, JEONG and CHANG [1999] and ANCILOTTO, BLANDIN and TOIGO [1999] who omit in particular the projector in Eq. (2.24). These approaches have to deal with the difficult question of how to define meaningful potentials – and sometimes eigenfunctions – on very coarse grids. BRIGGS, SULLIVAN and BERNHOLC [1996] proposed to keep only the Laplacian in the Hamiltonian operator on the coarse grids. In this approach, the multigrid V-cycles can be seen as a preconditioner or inexact solver. Such an approach can be very efficient for self-consistent Hamiltonians, since an accurate solution for Eq. (2.24) is not always useful when the operator changes at each iteration. This point of view has been more precisely formulated in FATTEBERT and BERNHOLC [2000] where the potential operator is used only once at the beginning of the multigrid cycle to compute the residual. The main advantage of the latter approach is that the operator in Eq. (2.24) then does not depend on j and can be used in any nonorthogonal representation Φ of the trial subspace.

Let us now focus on the preconditioning approach. Looking at the correction Eq. (2.24), we note that the right hand side is the SD direction for the minimization problem with orthonormality constraints associated with the KS eigenvalue problem (2.3). We note it $(-\vec{r}_j^{(k)})$, $\vec{r}_j^{(k)}$ being the residual of the eigenvalue problem. In a SD approach, $\delta\vec{\psi}_j^{(k)}$ would be given by $-\vec{r}_j^{(k)}/\varepsilon_{max}$ where ε_{max} is the largest eigenvalue of H. However, from the point of view of the inverse iteration method, an *optimal* correction is given by

$$\delta\vec{\psi}_j^{(k)} = -((I - \Psi^{(k)}\Psi^{(k)^T})(H - \varepsilon_j^{(k)})|_{\Psi^{(k)\perp}})^{-1}\vec{r}_j^{(k)}. \tag{2.25}$$

Thus one can consider a preconditioner K that approximates the operator

$$(I - \Psi^{(k)}\Psi^{(k)^T})(H - \varepsilon_j^{(k)})|_{\Psi^{(k)\perp}}. \tag{2.26}$$

A close look at the Hamiltonian operator shows that for high energy states, the Laplacian is the dominant part, and the corresponding eigenfunctions are essentially similar to those of the Laplacian, that is, PW perturbed by a relatively weak potential. It means that the operator $-\frac{1}{2}\nabla^2$ is a good approximation of $(H - \varepsilon_j^{(k)})$ in $\Psi^{(k)\perp}$ for $1 \le j \le N$, at least close to convergence, so that one can choose

$$K \sim (I - \Psi^{(k)}\Psi^{(k)^T})\left(-\frac{1}{2}\nabla^2\right)\Big|_{\Psi^{(k)\perp}}$$

$$= (I - \Phi^{(k)}S^{-1}\Phi^{(k)^T})\left(-\frac{1}{2}\nabla^2\right)\Big|_{\Phi^{(k)\perp}}. \tag{2.27}$$

In real-space, one can associate frequencies with grid resolution. Applying a single grid iterative method – like Jacobi or Gauss–Seidel – to solve a Poisson problem, one essentially obtains the high frequency components of the solution, the one that we cannot represent on a coarser grid. Using multigrid V-cycles based on such a *smoother*, we can solve the problem for components of lower frequencies by visiting coarser grids (BRANDT [1977]). Furthermore, by choosing a limited number of grids, one can *select* the components that we want to solve for. Following this heuristic argument, we define the application of the preconditioner K^{-1} to $\vec{r}_j^{(k)}$ as an iterative multigrid solver for the Poisson problem (FATTEBERT and BERNHOLC [2000])

$$-\frac{1}{2}\nabla^2 \delta\vec{\psi}_j^{(k)} = -\vec{r}_j^{(k)}$$

(2.28)

limited to the finest grids. For practical calculations, using 2 coarse grids is often optimal. We start the process with an initial trial solution $\delta\vec{\psi}_j^{(k)} = -\alpha\vec{r}_j^{(k)}$. Its main goal is to introduce some low frequency components in the correction $\delta\vec{\psi}_j^{(k)}$. The coefficient α is defined by looking at the initial guess as the SD correction one would make if the whole calculation was done on the coarse grids not visited during the V-cycles. As a smoother in the V-cycles, the Jacobi method is appropriate because of its inherent parallelism.

In Fig. 2.3, we present the convergence history of the error on the total energy for various discretization grids for an 8 atoms diamond cell self-consistent calculation.

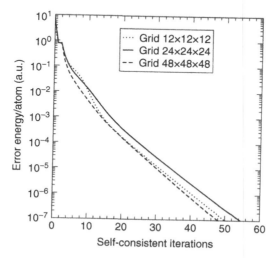

FIG. 2.3 Convergence rate for a diamond cell calculation (eight carbon atoms) for three different discretization grids.

We use the PSD algorithm with the multigrid preconditioner described above, doing for each correction $\vec{\delta\psi}_j^{(k)}$ and at each SC iteration 1 V-cycle with 2 presmoothing and 2 postsmoothing. The grid-independence of the convergence rate is observed. All the calculations use the same coarsest grid $6 \times 6 \times 6$ for the multigrid preconditioning, and the same total number of states (16 occupied + 8 unoccupied) $N = 24$. The initial trial functions are random functions.

REMARK 2.9. Preconditioners based on a similar idea have been developed for PW calculations (TETER, PAYNE and ALLAN [1989], FERNANDO, QUIAN, WEINERT and DAVENPORT [1989], CHETTY, WEINERT, RAHMAN and DAVENPORT [1995]). Since the numerical basis functions in PW are eigenfunctions of the Laplacian operator, efficient simple diagonal preconditioners can be designed for PW.

A different preconditioner was proposed by SAAD, STATHOPOULOS, CHELIKOWSKY, WU and OGUT [1996] in conjunction with a Lanczos algorithm. Realizing that the eigenfunctions corresponding to the lowest eigenvalues are in general smoother than the others, they proposed to apply a low frequency filter directly to the trial eigenfunctions. This is done on a single grid in real-space by an averaging of the value of a function at every grid point with the values at its neighboring points. They note however that this preconditioner is probably not always sufficient, in particular when a large number of eigenfunctions is required and the highest eigenvalues of interest correspond to eigenfunctions presenting a lower degree of smoothness.

2.6. *Energy and forces*

To optimize molecular geometries, or run molecular dynamics and measure physical quantities at finite temperature, it is important to be able to compute the forces acting on the atoms in any configuration. To derive expressions for these forces, we start by some considerations on the total energy of a physical system. In DFT, the total energy of a system can be expressed as the sum of three terms

$$E_t = E_{KS}[\Phi, \{\vec{R}_a\}_{a=1}^{N_a}] + E_{ions}[\{\vec{R}_a\}_{a=1}^{N_a}] + \frac{1}{2}\sum_{a=1}^{N_a} M_a \vec{\dot{R}}_a^2, \tag{2.29}$$

where E_{ions}, the electrostatic energy between ions of charges Z_i, is given by

$$E_{ions} = \frac{1}{2}\sum_{a,b=1,a\neq b}^{N_a} \frac{Z_a Z_b}{|\vec{R}_a - \vec{R}_b|}, \tag{2.30}$$

and M_a denotes the mass of the ion in $\vec{R}_a$. In the KS energy E_{KS}, the contribution due to the interaction between electrons and ions is given by the sum of two terms associated to the local and nonlocal parts of the pseudopotential

$$E_{ps} = \int_\Omega v_{ps}^{local}(\vec{r})\rho_e(\vec{r})d\vec{r} + \text{Tr}(\Phi^T V_{nl}\Phi_\rho^{(\Phi)}) = E_{ps}^{local} + E_{ps}^{nl}. \tag{2.31}$$

It is computationally more efficient to compute the electrostatic term E_{es} by solving a Poisson problem. In order to deal with a neutral charge, it is a standard procedure to add to the system smeared core charges centered at atomic sites,

$$\rho_a(\vec{r}) = -\frac{Z_a}{(\sqrt{\pi}r_c^a)^3}\exp\left(-\frac{|\vec{r} - \vec{R}_a|^2}{(r_c^a)^2}\right).$$ (2.32)

The sum of these charges, ρ_s, neutralizes the electronic charge by generating a total potential

$$v_s(\vec{r}) = \sum_{a=1}^{N_a} \frac{-Z_a}{|\vec{r} - \vec{R}_a|}\mathrm{erf}\left(\frac{|\vec{r} - \vec{R}_a|}{r_c^a}\right).$$ (2.33)

We then compute the Hartree potential v_H as the solution of a Poisson problem for a neutral total charge $\rho_e + \rho_s$,

$$-\nabla^2(v_H + v_s)(\vec{r}) = 4\pi(\rho_e + \rho_s)(\vec{r}),$$ (2.34)

with periodic or Dirichlet boundary conditions. This problem can be efficiently solved on the discretization grid in $O(N)$ operations by the multigrid method (BRANDT [1977]).

With the introduction of smeared neutralizing core charges, one can write

$$E_{ions} = \frac{1}{2}\sum_{a,b=1}^{N_a} \int_{\mathbb{R}^3} \frac{\rho_a(\vec{r} - \vec{R}_a)\rho_b(\vec{r}' - \vec{R}_b)}{|\vec{r} - \vec{r}'|}\,\mathrm{d}\vec{r}\mathrm{d}\vec{r}' - E_{self} + E_{diff},$$ (2.35)

where E_{self} is the self-interaction of the core charges,

$$E_{self} = \frac{1}{2}\sum_{a=1}^{N_a} \int_{\mathbb{R}^3} \frac{\rho_a(\vec{r} - \vec{R}_a)\rho_a(\vec{r}' - \vec{R}_a)}{|\vec{r} - \vec{r}'|}\,\mathrm{d}\vec{r}\mathrm{d}\vec{r}' = \frac{1}{\sqrt{2\pi}}\sum_{a=1}^{N_a}\frac{Z_a^2}{(r_c^a)^2}$$ (2.36)

and

$$E_{diff} = \frac{1}{2}\sum_{a,b=1,a\neq b}^{N_a}\left[\frac{Z_aZ_b}{|\vec{R}_a - \vec{R}_b|} - \int_{\mathbb{R}^3}\frac{\rho_a(\vec{r} - \vec{R}_a)\rho_b(\vec{r}' - \vec{R}_b)}{|\vec{r} - \vec{r}'|}\,\mathrm{d}\vec{r}\mathrm{d}\vec{r}'\right]$$

$$= \sum_{a,b=1,a<b}^{N_a}\frac{Z_aZ_b}{|\vec{R}_a - \vec{R}_b|}\mathrm{erfc}\left(\frac{|\vec{R}_a - \vec{R}_b|}{\sqrt{(r_c^a)^2 + (r_c^b)^2}}\right).$$ (2.37)

We then have, for r_c^a sufficiently small compared to Ω,

$$E_{es} + E_{ps}^{local} + E_{ions}$$

$$= \frac{1}{2} \int_\Omega \frac{\rho_e(\vec{r})\rho_e(\vec{r}')}{|\vec{r} - \vec{r}'|} d\vec{r} d\vec{r}' + \int_\Omega v_{ps}^{local} \rho_e(\vec{r}) d\vec{r}$$

$$+ \frac{1}{2} \sum_{a,b=1}^{N_a} \int_{\mathbb{R}^3} \frac{\rho_a(\vec{r} - \vec{R}_a)\rho_b(\vec{r}' - \vec{R}_b)}{|\vec{r} - \vec{r}'|} d\vec{r} d\vec{r}' - E_{self} + E_{diff}$$

$$\approx \frac{1}{2} \int_\Omega \frac{(\rho_e(\vec{r}) + \rho_s(\vec{r}))(\rho_e(\vec{r}') + \rho_s(\vec{r}'))}{|\vec{r} - \vec{r}'|} d\vec{r} d\vec{r}' - \int_\Omega \frac{\rho_e(\vec{r})\rho_s(\vec{r}')}{|\vec{r} - \vec{r}'|} d\vec{r} d\vec{r}'$$

$$+ \int_\Omega v_{ps}^{local}(\vec{r})\rho_e(\vec{r}) d\vec{r} - E_{self} + E_{diff}$$

$$= \frac{1}{2} \int_\Omega (\rho_e(\vec{r}) + \rho_s(\vec{r}))(v_H + v_s)(\vec{r}) d\vec{r}$$

$$+ \int_\Omega (v_{ps}^{local} - v_s)(\vec{r})\rho_e(\vec{r}) d\vec{r} - E_{self} + E_{diff}.$$

$$(2.38)$$

Knowing the ground state electronic structure for a given atomic configuration $\{\vec{R}_a\}_{a=1}^{N_a}$, one can compute the internal force acting on the ion I by deriving the total energy with respect to the atomic coordinates $\vec{R}_I$,

$$\vec{F}_I = -\frac{d}{d\vec{R}_I E_t(\Phi, \{\vec{R}_a\}_{a=1}^{N_a})}. \qquad (2.39)$$

Using the property that Φ is the minimum of the functional E, one shows that

$$\vec{F}_I = -\frac{\partial}{\partial \vec{R}_I} E_t(\Phi, \{\vec{R}_a\}_{a=1}^{N_a}) \qquad (2.40)$$

(Hellmann–Feynman forces, FEYNMAN [1939]). Since the electronic structure does not explicitly depend on the atomic positions, Eq. (2.40) means that the forces can be computed from a single ground state calculation, by deriving the atomic potentials only.

REMARK 2.10. To obtain Eq. (2.40), we also use the fact that the numerical representation of Ψ does not explicitly depend on the atomic positions since the grid is atom independent. For atom-centered orbitals moving with the atoms, this is not true anymore and additional terms (Pulay forces) have to be included.

From Eqs. (2.38) and (2.40), using Eq. (2.34), we obtain the total force acting on atom I in the form

$$\vec{F}_I = \int_\Omega (v_H(\vec{r}) + v_s(\vec{r})) \frac{d}{d\vec{R}_I} \rho_s d\vec{r} + \int_\Omega \frac{d}{d\vec{R}_I} (v_{ps}^{local} - v_s)\rho_e(\vec{r}) d\vec{r}$$

$$+\frac{\partial}{\partial\vec{R}_I}\mathrm{Tr}(\Phi^T V_{\mathrm{nl}}\Phi_\rho^{(\Phi)})+\frac{d}{d\vec{R}_I}E_{\mathrm{diff}}. \tag{2.41}$$

Writing the forces in this form lets appear the functions $(v_{\mathrm{ps}}^{\mathrm{local}}-v_s)$ and ρ_s which are localized in real-space. This can be directly used to reduce the complexity of the computation of the forces on a grid. In principle all the derivatives with respect to $\hat{R}_I$ in Eq. (2.41) can be computed analytically. In practice, because of the filtering of the pseudopotentials, the derivatives have to be evaluated numerically on the filtered pseudopotentials.

REMARK 2.11. The sum of the forces over all the atoms should be zero if no external force is applied. In practice, the use of a finite grid introduces small errors (see Section 2.4) and provides an estimate of the accuracy of the forces.

2.7. Born–Oppenheimer molecular dynamics

To perform Born–Oppenheimer molecular dynamics simulations of quantum systems described by the KS equations, we compute the forces acting on the ions according to Eq. (2.41) and let the system evolve accordingly. The ions evolve like classical particles surrounded by quantum electrons (Born–Oppenheimer approximation). The error in the energy is second order with respect to the error in the electronic wave functions, but the error in the forces is first order. It means that one should be particularly careful in the computation of the ground state of the KS energy functional for the each atomic configuration at each iteration. It is particularly important to have accurate forces to ensure a perfect conservation of the total energy of the system in a microcanonical simulation.

As shown by JING, TROULLIER, DEAN, BINGGELI and CHELIKOWSKY [1994] and BRIGGS, SULLIVAN and BERNHOLC [1996], the computation of the forces in FD methods is accurate enough to allow for energy conserving microcanonical *ab initio* simulations. This is illustrated in Fig. 2.4 where we show the evolution of the energy during a molecular dynamics simulation of a Si_5 cluster. To avoid any

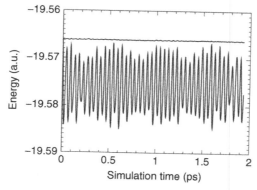

FIG. 2.4 Molecular dynamics simulation of a Si_5 cluster. The total energy (top) shows fluctuations of the order 10^{-4} a.u., but no systematic drift. The KS energy is also plotted (bottom).

systematic drift of the energy due to the integration scheme, the equations of motions were integrated numerically using the time reversible Verlet's second order algorithm (e.g., HEERMANN [1990], Chapter 3)

$$\vec{R}_I^{(n+1)} = 2\vec{R}_I^{(n)} - \vec{R}_I^{(n-1)} + \vec{F}_I^{(n)} (\Delta t)^2 / M_I$$

A grid spacing $h = 0.56$ Bohr and a time step $\Delta t = 80$ a.u. were used for this simulation.

2.8. Localized orbitals

Formulating the minimization problem in terms of nonorthogonal functions Φ instead of the Ritz functions Ψ, one can formally impose localization constraints on Φ to reduce the cost of the calculation. This is one of the most popular methods to obtain a linear scaling of the computational cost with respect to the size of the system. These so called Order N methods are discussed in another chapter (e.g., GOEDECKER [2003], this Handbook) and we will limit here the discussion to the grid-based method exposed in this chapter.

On a real-space grid, spatial localization can be imposed by forcing each orbital to be zero outside of a spherical region centered on a particular ion (HOSHI and FUJIWARA [1997], FATTEBERT and BERNHOLC [2000]). Such a truncation will linearize the computational cost of $D^{(\Phi)}$ in Eq. (2.16) and ρ_e in Eq. (2.11), the most expensive operations in the minimization algorithm. It also reduces to O(N) the storage requirements for the wave functions. Of course, this reduction in computational cost does not come for free. It introduces some approximation error that one expects to keep within a certain tolerance. It means in particular that one cannot choose the localization regions arbitrarily small.

The application of the compact FD Laplacian operator to a wave function localized in a sphere of radius R_c generates a function localized in a sphere of radius $R_c + h\sqrt{2}$, which is used as the localization radius for $H\phi_j$. This truncation suppresses some components of $H\phi_j$ that are generated by the nonlocal, short-range projectors of the pseudopotential operator. These components are lost in the correction of the wave functions since the latter should remain localized. However, they are included exactly in the computation of the matrix Θ and the total energy by writing $H^{(\Phi)}$ as the sum of two matrices

$$H^{(\Phi)} = \Phi^T(H - V_{\mathrm{nl}})\Phi + \Phi^T V_{\mathrm{nl}}\Phi,$$

which isolates the nonlocal potential V_{nl} in the second term. This second term is easily computed in O(N) operations using only the nonzero terms $\langle \phi_j | v_i \rangle$. Therefore, the only approximation in this approach is the use of a localization radius to limit the spatial extent of each nonorthogonal orbital.

Since the eigenfunctions are in general not localized, the matrix C that solves Eq. (2.8) is not sparse and the computation of C requires O(N^3) operations. To linearize the cost of the whole calculation, one could impose localization constraints on the density matrix, requiring $\bar{\rho}_{ij}^{(\Phi)} = 0$ if the localization regions of orbitals i and j are separated by a distance larger than a truncation radius R_ρ as in HERNANDEZ and GILLAN [1995]. This can be imposed at each step of the iterative minimization in order to achieve linear scaling. Such a truncature is justified by the exponential decay of $\rho(\vec{r}, \vec{r}')$ as $|\vec{r} - \vec{r}'| \to \infty$ in insulators or metals at a finite temperature (ISMAEL-BEIGI and ARIAS [1999]). However, for $M \gg N$, the full evaluation of S^{-1}, Θ or $\bar{\rho}^{(\Phi)}$, even with an order N^3 algorithm, constitutes a small fraction of the total calculations for a large range of system sizes. Since a good accuracy can be obtained only by keeping the number of nonzero elements in $\bar{\rho}$ much larger than in S (MILLAM and SCUSERIA [1997]), using the sparsity of $\bar{\rho}$ does not lead to much gain in this context.

REMARK 2.12. If an exact and explicit $O(N^3)$ diagonalization is performed, in Eq. (2.8), partially occupied and unoccupied orbitals can be used, which permits calculations for metallic as well as semiconducting systems. However, calculations for metals may require more localized orbitals or larger localization radii for an accurate description of their electronic structure.

For systems with $N > 1000$, solving Eq. (2.8) on a single processor becomes very expensive, if required at each self-consistent iteration. However, for fully parallel calculations, it is natural to also parallelize the $N \times N$ submatrices operations. This can be done using standard libraries, such as PBLAS (Parallel Basic Linear Algebra Subprograms) and ScaLapack (BLACKFORD, CHOI, CLEARY, D'AZEVEDO, DEMMEL, DHILLON, DONGARRA, HAMMARLING, HENRY, PETITET, STANLEY, WALKER and WHALEY [1997]).

REMARK 2.13. According to PBLAS and ScaLapack requirements, S, $H^{(\Phi)}$, Θ and $\bar{\rho}^{(\Phi)}$ are stored as full $N \times N$ matrices, distributed among the processors. Although most of the operations on these matrices can be optimized using their sparsity – except the diagonalization in Eq. (2.9) – the full storage approach is adequate for a substantial range of calculations. It is also the easiest implementation given the available standard numerical libraries for distributed memory multiprocessors computers. The solution of the eigenvalue problem (2.9) is clearly the dominant part of these $O(N^3)$ operations.

In the iterative minimization of the KS energy functional, the truncature of the orbitals modifies the correction directions in a way that can slow down the convergence process. On the other hand, the localization constraints break the invariance in the representation of the occupied subspace and may generate multiple local minima for E_{KS} (GOEDECKER [2003], this Handbook). One way to deal with these issues is to choose localization radii large enough so that one can easily end up in a minima of energy close enough to the *true* global minima – the one obtained without localization constraints. For example, FATTEBERT and BERNHOLC [2000] were able to compute accurately energy differences in a carbon nanotube using a localization radius of 8 Bohr.

Since the method described above allows to determine the eigenfunctions of the KS equations in a basis of localized functions – according to Eq. (2.7) –, it can be considered as a generalization of *ab initio* methods that use an LCAO to expand the eigenfunctions: $\psi_j = \sum_i c_i \phi_i$. The main difference is that grid-based local functions ϕ_i are defined by their values on a grid and are variationally optimized according to their environment. In particular, the functions ϕ_i have many more degrees of freedom and one can systematically increase the accuracy of the calculations by mesh refinement or expansion of the localization domain. An example of such an orbital

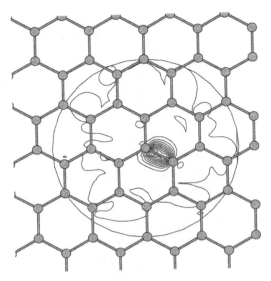

FIG. 2.5 Contour plot of the square of a typical localized orbital in the plane defined by the cylindrical surface of a (5, 5) nanotube. The external circle shows the localization region (radius 8 Bohr).

computed for a (5, 5) carbon nanotube is plotted in Fig. 2.5. The total number of basis functions in high precision calculations is thus much smaller than in LCAO approaches and minimizes the $O(N^3)$ part.

3. Quantum transport

3.1. Electron transmission and Green's functions

Let us consider a system composed of a conductor, C, connected to two semi-infinite leads, R and L, as in Fig. 3.1. A fundamental result in the theory of electronic transport is that the conductance through a region of interacting electrons (the C region in Fig. 3.1) is related to the scattering properties of the region itself via the Landauer formula (LANDAUER [1970])

$$C = \frac{2e^2}{h} \mathcal{T}, \tag{3.1}$$

where $\mathcal{T}$ is the transmission function and C is the conductance. The former represents the probability that an electron injected at one end of the conductor will transmit to the other end. In principle, we can compute the transmission function for a coherent conductor[1] starting from the knowledge of the scattering matrix, S. The latter is the mathematical quantity that describes the response at one lead due an excitation at another. In principle, the scattering matrix can be uniquely computed from the

[1] A conductor is said to be coherent if it can be characterized by a transmission matrix that relates each of the outgoing wave amplitudes to the incoming wave amplitudes at a given energy

FIG. 3.1 A conductor described by the Hamiltonian H_C, connected to two semi-infinite leads L and R, through the coupling matrices h_{LC} and h_{CR}.

solution of the Schroedinger equation and would suffice to describe the transport processes we are interested in this work. However, it is a general result of conductance theory that the elements of the $\mathcal{S}$-matrix can be expressed in terms of the Green's function of the conductor (DATTA [1995], FISHER and LEE [1981], MEIR and WINGREEN [1992]) which, in practice, can be sometimes simpler to compute.

Let us consider a physical system represented by an Hamiltonian H. Its Green's function for an energy E is defined by the equation

$$(E \pm i\eta - H)G(\vec{r}, \vec{r}') = \delta(\vec{r}, \vec{r}'), \tag{3.2}$$

where $i\eta > 0$ is an infinitesimal imaginary part added to the energy to incorporate the boundary conditions into the equation. The solution with $+$ sign is the retarded Green's function G^r, while the solution with $-$ sign is called advanced Green's function G^a. The transmission function can then be expressed in terms of the Green's functions of the conductors and the coupling of the conductor to the leads in a simple manner (see DATTA [1995], p. 141 and ff.)

$$\mathcal{T} = \text{Tr}(\Gamma_L G_C^r \Gamma_R G_C^a), \tag{3.3}$$

where $G_C^{\{r,a\}}$ are the retarded and advanced Green's functions of the conductor, and $\Gamma_{\{L, R\}}$ are functions that describe the coupling of the conductor to the leads.

In the following we are going to restrict the discussion to discrete systems that we can describe by ordinary matrix algebra. More precisely, we are going to work with matrices representing a physical system in a basis of localized electronic orbitals centered on the atoms constituting the system. It includes in particular the tight-binding model.

For a discrete media, the Green's function is then solution of a matrix equation

$$(\varepsilon - H)G = I, \tag{3.4}$$

where $\varepsilon = E \pm i\eta$ with η arbitrarily small and I is the identity matrix. To simplify the notations, we drop the exponent $\{a,r\}$ referring to advanced and retarded functions when implicitly defined by ε. For an open system, consisting of a conductor and two semi-infinite leads (see Fig. 3.1), the above Green's function can be partitioned into submatrices that correspond to the individual subsystems

$$\begin{pmatrix} g_L & g_{LC} & g_{LCR} \\ g_{CL} & G_C & g_{CR} \\ g_{LRC} & g_{RC} & g_R \end{pmatrix} = \begin{pmatrix} (\varepsilon - h_L) & -h_{LC} & 0 \\ -h_{LC}^* & (\varepsilon - H_C) & -h_{CR} \\ 0 & -h_{CR}^* & (\varepsilon - h_R) \end{pmatrix}^{-1}, \tag{3.5}$$

where the matrix $(\varepsilon - H_C)$ represents the finite "isolated" conductor (with no coupling elements to the leads), $(\varepsilon - h_{\{R, L\}})$ represent the semi-infinite leads, and h_{CR} and h_{LC} are the coupling matrices between the conductor and the leads, and h^* denotes the familiar conjugate transpose of h. As a convention, we use lower case letters for (semi-)infinite matrices and upper case for finite dimension matrices. In Eq. (3.5) we have made the assumption that there is no direct interaction between the left and right leads. From this equation it is straightforward to obtain an explicit expression for G_C

$$G_C = (\varepsilon - H_C - \Sigma_L - \Sigma_R)^{-1}, \tag{3.6}$$

where the finite-dimension matrices

$$\Sigma_L = h_{LC}^*(\varepsilon - h_L)^{-1}h_{LC}; \Sigma_R = h_{RC}(\varepsilon - h_R)^{-1}h_{RC}^* \tag{3.7}$$

are defined as the self-energy terms due to the semi-infinite leads. The self-energy terms can be viewed as effective Hamiltonians that arise from the coupling of the conductor with the leads. The coupling functions $\Gamma_{\{L,R\}}$ can then be obtained as (DATTA [1995])

$$\Gamma_{\{L,R\}} = i\left[\Sigma_{\{L,R\}}^r - \Sigma_{\{L,R\}}^a\right], \tag{3.8}$$

where the advanced self-energy $\Sigma_{\{L,R\}}^a$ is the conjugate transpose of the retarded self-energy $\Sigma_{\{L,R\}}^r$. The core of the problem lies in the calculation of the self-energies of the semi-infinite leads.

It is well known that any solid (or surface) can be viewed as an infinite (semi-infinite in the case of surfaces) stack of principal layers with nearest-neighbor interactions (LEE and JOANNOPOULOS [1981a], LEE and JOANNOPOULOS [1981b]). This corresponds to transforming the original system into a linear chain of principal layers. For a lead-conductor-lead system, the conductor can be considered as one principal layer sandwiched between two semi-infinite stacks of principal layers. The next three sections are devoted to the computation of the self-energies using the principal layers approach for different geometries.

3.2. Transmission through a bulk system

Within the principal layer approach, the matrix elements of Eq. (3.4) between layer orbitals yield a series of matrix equations for the Green's functions

$$\begin{aligned}(\varepsilon - H_{00})G_{00} &= I + H_{01}G_{10}, \\ (\varepsilon - H_{00})G_{10} &= H_{01}^*G_{00} + H_{01}G_{20}, \\ &\vdots \\ (\varepsilon - H_{00})G_{n0} &= H_{01}^*G_{n-1,0} + H_{01}G_{n+1,0},\end{aligned} \tag{3.9}$$

where the finite dimension matrices H_{nm} and G_{nm} are formed by the matrix elements of the Hamiltonian and Green's function between the layer orbitals. We assume that in a bulk system $H_{00} = H_{11} = \ldots$ and $H_{01} = H_{12} = \ldots$. Following LOPEZ-SANCHO, LOPEZ-SANCHO and RUBIO [1984], LOPEZ-SANCHO, LOPEZ-SANCHO and RUBIO [1985], this chain can be transformed in order to express the Green's function of one individual layer in terms of the Green's function of the preceding (or following) one. This is done via the introduction of the transfer matrices T and $\bar{T}$, defined such that

$$G_{10} = TG_{00}$$

and

$$G_{00} = \bar{T}G_{10}.$$

Using these definitions, we can write the bulk Green's function as (GARCIA-MOLINER and VELASCO [1992])

$$G(E) = (\varepsilon - H_{00} - H_{01}T - H_{01}^*\bar{T})^{-1}. \tag{3.10}$$

The transfer matrix can be easily computed from the Hamiltonian matrix elements via an iterative procedure, as outlined in LOPEZ-SANCHO, LOPEZ-SANCHO and RUBIO [1984]. In particular T and $\bar{T}$ can be written as

$$T = t_0 + \tilde{t}_0 t_1 + \tilde{t}_0 \tilde{t}_1 t_2 + \ldots + \tilde{t}_0 \tilde{t}_1 \ldots \tilde{t}_{n-1} t_n,$$
$$\bar{T} = \tilde{t}_0 + t_0 \tilde{t}_1 + t_0 t_1 \tilde{t}_2 + \ldots + t_0 t_1 \ldots t_{n-1} \tilde{t}_n,$$

where t_i and $\sim t_i$ are defined via the recursion formulas

$$t_i = (I - t_{i-1}\tilde{t}_{i-1} - \tilde{t}_{i-1}t_{i-1})^{-1} t_{i-1}^2,$$
$$\tilde{t}_i = (I - t_{i-1}\tilde{t}_{i-1} - \tilde{t}_{i-1}t_{i-1})^{-1} \tilde{t}_{i-1} 2$$

and

$$t_0 = (\varepsilon - H_{00})^{-1} H_{01}^*,$$
$$\tilde{t}_0 = (\varepsilon - H_{00})^{-1} H_{01}.$$

The process is repeated until $t_n, \tilde{t}_n \leq \delta$ with δ arbitrarily small. Usually no more than 5 or 6 terms are required to converge the above sums.

In the hypothesis of leads and conductors being of the same material (bulk conductivity), we can identify one principal layer of the bulk system with the conductor C, so that $H_{00} \equiv H_C$. If we compare Eq. (3.10) with Eq. (3.6), we obtain the expression for the self-energies of the conductor-leads system

$$\Sigma_L = H_{01}^*\bar{T}; \Sigma_R = H_{01}T. \tag{3.11}$$

The coupling functions are then obtained from the sole knowledge of the transfer matrices and the coupling Hamiltonian matrix elements: $\Gamma_L = -\text{Im}(H_{01}^*\bar{T})$ and $\Gamma_R = -\text{Im}(H_{01}\bar{T})$ (BUONGIORNO NARDELLI [1999]).

REMARK 3.1. In the application of the Landauer formula, it is customary to compute the transmission probability from one lead to the other assuming that the leads are connected to a reflectionless contact whose electron energy distribution is known (see for instance DATTA [1995], p. 59 and ff.).

3.3. Transmission through an interface

The procedure outlined above can also be applied in the case of electron transmission through an interface between two different media A and B. To study this case we

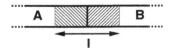

FIG. 3.2 Sketch of a system containing an interface between two media A and B. I is the interface region for which we need to compute the Green's function G_I. I is composed of two principal layers, one in each side of the interface (dashed area).

make use of the Surface Green's Function Matching (SGFM) theory, pioneered by GARCIA-MOLINER and VELASCO [1991], GARCIA-MOLINER and VELASCO [1992].

We have to compute the Green's function G_I, where the subscript I refers to the interface region composed of two principal layers – one in each media – (see Fig. 3.2). Using the SGFM method, G_I is calculated from the bulk Green's function of the isolated systems, G_A and G_B, and the coupling between the two principal layers at the two sides of the interface, H_{AB} and H_{BA}. Via the calculation of the transmitted and reflected amplitudes of an elementary excitation that propagates from medium A to medium B, it can be shown that the interface Green's function obeys the equation (GARCIA-MOLINER and VELASCO [1992], Chapter 4)

$$G_I = \begin{pmatrix} G_{AA} & G_{AB} \\ G_{BA} & G_{BB} \end{pmatrix} = \begin{pmatrix} \varepsilon - H_{00}^A - (H_{01}^A)^* \bar{T} & -H_{AB} \\ -H_{BA} & \varepsilon - H_{00}^B - H_{01}^B T \end{pmatrix}^{-1}. \tag{3.12}$$

Once the interface Green's function is known, we can compute the transmission function in terms of block super-matrices

$$\mathcal{T}(E) = \mathrm{Tr}(\Gamma_A G_{AB}^r \Gamma_B G_{BA}^a)$$

with $\Gamma_{\{A,B\}} = i\left[\Sigma_{\{A,B\}}^r - \Sigma_{\{A,B\}}^a\right]$, $\Sigma_{\{A,B\}}$ given by the analogous of Eq. (3.11) for the two semi-infinite sections, and $G_{BA}^a = (G_{AB}^r)^*$ (BUONGIORNO NARDELLI [1999]).

3.4. Transmission through a left lead-conductor-right lead (LCR) system

Within the SGFM framework, the approach described in the previous section can be extended to the case of multiple interfaces and superlattices (GARCIA-MOLINER and VELASCO [1991], GARCIA-MOLINER and VELASCO [1992]) with little complication. For the calculation of conductances in realistic experimental geometry, the method can be expanded to the general configuration of a Left-lead-Conductor-Right-lead (LCR) systems – as displayed in Fig. 3.1. In the language of block matrices and principal layers, outlined in the previous sections, the LCR Green's function obeys the equation

$$\begin{aligned} G_{LCR} &= \begin{pmatrix} G_L & G_{LC} & G_{LR} \\ G_{CL} & G_C & G_{CR} \\ G_{RL} & G_{RC} & G_R \end{pmatrix} \\ &= \begin{pmatrix} \varepsilon - H_{00}^L - (H_{01}^L)^* \bar{T} & -H_{LC} & 0 \\ -H_{CL} & \varepsilon - H_C & -H_{CR} \\ 0 & -H_{RC} & \varepsilon - H_{00}^R - H_{01}^R T \end{pmatrix}^{-1}, \end{aligned} \tag{3.13}$$

where $H_{nm}^{\{L,R\}}$ are the block matrices of the Hamiltonian between the layer orbitals in the left and right leads respectively, and $T_{\{L,R\}}$ and $\bar{T}_{\{L,R\}}$ are the appropriate transfer matrices. The latter are easily computed from the Hamiltonian matrix elements via the iterative procedure already described in the bulk case (Section 3.2). Correspondingly, H_{LC} and H_{CR} are the coupling matrices between the conductor and the leads principal layers in contact with the conductor.

As in Section 3.1, it is straightforward to obtain in the form of Eq. (3.6), $G_C = (\varepsilon - H_C - \Sigma_L - \Sigma_R)^{-1}$, where Σ_L and Σ_R are the self-energy terms due to the semi-infinite leads, and identify (BUONGIORNO NARDELLI and BERNHOLC [1999])

$$\Sigma_L = H_{LC}^*(\varepsilon - H_{00}^L - (H_{01}^L)^* \bar{T}_L)^{-1} H_{LC},$$
$$\Sigma_R = H_{CR}(\varepsilon - H_{00}^R - H_{01}^R T_R)^{-1} H_{CR}^*. \tag{3.14}$$

The transmission function in the LCR geometry can then be derived from Eqs. (3.3) and (3.8).

REMARK 3.2. The knowledge of the conductor's Green's function G_C gives also direct information on the electronic spectrum of the system via the spectral density of electronic states

$$N(E) = -(1/\pi)\mathrm{Im}[\mathrm{Tr}(G_C(E))].$$

REMARK 3.3. We have assumed a truly one-dimensional chain of principal layers, which is physical only for systems like nanotubes or quantum wires that have a definite quasi-one-dimensional character. The extension to a truly three-dimensional case is straightforward using Bloch functions wave vectors $k_\parallel$ parallel to the layers (in the directions perpendicular to the conduction). The introduction of the principal layer concept implies that along the direction of the conduction the system is described by an infinite set of wave vectors $\hat{k}_\perp$. The above procedure effectively reduces the three-dimensional system to a set of noninteracting linear-chains, one for each $\hat{k}_\parallel$ (LEE and JOANNOPOULOS [1981a], LEE and JOANNOPOULOS [1981b]). We can then use the usual k-point summation techniques to evaluate, for instance, the quantum conductance

$$T(E) = \sum_{\vec{k}_\parallel} w_{\vec{k}_\parallel} T_{\vec{k}_\parallel}(E),$$

where $w_{\vec{k}_\parallel}$ are the relative weights of the different wave vectors $\vec{k}_\parallel$ in the irreducible wedge of the surface Brillouin zone (BALDERESCHI [1973]).

3.5. Generalization to nonorthogonal orbitals

In the previous sections we have assumed to have a Hamiltonian representation in terms of orthogonal orbitals. The expression for the Green's and transmission functions of a bulk system described by a general nonorthogonal localized-orbital Hamiltonian follows directly from the procedure outlined in Section 3.2. All the quantities can be obtained making the substitutions $(\varepsilon - H_{00}) \rightarrow (\varepsilon S_{00} - H_{00})$ and $H_{01} \rightarrow -(\varepsilon S_{01} - H_{01})$. Here, we introduce the matrices S that represent the overlap between the localized orbitals. With this recipe, the equation chain (3.9) now reads

$$(\varepsilon S_{00} - H_{00})G_{10} = -(\varepsilon S_{01}^* - H_{01}^*)G_{00} - (\varepsilon S_{01} - H_{01})G_{20},$$
$$(\varepsilon S_{00} - H_{00})G_{n0} = -(\varepsilon S_{01}^* - H_{01}^*)G_{n-1,0} - (\varepsilon S_{01} - H_{01})G_{n+1,0}.$$
$$\vdots$$
$$(\varepsilon S_{00} - H_{00})G_{00} = I - (\varepsilon S_{01} - H_{01})G_{10},$$

From here, via the same series of algebraic manipulations as in the orthogonal case, we obtain the Green's function

$$G = [(\varepsilon S_{00} - H_{00}) + (\varepsilon S_{01} - H_{01})T + (\varepsilon S_{01}^* - H_{01}^*)\bar{T}]^{-1},$$

and from the latter we can identify the self-energies

$$\Sigma_L = -(\varepsilon S_{01}^* - H_{01}^*)\bar{T}; \Sigma_R = -(\varepsilon S_{01} - H_{01})T.$$

The above procedure can be extended to the case of the transmission through an interface or an LCR junction. For the latter case, we obtain

$$\Sigma_L = (\varepsilon S_{LC} - H_{LC})^* \times \left[\varepsilon S_{00}^L - H_{00}^L + (\varepsilon S_{01}^L - H_{01}^L)^* \bar{T}_L\right]^{-1}(\varepsilon S_{LC} - H_{LC}),$$
$$\Sigma_R = (\varepsilon S_{CR} - H_{CR}) \times \left[\varepsilon S_{00}^R - H_{00}^R + (\varepsilon S_{01}^R - H_{01}^R)T_R\right]^{-1}(\varepsilon S_{CR} - H_{CR})^*,$$

$$(3.15)$$

where $H_{nm}^{\{L,R\}}$ are the matrix elements of the Hamiltonian between layer orbitals in the left and right leads, respectively. $S_{nm}^{\{L,R\}}$ are the corresponding overlap matrices and $T_{\{L,R\}}$ and $\bar{T}_{\{L,R\}}$ are the appropriate transfer matrices. The latter are easily computed from the Hamiltonian and overlap matrix elements via the usual iterative procedure (see Section 3.2). Correspondingly, $H_{LC}, H_{CR}, S_{LC},$ and S_{CR} are the coupling and overlap matrices for the conductor-leads assembly.

4. Applications: conductivity from *ab initio* local orbital Hamiltonian

4.1. Methodology

The procedure described in Section 3 requires the knowledge of the Hamiltonian and overlap matrix elements between layer orbitals of the conductor, and the left and right leads. In *ab initio* density-functional calculations, such matrix elements can be computed using the $O(N)$-like algorithm described in Section 2.8. In this context the numerical orbitals – defined on a uniform grid in real-space – are centered on atoms and localized in spherical regions of radius R_L around the respective atoms. Since the orbitals are variationally optimized on the grid according to their environment until they accurately describe the ground state of the system, it allows us to use only a small number of orbitals per atom, much smaller than in LCAO-based calculations. The size of the matrices that enter in the quantum conductance calculation and the computational cost of the whole procedure are thus minimized. In order to ensure fast convergence and accuracy – even for metallic systems – we use both occupied and unoccupied orbitals.

The matrices that enter the electronic transport calculation of an LCR system are computed in two steps. In the first calculation, we compute the ground state of the bare leads in a supercell with periodic boundary conditions. From this calculation we extract the Hamiltonian in the basis of the localized nonorthogonal orbitals and the overlap matrices required for the computation of the self-energies. We then perform a second ground state calculation in a supercell with periodic boundary conditions containing the conductor and one principal layer of the leads. In this calculation, the orbitals in the leads are kept the same as in the bare lead calculation, in order to extract the matrix elements describing the coupling between the conductor and the leads. This procedure fully accounts for the electronic structure of the conductor and the interaction between the conductor and the leads, provided that the lead region is large enough to avoid spurious interactions between periodic images of the contacts. In order to have interactions between the nearest-neighbor principal layers only, the width of the layers has to be sufficiently large compared to the localization regions. On the other hand, the localization regions have to be large enough to ensure an accurate solution of the density-functional equations. Moreover, in the Green's function matching procedure one has to carefully align the Fermi levels of both systems in order to avoid spurious bias effects. Provided that in the conductor-lead calculation the lead region is large enough to recover bulk-like behavior far from the interfaces, we align the macroscopic average of the electrostatic potentials in the bare lead and in the conductor-lead geometry. This ensures a seamless conductor-lead geometry and prevents the spurious bias. An equivalent procedure is often used to extract band offsets in superlattice calculations (BALDERESCHI, BARONI and RESTA [1988], BUONGIORNO NARDELLI, RAPCEWICZ and BERNHOLC [1997]).

REMARK 4.1. If a principal layer is composed of N orbitals, the calculation of the Green's function requires a matrix inversion that scales as $O(N^3)$. However, for very large systems, the localization of the orbitals allows us to divide a principal layer into thiner layers and compute the quantities of interest in largely $O(N)$ fashion (ANANTRAM and GOVINDAN [1998]).

4.2. Example: carbon nanotube on metallic contacts

To illustrate the above *ab initio* methodology we use the example of transport behavior of nanotube-metal contacts studied by BUONGIORNO NARDELLI, FATTEBERT and BERNHOLC [2001]. The problem of contacts in metal-carbon nanotubes assemblies is a crucial issue for technological development, and determines much of the nanoscale device characteristics. A perfect metallic nanotube behaves like a ballistic conductor: every electron injected into the nanotube at one end should come out at the other end. The basic electronic properties of metallic nanotubes imply the existence of two propagating modes for electronic transmission, independent of the diameter (BERNHOLC, BRENNER, BUONGIORNO NARDELLI, MEUNIER and ROLAND [2002]). The electronic conductance is then expected to be twice the fundamental quantum of conductance, $G_0 = 2e^2/h = 1/12.9$ $(k\Omega)^{-1}$. At higher energies, the electrons are able to probe different subbands, which gives rise to an increase in G that is proportional to the number of additional bands available for transport. Hence, G for ideal nanotubes is expected to consist of a series of "down-and-up" steps as a function of the electron energy, in which the position of the steps correlate with the band edges. An illustration of this behavior is displayed in Fig. 4.1. These considerations would suggest that nanotubes should behave as ideal device elements because of their electrical properties.

However, one of the fundamental problems that hinder a broader technological application of carbon nanotubes is the observation that most carbon nanotube devices display contact resistances of the order of $M\Omega$ (TANS, DEVORET, DAI, THESS, SMALLEY, GEERLIGS and DEKKER [1997], TANS, VERSCHUEREN and DEKKER [1998], MARTEL, SCHMIDT, SHEA, HERTEL and AVOURIS [1998], BACHTOLD, HENNY, TARRIER, STRUNK, SCHONENBERGER, SALVETAT, BONARD and FORRO [1998]), rather than $k\Omega$, as one would expect.

What is the physical origin behind the very high contact resistance for carbon nanotube systems? As a prototypical example, we consider the transport properties of a metallic (5, 5) nanotube deposited on an Al (111) surface in an idealized side-contact geometry, as shown in the inset of Fig. 4.2. In order to accurately account for the highly inhomogeneous environment of the nanowire-metal junction, and to account for the charge transfer occurring at the interface between these two dissimilar materials, it is important to use the accurate and self-consistent *ab initio* description we have previously discussed. The main characteristics of the electronic response of the system is a marked transfer of charge from the nanotube to the metal that allows the

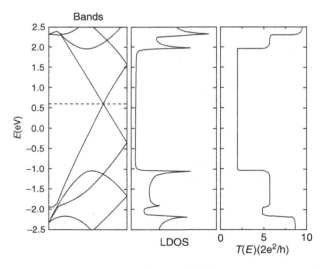

FIG. 4.1 Left: Electronic band structure of a metallic nanotube. Note the crossing of the bands at the Fermi energy. Middle: corresponding Density of States. Right: Quantum conductance spectrum. Note the metallic plateau of conductance equal to $2G_0$.

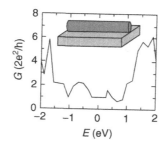

FIG. 4.2 The geometry and conductance spectrum of an infinite (5, 5) nanotube deposited on Al(111). Adapted from BUONGIORNO NARDELLI, FATTEBERT and BERNHOLC [2001].

valence band edge of the nanotube to align with the Fermi level of the metal electrode (XUE and DATTA [1999]).

This charge transfer, which has been already observed for other experimental systems (TANS, VERSCHUEREN and DEKKER [1998], WILDOER, VENEMA, RINZLER, SMALLEY and DEKKER [1998], MARTEL, SCHMIDT, SHEA, HERTEL and AVOURIS [1998]) and calculations (XUE and DATTA [1999], RUBIO, SANCHEZ-PORTAL, ARTACHO, ORDEJON and SOLER [1999], KONG, HAN and IHM [1999]), leads to enhanced conductivity along the tube axis and gives rise to a weak ionic bonding between the tube and the metal. The conductance spectrum for the coupled nanotube is displayed in Fig. 4.2. Although the metal contact increases the resistance by a factor of two as compared to ideal isolated tubes, the transmission through the system is still substantial. To further analyze the contact resistance, we have calculated the *eigenchannels* (BRANDBYGE, SORENSEN and JACOBSEN [1997]). Among the conducting channels, that is, those with a significant nonzero transmission coefficient, we observe a clear distinction between channels that are localized in the metal and those on the nanotube itself. This result reflects the clear separation of the individual electronic wave functions of each of the components of the system. In particular, the eigenchannel corresponding to the plateau of conductance around the Fermi energy corresponds to an individual wave function, reproduced in Fig. 4.3, almost fully localized on the nanotube. This implies that there is very little hybridization and intermixing between the nanotube and the metal in the channel responsible for conduction at the Fermi level. Thus, the conduction electron transfer between the tube and the metal in the idealized side-contact geometry considered here is very inefficient, which can explain the high contact resistance observed in nanotube-metal contacts.

This initial investigation has been extended to a geometry that more closely resembles an experimental two-terminal device, with two semi-infinite contacts connected by a nanotube bridge, 1.5 nm long. In this geometry, the system recovers the ideal conductance of an isolated tube with two conductance channels at the Fermi energy, as shown in Fig. 4.4. This behavior is induced by the alignment of the valence band edge of the nanotube with the Fermi energy of the metal contacts, triggered by the charge transfer in the lead regions. In this particular geometry, these conditions restore the two original eigenchannels of the nanotube and thus conserve the number

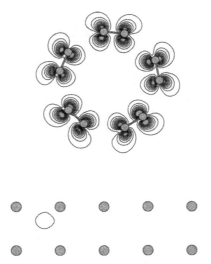

FIG. 4.3 Cross-section of the square modulus of the electronic wave function corresponding to the only open eigenchannel at the Fermi level that has a sizable component on the nanotube for the system represented in Fig. 4.2. The other wave functions at the Fermi level are mostly localized on the metal. Adapted from BUONGIORNO NARDELLI, FATTEBERT and BERNHOLC [2001].

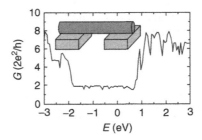

FIG. 4.4 The conductance spectrum of an ideal two-terminal device, as shown in the inset. The Fermi level is taken as a reference. From BUONGIORNO NARDELLI, FATTEBERT and BERNHOLC [2001].

of conducting channels throughout the system. It is important to note that the weak nanotube-metal interaction, responsible for the pathologically high resistance of the nanotube-metal assembly, is not strengthened.

REMARK 4.2. In this calculation, the conductor C is made of a $(5, 5)$ carbon nanotube composed of 120 atoms, while a principal layer of the leads is made of the same 120 atoms carbon nanotube and of an aluminium surface of 100 atoms. The width of a principal layer is 14.7 Å and a localization radius of 5.3 Å was used for the orbitals. The LDA exchange and correlation functional was used with the pseudopotentials by HAMANN [1989].

REMARK 4.3. The occurrence of a single channel in the first case is due to the idealized geometry of an infinite nanotube on an infinite metallic surface. The conservation of the total number of channels is ensured by the channels localized on the metal.

These examples clearly demonstrate that the weak nanotube-metal coupling is mostly responsible for the weak electron transport in the combined system, and that wave vector conservation is not a significant factor (TERSOFF [1999], DELANEY, DI VENTRA and PANTELIDES [1999]). The weak distributed coupling also explains why the measured contact resistance is inversely proportional to the contact length (TANS, DEVORET, DAI, THESS, SMALLEY, GEERLIGS and DEKKER [1997], FRANK, PONCHARAL, WANG and HEER [1998], ANANTRAM, DATTA and XUE [2000]). Although the nanotube behaves as an ideal ballistic conductor, the bonding characteristics of the nanotube-metal system prevent an efficient electron transfer mechanism from the nanotube to the Al contact. Indeed, inducing defects in the contact region, for example, by localized electron bombardment (BACHTOLD, HENNY, TARRIER, STRUNK, SCHONENBERGER, SALVETAT, BONARD and FORRO [1998]), drastically increases the bonding strength of the nanotube-metal assembly and greatly improve the performance of the device. Alternatively, mechanically pushing the nanotube closer to the Al surface by a small amount (≈ 1 Å, with an energy cost of ≈ 10 meV/atom) more than doubles the transmission efficiency between the metal and the nanotube. The mechanical deformation induces a small inward relaxation of the Al surface in the contact region, facilitating stronger hybridization between the nanotube and the metal contact in the conducting channels and therefore leading to a higher transmission rate.

Acknowledgments

The authors are very grateful to Prof. J. Bernholc for pointing them toward the problems of linear scaling and electronic transport, and for numerous stimulating discussions about these problems. A portion of the work was performed under the auspices of the U.S. Department of Energy by University of California Lawrence Livermore National Laboratory under contract No. W-7405-Eng-48.

References

ANANTRAM, M. and T. GOVINDAN (1998), Conductance of carbon nanotubes with disorder: A numerical study, *Phys. Rev. B* **58** (8), 4882–4887.

ANANTRAM, M.P., S. DATTA and Y.Q. XUE (2000), Coupling of carbon nanotubes to metallic contacts, *Phys. Rev. B* **61** (20), 14219–14224.

ANCILOTTO, F., P. BLANDIN and F. TOIGO (1999), Real-space full-multigrid study of the fragmentation of Li_{11}^{+} clusters, *Phys. Rev. B* **59** (12), 7868–7875.

ASHCROFT, N. and N. MERMIN (1976), *Solid State Physics* (Holt–Saunders).

BACHTOLD, A., M. HENNY, C. TARRIER, C. STRUNK, C. SCHONENBERGER, J.P. SALVETAT, J.M. BONARD and L. FORRO (1998), Contacting carbon nanotubes selectively with low-ohmic contacts for four-probe electric measurements, *Appl. Phys. Lett.* **73** (2), 274–276.

BALDERESCHI, A. (1973), Mean-value point in the Brillouin zone, *Phys. Rev. B* **7**, 5212.

BALDERESCHI, A., S. BARONI and R. RESTA (1988), Band offsets in lattice-matched heterojunctions: A model and first-principles calculations for GaAs/AlAs, *Phys. Rev. Lett.* **61**, 734.

BECK, T. (2000), Real-space mesh techniques in density-functional theory, *Rev. Mod. Phys.* **72** (4), 1041–1080.

BEENAKKER, C. and H. VAN HOUTEN (1991), Quantum transport in semiconductor nanostructures, *Solid State Phys.* **44**, 1–228.

BERNHOLC, J., D. BRENNER, M. BUONGIORNO NARDELLI, V. MEUNIER and C. ROLAND (2002), Mechanical and electrical properties of nanotubes, *Ann. Rev. Mater. Sci.*

BLACKFORD, L.S., J. CHOI, A. CLEARY, E. D'AZEVEDO, J. DEMMEL, I. DHILLON, J. DONGARRA, S. HAMMARLING, G. HENRY, A. PETITET, K. STANLEY, D. WALKER and R.C. WHALEY (1997), *SCALAPACK User's Guide* (SIAM, Philadelphia).

BRANDBYGE, M., M. SORENSEN and K. JACOBSEN (1997), Conductance eigenchannels in nanocontacts, *Phys. Rev. B* **56**, 14956.

BRANDT, A. (1977), Multilevel adaptative solutions to boundary-value problems, *Math. Comp.* **31** (138), 333–390.

BRIGGS, E.L., D.J. SULLIVAN and J. BERNHOLC (1995), Large scale electronic structure calculations with multigrid acceleration, *Phys. Rev. B* **52** (8), R5471–R5474.

BRIGGS, E.L., D.J. SULLIVAN and J. BERNHOLC (1996), Real-space multigrid-based approach to large-scale electronic structure calculations, *Phys. Rev. B* **54** (20), 14362–14375.

BUNGARO, C., K. RAPCEWICZ and J. BERNHOLC (1999), Surface sensitivity of impurity incorporation: Mg at GaN(0001) surfaces, *Phys. Rev. B* **59** (15), 9771–9774.

BUONGIORNO NARDELLI, M. (1999), Electronic transport in extended systems: Application to carbon nanotubes, *Phys. Rev. B* **60** (11), 7828–7833.

BUONGIORNO NARDELLI, M. and J. BERNHOLC (1999), Mechanical deformations and coherent transport in carbon nanotubes, *Phys. Rev. B* **60** (24), R16338–R16341.

BUONGIORNO NARDELLI, M., J.-L. FATTEBERT and J. BERNHOLC (2001), An O(N) real-space method for ab initio quantum transport calculations: Application to carbon nanotube-metal contacts, *Phys. Rev. B* **64**, 245423.

BUONGIORNO NARDELLI, M., K. RAPCEWICZ and J. BERNHOLC (1997), Strain effects on the interface properties of nitride semiconductors, *Phys. Rev. B* **55**, R7323.

BUONGIORNO NARDELLI, M., B. YAKOBSON and J. BERNHOLC (1998), Mechanism of strain release in carbon nanotubes, *Phys. Rev. B* **57** (8), R4277–R4280.

CANCÈS, E., M. DEFRANCESCHI, W. KUTZELNIGG, C. LE BRIS and Y. MADAY (2003), Computational quantum chemistry: A primer, in: P.G. Ciarlet and C. Le Bris, eds., *Computational Chemistry, Handbook of Numerical Analysis*, Vol. X (Elsevier, Amsterdam) 3–270.

CHELIKOWSKY, J.R., N. TROUILLER and Y. SAAD (1994), Finite-difference-pseudopotential method: Electronic structure calculations without a basis, *Phys. Rev. Lett.* **72** (8), 1240–1243.

CHETTY, N., M. WEINERT, T.S. RAHMAN and J.W. DAVENPORT (1995), Vacancies and impurities in aluminium and magnesium, *Phys. Rev. B* **52** (9), 6313–6326.

CHOI, H. and J. IHM (1999), *Ab initio* pseudopotential method for the calculation of conductance in quantum wires, *Phys. Rev. B* **59**, 2267.

CHOI, H., J. IHM, S. LOUIE and M. COHEN (2000), Defects, quasibound states, and quantum conductance in metallic carbon nanotubes, *Phys. Rev. Lett.* **84**, 2917.

COLLATZ, L. (1966), *The Numerical Treatment of Differential Equations* (Springer, Berlin).

COSTINER, S. and S. TAASAN (1995), Simultaneous multigrid techniques for nonlinear eigenvalue problems: Solutions of the nonlinear Schrödinger–Poisson eigenvalue problem in two and three dimensions, *Phys. Rev. E* **52** (1), 1181–1192.

DATTA, S. (1995), *Electronic Transport in Mesoscopic Systems* (Cambridge University Press).

DELANEY, P., M. DI VENTRA and S.T. PANTELIDES (1999), Quantized conductance of multiwalled carbon nanotubes, *Appl. Phys. Lett.* **75** (24), 3787–3789.

DESCLOUX, J., J.-L. FATTEBERT and F. GYGI (1998), RQI (Rayleigh Quotient Iteration), an old recipe for solving modern large scale eigenvalue problems, *Comput. Phys.* **12** (1), 22–27.

DI VENTRA, M., S. PANTELIDES and N. LANG (2000), First-principles calculation of transport properties of a molecular device, *Phys. Rev. Lett.* **84**, 979.

EDELMAN, A., T. ARIAS and S. SMITH (1998), The geometry of algorithms with orthogonality constraints, *SIAM J. Matrix Anal. Appl.* **20** (2), 303–353.

FATTEBERT, J.-L. (1996), An inverse iteration method using multigrid for quantum chemistry, *BIT* **36** (3), 509–522.

FATTEBERT, J.-L. (1998), A block Rayleigh Quotient Iteration with local quadratic convergence, *ETNA* **7**, 56–74.

FATTEBERT, J.-L. (1999), Finite difference schemes and block Rayleigh Quotient Iteration for electronic structure calculations on composite grids, *J. Comput. Phys.* **149**, 75–94.

FATTEBERT, J.-L. and J. BERNHOLC (2000), Towards grid-based O(N) density-functional theory methods: Optimized non-orthogonal orbitals and multigrid acceleration, *Phys. Rev. B* **62** (3), 1713–1722.

FATTEBERT, J.-L. and F. GYGI (2002), Density functional theory for efficient ab initio molecular dynamics simulations in solution, *J. Comput. Chem.* **23**, 662–666.

FERNANDO, G.W., G.-X. QUIAN, M. WEINERT and J.W. DAVENPORT (1989), First-principles molecular dynamics for metals, *Phys. Rev. B* **40** (11), 7985–7988.

FERRY, F. and S. GOODNICK (1997), *Transport in Nanostructures* (University Press, Cambridge).

FEYNMAN, R. (1939), Forces in molecules, *Phys. Rev.* **56**, 340–343.

FISHER, D. and P. LEE (1981), Relation between conductivity and transmission matrix, *Phys. Rev. B* **23**, 6851.

FRANK, S., P. PONCHARAL, Z.L. WANG and W.A.D. HEER (1998), Carbon nanotube quantum resistors, *Science* **280** (5370), 1744–1746.

GALLI, G. and M. PARRINELLO (1992), Large scale electronic structure calculations, *Phys. Rev. Lett.* **69** (24), 3547–3550.

GALLI, G. and A. PASQUARELLO (1993), First-principles molecular dynamics, in: M. Allen and D. Tildesley, eds., *Computer Simulation in Chemical Physics* (Kluwer Academic Publishers) 261–313.

GARCIA-MOLINER, F. and V. VELASCO (1991), Matching methods for single and multiple interfaces: Discrete and continuous media, *Phys. Rep.* **200** (3), 83–125.

GARCIA-MOLINER, F. and V. VELASCO (1992), *Theory of Single and Multiple Interfaces* (World Scientific, Singapore).

GOEDECKER, S. (2003), Linear scaling methods for the solution of Schrödinger's equation, in: P.G. Ciarlet and C. Le Bris, eds., *Computational Chemistry, Handbook of Numerical Analysis*, Vol. X (Elsevier, Amsterdam) 537–570.

GYGI, F. and G. GALLI (1995), Real-space adaptive-coordinate electronic-structure calculations, *Phys. Rev. B* **52** (4), 2229–2232.

HAMANN, D.R. (1989), Generalized norm-conserving pseudopotentials, *Phys. Rev. B* **40** (5), 2980–2987.

HEERMANN, D. (1990), *Computer Simulation Methods*, (Springer-Verlag, Berlin Heidelberg), 2nd edn.

HEISKANEN, M., T. TORSTI, M. PUSKA and R. NIEMINEN (2001), Multigrid method for electronic structure calculations, *Phys. Rev. B* **63**, 245106.

HERNANDEZ, E. and M.J. GILLAN (1995), Self-consistent first-principles technique with linear scaling, *Phys. Rev. B* **51** (15), 10157–10160.

HIROSE, K. and M. TSUKADA (1995), First-principles calculation of the electronic structure for a bielectrode junction system under strong field and current, *Phys. Rev. B* **51**, 5278.

HOSHI, T. and T. FUJIWARA (1997), Fully-selfconsistent electronic-structure calculation using nonorthogonal localized orbitals within a finite-difference real-space scheme and ultrasoft pseudopotential, *J. Phys. Soc. Jpn.* **66** (12), 3710–3713.

ISMAEL-BEIGI, S. and T.A. ARIAS (1999), Locality of the density matrix in metals, semiconductors, and insulators, *Phys. Rev. Lett.* **82** (10), 2127–2130.

JIN, Y.-G. and K. CHANG (2001), Mechanism for the enhanced diffusion of charged oxygen ions in SiO_2, *Phys. Rev. Lett.* **86** (9), 1793–1796.

JIN, Y.-G., J.-W. JEONG and K. CHANG (1999), Real-space electronic structure calculations of charged clusters and defects in semiconductors using a multigrid method, *Physica B* **273–274**, 1003–1006.

JING, X., N. TROULLIER, D. DEAN, N. BINGGELI and J. CHELIKOWSKY (1994), *Ab initio* molecular dynamics simulations of Si clusters using higher-order finite difference-pseudopotential method, *Phys. Rev. B* **50** (16), 12234–12237.

KLEINMAN, L. and D. BYLANDER (1982), Efficacious form for model pseudopotentials, *Phys. Rev. Lett.* **48** (20), 1425–1428.

KOBAYASHI, N., M. BRANDBYGE and M. TSUKADA (2000), Conduction channels at finite bias in singleatom gold contact, *Phys. Rev. B* **62**, 8430.

KOHN, S., J. WEARE, M. ONG and S. BADEN (1997), Software abstractions and computational issues in parallel structured adaptive mesh methods for electronic structure calculations, in: *Proc. of Workshop on Structured Adaptive Mesh Refinement Grid Methods* (Minneapolis).

KOHN, W. and L.J. SHAM (1965), Self-consistent equations including exchange and correlation effects, *Phys. Rev. A* **140**, 1133–1138.

KONG, K., S. HAN and J. IHM (1999), Development of an energy barrier at the metal-chain-metallic-carbonnanotube nanocontact, *Phys. Rev. B* **60** (8), 6074–6079.

LANDAUER, R. (1970), Electrical resistance of disordered one-dimensional lattices, *Philos. Mag.* **21**, 863–867.

LANDMAN, U., R. BARNETT, A.G. SCHERBAKOV and P. AVOURIS (2000), Metal-semiconductor nanocontacts: silicon nanowires, *Phys. Rev. Lett.* **85**, 1958.

LANG, N. (1995), Resistance of atomic wires, *Phys. Rev. B* **52**, 5335.

LEE, D. and J. JOANNOPOULOS (1981a), Simple scheme for surface-band calculations. I, *Phys. Rev. B* **23** (10), 4988.

LEE, D. and J. JOANNOPOULOS (1981b), Simple scheme for surface-band calculations. II. The Green's function, *Phys. Rev. B* **23** (10), 4997.

LOPEZ-SANCHO, M., J. LOPEZ-SANCHO and J. RUBIO (1984), Quick iterative scheme for the calculation of transfer matrices: application to Mo(100), *J. Phys. F: Metal Phys.* **14**, 1205–1215.

LOPEZ-SANCHO, M., J. LOPEZ-SANCHO and J. RUBIO (1985), Highly convergent schemes for the calculation of bulk and surface Green functions, *J. Phys. F: Metal Phys.* **15**, 851–858.

MANDEL, J. and S. MCCORMICK (1989), A multilevel variational method for $Au = \lambda Bu$ on composite grids, *J. Comput. Phys.* **80** (2), 442–452.

MARTEL, R., T. SCHMIDT, H.R. SHEA, T. HERTEL and P. AVOURIS (1998), Single- and multi-wall carbon nanotube field-effect transistors, *Appl. Phys. Lett.* **73** (17), 2447–2449.

MEIR, Y. and N. WINGREEN (1992), Landauer formula for the current through an interacting electron region, *Phys. Rev. Lett.* **68** (16), 2512–2515.

MILLAM, J.M. and G.E. SCUSERIA (1997), Linear scaling conjugate gradient density matrix search as an alternative to diagonalization for first principles electronic structure calculations, *J. Chem. Phys.* **506** (13), 5569–5577.

MODINE, N.A., G. ZUMBACH and E. KAXIRAS (1996), Adaptive-coordinate real-space electronic-structure calculations on parallel computers, *Solid State Comm.* **99** (2), 57–61.

MURAKAMI, H., V. SONNAD and E. CLEMENTI (1992), A three-dimensional finite element approach towards molecular SCF computations, *Int. J. Quant. Chem.* **42**, 785–817.

PARLETT, B.N. (1998), *The Symmetric Eigenvalue Problem* (SIAM, Philadelphia).

PARRINELLO, M. (1997), From Silicon to RNA: The coming age of *ab initio* molecular dynamics, *Sol. State Comm.* **102** (2–3), 107–120.

PASK, J.E., B.M. KLEIN, C.Y. FONG and P.A. STERNE (1999), Real-space polynomial basis for solid-state electronic structure calculations: A finite-element approach, *Phys. Rev. B* **59** (19), 12352–12358.

PAYNE, M.C., M.P. TETER, D.C. ALLAN, T. ARIAS and J. JOANNOPOULOS (1992), Iterative minimization techniques for ab initio total-energy calculations: Molecular dynamics and conjugate gradients, *Rev. Mod. Phys.* **64** (4), 1045–1097.

PERDEW, J., K. BURKE and M. ERNZERHOF (1996), Generalized gradient approximation made simple, *Phys. Rev. Lett.* **77** (18), 3865–3868.

PFROMMER, B., J. DEMMEL and H. SIMON (1999), Unconstraint energy functionals for electronic structure calculations, *J. Comput. Phys.* **150**, 287–298.

RAMAMOORTHY, M., E. BRIGGS and J. BERNHOLC (1998), Chemical trends in impurity incorporation into Si (100), *Phys. Rev. Lett.* **81** (8), 1642–1645.

RUBIO, A., D. SANCHEZ-PORTAL, E. ARTACHO, P. ORDEJON and J.M. SOLER (1999), Electronic states in a finite carbon nanotube: A one-dimensional quantum box, *Phys. Rev. Lett.* **82** (17), 3520–3523.

SAAD, Y., A. STATHOPOULOS, J. CHELIKOWSKY, K. WU and S. OGUT (1996), Solution of large eigenvalue problems in electronic structure calculations, *BIT* **36** (3), 563–578.

SCHMIDT, W.G., F. BECHSTEDT and J. BERNHOLC (2001), Terrace and step contributions to the optical anisotropy of Si(001) surfaces, *Phys. Rev. B* **63** (4), 045322.

SLEIJPEN, G.L.G. and H.A. VAN DER VORST (1996), A generalized Jacobi–Davidson iteration method for linear eigenvalue problem, *SIAM Matrix Anal. Appl.* **17** (2), 401–425.

TAKAHASHI, H., T. HORI, H. HASHIMOTO and T. NITTA (2001), A hybrid QM/MM method employing real space grids for QM water in the TIP4P water solvent, *J. Comp. Chem.* **22** (12), 1252–1261.

TANS, S.J., M.H. DEVORET, H.J. DAI, A. THESS, R.E. SMALLEY, L.J. GEERLIGS and C. DEKKER (1997), Individual single-wall carbon nanotubes as quantum wires, *Nature* **386** (6624), 474–477.

TANS, S.J., A.R.M. VERSCHUEREN and C. DEKKER (1998), Room-temperature transistor based on a single carbon nanotube, *Nature* **393** (6680), 49–52.

TAYLOR, J., H. GUO and J. WANG (2000), *Ab initio* modeling of open systems: Charge transfer, electron conduction, and molecular switching of a C60 device, *Phys. Rev. B* **63**, 121104.

TERSOFF, J. (1999), Response to comment on "Contact resistance of carbon nanotubes [Appl. Phys. Lett. 75, 4028 (1999)]" *Appl. Phys. Lett.* **75** (25), 4030.

TETER, M.P., M.C. PAYNE and D.C. ALLAN (1989), Solution of Schrödinger's equation for large systems, *Phys. Rev. B* **40** (18), 12255–12263.

TSUCHIDA, E. and M. TSUKADA (1995), Electronic-structure calculations based on the finite-element method, *Phys. Rev. B* **52** (8), 5573–5578.

VASILIEV, I., S. OGUT and J. CHELIKOWSKY (1999), *Ab initio* excitation spectra and collective electronic response in atoms and clusters, *Phys. Rev. Lett.* **82** (9), 1919–1922.

WANG, J. and T. BECK (2000), Efficient real-space solution of the Kohn–Sham equations with multiscale techniques, *J. Chem. Phys.* **112** (21), 9223–9228.

WHITE, S.R., J.W. WILKINS and M.P. TETER (1989), Finite-element method for electronic structure calculations, *Phys. Rev. B* **39** (9), 5819–5833.

WILDOER, J.W.G., L.C. VENEMA, A.G. RINZLER, R.E. SMALLEY and C. DEKKER (1998), Electronic structure of atomically resolved carbon nanotubes, *Nature* **391** (6662), 59–62.

XUE, Y.Q. and S. DATTA (1999), Fermi-level alignment at metal-carbon nanotube interfaces: Application to scanning tunneling spectroscopy, *Phys. Rev. Lett.* **83** (23), 4844–4847.

YALIRAKI, S., A.E. ROITBERG, C. GONZALEZ, V. MUJICA and M. RATNER (1999), The injecting energy at molecule/metal interfaces: Implications for conductance of molecular junctions from an ab initio molecular description, *J. Chem. Phys.* **111**, 6997.

YOON, Y.-G., M. MAZZONI, H. CHOI, J. IHM and S. LOUIE (2001), Structural deformation and intertube conductance of crossed carbon nanotube junctions, *Phys. Rev. Lett.* **86**, 688.

Simulating Chemical Reactions in Complex Systems

M.J. Field

Laboratoire de Dynamique Moléculaire, Institut de Biologie Structurale – Jean-Pierre Ebel, 41 rue Jules Horowitz, Grenoble cedex 1, France

1. Introduction

Chemistry is the central science, straddling as it does the physical and the biological sciences, and it has as its goal the study of the elements and the myriad compounds that they can form. Perhaps the most remarkable and the most interesting property of molecules is their ability to react or, in other words, to undergo transformations that change them into other molecular species. The evidence of chemical reaction surrounds us in our everyday life both in the inanimate – for example, in a flame – and in the animate. Reactions occur in all phases of matter, from the gas phase to the solid state, but it is in condensed phases, and particularly in solution and at surfaces, that the most diverse reactions occur.

The aim of this chapter is to give a review of modern simulation methods for the investigation of chemical reactions in the condensed phase with a bias, due to the interests of the author, toward systems of biological importance. As in many areas of molecular science, simulation approaches are proving to be essential complements to purely experimental or theoretical work but the simulation of reactions poses unique challenges because it requires a combination of a wide range of different theoretical methodologies.

The outline of this chapter is as follows. Section 2 introduces the standard classical mechanical framework for understanding reaction rate processes, Section 3 discusses the determination of potential energy surfaces (PES) for reacting systems and Section 4 describes how to use these surfaces for reaction studies. The chapter continues, in Section 5, with a brief outline of methods for including quantum dynamical effects

Computational Chemistry
Special Volume (C. Le Bris, Guest Editor) of
HANDBOOK OF NUMERICAL ANALYSIS
P.G. Ciarlet (Editor)

in simulations of reactions and concludes, in Section 6, with some perspectives for the future.

2. Classical theories of reaction rates

This section briefly describes the theory that is necessary to calculate rate constants within the classical reactive flux formalism. It starts with a discussion of the nature of the phenomenological rate constant before going on to outline how this macroscopic quantity can be determined from microscopic theories. The arguments in this section follow closely those presented by CHANDLER [1978], CHANDLER [1987] and FIELD [1993]. Complementary reviews are given by BERNE [1985], HYNES [1985], BERNE, BORKOVEC and STRAUB [1988], ANDERSON [1995], KARPLUS [2000] and STRAUB [2001]. Due to space limitations, the discussion of this section is necessarily cursory and so readers are urged to consult these reviews for further details.

2.1. The phenomenological rate constant

The standard approach that chemists use when they want to predict the behavior of a set of chemical reactions is to formulate a series of phenomenological differential equations which describe how the concentrations of each reactive species change as a function of time. For simplicity, consider a unimolecular reaction that interconverts between two states, A and B, of a system:

$$A \rightleftharpoons B. \tag{2.1}$$

Reasonable rate equations that describe the change in the concentrations of the two states, c_A and c_B are:

$$\begin{aligned}
\frac{dc_A}{dt} &= -k_{A \to B} c_A(t) + k_{B \to A} c_B(t), \\
\frac{dc_B}{dt} &= -k_{B \to A} c_B(t) + k_{A \to B} c_A(t),
\end{aligned} \tag{2.2}$$

where t is the time and $k_{A \to B}$ and $k_{B \to A}$ are the forward and backward rate constants, respectively. There are two constraints on the concentration variables. First, the sum of the concentrations, $c_A(t) + c_B(t)$, must be constant at all times if the only reaction in the system is that given by Eq. (2.1) and, second, at long times the equilibrium values of the concentrations, $\langle c_A \rangle$ and $\langle c_B \rangle$, must obey the detailed balance condition:

$$k_{A \to B} \langle c_A \rangle = k_{B \to A} \langle c_B \rangle. \tag{2.3}$$

This says that the rate of transition from A to B must be equal to the rate from B to A.

The solution to these equations is:

$$\frac{c_A(t) - \langle c_A \rangle}{c_A(0) - \langle c_A \rangle} = \frac{c_B(t) - \langle c_B \rangle}{c_B(0) - \langle c_B \rangle} = \exp\left(-\frac{t}{\tau_{AB}}\right), \tag{2.4}$$

where $\tau_{AB}^{-1} = k_{A \to B} + k_{B \to A}$.

The scheme outlined above, and others like it for more complicated reactions, have been observed experimentally to work well in many cases. The problem then arises of how to relate the macroscopic parameters describing the reaction, τ_{AB}, $k_{A \to B}$ and $k_{B \to A}$, to microscopic quantities that can be calculated by simulation.

2.2. Potential energy surfaces and molecular dynamics

Before going on to discuss microscopic rate theories, it will be necessary to recapitulate some basic concepts concerned with how systems are described theoretically at an atomic level. As is well-known, quantum mechanics (QM) is the appropriate theory for describing molecules and one of its most fundamental equations is the time-dependent Schrödinger equation. This implies that, in principle, the behavior of any system could be determined by setting up its Schrödinger equation and then solving it. Unfortunately, this is very difficult, except in the simplest of cases, and so approximations have to be made.

The normal approach is twofold. First of all, the time-dependent problem is transformed into a time-independent one by considering only the stationary states of the system and second, the Born–Oppenheimer approximation is invoked which has the effect of being able to treat the dynamics of the electrons and nuclei separately due to their large differences in mass. This leads to a time-independent electronic Schrödinger equation which has the following form:

$$\hat{H}_{el}(r, R)\Psi_{el}(r, R) = E_{el}(R)\Psi_{el}(r, R). \tag{2.5}$$

Here r and R are vectors of coordinates for the electrons and the nuclei, respectively, Ψ is the wavefunction that gives the distribution of electrons in the system and E_{el} is the electronic energy. $\hat{H}_{el}$, is the nonrelativistic, electrostatic Hamiltonian for the electrons in the system, which in atomic units has the form:

$$\hat{H}_{el} = -\frac{1}{2}\sum_i \nabla_i^2 - \sum_{i\alpha} \frac{Z_\alpha}{r_{\alpha i}} + \sum_{ij} \frac{1}{r_{ij}} + \sum_{\alpha\beta} \frac{Z_\alpha Z_\beta}{r_{\alpha\beta}}, \tag{2.6}$$

where the subscripts i and j and α and β refer to electrons and nuclei, respectively, Z is a nuclear charge and r_{xy} is the distance between particles x and y.

If the wavefunction for the system is known, the electronic energy can be written as an expectation value over the electronic Hamiltonian:

$$E_{el} = \frac{\langle \Psi|\hat{H}_{el}|\Psi\rangle}{\langle \Psi|\Psi\rangle}. \tag{2.7}$$

In this equation the expectation value is performed with respect to the electronic variables only and not those of the nuclei because the nuclear positions are fixed and enter into Eqs. (2.5) and (2.6) parametrically. This means that the electronic energy must be solved for each distinct configuration of the nuclei of the system (defined by the vector R). As this energy is a function of the nuclear coordinates, it defines a multidimensional surface, the PES, for the system which represents the effective potential energy of interaction between the nuclei. An accurate determination of the PES of a system is a critical aspect of the study of reactions and will be discussed in Section 3.

Once the electronic problem has been solved or, in other words, once a method of obtaining the PES for the system exists, the dynamics of the nuclei can be treated. Quantum approaches do exist to do this – indeed, some will be discussed in Section 5 – but it is more usual, and certainly more straightforward, to employ classical mechanics. Probably the simplest classical mechanical algorithms start by considering the classical Hamiltonian, H, which can be written as the sum of a kinetic energy contribution, T, and a potential energy term, U. Thus:

$$H = T + U. \tag{2.8}$$

The kinetic energy term is written as:

$$T = \frac{1}{2} P^{\mathrm{T}} M^{-1} P, \tag{2.9}$$

where P is the vector of particle momenta and M is the diagonal matrix of particle masses. The potential energy term is simply the electronic energy, that is,

$$U = E_{\mathrm{el}}. \tag{2.10}$$

The equations of motion arising from this Hamiltonian are:

$$\frac{\mathrm{d}R}{\mathrm{d}t} = M^{-1} P, \tag{2.11}$$

$$\frac{\mathrm{d}P}{\mathrm{d}t} = -\frac{\partial U}{\partial R}, \tag{2.12}$$

and are equivalent to Newton's equations:

$$F = M \frac{\mathrm{d}^2 R}{\mathrm{d}t^2}, \tag{2.13}$$

where the forces, F, are defined as:

$$F = \frac{\partial U}{\partial R}. \tag{2.14}$$

The solution of these equations, or of similar ones, for the particles in the system constitutes the classical molecular dynamics method.

2.3. Microscopic states

The first step in describing the macroscopic process of Eqs. (2.1) and (2.2) in terms of microscopic variables is to define what the states A and B are at the atomic level. From the discussion of the preceding section, it is clear that this can be done informally by identifying which nuclear configurations or which regions of the system's PES correspond to state A and which to state B. More rigorously, the formulation is done in terms of a surface, the transition state (TS) surface, that separates configurations of state A from those of state B.

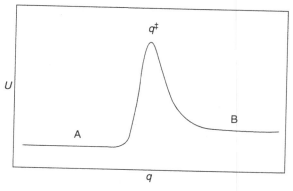

Fig. 2.1 A one-dimensional potential energy surface with two states A and B.

Consider the simple case of a one-dimensional potential shown in Fig. 2.1. It is possible to choose the TS (in this case a point) anywhere along the curve that divides the regions of the surface corresponding to states A and B. For reasons that will be apparent later, though, it is customary to place the TS at the saddle point. With this definition, characteristic functions, χ_A and χ_B, may be defined for each state as:

$$\chi_A(q) = \begin{cases} 1, & q < q^{\ddagger}, \\ 0, & q \geqslant q^{\ddagger}, \end{cases} \tag{2.15}$$

$$\chi_B(q) = 1 - \chi_A(q) = \begin{cases} 0, & q \leqslant q^{\ddagger}, \\ 1, & q > q^{\ddagger}, \end{cases} \tag{2.16}$$

where q is the coordinate, the reaction coordinate, whose value distinguishes the two states and $q^{\ddagger}$ is the value of the reaction coordinate at the TS.

The principles underlying the definition of the TS surface for multidimensional systems are the same as for the one-dimensional case except that the surface will have $D - 1$ dimensions if the system itself is D-dimensional. The choice of such a surface or, alternatively, of the reaction coordinate, q, that complements it, is often far from obvious, even for the simplest systems, and will be discussed in more detail in Section 4.

2.4. A rate constant expression

The connection between the macroscopic description of Section 2.1 and the microscopic states just defined is made using the equations of statistical mechanics. Suppose, for concreteness, that we consider the case of a particular thermodynamic ensemble, the canonical ensemble, in which the thermodynamic state variables are the particle number, N_{atoms}, the absolute temperature, T, and the volume, V. The ensemble average of the characteristic function for the state A is given by the following integral over the $3N_{atoms}$ momentum and $3N_{atoms}$ position variables for the system:

$$\langle \chi_A \rangle \propto \int d\boldsymbol{P} \int d\boldsymbol{R} \, \exp\left[-\frac{H(\boldsymbol{P}, \boldsymbol{R})}{k_B T}\right] \chi_A(\boldsymbol{R}), \tag{2.17}$$

where H is the classical Hamiltonian of Eq. (2.8) and k_B is Boltzmann's constant. At thermodynamic equilibrium, the value of this integral is independent of time and will be proportional to the equilibrium concentration of species A, $\langle c_A \rangle$. In fact, if A and B are the only states of the system:

$$\langle \chi_A \rangle = \langle c_A \rangle / (\langle c_A \rangle + \langle c_B \rangle). \tag{2.18}$$

Suppose that the system is perturbed a little due to a change in the Hamiltonian to, say, H', and that we want to determine how the ensemble average of the characteristic function decays back to its equilibrium value. The ensemble average of Eq. (2.17) will now depend on time so that:

$$\tilde{\chi}_A(t) \propto \int dP \int dR \exp\left[-\frac{H'}{k_B T}\right] \chi_A(t; R). \tag{2.19}$$

Expanding this equation to first order in the perturbation, $H' - H$, gives the following relation:

$$\tilde{\chi}_A(t) - \langle \chi_A \rangle \propto \langle (\chi_A(t) - \langle \chi_A \rangle)(\chi_A(0) - \langle \chi_A \rangle) \rangle, \tag{2.20}$$

where the right-hand side of the equation is the time autocorrelation function for the characteristic function of state A. Comparison of this equation with the solution of the macroscopic rate equations in Eq. (2.4), shows the following relation to be valid:

$$\exp\left(-\frac{t}{\tau_{AB}}\right) = \frac{\langle (\chi_A(t) - \langle \chi_A \rangle)(\chi_A(0) - \langle \chi_A \rangle) \rangle}{\langle (\chi_A(0) - \langle \chi_A \rangle)^2 \rangle}. \tag{2.21}$$

This, greatly simplified, derivation is a consequence of the fluctuation–dissipation theorem of the linear response theory of statistical thermodynamics and allows the connection that we are seeking to be made between the macroscopic rate constant and the microscopic characteristic functions.

The expression of Eq. (2.21) is not especially convenient and it is normal to manipulate it further, the details of which will not be given here. This results in a quantity called the reactive flux correlation function, $k(t)$, whose form, for the single dimensional case of Fig. 2.1, is:

$$k(t) = \langle \dot{q}(0) \delta(q(0) - q^{\ddagger}) \chi_B(q(t)) \rangle \tag{2.22}$$

and in which $\dot{q}(0)$ denotes the reaction coordinate velocity at $t = 0$. The first two terms in the correlation function, $\dot{q}(0)\delta(q(0) - q^{\ddagger})$, give the flux across the TS surface at $t = 0$ whereas the characteristic function $\chi_B(t)$ is zero unless the system is in state B at time, t.

In terms of the correlation function, $k(t)$, the rate constant for the conversion of state A to state B is:

$$k_{A \to B} \exp\left(-\frac{t}{\tau_{AB}}\right) = \frac{k(t)}{\langle \chi_A \rangle}. \tag{2.23}$$

Is this equation correct or, in other words, does $k(t)$ exhibit a simple exponential dependence? In many instances the correlation function has the form shown

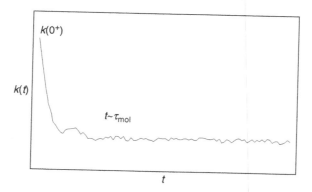

FIG. 2.2 The reactive flux correlation function.

schematically in Fig. 2.2. At short times there is a rapid decrease in the value of the function because it takes the system a certain time, $t \sim \tau_{\text{mol}}$, to commit itself to one of the regions of the PES corresponding to the states A or B. At times shorter than τ_{mol}, the system can "recross" the TS surface one or more times before the motions of the other degrees of freedom in the system can remove energy from the reactive motion and trap it in one of the states. At times longer than τ_{mol}, but much shorter than τ_{AB}, for which $\exp(-t/\tau_{\text{AB}}) \sim 1$, $k(t)$ appears to reach a plateau, the value of which is proportional to the rate constant $k_{\text{A} \rightarrow \text{B}}$.

The fact that the correlation function, $k(t)$, should have a plateau value for $\tau_{\text{mol}} \leqslant t \ll \tau_{\text{AB}}$ provides a criterion for the existence of a rate constant of the form given by the phenomenological model. It shows that there must be a separation of timescales between the "normal," fast dynamics of the system (characterized by the time τ_{mol}) and the slower reactive dynamics (characterized by τ_{AB}). This implies that the reactive events should be rare or that the energy barrier separating the states A and B be high ($\gg k_{\text{B}}T$) so that the probability of the system existing at the TS be small. If this is not true in a particular case, it means that a phenomenological model for the reaction process of the type described in Section 2.1 is inappropriate.

2.5. Transition state theory

Consider the equation for the correlation function of Eq. (2.22) and suppose that the time-dependence is removed by replacing the term $\chi_{\text{B}}(q(t))$ by a term whose value is 1 if the initial velocity $\dot{q}(0)$ takes the system to state B and 0 otherwise. Denoting this modified term by $\chi_{\text{B}}(q(0^+))$ where $q(0^+)$ is the configuration of the system at an infinitesimal time after $t = 0$, gives the transition state theory (TST) rate constant, $k_{\text{A} \rightarrow \text{B}}^{\text{TST}}$:

$$k_{\text{A} \rightarrow \text{B}}^{\text{TST}} = \frac{k(0^+)}{\langle \chi_{\text{A}} \rangle}. \tag{2.24}$$

Physically what has been done by removing the time-dependence is to forbid recrossing of the TS due to the dynamics of the remaining degrees of freedom in the system.

It also means that the TST rate constant will always be an upper bound to the true rate constant (see Fig. 2.2). The ratio of the true and the TST rate constants is called the transmission coefficient, k, whence:

$$k_{A \to B} = \kappa k_{A \to B}^{TST}. \tag{2.25}$$

TST is useful because it provides a reasonably practicable way of calculating the rate constant for a rare event in a complex system. It does this by dividing the true rate constant into a statistical part, the TST rate constant, which depends uniquely upon the equilibrium properties of the system, and a dynamical part, the transmission coefficient, which is essentially equivalent to the proportion of trajectories, starting off at the TS, that stay in the state to which they were initially directed. Each of these two terms can be evaluated independently as we shall see in Section 4.

3. Calculating condensed-phase potential energy surfaces

The evaluation of the potential energy of a reacting system as a function of its nuclear geometry is a crucial part of any simulation study. Without an accurate or, at the very least, a qualitatively correct PES, the results of any calculation will, in most cases, be meaningless.

There are several different ways to calculate PESs. The most fundamental are the *ab initio* quantum chemical methods which try to determine the electronic energy for a system at a given nuclear configuration by solving the Schrödinger equation. At the other extreme are empirical methods that use experimental, and sometimes theoretical, data to fit a function that approximates the PES for a system in a certain region of interest. And, in between, there are methods which blend elements of both the quantum chemical and empirical approaches.

Quantum chemical methods of some type are essential for the investigation of the energetics of reacting systems because it is normally difficult to obtain sufficient empirical information to give an adequate description of a PES in the region of a reaction. It is for this reason that quantum chemical methods will be given prominence in what follows below although empirical methods, when employed in conjunction with quantum chemical potentials, are also useful tools for studying reacting systems.

3.1. Ab initio quantum chemical methods

Ab initio quantum chemical methods aim to solve the time-independent Schrödinger equation (Eq. (2.5)) directly and precisely by making as few approximations as possible. The principal methods in use are based upon density functional theory (DFT) and molecular orbital (MO) theory although valence bond (VB) and quantum Monte Carlo (QMC) approaches also exist. For many applications, it will be possible to choose a technique that will give results of the desired accuracy but for others – particularly those that involve complicated electronic structures or changes in electronic structure, such as radicals, transition metal compounds and many chemical reactions – obtaining precise results can be a challenging task.

DFT and MO-based techniques are the most common contemporary *ab initio* methods in quantum chemistry. They have been covered extensively in other chapters in this volume (see, e.g., CANCÈS, DEFRANCESCHI, KUTZELNIGG, LE BRIS and MADAY [2003]), and in other references (KOCH and HOLTHAUSEN [2000], PARR and YANG [1989], KOHN [1999], POPLE [1999], SZABO and OSTLUND [1989]), and so only a few salient points concerning them will be made here.

DFT methods have revolutionized quantum chemistry over the past 10 years or so and currently provide the most cost-effective means of determining the energetics and structures of reacting groups in large molecular systems. They can be implemented with efficiencies that are similar to those of the fastest MO methods based upon Hartree–Fock (HF) theory but they provide energetic and structural results for ground and excited state systems that are often as precise as much more expensive MO-based methods. They do have some limitations, notably for applications to hydrogen-bonded and weakly bound systems and for the determination of the barrier heights for certain reactions. These problems are likely to become less marked though as better exchange-correlation functionals are developed (KOCH and HOLTHAUSEN [2000]). Another important development that should be mentioned, and which will increase the usefulness of DFT calculations, is the introduction of linear-scaling or $O(n)$ algorithms (GOEDECKER [1999]). Calculations with most existing DFT implementations have costs that scale somewhere between $O(n^2)$ and $O(n^4)$ where n is a measure of the size of the system (such as the number of electrons).

Although calculations relying upon DFT have come to dominate quantum chemistry, MO-based methods still have a role to play as they are, in principle, capable of providing results of arbitrary accuracy by systematically enhancing the quality of the basis set and including more Slater determinants in the wavefunction expansion. The simplest MO HF methods are not competitive with DFT techniques in terms of accuracy and, in any case, are often not appropriate for describing the electronic structure of a system in which bonds are being broken and formed. In contrast, the more accurate MO methods have costs that scale as $O(n^5)$ or higher which means that they are limited to systems of small size ($N_{atoms} \sim 10$).

Two other approaches deserve comment in this brief survey. The first are methods based upon VB theory in which the wavefunction for the system is constructed as a linear combination of "resonant" structures that represent particular, idealized electronic configurations of the system. The wavefunctions of the resonant states are normally constructed from nonorthogonal spin-orbitals. *Ab initio* VB methods give results that are of similar quality to MO calculations but they are not often employed because of the extra complexity arising from the use of nonorthogonal orbitals (GALLUP, VANCE, COLLINS and NORBECK [1982], COOPER, GERRATT and RAIMONDI [1987]). The second type of approach comprise the QMC methods which make use of the similarity between the diffusion equation and the Schrödinger equation in imaginary time to statistically estimate the electronic energy of the ground state of a system (CEPERLEY and MITAS [1996]). QMC methods have the ability to produce results of very high accuracy at a much lower cost than the corresponding MO methods of equivalent precision (GROSSMAN, LESTER and LOUIE [2000]) but their use has been hampered by the absence of any widely available general-purpose QMC programs.

To terminate this section, it should be pointed out that calculations with the quantum chemical methods of the type described here can be applied in two distinct ways. Either the calculations can be done "on-the-fly," which means that the quantum chemical method is used directly during a simulation to obtain the electronic energy for each new configuration of a system, or the quantum chemical calculations can be done beforehand and the results employed to parametrize an analytical function that reproduces the reactive part of the PES. The latter, indirect, approach has the advantage that simulations with the parametrized surface are very fast but the disadvantage that parametrization of a surface for a reaction becomes progressively more difficult as the number of configurational degrees of freedom increases. It is for this reason that the former, direct, method is nowadays generally preferred for studies of reactions in complex systems.

3.2. Semiempirical quantum chemical methods

Ab initio quantum chemical methods are preferable for the investigation of chemical reactions and other aspects of the electronic structure of molecules but, due to their computational expense, faster, albeit less precise, quantum chemical methods have been developed. In general, these semiempirical methods have similar formalisms to their *ab initio* counterparts but differ in that simplifications are made that greatly speed up the more time-consuming parts of the calculation. These simplifications normally entail the replacement of analytically determinable quantities, such as certain Hamiltonian matrix elements, by empirical functions that must be parametrized against experimental or *ab initio* data.

The great majority of semiempirical methods in chemistry are based upon MO theory. The Hückel method was an early method that is still widely used for qualitative calculations (ALBRIGHT, BURDETT and WHANGBO [1985]). Equally successful have been the semiempirical MO approaches developed by Dewar and coworkers. These include the AM1 and MNDO methods of Dewar himself (DEWAR and THIEL [1977], DEWAR, ZOEBISCH, HEALY and STEWART [1985]) and the PM3 parametrization of the AM1 Hamiltonian of Stewart (STEWART [1989a], STEWART [1989b]). The AM1 family of techniques are now dated and insufficiently accurate for certain problems, including many chemical reactions (THOMAS, JOURAND, BRET, AMARA and FIELD [1999]). Even so, in favorable cases, they have a precision comparable to or better than *ab initio* DFT or HF calculations performed with small basis sets (up to double-ζ) despite being substantially less expensive (DEWAR and O'CONNOR [1987]).

Given the recent success of *ab initio* DFT methods in quantum chemistry, it is perhaps not surprising that semiempirical versions of the theory have been introduced. One promising avenue concerns the application of tight-binding methods to chemistry. These methods, commonly used in solid-state physics, can be shown to arise from DFT by expanding the Kohn–Sham equations in terms of fluctuations in the atomic density (FOULKES and HAYDOCK [1989]). In addition to providing a firm theoretical foundation, such a link gives expressions for many of the parameters appearing in tight-binding models in terms of quantities that can be determined from *ab initio* DFT calculations. This, in principle, greatly simplifies the parametrization procedure.

An example of this type of model is the self-consistent charge tight-binding (SCCTB) method developed by Elstner, Porezag, Frauenheim and coworkers and which has been claimed to give improvements over the Dewar MO methods (ELSTNER, POREZAG, JUNGNICKEL, ELSNER, HAUGK, FRAUENHEIM, SUHAI and SEIFERT [1998], FRAUENHEIM, SEIFERT, ELSTNER, HAJNAL, JUNGNICKEL, POREZAG, SUHAI and SCHOLZ [2000], POREZAG, FRAUENHEIM, KÖHLER, SEIFERT and KASCHNER [1995]). In this method, the energy of the system, E, can be written as the sum of two terms. The first, E_0, is the typical tight-binding energy expression:

$$E_0 = \sum_i n_i \langle \psi_i | \hat{H} | \psi_i \rangle + E_{\text{rep}}, \tag{3.1}$$

where $\hat{H}$ is the tight-binding Hamiltonian for the system, ψ_i is the ith orbital with occupation number n_i and E_{rep} is the repulsion energy between atomic cores. The second term, E_1, is:

$$E_1 = \frac{1}{2} \sum_{\alpha\beta} \gamma_{\alpha\beta} q_\alpha q_\beta, \tag{3.2}$$

where q_α is the partial charge on atom α and $\gamma_{\alpha\beta}$ is a matrix element for the interaction between two charges. The orbitals for the system are expanded in terms of atomcentered basis functions and the optimum orbitals are obtained by minimizing the total energy expression with respect to the orbital expansion coefficients. As the atomic charges are derived from a population analysis of the system's wavefunction, they depend upon the orbitals and so solution of the equations must be performed in a self-consistent fashion.

3.3. Empirical potentials

With quantum chemical methods, there are two possible strategies for the use of data resulting from the solution of the Schrödinger equation at a particular nuclear configuration. Either it can be used directly in a simulation or it can be employed to fit an intermediate function of arbitrary form that represents the PES of the system in the regions of interest. Once fitting is complete, this function will be used as the potential energy function in subsequent simulation. The rationale underlying empirical potentials (or, as they are also called, force fields or molecular mechanical (MM) potentials) is different. Instead, an empirical form for the energy function is chosen, each term of which encapsulates a particular aspect of the bonding properties or the interatomic interactions in the system. Once, the form of the potential has been decided, it must be parametrized so that it reproduces experimentally observable quantities.

There are many different types of empirical force field and they are often tailored for simulations of different classes of molecule. One of the more common types and the type that is most frequently employed for simulations of biomolecular systems, such as proteins, writes the total potential energy of the system, E, as the sum of two energies, one for the covalent or bonding interactions, E_{bonding}, and one for the noncovalent or nonbonding interactions, $E_{\text{nonbonding}}$ (FIELD [1999]):

$$E = E_{\text{bonding}} + E_{\text{nonbonding}}. \tag{3.3}$$

It is normal to subdivide the bonding and nonbonding energies further. Thus, the bonding energy often consists of a sum of bond, angle, dihedral, and out-of-plane terms:

$$E_{\text{bonding}} = E_{\text{bond}} + E_{\text{angle}} + E_{\text{dihedral}} + E_{\text{out-of-plane}}, \tag{3.4}$$

whereas the nonbonding energies are the sum of a electrostatic term and a Lennard-Jones term:

$$E_{\text{nonbonding}} = E_{\text{elect}} + E_{\text{LJ}}. \tag{3.5}$$

A common form for the bond energy is a sum of harmonic terms, one for each bond in the system:

$$E_{\text{bond}} = \sum_{\text{bonds}} \frac{1}{2} k_b (b - b_0)^2. \tag{3.6}$$

In this equation, b is the actual distance between the two atoms involved in the bond, b_0 is an equilibrium distance characteristic of the bond and k_b is the force constant for the bond which determines the steepness of the potential well and, hence, the bond's frequency of oscillation.

The angle energy is usually similar except that energy terms are functions of the angle, θ, subtended by three atoms:

$$E_{\text{angle}} = \sum_{\text{angles}} \frac{1}{2} k_\theta (\theta - \theta_0)^2. \tag{3.7}$$

For the dihedral or torsional energy, a harmonic form is not appropriate as the energy must be a periodic function of the torsion angle about the bond. A suitable form is an expansion in terms of trigonometric functions, that is,

$$E_{\text{dihedral}} = \sum_{\text{dihedrals}} \frac{1}{2} V_n (1 + \cos(n\phi - \delta)). \tag{3.8}$$

Here, V_n is the height of the torsional barrier, n is the periodicity of the term and δ is its phase.

The out-of-plane or improper torsional energy, $E_{\text{out-of-plane}}$, is used to keep atoms, such as those which are sp^2 hybridized, planar. In some force fields, a harmonic form for this energy term is employed whereas, in others, terms reminiscent of the dihedral energy are preferred.

In many force fields, the electrostatic energy, E_{elect}, is given by a simple Coulomb-type expression because the charge distribution of a molecule is represented by fixed partial charges centered on the nuclei of the atoms. The energy then has the form:

$$E_{\text{elect}} = \sum_{\alpha\beta \text{pairs}} \frac{q_\alpha q_\beta}{\varepsilon r_{\alpha\beta}}, \tag{3.9}$$

where q_α and q_β are the partial charges on atoms α and β and $r_{\alpha\beta}$ is the distance between them. ε is the dielectric constant for the interaction which will be unity for two atoms in vacuum.

The Lennard-Jones energy mimics the quantum mechanical exchange-repulsion interaction arising when two charge clouds overlap at short-range and the attractive dispersive inverse sixth power interaction at longer range. It has the form:

$$E_{LJ} = \sum_{\alpha\beta \text{pairs}} \frac{A_{\alpha\beta}}{r_{\alpha\beta}^{12}} - \frac{B_{\alpha\beta}}{r_{\alpha\beta}^{6}}, \tag{3.10}$$

where $A_{\alpha\beta}$ and $B_{\alpha\beta}$ are constants whose values depend upon the nature of the atoms α and β. The Lennard-Jones energy between two atoms will be nonzero even when they have no net charge.

The nonbonding interaction energies of Eqs. (3.9) and (3.10) are normally calculated for all pairs of atoms in the system but it is usual to exclude pairs of atoms from the sum which are separated by only one or two covalent bonds so as to avoid the overcounting that would result if both bonding and nonbonding terms were calculated for these atoms.

The calculation of the nonbonding interactions, and particularly the longer range electrostatics terms, is invariably the most expensive part of an energy calculation. This is because the number of interactions scales as the square of the number of particles whereas the number of bonding terms scales roughly linearly with the size of the system. There are a number of ways in which the cost of the nonbonding energy calculation can be reduced. One of these is the approximate truncation technique in which interactions are either neglected or tapered to zero beyond a certain cutoff distance. Other methods, which are to be preferred, attempt to calculate the full nonbonding energy of a system to a certain precision but with a cost that scales linearly with the size of the system.

The terms listed for the "typical" force field above are by no means the only ones in use. Thus, for example, in more complicated functions there will be bonding cross-terms which couple various internal coordinate deformations as well as nonbonding polarization terms which describe the interactions due to the changes in the charge distribution of a molecule in different environments.

An advantage of empirical force fields of the type discussed here is that they are computationally efficient and can be used for the simulation of systems comprising many thousands of atoms. In addition, their analytic form is such that it is straightforward to calculate the derivatives of the energy with respect to various atomic quantities. In practical applications, the most important of these is the first derivative of the potential energy with respect to the atomic coordinates which is proportional to the force (see Eq. (2.14)). A disadvantage of force fields is that they must be parametrized to obtain values for the many parameters (b_0, k_θ, V_n, q_i, etc.) that they contain. Some applications do not require a very precise parametrization but, in most cases, it will be necessary to parametrize the force field against large amounts of data from experiments or from high quality QM calculations. Such parametrizations are laborious and can demand great effort if force fields of reasonable precision are to be obtained. Another disadvantage is that force fields are of limited flexibility as they are conceived for the simulation of particular systems in particular circumstances and will be unsuitable for studying processes outside their range of applicability. Thus, for example, it will not be possible to study reactions in which bonds are broken and formed with the force field described above because the harmonic bond term of Eq. (3.6) does not allow dissociation.

3.4. Hybrid potentials

As we have seen, the investigation of reactions in condensed-phase systems with purely quantum chemical methods is currently impractical if the system is of any reasonable size and the use of empirical force fields for studying reactions is problematical because special functions would have to be devised and parametrized for each reaction under study. To circumvent these problems, hybrid potentials have been developed in which potentials of differing accuracy are used to treat different regions of the system. For example, a chemical reaction could be studied by treating the reacting atoms and those immediately surrounding them with a QM potential and using a simpler method for the atoms of the remainder of the system. When partitioning the system in this way, the assumption is made that the reactive process is localized in the QM region – this will be reasonable for many reactions but it will not be valid in some instances, such as when there is a long-range electron-transfer event.

Hybrid or QM/MM potentials were first introduced for the treatment of reactions in enzymes by WARSHEL and LEVITT [1976]. Since then, many types of hybrid potential have been implemented (for reviews, see GAO [1995], AMARA and FIELD [1999], MONARD and MERZ [1999]). They differ in the number of regions into which the system is divided, the types of potential used to treat the different regions and the ways in which the interfaces between the potentials are handled. Although diverse in nature, the basic principles underlying most hybrid potentials are similar and are best illustrated by a simple example.

Consider a system divided into two regions (see Fig. 3.1), one of which is treated by an MO QM method and the other with an MM potential. The first steps in the

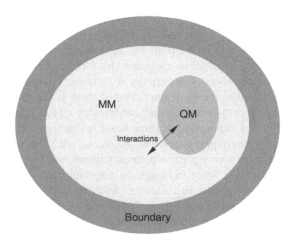

FIG. 3.1 Partitioning a system into a QM and an MM region for use with a hybrid potential.

formulation of the hybrid potential in this case are to invoke the Born–Oppenheimer approximation and to construct an effective electronic Hamiltonian, $\hat{H}_{\text{Eff}}$, which will be the sum of three terms:

$$\hat{H}_{\text{Eff}} = \hat{H}_{\text{QM}} + \hat{H}_{\text{MM}} + \hat{H}_{\text{QM/MM}}, \tag{3.11}$$

where $\hat{H}_{\text{QM}}$ is the Hamiltonian for the particles (electrons and nuclei) in the QM region of the system, $\hat{H}_{\text{MM}}$ is the Hamiltonian for the atoms in the MM region and $\hat{H}_{\text{QM/MM}}$ is the Hamiltonian for the interaction between the two regions (FIELD, BASH and KARPLUS [1990]).

The exact form for the terms in the Hamiltonian will depend upon the QM and MM potentials of the individual regions but common examples are:

$\hat{H}_{\text{QM}}$ The form for the QM Hamiltonian depends upon the approximation chosen to solve the time-independent Schrödinger equation. For an *ab initio* method, this would be the standard nonrelativistic electronic Hamiltonian whereas for a semiempirical method it would be the Hamiltonian appropriate to the method.

$\hat{H}_{\text{MM}}$ The MM Hamiltonian describes the potential energy of the MM atoms, E_{MM}, and is independent of the electronic coordinates. Any molecular mechanics force field is suitable that includes covalent terms that describe the bond, angle, and dihedral energies of the molecules and nonbonding terms that account for electrostatic and dispersion/repulsion effects between atoms that are further apart.

$\hat{H}_{\text{QM/MM}}$ The Hamiltonian for the interactions between the QM and MM regions consists of a sum of terms which represent the electrostatic interactions between the charges of the MM atoms and the electrons and nuclei of the QM region and the Lennard-Jones interactions between the MM atoms and the QM nuclei. If an *ab initio* method is being used for the QM region, the appropriate form, in atomic units, would be:

$$\hat{H}_{\text{QM/MM}} = -\sum_{iM} \frac{qM}{r_{iM}} + \sum_{\alpha M} \frac{Z_\alpha qM}{r_{\alpha M}} + \sum_{\alpha M} \left\{ \frac{C_{\alpha M}}{r_{\alpha M}^{12}} - \frac{D_{\alpha M}}{r_{\alpha M}^{6}} \right\}, \tag{3.12}$$

where the subscripts i, α, and M refer to electrons, QM nuclei and MM atoms respectively, Z is the nuclear charge and C and D are parameters for the Lennard-Jones interaction.

The energy of the system, E, and the wavefunction, Ψ, for the electrons of the QM region are determined by solving the time-independent Schrödinger equation (Eq. (2.5)) with the effective Hamiltonian defined in Eq. (3.11) as the operator. As the wavefunction and the effective Hamiltonian depends parametrically upon the positions of both the QM nuclei and the MM atoms, the Schrödinger equation must be solved at each different configuration of the QM nuclei and the MM atoms.

The hybrid potential formulated above was for an MO QM method and was done in terms of an effective Hamiltonian. For a DFT QM method, it is more appropriate to formulate the potential in terms of the electron density in the QM region but as the procedure is essentially equivalent no details will be given here. Instead, readers are referred to AMARA, VOLBEDA, FONTECILLA-CAMPS and FIELD [1999].

The treatment of the junction between the different regions is the crucial aspect of the definition of a hybrid potential and remains, in many respects, an outstanding problem. The interaction Hamiltonian of Eq. (3.12) was for the straightforward case in which there are only nonbonding interactions between the atoms of the different regions. For many applications, however, it will be necessary for a single molecule to be split between different regions which will result in there being covalent bonds between QM and MM atoms. These "dangling" bonds must be treated in some way as otherwise the presence of broken bonds and unpaired electrons at the boundary of the QM region will dramatically change the electronic structure of the QM system. The commonest and also the simplest way of tackling this problem is to use the link-atom approximation in which a single unphysical atom is introduced into the QM region for each dangling bond (see Fig. 3.2). These atoms, which are normally hydrogens, enter into the QM calculation and serve to sate the unsatisfied valencies of the QM atoms with broken bonds. How their interactions with the MM regions are handled varies according to the implementation of the method. Apart from the link-atom method, alternatives, such as those based upon hybrid orbitals, have been proposed but all these methods are significantly more complicated to implement and, in the author's opinion, have not been shown to be consistently more accurate. In addition to the problem of covalent bonds between QM and MM atoms, the treatment of the interactions between the atoms of different regions needs improvement, especially for hybrid potentials which use *ab initio* QM methods. This is an area of on-going research.

The hybrid potentials described above have been implemented with both sophisticated *ab initio* and the simpler semiempirical QM methods. Hybrid potentials have also been developed with even simpler QM potentials, the most notable examples being the empirical valence bond (EVB) potentials of Warshel and coworkers (WARSHEL [1991]). The EVB method is like a normal VB method in that the total wave-function for the system is formed as a linear combination of wavefunctions of resonant states but differs because the matrix elements between the resonant states are parametrized using simple functional forms. Although computationally efficient and applicable to reactions, EVB potentials must be redesigned and reparametrized every time a new reaction is to be studied. Unfortunately, this requires considerable

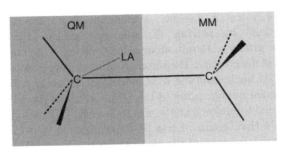

FIG. 3.2 Treating covalent bonds between atoms of the QM and MM regions of a hybrid potential using the "link-atom" approximation.

experience as there are systematic ways neither for the generation of the important quantum states that partake in the reaction nor for their parametrization. Consequently, only a relatively small number of research groups worldwide routinely apply these methods.

3.5. Extended systems

Experiments are usually done on systems with $O(N_{Avogadro})$ ($\sim 10^{23}$) particles which is far from the size of the system that can be studied computationally. Thus, to be feasible, a simulation of a condensed-phase system must select a certain group of atoms to treat explicitly and then approximate the effect of the remaining, neglected atoms which act as the environment to the central simulation system. There are two main ways of dealing with extended systems (apart from the easiest one of neglecting the environment entirely). The first is to use the method of periodic boundary conditions in which the entire system is taken to be a periodic array (or a "crystal") of copies of the central simulation system (ALLEN and TILDESLEY [1987]). This method is probably the most exact as all the atoms in the system are treated explicitly. The second class of methods is to dispense with an atomic model of the environment and use some sort of approximate description. There are many of these, including the widely used reaction field methods in which the simulation system is immersed in a medium whose dielectric constant mimics that of the surrounding atoms (see TOMASI and PERSICO [1994] and SIMONSON [2001] for nice reviews). None of the methods described in this section are specific to studies of reacting systems but no discussion of how to calculate the PES for a condensed-phase system would be complete without mentioning them.

4. Simulation methods for investigating chemical reactions

This section assumes that a method for the calculation of the PES for a reacting system is available and outlines the various simulation approaches that can be employed to obtain information about a chemical reaction. The discussion is focused on methods for investigating reactions within the classical reactive flux framework but a more recent development is mentioned at the end.

4.1. The classical reactive flux formalism

Probably the most practical way of determining the rate constant within the reactive flux formalism is based upon Eq. (2.25) which expresses the rate constant as the product of two quantities, the TST rate constant and the transmission coefficient, which can be evaluated independently. Discussion of the transmission coefficient will be left until later, but it is convenient to manipulate the TST rate constant expression of Eq. (2.24) to put it into a more suitable form for computation. A little rearrangement allows it to be rewritten as:

$$k_{A \to B}^{TST} = \frac{\langle \dot{q} \Theta(\dot{q}) \delta(q - q^\ddagger) \rangle}{\langle \delta(q - q^\ddagger) \rangle} \times \frac{\rho(q^\ddagger)}{\int_R dq \rho(q)}. \tag{4.1}$$

Here Θ is the Heaviside function which takes values of 1 for positive arguments and zero otherwise, $\rho(q)$ represents the probability density for the system as a function of the reaction coordinate and the integration in the denominator of the second term on the right-hand side is done over the values of the reaction coordinate that correspond to the reactant well. Not only is Eq. (4.1) easier for calculation, as we shall see below, but it also takes the same form as the well-known Arrhenius rate constant expression:

$$k_{A \to B}^{Arrhenius} = v \exp\left[-\frac{\Delta G^{\ddagger}}{k_B T}\right], \tag{4.2}$$

where v is a frequency prefactor and $\Delta G^{\ddagger}$ is the activation free energy which is the free energy difference between the reactant and the TS structures.

Inspection of Eq. (2.25), together with the fact that the transmission coefficient takes values between 0 and 1, indicates that the TST rate constant will always overestimate the true rate constant. As the transmission coefficient can be expensive to evaluate, especially when it takes very small values, it is most efficient to select a TS surface dividing reactants from products that minimizes the value of the TST rate constant and, hence, maximizes the value of the transmission coefficient. As Chandler has shown, this is equivalent to finding the surface with the highest free energy (CHANDLER [1978]).

Unfortunately, the selection of the TS surface of most general form and of highest free energy is a very difficult problem even for small systems. Marked simplifications arise when it is assumed that the surface dividing states A and B is a hyperplane, as in this case it is only necessary to choose the position of the hyperplane and its orientation (SCHENTER, MILLS and JÓNSSON [1994], JÓHANNESSON and JÓNSSON [2001]). Even so, calculations that make use of algorithms for finding this optimum hyperplane have been limited due to their expense. An even simpler approximation, and the one that is usually employed, is to choose a reaction coordinate that describes the reaction process between the two states, calculate the free-energy as a function of this coordinate and then take the TS as being the point of highest free-energy along the path. This approach also assumes that the dividing surface between states A and B is a hyperplane but only the location of the hyperplane is optimized – its orientation is fixed as being normal to the tangent of the reaction coordinate.

To summarize, a standard approach when using the reactive flux formalism consists of three steps. The first is the determination of a reaction coordinate that somehow represents the transition process between the reactant and product states, the second is the calculation of the TST rate constant by the estimation of the free-energy as a function of the reaction coordinate and the third is the correction of the TST rate constant by evaluating the transmission coefficient. In very general terms, each of these steps is more expensive than the previous one, so that many studies stop after only the first or second steps. This is particularly true for simulations with *ab initio* potentials as these most frequently locate a suitable TS structure for the reaction and then, at best, make a crude estimate of the TST rate constant. It should be emphasized, however, that much useful chemical information can still be gained by limited studies of this type.

4.2. Determining a reaction coordinate

The ideal reaction path is the one which minimizes the TS rate constant. Unfortunately, such paths are difficult to determine directly and so it is usual to make assumptions about what such paths look like. For the simplest reactions, it may be possible to choose a reaction path by inspection. Thus, for example, the dissociation of a diatomic molecule obviously depends upon the distance between the two atoms whereas the rotation of a methyl group or an aromatic ring within an organic molecule could be reasonably described by a torsional angle. Such choices are, however, fraught with danger as the example of the diatomic makes clear. Whereas in the gas phase the reaction coordinate is obvious, in solution it is highly probable that various solvent degrees of freedom are important for the definition of the reaction coordinate, particularly if the dissociation involves a separation of charge.

A more rigorous definition of a path is one that says that the optimum path is the "lowest-energy" path on the PES that connects in a continuous fashion the potential wells (or minima) that define the reactant and product states. This requires that the path has, as its highest-energy point, the first-order saddle point of lowest energy between the two stable states. For reactions where there are intermediate stable states between reactants and products, it is normal to define the complete path as the one made up of the individual paths between each of the intervening minima. An example is shown in Fig. 4.1.

The location of a reaction path going through saddle points is most conveniently done for small systems by first finding the saddle points and then tracing out the path downhill from the saddle point to the stable states that lie on either side. The location

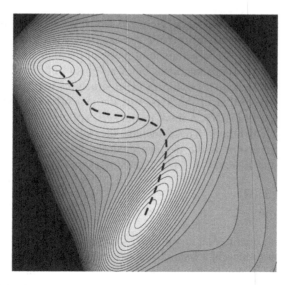

FIG. 4.1 A two-dimensional potential energy surface and a possible reaction path between two states. There is one intermediate.

of saddle points is a geometry-optimization procedure with the function determining the PES as the target function and the atomic coordinates as its variables. Saddle point location is more difficult than finding minima because instead of searching for a direction which reduces the energy in every direction, it is necessary to maximize the energy in one direction and minimize in all the rest. An early, but nice introductory review of saddle-point-search algorithms has been given by BELL and CRIGHTON [1984].

Probably the most successful algorithms are the mode-walking algorithms that start off at a particular point on the surface, often a minimum, and move uphill along one of the normal modes for the system and downhill along all the rest. These algorithms were introduced by CERJAN and MILLER [1981] and have been refined by other workers, including SIMONS, JORGENSON, TAYLOR and OZMENT [1983], BAKER [1986] and WALES [1994]. The determination of the normal modes along which to walk requires the calculation or the estimation and the diagonalization of the second derivative matrix of the energy with respect to the coordinate variables. This limits the use of this algorithm to systems with a relatively small number of degrees of freedom as the calculation and storage of the second derivative matrix and its diagonalization are computationally demanding tasks.

Once the saddle point between two minima has been identified, the full path can be generated. A common definition is to follow the steepest-descent direction down from the saddle point either in Cartesian coordinates, R, or in mass-weighted Cartesians, $M^{-1/2}R$ (FUKUI [1981]). A straightforward steepest-descent algorithm is often sufficient (MILLER, HANDY and ADAMS [1980], MILLER [1983], FIELD [1999]) but other algorithms have been devised if very smooth paths are desired (see, e.g., DAVIS, WALES and BERRY [1990] and GONZALEZ and SCHLEGEL [1989]).

While the two-step procedure described above is suitable for finding the reaction paths for small systems, other methods, most of which do not require the use of second derivatives, have been developed for locating transition paths in systems with many degrees of freedom. A characteristic of many of these methods is that they take as input reactant and product structures, build a "chain" of intermediate structures between them and then geometry optimize these structures until they lie on the reaction path. An early algorithm of this type was the synchronous transit algorithm of HALGREN and LIPSCOMB [1977]. A more recent series of algorithms have been developed by Elber and coworkers. These include the self-avoiding walk algorithm in which a discretized version of the average potential energy along the path, $\langle U \rangle$, is optimized (ELBER and KARPLUS [1987], CZERMINSKI and ELBER [1990]):

$$\langle U \rangle = \frac{\int_R^P \mathrm{d}l(R)U(R)}{\int_R^P \mathrm{d}l(R)}. \tag{4.3}$$

Here R and P refer to the reactant and product structures, respectively, and l is the distance along the path (which is a function of the atomic coordinates, R).

The early chain-type algorithms were reasonably effective but they had some problems, including limited radii of convergence and the tendency to produce nonsmooth paths. To alleviate these problems, Elber and coworkers introduced a couple of complementary algorithms to further refine the path once an initial reaction path had been determined (ULITSKY and ELBER [1990], CHOI and ELBER [1991]). A notable recent development is the nudged-elastic band algorithm that takes many elements of the Elber algorithms but considerably enhances the stability and the convergence rate of the algorithm (HENKELMAN, JÓHANNESSON and JÓNSSON [2000], HENKELMAN and JÓNSSON [2000]). The objective function of Eq. (4.3) is also not the only one that leads to reasonable reaction paths. Thus, for example, both HUO and STRAUB [1997] and ELBER and SHALLOWAY [2000] have introduced functions that allow average paths at nonzero temperature to be determined. In the work of Huo and Straub, the potential energy, U, in the integral of Eq. (4.3) is replaced by the quantity $\exp(U/k_B T)$.

4.3. Calculating the TST rate constant

As we saw in Section 4.1, the TST rate constant (Eq. (4.1)) can be expressed as the product of two factors, both of which is a configurational average. The first term is the average forward flux at the TS surface. It can be evaluated as it appears in Eq. (4.1) by fixing the system at the TS and then evaluating the average from a molecular dynamics simulation. Alternatively, the expression can be manipulated further by integrating over the momentum degrees of freedom to give the expression $\sqrt{k_B T / 2\pi m_{\text{eff}}}$ where m_{eff} is an effective or reduced mass for motion along the reaction coordinate:

$$
m_{\text{eff}} = \left[\frac{\left\langle \sqrt{\sum_{\alpha=1}^{N_{\text{atoms}}} \frac{1}{M_\alpha} \left(\frac{\partial q}{\partial R_\alpha}\right)^2} \delta(q - q^\ddagger) \right\rangle}{\langle \delta(q - q^\ddagger) \rangle} \right]^{-2}.
\tag{4.4}
$$

The effective mass can be evaluated as a configurational average too but the assumption is often made that its value is only weakly dependent on configuration and so it can be determined for a single structure that is representative of the TS, such as a saddle point.

The second term in Eq. (4.1) is the ratio of the equilibrium populations of the system in the transition and the reactant states. It corresponds to the free-energy term of Eq. (4.2) and so allows the free-energy of activation of a reaction to be written as:

$$
G^\ddagger = -k_B T \ln\left[\frac{\rho(q^\ddagger)}{\int_R dq \rho(q)}\right].
\tag{4.5}
$$

The quantity $\mathcal{W}(q) = -k_B T \ln[\rho(q)]$ is called the potential of mean force (PMF) and gives the reversible work (or free-energy) required to displace the system along the reaction coordinate. In principle, the PMF could be calculated directly from a molecular dynamics or Monte Carlo simulation that samples the correct thermodynamic ensemble. However, in the cases that we are considering for which the energetic barrier to reaction is high, the population of the TS relative to that of the reactant state will be extremely small. This means that very long simulations, impractically long in most cases, would be needed to adequately sample the TS region to obtain a statistically significant estimate of the PMF.

To overcome this problem a number of techniques exist which either enhance sampling in various regions of phase space or allow the determination of the free energy of the system as a function of certain system parameters, such as the geometrical variables that define the reaction coordinate. Umbrella sampling is an example of the first type of technique and involves performing a series of simulations with constraint potentials that confine the system to different regions of configuration space. The bias introduced by the constraint potentials is then removed after the simulations when the probability density, $p(q)$, is constructed from the individual, biased probability densities obtained from the separate simulations. Examples of the second type of technique are the thermodynamic integration and statistical perturbation methods that rigidly constrain the system to be at a certain value of the reaction coordinate and then determine the free-energy difference as the value is changed. All these methods work well but they must be applied with care – in particular, simulations of sufficiently long duration must be performed for accurate values of the free energies to be determined. For fuller details about these techniques, readers are referred to the references by BEVERIDGE and DI CAPUA [1989], FIELD [1999], JORGENSEN [1989], KOLLMAN [1993], MARK [1998], ROUX [1995] and STRAATSMA and MCCAMMON [1992].

4.4. *Evaluating the transmission coefficient*

The transmission coefficient is the ratio of the true rate constant to the TST rate constant which, in terms of correlation functions of Eqs. (2.22) and (2.24), can be written as:

$$\kappa \sim \frac{k(\tau_{\mathrm{mol}})}{k(0^+)}, \tag{4.6}$$

where the plateau value of the correlation function is assumed to occur at $t \sim \tau_{\mathrm{mol}}$.

Calculation of κ requires finding the value of the plateau for the (normalized) reactive flux correlation function displayed in Fig. 2.2 and is equivalent to determining the fraction of trajectories, originating in the reactant well, that cross the TS and remain in the product well. There are a number of different approaches for doing this, but as they are mostly similar, only one will be outlined here. It is possible to rewrite the expression for the correlation function as:

$$k(t) \propto \langle \chi_{\mathrm{B}}(q(t)) \rangle_+ - \langle \chi_{\mathrm{B}}(q(t)) \rangle_-. \tag{4.7}$$

The subscripts $+$ and $-$ on the angled brackets imply an averaging but using initial conditions for the position and velocity (or momentum) degrees of freedom in the system drawn from the probability distributions, P_+ and P_-, defined as follows:

$$P_\pm \propto \dot{q}\delta(q - q^\ddagger)|\dot{q}|\Theta(\pm\dot{q}) \exp[-H/k_{\mathrm{B}}T]. \tag{4.8}$$

The function P_+ forces the system to be at the TS initially with a positive velocity that points toward the product well. The function P_- acts similarly except that the velocity points toward the reactant well.

Practically, the averaging of Eq. (4.7) is performed by starting off a large number of trajectories for the system at the TS and with momenta that satisfy the functions of

Eq. (4.8). The equations of motion for each trajectory are then integrated for a length of time that it is estimated will be sufficient for each trajectory to have committed itself to either the reactant or the product well. The average is then simply the fraction of trajectories with velocities pointing to the product well that remain in the product well minus the fraction of trajectories with velocities pointing to the reactant well that end up in the product well. This scheme works well and has been used extensively for studying complex systems (see, e.g., the application to an enzyme by NERIA and KARPLUS [1997]). It can, however, be expensive to implement as many trajectories can be needed to obtain a statistically reliable estimate of k. In certain regimes, where there are many recrossings or where it takes a long time for the trajectories to become trapped on one side of the TS, the method can become impractical. Alternative algorithms have been developed to cope with such circumstances but due to space limitations readers are urged to refer to STRAUB [2001] for further details.

4.5. Transition path sampling

The discussion of Section 2 and of the preceding part of this section has focused on the calculation of reaction rates using the reactive flux formalism in which a reaction coordinate is defined for the reaction process, a TS is identified at some point along the reaction coordinate and, finally, the various terms of Eqs. (4.1) and (4.6) are calculated. Such a procedure is satisfactory as long as a suitable set of variables can be found to describe the reaction coordinate and TS. This is quite readily done, to reasonable accuracy, for simple reactions or for systems with only a few degrees of freedom by using the methods described in Section 4.2 which locate the saddle points and the associated paths that lead between reactants and products. In complex systems, however, where there can be many saddle points and paths of similar energy, these approaches are much less effective and, hence, determination of the rate constant becomes much more difficult.

To circumvent these problems, a new class of methods for investigating transitions in complex systems has recently been developed. These methods, or ones like them, are very exciting and likely to change radically the way reactions are studied in the future. The original idea was probably due to PRATT [1986] but most of the recent advances have been made by BOLHUIS, CHANDLER, DELLAGO and GEISSLER [2002]. These methods dispense with the notion of a TS and, instead, aim to generate a representative set of transition paths that go between reactant and product configurations. Assuming that it is possible to sample a statistically significant number of trajectories and that these trajectories are consistent with a particular dynamics and thermodynamic ensemble, then dynamical quantities, such as rate constants, may be determined directly.

To give a flavor of what the methods entail, consider a transition path going between states A and B and consisting of $L + 1$ different configurations of the system. The Ith structure is specified by values of its position, R_I, and its momenta, P_I, and the time at which it occurs, τ_I. τ_I is equal to ΔI if the timestep between configurations is Δ. If neighboring configurations are linked by a Markovian transition probability, $p(I \rightarrow I + 1)$, the probability for the whole path is given by:

$$\exp\left[-\frac{E_0}{k_{\mathrm{B}}T}\right] \prod_{I=0}^{L-1} p(I \to I+1), \tag{4.9}$$

where E_0 is the total energy of the first structure. Note that it has been assumed that the structures are drawn from the canonical ensemble (hence, the Boltzmann factor) and that the transition probability conserves this distribution. To ensure that the trajectories are transition paths between the two states, it is necessary to add terms corresponding to the characteristic functions for each of the states (Eqs. (2.15) and (2.16)). Thus, the probability of a path forced to start in state A is:

$$\mathcal{P}_{\mathrm{A}}(\mathbf{R}_L) = \chi_{\mathrm{A}}(\mathbf{R}_0) \exp\left[-\frac{E_0}{k_{\mathrm{B}}T}\right] \prod_{I=0}^{L-1} p(I \to I+1), \tag{4.10}$$

whereas the probability of a path starting in state A and finishing in state B is:

$$\mathcal{P}_{\mathrm{AB}}(\mathbf{R}_L) = \chi_{\mathrm{A}}(\mathbf{R}_0) \exp\left[-\frac{E_0}{k_{\mathrm{B}}T}\right] \prod_{I=0}^{L-1} p(I \to I+1)\chi_{\mathrm{B}}(\mathbf{R}_L). \tag{4.11}$$

Once these probabilities have been defined, it is possible to generate a "statistical mechanics" of transition paths, that is analogous in many respects to that for individual configurations in phase space, and from which expressions for time correlation functions (such as that in Eq. (2.21)) arise naturally.

Chandler and coworkers have formulated this method for different ensembles and for systems with different dynamics and they have discussed in detail how to sample effectively the transition path ensemble and calculate various dynamical quantities, such as rate constants, from it (DELLAGO, BOLHUIS, CSAJKA and CHANDLER [1998], BOLHUIS, DELLAGO and CHANDLER [1998], DELLAGO, BOLHUIS and CHANDLER [1999]). They have also applied the method to a number of different transition processes including a conformational isomerization reaction in the alanine dipeptide, which is a molecule that is often used to model the peptide bond in proteins (BOLHUIS, DELLAGO and CHANDLER [2000]), and autoionization processes in liquid water (GEISSLER, DELLAGO, CHANDLER, HUTTER and PARRINELLO [2001]). This last study was particularly interesting because it combined two of the most "advanced" technologies available at the moment for studying reactions – molecular dynamics simulations with *ab initio* DFT QM potentials and transition path sampling. Work on similar approaches has also been done by other workers including by ZUCKERMAN and WOOLF [1999].

5. Quantum algorithms

The techniques that have been discussed so far in this chapter have assumed that the dynamics of the nuclei can be dealt with classically and QM theory has been employed solely for the calculation of the PES of the reacting system. At present, QM algorithms for the determination of rate constants in condensed-phase systems are less well developed and more complicated than their classical counterparts, but it is known that quantum dynamical effects are important and so some effort needs to be made to estimate them. This section highlights some currently active areas of research into quantum dynamical algorithms but makes no attempt at an exhaustive survey. Instead, interested readers are encouraged to consult some of the references cited below.

Perhaps the most straightforward way of including quantum effects (but not those due to spin) is with the path-integral method that can be used to calculate the equilibrium properties of a system. It is based on the observation that there is an isomorphism between a discretized Feynman path integral representation of the equilibrium QM density operator and a classical system in which the quantum particles are represented by polymers of classical particles (FEYNMAN and HIBBS [1965], CHANDLER and WOLYNES [1981]). This makes it a relatively easy method to implement in an existing classical molecular dynamics program because the simulation procedure is the same but the potential is changed as follows:

$$U_{\text{PI}} = \sum_{\alpha=1}^{N_{\text{atoms}}} \frac{Pm_\alpha(k_BT)^2}{2\hbar^2} \sum_{p=1}^{P} (R_\alpha^{(p)} - R_\alpha^{(p+1)})^2 + \frac{1}{P}\sum_{p=1}^{P} U_{\text{classical}}\left(R_1^{(p)}, \ldots, R_{N_{\text{atoms}}}^{(p)}\right).$$

$$(5.1)$$

Here P is the number of beads in each polymer, m_α is the mass of atom α, and $R_\alpha^{(p)}$ is the position of the pth bead of atom α (noting that $R_\alpha^{(P+1)} = R_\alpha^{(1)}$). $U_{\text{classical}}$ is the normal, classical potential of the system which will be a function of the pth copy of all the quantum particles. The path-integral approach has been used successfully in many cases (see, e.g., FIELD [2002] for references to applications to enzymatic reactions) but it is only valuable as a way of calculating the ensemble averages for the corresponding quantum system. Extensions of the approach to allow approximate dynamical quantities to be determined from simulations have been developed, the most notable being the path-integral centroid molecular dynamics method (see, e.g., WARSHEL and CHU [1990] and VOTH [1996]).

The quantum mechanical equivalent of the classical reactive flux correlation function (Eq. (2.23)) has been known for some time. It was first derived by YAMAMOTO [1960] with later developments by a number of workers (MILLER [1974], MILLER, SCHWARTZ and TROMP [1983]). Unfortunately, applications of the formalism to real condensed-phase systems have been very limited due to its complexity although some recent work may change this (CHAKRABARTI, CARRINGTON and ROUX [1998], KOHEN and TANNOR [2000]). In the light of these problems, a very large range of approximate methodologies have been devised. These include the centroid path-integral methods mentioned above as well as methods that add quantum corrections to classical rate constants (see, e.g., TRUHLAR [1998] and the references therein). Another area of active research are the surface-hopping methods that aim to study the dynamics of a system in which transitions between different electronic states can occur. Recent work in this area has been done by SHOLL and TULLY [1998], FANG and HAMMES-SCHIFFER [1999] and HACK, WENSMANN, TRUHLAR, BEN-NUN and MARTÍNEZ [2001].

6. Challenges and perspectives

The simulation of chemical reactions is a very challenging task that requires the synthesis of a number of simulation technologies from fields as diverse as molecular modeling, quantum chemistry and statistical physics. Much has been achieved and it is now possible to obtain at least qualitative insights into how many reactions occur in the condensed phase. Much, however, remains to be done. Areas of improvement are easy to identify and include (i) more accurate quantum chemical descriptions of reacting systems, whether through more precise DFT methods or faster wavefunction-based approaches, (ii) better sampling of configuration space for free-energy and transition-path-sampling simulations and (iii) effective quantum-dynamical simulation algorithms. All-in-all, though, the outlook is bright – algorithmic advances combined with increased computer power mean that precise simulation of reactions will become more straightforward. This is indeed fortunate if molecular simulation approaches are to contribute fully in the design of new materials and molecules.

Acknowledgments

The author would like to thank the Institut de Biologie Structurale – Jean-Pierre Ebel, the Commissariat à l'Energie Atomique and the Centre National de la Recherche Scientifique for financial support.

References

ALBRIGHT, T.A., J.K. BURDETT, and M.H. WHANGBO (1985), *Orbital Interactions in Chemistry* (Wiley-Interscience, New York).

ALLEN, M.P. and D.J. TILDESLEY (1987), *Computer Simulations of Liquids* (Oxford University Press, Oxford).

AMARA, P. and M.J. FIELD (1999), Hybrid potentials for large molecular systems, in: J. Leszczynski, ed., *Computational Molecular Biology* (Elsevier, Amsterdam) pp. 1–33.

AMARA, P., A. VOLBEDA, J. FONTECILLA-CAMPS and M.J. FIELD (1999), A hybrid density functional theory/molecular mechanics study of nickel–iron hydrogenase: investigation of the active site redox states, *J. Am. Chem. Soc.* **121**, 4468–4477.

ANDERSON, J.B. (1995), Predicting rare events in molecular dynamics, *Adv. Chem. Phys.* **91**, 381–431.

BAKER, J. (1986), An algorithm for the location of transition states, *J. Comput. Chem.* **7**, 385–395.

BELL, S. and J.S. CRIGHTON (1984), Locating transition states, *J. Chem. Phys.* **80**, 2464–2475.

BERNE, B.J. (1985), Molecular dynamics and Monte Carlo simulations of rare events, in: J.U. Brackhill and B.I. Cohen, eds., *Multiple Time Scales* (Academic Press, New York) pp. 419–436.

BERNE, B.J., M. BORKOVEC and J.E. STRAUB (1988), Classical and modern methods in reaction rate theory, *J. Phys. Chem.* **92**, 3711–3725.

BEVERIDGE, D.L. and F.M. DI CAPUA (1989), Free energy via molecular simulation: applications to biomolecular systems, *Annu. Rev. Biophys. Biophys. Chem.* **18**, 431–492.

BOLHUIS, P.G., C. DELLAGO, and D. CHANDLER (1998), Sampling ensembles of deterministic transition pathways, *Faraday Discuss.* **110**, 421–436.

BOLHUIS, P.G., C. DELLAGO, and D. CHANDLER (2000), Reaction coordinates of biomolecular isomerization, *Proc. Natl. Acad. Sci.* **97**, 5877–5882.

BOLHUIS, P.G., D. CHANDLER, C. DELLAGO, and P.L. GEISSLER (2002), Transition path sampling: throwing ropes over rough mountain passes, in the dark, *Ann. Rev. Phys. Chem.* **53**, 291–318.

CANCÈS, E., M. DEFRANCESCHI, W. KUTZELNIGG, C. LE BRIS and Y. MADAY (2003), Computational quantum chemistry: a primer, in: P.G. Ciarlet and C. Le Bris, eds., *Computational Chemistry, Handbook of Numerical Analysis*, vol. X (Elsevier, Amsterdam) pp. 3–270.

CEPERLEY, D.M. and L. MITAS (1996), Quantum Monte Carlo methods in chemistry, *Adv. Chem. Phys.* **93**, 1–38.

CERJAN, C.J. and W.H. MILLER (1981), On finding transition states, *J. Chem. Phys.* **75**, 2800–2806.

CHAKRABATI, N., T. CARRINGTON JR., and B. ROUX (1998), Rate constants in quantum mechanical systems: a rigorous and practical path-integral formulation for computer simulations, *Chem. Phys. Lett.* **293**, 209–220.

CHANDLER, D. (1978), Statistical mechanics of isomerization dynamics in liquids and the transition state approximation, *J. Chem. Phys.* **68**, 2959–2970.

CHANDLER, D. (1987), *Introduction to Modern Statistical Mechanics* (Oxford University Press, Oxford).

CHANDLER, D. and P.G. WOLYNES (1981), Exploiting the isomorphism between quantum theory and classical statistical mechanics of polyatomic fluids, *J. Chem. Phys.* **74**, 4078–4095.

CHOI, C. and R. ELBER (1991), Reaction path study of helix formation in tetrapeptides: effect of side chains, *J. Chem. Phys.* **94**, 751–760.

COOPER, D.L., J. GERRATT and M. RAIMONDI (1987), Modern valence bond theory, *Adv. Chem. Phys.* **69**, 319–397.

CZERMINSKI, R. and R. ELBER (1990), Self-avoiding walk between two fixed points as a tool to calculate reaction paths in large molecular systems, *Int. J. Quant. Chem.* **24**, 167–186.

DAVIS, H.L., D.J. WALES, and R.S. BERRY (1990), Exploring potential energy surfaces via transition state calculations, *J. Chem. Phys.* **92**, 4308–4319.

DELLAGO, C., P.G. BOLHUIS and D. CHANDLER (1999), On the calculation of rate constants in the transition path ensemble, *J. Chem. Phys.* **110**, 6617–6625.

DELLAGO, C., P.G. BOLHUIS, F.S. CSAJKA and D. CHANDLER (1998), Transition path sampling and the calculation of rate constants, *J. Chem. Phys.* **108**, 1964–1977.

DEWAR, M.J.S. and B.M. O'CONNOR (1987), Testing *ab initio* procedures: the 6–31G* model, *Chem. Phys. Lett.* **138**, 141–145.

DEWAR, M.J.S. and W. THIEL (1977), Ground states of molecules. 38. The MNDO method. Approximations and parameters, *J. Am. Chem. Soc.* **99**, 4899–4907.

DEWAR, M.J.S., E.G. ZOEBISCH, E.F. HEALY and J.J.P. STEWART (1985), AM1: a new general purpose quantum mechanical molecular model, *J. Am. Chem. Soc.* **107**, 3902–3909.

ELBER, R. and M. KARPLUS (1987), A method for determining reaction paths in large molecules: application to myoglobin, *Chem. Phys. Lett.* **139**, 375–380.

ELBER, R. and D. SHALLOWAY (2000), Temperature dependent reaction coordinates, *J. Chem. Phys.* **112**, 5539–5545.

ELSTNER, M., D. POREZAG, G. JUNGNICKEL, J. ELSNER, M. HAUGK, T.H. FRAUENHEIM, S. SUHAI and G. SEIFERT (1998), Self-consistent-charge density-functional tight-binding method for simulations of complex materials properties, *Phys. Rev. B* **58**, 7261–7268.

FANG, J.Y. and S. HAMMES-SCHIFFER (1999), Comparison of surface hopping and mean-field approaches for model proton transfer reactions, *J. Chem. Phys.* **110**, 11166–11175.

FEYNMAN, R.P. and A.R. HIBBS (1965), *Quantum Mechanics and Path Integrals* (McGraw-Hill, New York).

FIELD, M.J. (1993), The simulation of chemical reactions, in: W.F. van Gunsteren, P.K. Weiner and A.J. Wilkinson, eds., *Computer Simulation of Biomolecular Systems*, vol. 2 (ESCOM, Leiden) pp. 82–123.

FIELD, M.J. (1999), *A Practical Introduction to the Simulation of Molecular Systems* (Cambridge University Press, Cambridge).

FIELD, M.J. (2002), Simulating enzyme reactions: challenges and perspectives, *J. Comput. Chem.* **23**, 48–58.

FIELD, M.J., P.A. BASH and M. KARPLUS (1990), A combined quantum mechanical and molecular mechanical potential for molecular dynamics simulations, *J. Comput. Chem.* **11**, 700–733.

FOULKES, W.M.C. and R. HAYDOCK (1989), Tight-binding models and density-functional theory, *Phys. Rev. B* **39**, 12520–12536.

FRAUENHEIM, T.H., G. SEIFERT, M. ELSTNER, Z. HAJNAL, G. JUNGNICKEL, D. POREZAG, S. SUHAI and R. SCHOLZ (2000), A self-consistent charge density-functional based tight-binding method for predictive materials simulations in physics, chemistry and biology, *Phys. Stat. Sol. B* **217**, 41–62.

FUKUI, K. (1981), The path of chemical reactions—The IRC approach, *Acc. Chem. Res.* **14**, 363–368.

GALLUP, G.A., R.L. VANCE, J.R. COLLINS and J.M. NORBECK (1982), Practical valence bond calculations, *Adv. Quant. Chem.* **16**, 229–272.

GAO, J. (1995), Methods and applications of combined quantum mechanical and molecular mechanical potentials, in: K.B. Lipkowitz and D.B. Boyd, eds., *Reviews in Computational Chemistry*, vol. 7 (VCH, New York) pp. 119–185.

GEISSLER, P.L., C. DELLAGO, D. CHANDLER, J. HUTTER and M. PARRINELLO (2001), Autoionization in liquid water, *Science* **291**, 2121–2124.

GOEDECKER, S. (1999), Linear-scaling electronic structure methods, *Rev. Mod. Phys.* **71**, 1085–1123.

GONZALEZ, C. and H.B. SCHLEGEL (1989), An improved algorithm for reaction path following, *J. Chem. Phys.* **90**, 2154–2161.

GROSSMAN, J.C., W.A. LESTER JR., and S.G. LOUIE (2000), Quantum Monte Carlo and density functional theory characterization of the 2-cyclopentenone and 3-cyclopentenone formation from $O(^3P)$ + cyclopentadiene, *J. Am. Chem. Soc.* **122**, 705–711.

HACK, M.D., A.M. WENSMANN, D.G. TRUHLAR, M. BEN-NUN, and T.J. MARTÍNEZ (2001), Comparison of full multiple spawning, trajectory surface hopping, and converged quantum mechanics for electronically nonadiabatic dynamics, *J. Chem. 0050hys.* **115**, 1172–1186.

HALGREN, T.A. and W.N. LIPSCOMB (1977), The synchronous-transit method for determining reaction pathways and locating molecular transition states, *Chem. Phys. Lett.* **19**, 225–232.

HENKELMAN, G., JÓHANNESSON, G., and JÓNSSON, H. (2000), Methods for finding saddle points and minimum energy paths, in: S.D. Schwartz, ed., *Progress on Theoretical Chemistry and Physics* (Kluwer Academic Publishers).

HENKELMAN, G. and H. JÓNSSON (2000), Improved tangent estimate in the NEB method for finding minimum energy paths and saddle points, *J. Chem. Phys.* **113**, 9978–9985.

HUO, S. and J.E. STRAUB (1997), The MaxFlux algorithm for calculating variationally optimized reaction paths for conformational transitions in many body systems at finite temperature, *J. Chem. Phys.* **107**, 5000–5006.

HYNES, J.T. (1985), Chemical reaction dynamics in solution, *Ann. Rev. Phys. Chem.* **36**, 573–597.

JÓHANNESSON, G.H. and H. JÓNSSON (2001), Optimization of hyperplanar transition states, *J. Chem. Phys.* **115**, 9644–9656.

JORGENSEN, W.L. (1989), Free energy calculations: a breakthrough for modeling organic chemistry in solution, *Acc. Chem. Res.* **22**, 184–189.

KARPLUS, M. (2000), Aspects of protein reaction dynamics: deviations from simple behavior, *J. Phys. Chem. B* **104**, 11–27.

KOCH, W. and M.C. HOLTHAUSEN (2000), *A Chemist's Guide to Density Functional Theory* (Wiley VCH, New York).

KOHEN, D. and D.J. TANNOR (2000), Phase space approach to dissipative molecular dynamics, *Adv Chem. Phys.* **111**, 219–398.

KOHN, W. (1999), Noble lecture: electronic structure of matter—Wavefunctions and density functionals, *Rev. Mod. Phys.* **71**, 1253–1266.

KOLLMAN, P.A. (1993), Free energy calculations: applications to chemical and biochemical phenomena, *Chem. Rev.* **93**, 2395–2417.

MARK, A.E. (1998), Free energy perturbation calculations, in: P. von Rague Schleyer, ed., *Encyclopedia of Computational Chemistry* (Wiley, New York) pp. 1211–1216.

MILLER, W.H. (1974), Quantum mechanical transition state theory and a new semiclassical model for reaction rate constants, *J. Chem. Phys.* **61**, 1823–1834.

MILLER, W.H. (1983), Reaction-path dynamics for polyatomic systems, *J. Phys. Chem.* **87**, 3811–3819.

MILLER, W.H., N.C. HANDY and J.E. ADAMS (1980), Reaction path Hamiltonian for polyatomic molecules, *J. Chem. Phys.* **72**, 99–112.

MILLER, W.H., S.D. SCHWARTZ, and J.W. TROMP (1983), Quantum mechanical rate constants for biomolecular reactions, *J. Chem. Phys.* **79**, 4889–4898.

MONARD, G. and K.M. MERZ (1999), Combined quantum mechanical/molecular mechanical methodologies applied to biomolecular systems, *Acc. Chem. Res.* **32**, 904–911.

NERIA, E. and M. KARPLUS (1997), Molecular dynamics of an enzyme reaction: proton transfer in TIM, *Chem. Phys. Lett.* **267**, 23–30.

PARR, R.G. and W. YANG (1989), *Density-Functional Theory of Atoms and Molecules* (Oxford University Press, Oxford).

POPLE, J.A. (1999), Noble lecture: quantum chemical models, *Rev. Mod. Phys.* **71**, 1267–1274.

POREZAG, D., T.H. FRAUENHEIM, T.H. KÖHLER, G. SEIFERT and R. KASCHNER (1995), Construction of tight-binding-like potentials on the basis of density-functional theory: application to carbon, *Phys. Rev. B* **51**, 12947–12957.

PRATT, L.R. (1986), A statistical method for identifying transition states in high dimensional problems, *J. Chem. Phys.* **85**, 5045–5048.

ROUX, B. (1995), The calculation of the potential of mean force using computer simulations, *Comput. Phys. Commun.* **91**, 275–282.

SCHENTER, G.K., G. MILLS and H. JÓNSSON (1994), Reversible work based quantum transition state theory, *J. Chem. Phys.* **101**, 8964–8971.

SHOLL, D.S. and J.C. TULLY (1998), A generalized surface-hopping method, *J. Chem. Phys.* **109**, 7702–7710.

SIMONS, J., P. JORGENSON, H. TAYLOR and J. OZMENT (1983), Walking on potential energy surfaces, *J. Chem. Phys.* **87**, 2745–2753.

SIMONSON, T. (2001), Macromolecular electrostatics: continuum models and their growing pains, *Curr. Opin. Struct. Biol.* **11**, 243–252.

STEWART, J.J.P. (1989a), Optimization of parameters for semiempirical methods I. Method, *J. Comput. Chem.* **10**, 209–220.

STEWART, J.J.P. (1989b), Optimization of parameters for semiempirical methods I. Applications, *J. Comput. Chem.* **10**, 221–264.

STRAATSMA, T.P. and J.A. MCCAMMON (1992), Computational alchemy, *Annu. Rev. Phys. Chem.* **43**, 407–435.

STRAUB, J. (2001), Reaction rates and transition pathways, in: O.M. Becker, A.D. MacKerell, Jr., B. Roux, M. Watanabe, eds., *Computational Biochemistry and Biophysics* (Marcel Dekker, New York) pp. 199–220.

SZABO, A. and N. OSTLUND (1989), *Modern Quantum Chemistry: Introduction to Advanced Electronic Structure Theory* (McGraw-Hill, New York).

THOMAS, A., D. JOURAND, C. BRET, P. AMARA and M.J. FIELD (1999), Is there a covalent intermediate in the viral neuraminidase reaction? A hybrid-potential free-energy study, *J. Am. Chem. Soc.* **121**, 9693–9702.

TOMASI, J. and M. PERSICO (1994), Molecular interactions in solution: an overview of methods based on continuous distributions of the solvent, *Chem. Rev.* **94**, 2027–2094.

TRUHLAR, D.G. (1998), Chemical reaction theory: summarizing remarks, *Faraday Discuss.* **110**, 521–535.

ULITSKY, A. and R. ELBER (1990), A new technique to calculate steepest descent paths in flexible polyatomic systems, *J. Chem. Phys.* **92**, 1510–1511.

VOTH, G.A. (1996), Path-integral centroid methods in quantum statistical mechanics and dynamics, *Adv. Chem. Phys.* **93**, 135–218.

WALES, D.J. (1994), Rearrangements of 55-atom Lennard-Jones and $(C_{60})_{55}$ clusters, *J. Chem. Phys.* **101**, 3750–3762.

WARSHEL, A. (1991), *Computer Modeling of Chemical Reactions in Enzymes and Solutions* (Wiley, New York).

WARSHEL, A. and Z.T. CHU (1990), Quantum corrections for rate constants of diabatic and adiabatic reactions in solutions, *J. Chem. Phys.* **93**, 4003–4015.

WARSHEL, A. and M. LEVITT (1976), Theoretical studies of enzymic reactions: dielectric, electrostatic and steric stabilization of the carbonium ion in the reaction of lysozyme, *J. Mol. Biol.* **103**, 227–249.

YAMAMOTO, T. (1960), Quantum statistical mechanical theory of the rate of exchange chemical reactions in the gas phase, *J. Chem. Phys.* **33**, 281–289.

ZUCKERMAN, D.M. and T.B. WOOLF (1999), Dynamic reaction paths and rates through importance-sampled stochastic dynamics, *J. Chem. Phys.* **111**, 9475–9484.

Biomolecular Conformations Can Be Identified as Metastable Sets of Molecular Dynamics

Christof Schütte

Institute of Mathematics II, Department of Mathematics and Computer Science,
Free University (FU) Berlin, Germany
E-mail: schuette@math.fu-berlin.de

Wilhelm Huisinga

Institute of Mathematics II, Department of Mathematics and Computer Science,
Free University (FU) Berlin, Germany
E-mail: huisinga@math.fu-berlin.de

1. Introduction

The biochemical functions of many important biomolecules result from their *dynamical* properties, particularly from their ability to undergo so-called *conformational transitions* (cf. ZHOU, WLODEC and MCCAMMON [1998]). In a conformation, the large scale geometric structure of the molecule is understood to be conserved, whereas on smaller scales the system may well rotate, oscillate, or fluctuate. Furthermore, transitions between conformations are rare events or, in other words, a typical trajectory of a molecular system stays for long periods of time within the conformation, while exits are long-term events. Hence, the term conformation includes both geometric and dynamical aspects. From the geometrical point of view, conformations are understood to represent all molecules with the same large scale geometric structure and may thus be identified with a subset of the state space. From the dynamical point of view, a conformation typically persists for long periods of time (compared to the fastest molecular motions) such that the associated subset of the state space is *metastable*

Computational Chemistry
Special Volume (C. Le Bris, Guest Editor) of
HANDBOOK OF NUMERICAL ANALYSIS
P.G. Ciarlet (Editor)

and the resulting *macroscopic dynamical behavior* can be described as a flipping process between the metastable subsets.

Understanding *conformation dynamics* – that is the statistics of the flipping process and the corresponding exit times as well as the actual transition paths between different conformations – is crucial to the understanding of biomolecular flexibility and activity. Prominent examples of conformation dynamics are the conformational changes accompanying the action of the muscle protein myosin, the light-induced conformational transition of the photo-receptor rhodopsin initializing the primary amplification cascade in vision, or the conformation conversion of human prions assumed to cause prion diseases.

The state-of-the-art biophysical explanation for the existence of conformations is as follows: The *free energy landscape* of a protein or peptide decomposes into particularly deep wells each containing huge numbers of local minima. These wells are separated by relatively large barriers – as measured on the scale of the thermal energy $k_B T$ – from each other and represent different metastable conformations. The hierarchy of barrier heights induces a hierarchy of metastable conformations (ELBER and KARPLUS [1987], FRAUENFELDER, STEINBACH and YOUNG [1989], FRAUENFELDER and McMAHON [2000]). The corresponding hierarchy of time scales observed for conformational transitions seems to confirm the biophysical explanation for the existence of conformations (NIENHAUS, MOURANT and FRAUENFELDER [1992]). However, the entire explanation depends on the concept of the free energy landscape, whose definition is typically based on the assumption that the conformational degrees of freedom are already known in advance. In other words, the model is of minor use if conformations and conformational degrees of freedom still have to be identified by simulations.

Mathematically, the dynamical aspect of conformations is based on the concept of metastability. In this paper we will pursue two characterizations of metastability. The *exit time approach* based on *exit rates* characterizes metastability of some subset by the property that a typical trajectory will only exit the subset on macroscopically long time scales. The *ensemble dynamics approach* based on *transition probabilities* characterizes metastability of some subset in statistical terms in the following sense: the fraction of systems in an ensemble that exit from the subset during some (not necessarily long) given time span is significantly small in comparison to other subsets. It is one of the main goals of this article to discuss and compare the similarities and distinguishing aspects of these two concepts in detail (the entire Section 2 will be devoted to the conceptual differences).

We will see that the two characterizations of metastability can be formalized and studied within the unified mathematical framework of the *transfer operator approach* to metastability. This approach originated from the work of Dellnitz et al. on the identification of almost invariant sets of discrete dynamical systems with small random perturbations (DEUFLHARD, DELLNITZ, JUNGE and SCHÜTTE [1999], DELLNITZ and JUNGE [1999]) and it has been successfully applied to examine metastable behavior of deterministic Hamiltonian systems by DEUFLHARD, DELLNITZ, JUNGE and SCHÜTTE [1999]. By reformulating this idea in the context of biophysical models of molecular motion, Schütte et al. showed that biomolecular conformations can be identified via the "dominant" eigenvectors of the transfer operator associated with the dynamical model used (SCHÜTTE, FISCHER, HUISINGA and DEUFLHARD [1999], SCHÜTTE, HUISINGA and DEUFLHARD [2001], SCHÜTTE [1998], HUISINGA [2001]). It has been demonstrated that, for moderate size (bio)molecules, the eigenvectors of interest can be computed efficiently and allow to identify the desired conformations and the associated conformation dynamics in a unified setting based on simulation of the dynamical behavior of molecular systems (FISCHER, SCHÜTTE, DEUFLHARD and CORDES [2002], DEUFLHARD, HUISINGA, FISCHER and SCHÜTTE [2000], HUISINGA and SCHMIDT [2002]).

The literature on conformation dynamics is enormously rich, see, for example, DEUFLHARD, HERMANS, LEIMKUHLER, MARK, REICH and SKEEL [1999], BERNE, CICCOTTI and COKER [1998]. However, the branch that deals with the *dynamical* aspects of conformations mainly contains approaches to the computational detection of transition paths between conformations and of the associated main transition coordinates, see, E, RAN and VANDEN-EIJNDEN [2002], FRAUENFELDER, STEINBACH and YOUNG [1989], BOLHUIS, DELLAGO, CHANDLER and GEISSLER [2001]. There are several approaches designed to bridge the time scale gap between realizably short trajectory simulations and significantly longer metastability periods of conformational substates. One example are approaches that exploit artificial accelerations of the dynamics, cf. GRUBMÜLLER [1995], HUBER, TORDA and VAN GUNSTEREN [1994]; another example is given by path integral approaches to long-term dynamics where transition paths are discretized in time using extremely large timesteps (OLENDER and ELBER [1996]).

The article will be organized as follows: First we will sharpen and conceptually complement the two characterizations of metastability (Section 2), then we will shortly summarize the different dynamical models designed to describe different aspects of the dynamical and statistical properties of molecules (Section 3). This will be followed by the presentation of the mathematical framework of the transfer operator approach to metastability (Sections 4 and 5). Within this framework we will reformulate the two concepts of metastability and justify in detail the key idea of the transfer operator approach. Moreover, it will be shown that the framework allows to incorporate almost all different dynamical models available. In Section 6 the theoretical level is left and the issues of practical realization are discussed: The concept of Galerkin discretization of transfer operators is studied which leads to the question of whether a discretization of the eigenvalue problem in huge dimensional state spaces will be possible without risking the increase of numerical effort beyond any tolerable amount. It is illustrated how this problem can be circumvented. Sections 7 and 8 conclude the article by demonstrating the application of the approach. In Section 7 the entire concept is illustrated by means of a simple but completely comprehensible test system whereas Section 8 is devoted to the application to a small oligonucleotide.

2. Conceptual preliminaries

Before we go into details about molecular dynamics, conformations, and metastability we first want to point out the fundamental principles of the approach to biomolecular conformations.

Let us assume for the remainder of this section that a mathematical model is available which, given an exact initial state, perfectly describes the true motion of the molecule under consideration in all necessary details. In general, this will be given by some (discrete or continuous, deterministic, or stochastic) dynamical system. In the deterministic setting, the corresponding initial value problem is thought to model the evolution of the state of a *single molecule*; its exact solution will be called "trajectory" in the following. In the stochastic setting, we use the same interpretation and wording for every single pathwise realization. Trajectories have to be distinguished from their numerical realization, which will be called "simulation" or "numerical integration."

In the context of biochemical applications one is generally not interested in single isolated molecules but in certain *molecular ensembles* that, for instance, model the collection of many identical molecules in a living cell or in a test tube with certain side conditions like, for example, constant temperature. The molecular ensemble is represented by a statistical distribution in molecular state space. If the ensemble is assumed to be stationary, the distribution does not change in time (see Fig. 2.1). Within this setting, the *transition probability* from some subensemble A to some subensemble B, both specified by some subset A and B of the state space, within a prescribed time span τ is given by the fraction of systems with initial state in A at $t = 0$ and final state in B at $t = \tau$. Built upon transition probabilities we may state the

Ensemble dynamics approach: Conformations are identified as subensembles/subsets, for which the fraction of systems that exit during a prescribed observation time τ is significantly smaller than for other subensembles/subsets.

In order to numerically *compute* the transition probabilities from subensemble A to B one has to generate (i) a sample that represents the *subensemble of initial states* in A, and (ii) a sample that represents the corresponding *subensemble of trajectories* starting from these initial states, as illustrated in Fig. 2.1. Within this statistical setting the ensemble dynamics approach is able to capture conformation dynamics by considering only *short-term* trajectories, since measurements on ensembles already contain

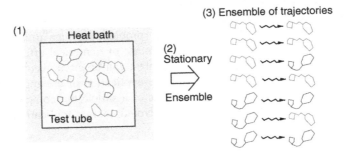

FIG. 2.1 Illustration of ensemble dynamics: (1) an ensemble of molecular systems embedded in a heat bath of constant temperature; (2) the usual assumption is that this can be modeled by some stationary probability distribution, for example, the canonical density; (3) the dynamical behavior of each single molecule in the ensemble (here assumed to be modeled accurately by some dynamical system with flow Φ^τ) induces dynamical fluctuations within the ensemble without effect on the stationary distribution.

information about all possible states, and short-term trajectories over time spans that are of the order of magnitude of the rapid conformational transition itself contain all transition paths from one conformation to another.

The ensemble dynamics approach has the advantage that it is based on a setting and requires information which is experimentally available: ensembles of short-term trajectories can be observed by means of femtochemistry (Nobel price 1999) (ZEWAIL [1995], ZEWAIL [1996]), a novel technique that permits us to observe the dynamical behavior of molecular systems in real-time. Boosted by the progress in laser technology, ultrashort light pulses can be generated with durations on the typical timescales of molecular vibrations, that is, from picoseconds down to femtoseconds. The prototypical experiment follows the *pump-probe scenario*: A molecular ensemble that is initially prepared in some stationary state is excited by a first laser pulse ("*pump*") thus lifting the system into an excited state. Subsequently, a second laser pulse ("*probe*") is used to stimulate another transition (emission or absorption) that serves to generate an observable signal. By measuring the observable signal as a function of the time delay between pump and probe pulse the evolution of the system in its excited state can be monitored. Hence, pump-probe measurements allow to experimentally realize the ensemble dynamics approach.

In this chapter we will present two different algorithmic approaches to generate the data needed in steps (i) and (ii) from above. The two concepts are sketched in Fig. 2.2. On the one hand, we may use *any* method available to compute a sample that appropriately represents the stationary distribution. Given this sample one then evaluates the trajectories starting from any sample point. We will see later that the two steps may be combined into a single procedure by introducing an appropriate Markov chain. On the other hand, whenever the dynamical system under consideration is *ergodic*,[1] a single long-term trajectory represents the average behavior of molecules in the ensemble. Then, chopping the long-term trajectory into pieces of length τ will also do the job.

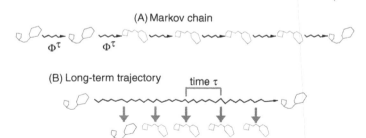

FIG. 2.2 Illustration of different algorithmic options for realizing a sample of the stationary ensemble under consideration and the induced sample the corresponding ensemble of trajectories: realization of e nsemble and trajectories can be done by means of (A) specially designed Markov chains, or (B) time-τ pieces from a single ergodic long-term simulation of the corresponding dynamical system.

---

[1]The notion of ergodicity has several different meanings in physics and mathematics. The typical rough "definition" states that "time average equals ensemble average". We will introduce the precise meaning used in this article in Section 5.2.

This algorithmic option via long-term simulation should not be confused with another approach to conformation analysis, the

Exit time approach: Conformations are identified as subsets for which the exit time of a typical trajectory is extraordinary large in comparison to other subsets.

This approach is build upon the belief that the description of conformational transitions requires to start a long-term simulation and wait through the in general tremendously long period of time until a transition takes place. However, as we will see in the following this is not necessarily the case, and the notion of conformations in the ensemble dynamics and exit time approach in some sense turn out to be very similar.

Both approaches may exploit long-term *simulations* in the ergodic interpretation in order to generate ensembles of short-term trajectories only. Yet, the authors want to emphasize that long-term *trajectories* of the dynamical system under consideration are not necessarily needed. This is of utmost importance because the literature on predictability and sensitivity w.r.t. perturbations states that for any dynamical system and prescribed accuracy there is a certain maximal time T, up to which initial value problems are make sense. For time spans longer than T the deviation between trajectories of the dynamical system caused by perturbations may exceed the accuracy requirement. The nature of the perturbations to be considered depends on the actual application context: one may have to take into account the uncertainty of the initial value, or perturbations due to the numerical realization. The actual value of T depends on the properties of the dynamical system and on the nature of the perturbations and can be characterized by means of different estimates (e.g., by Lyapunov exponents, so-called condition numbers (DEUFLHARD and BORNEMANN [1994]), or predictability analysis (KLEEMAN [2001])). However, in the context of biomolecular dynamics all available estimates indicate that for any tolerable accuracy, the time T is many orders of magnitude smaller than the expected exit times from typical conformations. The same situation is encountered in the above mentioned pump-probe experiments, where the time spans between the pump and probe pulse are orders of magnitude smaller than typical exit times from some conformation. However, experimental observations of conformational transitions in real-time over milliseconds or even on longer scales are very limited and possible only indirectly.

Before proceeding to the algorithmic realization of the ensemble dynamics or exit time approach we may first introduce the most prominent types of dynamical systems presently discussed in the context of molecular dynamics. They can be classified in two main categories:

(MC1) Dynamical systems that are designed to model the precise motion of some molecular system, at least on short time scales.

(MC2) Dynamical systems that are designed to sample the state space of some molecular system w.r.t. some prescribed statistical distribution.

3. Description of dynamical behavior

The literature on the description of the dynamical behavior of molecular systems is extremely rich; they range from classical deterministic Hamiltonian models that try to cover the actual motion of each single molecule in the system to stochastic descriptions like Langevin dynamics or iterative schemes that only model artificial dynamics like most Markov chain Monte Carlo approaches.

3.1. *Markov processes and transition functions*

We now introduce the mathematical framework that subsumes both approaches, whether stochastic or deterministic.

Consider the state space $\mathbf{X} \subset \mathbf{R}^m$ for some $m \in \mathbf{N}$ equipped with the Borel σ-algebra on $\mathbf{X}$.[2] The evolution of a single microscopic system is supposed to be given by a *homogeneous Markov process* $X_t = \{X_t\}_{t \in \mathbf{T}}$ in continuous or discrete time $\mathbf{T} = \mathbf{R}_0^+$ or $\mathbf{T} = \mathbf{N}$, respectively. We write $X_0 \sim \mu$, if the Markov process X_t is initially distributed according to the probability measure μ, that is, if $\mathbf{P}[X_0 \in A] = \mu(A)$ for every $A \subset \mathbf{X}$. We use $X_0 = x$, if $X_0 \sim \delta_x$, where δ_x denotes the Dirac measure at x. The motion of X_t is given in terms of the *stochastic transition function* $p \colon \mathbf{T} \times \mathbf{X} \times \mathcal{B}(\mathbf{X}) \to [0, 1]$ according to

$$p(t, x, A) = \mathbf{P}[X_{t+s} \in A | X_s = x] \tag{3.1.1}$$

for every $t, s \in \mathbf{T}$, $x \in \mathbf{X}$ and $A \subset \mathbf{X}$. Hence, $p(t, x, A)$ describes the probability that the system moves from state x into the subset A within time t. The relation between a stochastic transition function and a homogeneous Markov process is one-to-one (MEYN and TWEEDIE [1993], Chapter 3). In the special case, where $p(t, x, A) = \delta_{\phi(x, t)}(A)$, the Markov process is in fact a deterministic process, whose evolution is defined by the flow map $\Phi(x, t)$ in state space. Besides some more technical properties (see Appendix 9.1) the stochastic transition function fulfills the so-called Chapman–Kolmogorov equation

$$p(t + s, x, A) = \int_X p(t, x, dz) p(s, z, A), \tag{3.1.2}$$

that holds for every $t, s \in \mathbf{T}$, $x \in \mathbf{X}$ and $A \subset \mathbf{X}$ and represents the semigroup property of the Markov process. As a consequence, in the discrete time case $\mathbf{T} = \mathbf{N}$ it suffices to specify $p(x, dy) = p(1, x, dy)$, since the n-step transition probabilities $p^n(x, dy) = p(n, x, dy)$ are recursively determined by (3.1.2).

We say that the Markov process X_t admits an *invariant probability measure* μ, or μ is invariant w.r.t. X_t, if

$$\int_X p(t, x, A) \mu(dx) = \mu(A) \tag{3.1.3}$$

for every $t \in \mathbf{T}$ and $A \subset \mathbf{X}$ (MEYN and TWEEDIE [1993], Chapter 10). Note that the invariant probability measure needs not to be unique. A Markov process is called *reversible* w.r.t. an invariant probability measure μ if

$$\int_A p(t, x, B) \mu(dx) = \int_B p(t, x, A) \mu(dx) \tag{3.1.4}$$

---

[2]In the sequel every subset $C \subset \mathbf{X}$ is implicitly assumed to be measurable, i.e., we assume that additionally $C \in \mathcal{A}$. holds without further mentioning.

for every $t \in \mathbf{T}$ and $A, B \subset \mathbf{X}$. If μ is unique, X_t is simply called reversible. For the special case of a stochastic transition function being absolutely continuous w.r.t. μ, reversibility reads $p(t, x, y) = p(t, y, x)$ for every $t \in \mathbf{T}$ and μ-a.e. $x, y \in \mathbf{X}$.

3.2. Model systems

We now turn to the most prominent examples in the context of molecular dynamics.

Let N denote the number of atoms of the system and $\Omega = \mathbf{R}^{3N}$ the position space, that is, $q \in \Omega$ represents the vector of atomic position coordinates. Moreover, let $\xi \in \mathbf{R}^{3N}$ denote the vector of all conjugated momenta. Suppose that a differentiable potential energy function $V : \mathbf{R}^{3N} \to \mathbf{R}$ describing all interactions between the atoms is given. For each model system below we assume that the position space Ω belongs to one of the two fundamentally different cases:

Bounded systems: The potential energy function $V: \mathbf{R}^{3N} \to \mathbf{R}^{3N}$ is smooth, bounded from below, and satisfies $V \to \infty$ for $|q| \to \infty$. Such systems are called bounded, since the energy surfaces $\{(q, \xi) \in \mathbf{X} : H(q, \xi) = E\}$ are bounded subsets of Γ for every energy E.

Periodic systems: The position space Ω is some $3N$-dimensional torus and the potential energy function V is continuous on Ω and thus bounded. There is an intensive discussion concerning the question of whether V can also be assumed to be smooth as we will do herein, see SCHÜTTE [1998], Section 2, for details.

Both cases are typical for molecular dynamics applications. Periodic systems in particular include the assumption of periodic boundaries, which is by far the most popular modeling assumption for biomolecular systems.

Deterministic Hamiltonian system.   The most prominent model for the dynamical behavior of molecular systems exploits classical Hamiltonian mechanics, that is, atoms are described as mass points subject to forces that are generated by specified classical interaction potentials V. The dynamical behavior is described by some deterministic Hamiltonian system of the form

$$\dot{q} = M^{-1}\xi, \quad \dot{\xi} = -\nabla_q V(q), \tag{3.2.1}$$

defined on the state space $\mathbf{X} = \mathbf{R}^{3N} \times \mathbf{R}^{3N}$ and M denoting the diagonal mass matrix.

Eq. (3.2.1) models an energetically closed system, whose total energy, given by the Hamiltonian

$$H(q, \xi) = \frac{1}{2}\xi^T M^{-1}\xi + V(q), \tag{3.2.2}$$

is preserved under the dynamics. For the sake of simplicity, we assume in the following that M is the identity matrix. The deterministic Hamiltonian system is typically seen as the embodiment of our class (MC1) in the context of molecular dynamics.

Let Φ^t denote the flow associated with the Hamiltonian system (3.2.1), that is, the solution $x_t = (q_t, \xi_t)$ of (3.2.1) for the initial value $x_0 = (q_0, \xi_0)$ is given by $x_t = \Phi^t x_0$. Let $\mathbf{1}_C$ denote the characteristic function of the subset $C \subset \mathbf{X}$. Then, the stochastic transition function corresponding to (3.2.1) is given by

$$p(t, x, C) = 1_C(\Phi^t x) = \delta_{\Phi^t x}(C) \tag{3.2.3}$$

for every $t \in \mathbf{R}_0^+$ and $C \subset \mathbf{X}$. The Markov process $X_t = \{X_t\}_{t \in \mathbf{R}_0^+}$ induced by the stochastic transition function p coincides with the flow Φ^t; hence $X_t = \Phi^t x_0$ for the initial distribution $X_0 = x_0$.

It is well known that for every smooth function $\mathcal{F} : \mathbf{R} \to \mathbf{R}$ the probability measure $\mu(dx) \propto \mathcal{F}(H)(x)dx$ is invariant w.r.t. the Markov process X_t. The most prominent choice is the canonical density or *canonical ensemble*

$$f(x) \propto \exp(-\beta H(x))$$

for some constant $\beta > 0$ that can be interpreted as inverse temperature. The associated measure $\mu(dx) \propto f(x)dx$ is called the *canonical measure*. The canonical ensemble is often used in modeling experiments on molecular systems that are performed under the conditions of constant volume and temperature $\mathcal{T} = \frac{1}{k_B \beta}$, where k_B Boltzmann's constant. Obviously, a single solution of the Hamiltonian system (3.2.1) can never be ergodic w.r.t. the canonical measure, since it conserves the internal energy H, as defined in (3.2.2). Hence, w.r.t. the canonical measure, the deterministic Hamiltonian system is not in the class (MC2), while it might be w.r.t. to other measure such as, for example, the microcanonical measure.

Hamiltonian system with randomized momenta. Aiming at a conformational analysis of biomolecular systems in the context of the canonical ensemble, Schütte et al. introduced a specific stochastic Hamiltonian system (SCHÜTTE, FISCHER, HUISINGA and DEUFLHARD [1999]) as a discrete time Markov chain, defined solely on the position space and derived from the deterministic Hamiltonian system by "randomizing the momenta."

For some fixed observation time span $\tau > 0$ (for comments on the choice of τ see remark below) and some inverse temperature $\beta > 0$ the stochastic transition function for the Hamiltonian system with randomized momenta is given by

$$p(q, A) = \int_{\mathbf{R}^d} 1_A(\Pi_q \Phi^\tau(q, \xi)) \mathcal{P}(\xi) d\xi,$$

where $\Pi_q : (q, \xi) \mapsto q$ denotes the projection onto the position space $\Omega = \mathbf{R}^{3N}$ and $\mathcal{P}$ the canonical distribution of momenta $\mathcal{P} \propto \exp(-\beta \xi^t \xi / 2)$ (Fig. 3.1).

The associated discrete time Markov process $Q_n = \{Q_n\}_n \in \mathbf{N}$, defined on the state space $\mathbf{X} = \Omega$, satisfies

$$Q_{n+1} = \Pi_q \Phi^\tau(Q_n, \xi_n), \quad n \in \mathbf{N}, \tag{3.2.4}$$

where ξ_n is chosen randomly from $\mathcal{P}$ (SCHÜTTE [1998]). As it is shown in SCHÜTTE [1998] the positional canonical measure $\mu(dq) \propto \exp(-\beta V(q))dq$ is invariant w.r.t. Q_n and unique. Moreover, exploiting that Φ^τ is reversible and symplectic Q_n is shown to be reversible w.r.t. μ (SCHÜTTE [1998]).

The Hamiltonian system with randomized momenta generates an ensemble of time-τ trajectories such that each trajectory follows the deterministic Hamiltonian dynamics (3.2.1) starting at initial values distributed according to the positional canonical ensemble $f(q) \propto \exp(-\beta V(q))$ (see Fig. 2.2 for illustration). When the deterministic

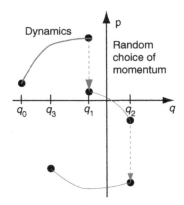

FIG. 3.1 Illustration of the Hamiltonian system with randomized momenta as defined in (3.2.4).

Hamiltonian system is believed to be contained in class (MC1), then the Hamiltonian system with randomized momenta is contained in the classes (MC1) and (MC2).

REMARK. For arbitrary, but fixed $\tau > 0$ we have defined the *one-step* transition function $p(q, D) = p^\tau (1, q, D)$. Changing the observation time to $\sigma > 0$ results in a new one-step transition function $p^\sigma (1, q, D)$. In general we will have $p^{2\tau} (1, q, D) \neq p^\tau (2, q, D)$; for an example see (SCHÜTTE [1998]).

Constant temperature molecular dynamics. One traditional aspect of molecular dynamics is the construction of (stochastic) dynamical systems that allow of sampling the canonical ensemble by means of long-term simulation. Several concepts have been discussed that all boil down to the idea to construct a Hamiltonian system in some extended state space $\hat{\mathbf{X}}$, whose projection onto the lower dimensional state space $\mathbf{X}$ of positions and momenta allows to generate such a sampling. One of the most prominent examples is defined in terms of the Nosé Hamiltonian

$$H_{\text{Nose}}(q, \xi, s, v) = \underbrace{\frac{1}{2s^2} \xi^T \xi + V(q)}_{= H(q, \xi)} + \frac{1}{2Q} v^2 + \frac{1}{\beta} \log s,$$

where s is called the thermostat with conjugated momentum π and associated artificial mass Q. Let the flow of the associated Nosé Hamiltonian system be denoted Ψ^t and let Π denote the projection $(q, \xi, s, v) \mapsto (q, \xi)$. If Ψ^t is ergodic w.r.t. the microcanonical measure on the associated energy cell of H_{Nose}, then $\Pi\Psi^t$ is ergodic w.r.t. the canonical measure $\mu(\mathrm{d}x) \propto \exp(-\beta H_{s=1}(x))\mathrm{d}x$, where $x = (q, \xi)$ (BOND, BENEDICT and LEIMKUHLER [1999]). Thus, the Nosé Hamiltonian system is contained in the class (MC2) but it is at least questionable whether it is part of (MC1).

Langevin system. The most popular model for an *open system* with stochastic interaction with its environment is the so-called Langevin System (RISKEN [1996]):

$$\dot{q} = \xi, \quad \dot{\xi} = -\nabla_q V(q) - \gamma \xi + \sigma \dot{W}_t, \tag{3.2.5}$$

defined on the state space $\mathbf{X} = \mathbf{R}^{6N}$. Here $\gamma > 0$ denotes some friction constant and $F_{\text{ext}} = \sigma \dot{W}_t$ the external forcing given by a $3N$-dimensional Brownian motion W_t. The external stochastic force is assumed to model the influence of the heat bath surrounding the molecular system. In this case, the internal energy given by the Hamiltonian H, as defined in (3.2.2), is not preserved, but the interplay between stochastic excitation and damping balances the internal energy. As a consequence, the canonical measure $\mu(\mathrm{d}x) \propto \exp(-\beta H(x))\mathrm{d}x$ with $x = (q, \xi)$ is invariant w.r.t. the Markov process corresponding to the Langevin system, where the noise and damping constants satisfy (RISKEN [1996]):

$$\beta = \frac{2\gamma}{\sigma^2}. \tag{3.2.6}$$

Thus, the Langevin system satisfies our expectation on (MC2) w.r.t. the canonical ensemble but simultaneously also allows to represent some essential aspects of (MC1), that is, of the true dynamical behavior of the molecular system.

Smoluchowski system. The Smoluchowski system can be understood as an approximation to the Langevin system in the limit of high friction $\gamma \to \infty$, see HUISINGA [2001], SCHÜTTE and HUISINGA [2000] for details. While the Langevin system gives a description of molecular motion in terms of positions and momenta of all atoms in the system, the Smoluchowski system is stated in the position space only. Moreover, in contrast to the Langevin equation it defines a *reversible* Markov process that is given by the equation

$$\dot{q} = -\frac{1}{\gamma}\nabla_q V(q) + \frac{\sigma}{\gamma}\dot{W}_t. \tag{3.2.7}$$

The stochastic differential equation (3.2.7) defines a continuous time Markov process Q_t on the state space $\mathbf{X} = \Omega$ with invariant probability measure $\mu(\mathrm{d}q) \propto \exp(-\beta V(q))\mathrm{d}q$ (RISKEN [1996]). Thus, this dynamical model satisfies our expectation on class (MC2) but should in general not be expected to satisfy those on (MC1). Nevertheless there is a long history of using it as a simple toolkit for investigation of dynamical behavior in complicated energy landscapes (CHANDLER [1998]). It is known that under weak conditions on the potential function V the Markov process is reversible (HUISINGA [2001]).

Markov chain Monte Carlo (MCMC). Markov chain Monte Carlo techniques are designed to sample a given probability density $f: \mathbf{R}^d \to \mathbf{R}$, particularly in highly dimensional state spaces. MCMC is an iterative realization of some specific Markov chain, whose stochastic transition function is given by

$$p(x, \mathrm{d}y) = q(x, y)\mu(\mathrm{d}y) + r(x)\delta_x(\mathrm{d}y).$$

That is, the stochastic transition function is composed of some transition kernel $q(x, y)$, which is assumed to be μ-integrable and some rejection probability

$$r(x) = 1 - \int_X q(x, y)\mu(\mathrm{d}y) \geqslant 0.$$

In almost all situations, the transition kernel q is chosen in such a way that the stochastic transition function is reversible w.r.t. μ.

In general MCMC is an artificial dynamical model that is in general understood as the embodiment of class (MC2), therefore being in general far from satisfying the properties of (MC1). However, there are MCMC methods like the popular hybrid Monte Carlo Method (HMC) that can be understood as a special realization of the Hamiltonian system with randomized momenta.

3.3. Summary

Concerning our main categories (MC1) and (MC2) the summary could be the following: constant temperature MD, MCMC, the Langevin and Smoluchowski systems clearly belong to (MC2), while the deterministic Hamiltonian system is supposed to be the incorporation of (MC1). However, this distinction is not sharp: the Langevin system is often also accepted as belonging to (MC1), while the deterministic Hamiltonian system is accepted for (MC1) only under the condition that enough details of the entire molecular system (including parts of the solute environment) are represented in atomic resolution and the interaction potential V is appropriate.

4. Metastability

Given a dynamical system, metastability of some subset of the state space is characterized by the property that the dynamical system is likely to remain within the subset for a long period of time, until it exits and hence a transition to some other region of the state space occurs. There is no unique but several definitions of metastability in literature (see, e.g., BOVIER, ECKHOFF, GAYRARD and KLEIN [2001], DAVIES [1982], SCHÜTTE, HUISINGA and DEUFLHARD [2001], SINGLETON [1984]); we will herein focus on two different concepts that are adapted to suit the ensemble dynamics and exit time approach, respectively, as discussed in Section 3.2 (Fig. 4.1).

Ensemble dynamics approach: A subset $C \subset \mathbf{X}$ of the entire state space is called metastable, if the fraction of systems in C, whose trajectory exits during some predescribed time span τ, is significantly small.

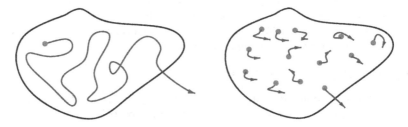

FIG. 4.1 Illustration of the different concepts of metastability: exit time approach on the left and ensemble dynamics approach on the right.

Exit time approach: A subset $C \subset \mathbf{X}$ of the entire state space is called metastable, if with high probability a typical long-term trajectory stays within C longer than some "macroscopic" time span.

Based on the discussion in section "Conceptual Preliminaries" we would expect to run into trouble when aiming at a "naive" numerical realization of the exit time approach while the ensemble dynamics approach seems to be numerically treatable for short observation time spans τ. We will come back to this question below.

In view of our biochemical application context we aim at the identification of a *decomposition of the state space into metastable subsets* and the corresponding "flipping dynamics" between these. In general, a *decomposition* $\mathcal{D} = \{\mathcal{D}_1, \ldots, \mathcal{D}_m\}$ of the state space $\mathbf{X}$ is a collection of subsets $D_k \subset \mathbf{X}$ with the properties:

positivity: $\mu(D_k) > 0$ for every k,
disjointness up to null sets: $\mu(D_k \cap D_l) = 0$ for $k \neq l$, and
covering property: $\cup_{k=1}^m \bar{D}_k = \mathbf{X}$.

The problem of identifying a decomposition into metastable subsets particularly poses the task of specifying the number m of subsets one is looking for. Within the transfer operator approach this is done via spectral analysis (see *key idea* on page 718).

4.1. Ensemble dynamics approach: transition probabilities

We aim at defining an (ensemble) transition probability from a subset B to C within some time span τ, denoted by $p(\tau, B, C)$, such that an invariant subensemble C is characterized by $p(\tau, C, C) = 1$, while a metastable subensemble can be characterized by $p(\tau, C, C) \approx 1$. We will see that within this approach metastability is measured w.r.t. the invariant probability measure μ of the dynamics; in biomolecular systems, the measure μ will often be defined in terms of the canonical ensemble.

We define the *transition probability* $p(t, B, C)$ from $B \subset \mathbf{X}$ to $C \subset \mathbf{X}$ within the time span t as the conditional probability

$$p(t, B, C) = \mathbf{P}_\mu[X_t \in C | X_0 \in B] = \frac{\mathbf{P}_\mu[X_t \in C \text{ and } X_0 \in B]}{\mathbf{P}_\mu[X_0 \in B]}, \quad (4.1.1)$$

where $\mathbf{P}_\mu$ indicates that the initially the Markov process X_t is distributed according to μ, hence $X_0 \sim \mu$. Exploiting the definition of the stochastic transition function $p(t, x, C)$ in (3.1.1) we rewrite (4.1.1) as

$$p(t, B, C) = \frac{1}{\mu(B)} \int_B p(t, x, C) \mu(dx). \quad (4.1.2)$$

In other words, the transition probability quantifies the dynamical fluctuations within the stationary ensemble μ. Due to the ensemble dynamics approach to metastability we call a subset $B \subset \mathbf{X}$ *metastable* on the time scale $\tau > 0$ if

$$p(\tau, B, B^c) \approx 0, \text{ or equivalently,} \quad p(\tau, B, B) \approx 1,$$

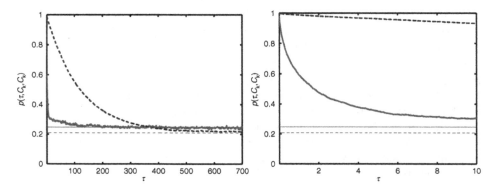

FIG. 4.2 Dependence of the transition probability $p(\tau, C, C)$ on the observation time span τ for $0 \leqslant \tau \leqslant 700$ (left) and zoomed into $0 \leqslant \tau \leqslant 10$ (right) w.r.t. two different subsets, one being metastable (dashed line) the other being much less metastable (solid line). The thick line corresponds to $p(\tau, C, C)$, while the corresponding predicted limit value $\mu(C)$ (see text) is indicated by a horizontal thin line. As can be seen from the graphics the distinction between metastable and not-metastable is clearly visible for mesoscopic observation time spans as, for example, $1/2 \leqslant \tau \leqslant 10$. (The data are based on the Smoluchowski dynamics w.r.t. the perturbed three-well potential as illustrated in Fig. 5.1.)

where $B^c = \mathbf{X} \backslash B$ denotes the complement of B. Obviously, the approximate equalities are not sharp enough for a rigorous definition of metastability. We will come back to this problem in Section 5.2.

REMARK. It is an intrinsic property of the ensemble transition probability to depend on the observation time span τ. It is obvious from its definition that $p(\tau, C, C)$ approaches 1 for $\tau \to 0$, while it decays to $\mu(C)$ for $\tau \to \infty$, see Fig. 4.2. The most interesting phenomena occur on mesoscopic time scales τ. In the biomolecular application context, the observation time span τ will be given by the experimental setting. In the earlier mentioned pump-and-probe experiments, the values of τ range in the subpicosecond regime.

4.2. *Exit time approach: exit rates*

The characterization of metastability within the exit time approach is related to the asymptotic decay of the distribution of exit times. Its precise formulation via exit (decay) rates requires some extended mathematical theory that hinders understanding at first reading. Therefore, we prefer to outline the fundamental idea rather than to give its mathematical justification, for which we refer to HUISINGA, MEYN and SCHÜTTE [2003].

Denote by $D \subset \mathbf{X}$ some connected open subset and consider some point $x \in D$. Then the *exit time* $\%_D(x)$ of the Markov process X_t from D started at $X_0 = x$ is defined as

$$\%_D(x) = \inf\left\{t \geqslant 0 : \int_0^1 \mathbf{1}_X(X_s \notin D)\mathrm{d}s > 0\right\} \tag{4.2.1}$$

and measures only exits that happen for some nonnull time interval neglecting exit events that are "singular" in time. Note that in general $\%_D$ is a random variable that depends on the realization of the Markov process X_t.

The fundamental idea is a characterization of metastability in terms of the asymptotic decay of the distribution of exit times

$$F_x(s) = \mathbf{P}_x[D(x) \geqslant s].$$

While for small values of s the function F_x may show complicated behavior, it asymptotically may decay almost exponentially, at least under certain well-established conditions. The decay rate of F_x can best be expressed by means of the conditional exit time distribution

$$F_x(s, t) = \mathbf{P}_x[D(x) \geqslant s + t | D(x) \geqslant t]$$

for $s, t \geqslant 0$ that describes the tail of the distribution, for which the exit time is larger than the so-called waiting time t. The decay rate is equal to Γ if the conditional distribution decays exponentially with rate $\Gamma > 0$, that is,

$$F_x(s, t) \propto \exp(-\Gamma s) \tag{4.2.2}$$

for $s \geqslant 0$ and $t \geqslant 0$. When aiming at a definition of metastability in terms of decay rates for entire *subsets*, there are two problems. Firstly, the relation in (4.2.2) will only hold for very special Markov processes (HUISINGA, MEYN and SCHÜTTE [2003]). Secondly, we would expect that the decay rate depends on the starting point, that is, $\Gamma = \Gamma_x$ (Fig. 4.3).

The approach presented herein is based on the fact that there do exist subsets C, for which the decay rate is basically independent for all states $x \in C$. In a more general

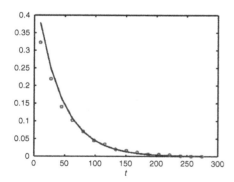
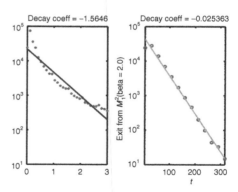

FIG. 4.3 Dependence of the exit time distribution $F_x(s)$ on the exit time s for some metastable subset D. The decay is shown for $0 \leqslant s \leqslant 300$ (left), and zoomed into $0 \leqslant s \leqslant 3$ (middle) and $10 \leqslant s \leqslant 300$ (right) on a semi-logarithmic plot. We observe regions of different decay (middle and right). Asymptotically, the decay rate of the exit time distribution is close to the predicted value of Γ (D), that is, $F_x(s)$ for $s > t$ decays approximately with rate Γ (D) for $t \to \infty$. Regions of different decay rate (like, e.g., initial rapid decay followed by a much slower decay) are typically due to the fact that with high probability the process exits very rapidly, while with almost vanishing, but existing probability the process moves into some much more metastable region contained in D such that asymptotically the exit rate becomes very small.

setting, but for a specific class of dynamical systems including, for example, the Smoluchowski dynamics, we are able to assign a so-called *exit rate* $\Gamma = \Gamma\,(C)$ to an entire subset $C \subset \mathbf{X}$ rather than to single points $x \in \mathbf{X}$, thus circumventing the two above mentioned problems with (4.2.2). This exit rates may be thought of as some generalization of decay rates; see Appendix 9.2, in particular Theorem 9.2.1.

Due to the exit time approach to metastability we call a subset $B \subset \mathbf{X}$ *metastable* with exit rate $\Gamma\,(B)$ if

$$\Gamma(A) > \Gamma(B), \quad \text{for all open, connected sets } A \subset B, A \neq B. \tag{4.2.3}$$

As in the ensemble approach there will be infinitely many metastable subsets and we expect to find a hierarchy of metastability.

5. Transfer operators

We now give the mathematical foundations and the algorithmic strategies to characterize and identify a decomposition of the state space into metastable subsets. It turns out that in either of the approaches transfer operators and their generators play a crucial role.

5.1. Transfer operators and generators

Based on the assumption that the dynamical description is given by a (homogeneous) Markov process X_t we now introduce a Markov operator that allows to describe the propagation of subensembles in time under the action of X_t. In so doing, we assume in the sequel that the probability measure μ is invariant w.r.t. the Markov process X_t.

The basic idea is the following: Consider all systems within the stationary ensemble, whose states are in some subset $C \subset \mathbf{X}$. This subensemble is distributed according to the probability measure

$$v_0(A) = \frac{1}{\mu(C)} \int_A \mathbf{1}_C(x)\mu(\mathrm{d}x) = \int_A v_0(x)\mu(\mathrm{d}x)$$

corresponding to the density $v_0 = \mathbf{1}_C/\mu(C)$ w.r.t. to μ. Since every single microscopic system evolves according to the Markov process defined by its stochastic transition function p, the distribution of the subensemble at time $t \in \mathbf{T}$ is given by the probability measure

$$v_t(A) = \int_{\mathbf{X}} v_0(x)p(t,x,A)\mu(\mathrm{d}x). \tag{5.1.1}$$

On the other hand, if v_t admits the density v_t, we have

$$v_t(A) = \int_A v_t(x)\mu(\mathrm{d}x). \tag{5.1.2}$$

Our interest is to define a transfer operator P^t that propagates subensembles in time according to

$$v_0 \mapsto v_t = P^t v_0. \tag{5.1.3}$$

This transfer operator is well-defined due to HUISINGA [2001], REVUZ [1975] and acts on the Banach spaces $L^r(\mu)$, $1 \leqslant r \leqslant \infty$, with corresponding norms $\| \cdot \|_r$, as defined in Appendix 9.1. Combining Eqs. (5.1.1) and (5.1.2), we define the *semigroup of propagators* or forward transfer operators $P^t: L^r(\mu) \to L^r(\mu)$ with $t \in \mathbf{T}$ and $1 \leqslant r < \infty$ as follows:

$$\int_A P^t v(y)\mu(\mathrm{d}y) = \int_{\mathbf{X}} v(x)p(t,x,A)\mu(\mathrm{d}x) \tag{5.1.4}$$

for $A \subset \mathbf{X}$. As a consequence of the invariance of μ, the characteristic function $\mathbf{1_X}$ of the entire state space is an invariant density of P^t, that is, $P^t \mathbf{1_X} = \mathbf{1_X}$. Furthermore, P^t is a Markov operator, that is, P^t conserves norm: $\| P^t v \|_1 = \| v \|_1$ and positivity: $P^t v \geqslant 0$ if $v \geq 0$, which is a simple consequence of the definition. Due to (5.1.3), the semigroup of propagators mathematically models the physical phenomena of evolution of subensembles in time.

In the theory of Markov processes another semigroup of operators is considered. We will call it the *semigroup of backward transfer operators* $T^t: L^r(\mu) \to L^r(\mu)$ with $t \in \mathbf{T}$ and $1 \leqslant r \leqslant \infty$, defined by

$$T^t u(x) = E_x[u(X_t)] = \int_{\mathbf{X}} u(y)p(t,x,\mathrm{d}y). \tag{5.1.5}$$

As a consequence of property (ii) in Appendix 9.1 of the stochastic transition function, we have $T^t \mathbf{1_X} = \mathbf{1_X}$ for every $t \in \mathbf{T}$. Both transfer operators are closely related via the duality bracket $\langle v, u \rangle_\mu = \int_{\mathbf{X}} v(x)u(x)\mu(\mathrm{d}x)$ for $v \in L^1(\mu)$ and $u \in L^\infty(\mu)$, namely

$$\langle P^t v, u \rangle_\mu = \langle v, T^t u \rangle_\mu. \tag{5.1.6}$$

In discrete time $t \in \mathbf{N}$ it is convenient to use the abbreviations $P = P^1$ and $T = T^1$ corresponding to the stochastic transition function $p(x, \mathrm{d}y) = p(1, x, \mathrm{d}y)$; compare Section 3.1. Propagators associated with *reversible* Markov processes are of particular interest, since they possess additional structure on the Hilbert space $L^2(\mu)$. Such propagators will be called reversible, too.

PROPOSITION 5.1.1 (HUISINGA [2001]). *Let $P^t: L^2(\mu) \subset L^1(\mu) \to L^2(\mu)$ denote the propagator corresponding to the Markov process X_t. Then P^t is self-adjoint w.r.t. the scalar product $\langle \cdot, \cdot \rangle_\mu$ in $L^2(\mu)$, that is,*

$$\langle u, P^t v \rangle_\mu = \langle P^t u, v \rangle_\mu \quad t \in \mathbf{T},$$

for every $u, v \in L^2(\mu)$, if and only if the Markov process X_t is reversible.

For the semigroup of propagators $P^t: L^r(\mu) \to L^r(\mu)$ with $1 \leqslant r \leqslant \infty$ define $\mathcal{D}(\mathcal{A})$ as the set of all $v \in L^r(\mu)$ such that the strong limit

$$\mathcal{A}v = \lim_{t \to \infty} \frac{P^t v - v}{t}$$

exists. Then, the operator $\mathcal{A} : \mathcal{D}(\mathcal{A}) \to L^r(\mu)$ is called the infinitesimal *generator* corresponding to the semigroup P^t (KARATZAS and SHREVE [1991], LASOTA and MACKEY [1994]).

REMARK. Physical experiments on molecular ensembles allow to measure relative frequencies in the canonical ensemble μ. Suppose again that μ has the form

$$\mu(\mathrm{d}x) = f(x)\mathrm{d}x,$$

that is, μ is absolutely continuous w.r.t. the Lebesgue measure $\mathrm{d}x$. Then physical experiments are related to densities of the form

$$v_{\mathrm{phys}}(x) = \hat{\mathbf{1}}_C(x)f(x) \in L^1(\mathrm{d}x)$$

w.r.t. the Lebesgue measure $\mathrm{d}x$. Whenever physicists use the phrase "probability density" they refer to v_{phys} rather than to densities

$$v_{\mathrm{math}}(x) = \hat{\mathbf{1}}_C(x) \in L^1(\mu)$$

w.r.t. probability measure μ, as we do. As it will become apparent later, it is mathematically advantageous to consider the semigroup of propagators acting on densities v_{math} rather than the semigroup of propagators acting on v_{phys}. In the former approach, we have $P^t \mathbf{1}_X = \mathbf{1}_X$, while in the latter this would read $P^t f = f$. However, it should be clear that results obtained in either of the two descriptions can be transformed into the other.

EXAMPLES 5.1.1. To be more specific, we now list the propagators for the different dynamical descriptions introduced in Section 3.2.

The propagator corresponding to the deterministic Hamiltonian system with flow Φ^t is known as the Frobenius–Perron operator (LASOTA and MACKEY [1994]) given by

$$P^t u(x) = u(\Phi^{-t}x).$$

For the Hamiltonian system with randomized momenta we have a kind of Frobenius–Perron operator averaged w.r.t. the momenta,

$$P^t u(q) = \int_{\mathbf{R}^d} u(\Pi_q \Phi^{-\tau}(q, \xi)) \mathcal{P}(\xi)\mathrm{d}\xi. \tag{5.1.7}$$

For MCMC the propagator is given by

$$Pu(y) = \int q(x, y)u(x)\mathrm{d}x + r(y)u(y). \tag{5.1.8}$$

For Langevin and Smoluchowski dynamics, the semigroups of propagators admit strong generators $\mathcal{A}_{\mathrm{Smo}}$ and $\mathcal{A}_{\mathrm{Lan}}$ in $L^r(\mu)$ for $1 \leqslant r < \infty$ such that the semigroups can be written as

$$P^t_{\mathrm{Smo}} = \exp(t\mathcal{A}_{\mathrm{Smo}}) \text{ and } P^t_{\mathrm{Lan}} = \exp(t\mathcal{A}_{\mathrm{Lan}}),$$

respectively. For twice continuously differentiable $u \in L^r(\mu)$ we have the identity

$$\mathcal{A}_{\mathrm{Smo}}u = \left(\frac{\sigma^2}{2\gamma^2}\Delta_q - \frac{1}{\gamma}\nabla_q V(q).\nabla_q \right) u,$$

$$\mathcal{A}_{\mathrm{Lan}}u = \left(\frac{\sigma^2}{2}\Delta_p - p.\nabla_q + \nabla_q V.\nabla_p - \gamma p.\nabla_p \right) u.$$

For details on $\mathcal{A}_{\text{Smo}}$ and $\mathcal{A}_{\text{Lan}}$ see the theory of Fokker–Planck equations and Kolmogoroff forward and backward equations (RISKEN [1996], SCHÜTTE, HUISINGA and DEUFLHARD [2001], HUISINGA [2001]).

5.2. Detecting metastability

The key idea of the transfer operator approach to metastability is to exploit the strong relation between stability properties of the Markov process and the presence of special eigenvalues in the spectrum.

Since propagators are Markov operators by definition, their spectrum is contained in the unit circle of the complex plane, that is, the modulus of every eigenvalue is smaller or equal to 1. Suppose that some proper subset $C \subset \mathbf{X}$ is invariant under the Markov process, that is, $p(t, x, C^c) = 0$ for all $x \in C$. Then we have:

Ensemble dynamics approach: the transition probability from C into its complement C^c is zero: $p(t, C, C^c) = 0$ for every $t \in \mathbf{T}$.

Exit rate approach: the exit rate from C is zero: $\Gamma(C) = 0$.

In literature on Markov and transfer operators, it is a well-known fact that the existence of invariant subsets has spectral consequences. Under well-established stability conditions (see (C1) and (C2) on page 719), we have:

Transfer operator: the propagator P^t exhibits an eigenvalue $\lambda_t \equiv 1$ with corresponding eigenfunction $\mathbf{1}_C$, hence $P^t \mathbf{1}_C = \mathbf{1}_C$ for every $t \in \mathbf{T}$.

Now suppose that the entire state space decomposes into exactly two invariant subsets, $X = B \cup C$. Then, the eigenvalue $\lambda = 1$ is two-fold, one corresponding to each invariant subset associated with the eigenfunctions $\mathbf{1}_B$ and $\mathbf{1}_C$. Introducing a weak coupling between the subsets B and C yields one invariant set, namely the entire state space $\mathbf{X}$, and two *weakly coupled* or *metastable* subsets, namely B and C. This has the following consequences:

Ensemble dynamics approach: the transition probability from B to C is almost zero: $p(\tau, B, C) \approx 0$ for $0 < \tau < T$ with large T. The same holds for the transition probability from C to B.

Exit rate approach: the exit rates from B and C are very small: $\Gamma(B) \approx 0$ and $\Gamma(C) \approx 0$.

Transfer operator: for $0 < \tau < T$ with large T, the propagator P^τ exhibits two dominant eigenvalues. More precisely, there exists $\eta_t \equiv 1$ corresponding to the invariant state space, and one eigenvalue $\lambda_\tau \approx 1$ corresponding to the weak coupling between the subsets B and C.

To the end, we fix some $\tau > 0$ and abbreviate $P = P^\tau$ and $p(x, C) = p(\tau, x, C)$. Hence, $(P)^n = P^{n\tau}$ corresponds to the Markov process sampled at rate τ with stochastic transition function given by $p^n(\cdot, \cdot) = p(n\tau, \cdot, \cdot)$.

The above considerations motivate the following *key idea of the transfer operator approach*:

> Metastable subsets can be detected via eigenvalues of the propagator P close to its maximal eigenvalue $\lambda = 1$; moreover they can be identified

by exploiting the corresponding eigenfunctions. In doing so, the number of metastable subsets is equal to the number of eigenvalues close to 1, including $\lambda = 1$ and counting multiplicity.

The strategy mentioned above has first been proposed by DELLNITZ and JUNGE [1999] for discrete dynamical systems with weak random perturbations and has been successfully applied to molecular dynamics in different contexts (SCHÜTTE, FISCHER, HUISINGA and DEUFLHARD [1999], SCHÜTTE and HUISINGA [2000], SCHÜTTE [1998]); its justification is given in Section 5.4. The key idea requires the following two *conditions on the propagator P* (for a definition of the essential spectral radius see Appendix 9.1):

(C1) The essential spectral radius of P is less than one, that is, $r_{\mathrm{ess}}(P) < 1$.

(C2) The eigenvalue $\lambda = 1$ of P is simple and dominant, that is, $\eta \in \sigma(P)$ with $|\eta| = 1$ implies $\eta = 1$.

While condition (C1) allows to ensure convergence results of the numerical discretization scheme, condition (C2) excludes modeling and interpretation problems; for more details see SCHÜTTE [1998], HUISINGA [2001]. In order to proceed along the way indicated by the key idea we have to check in which situations the two conditions (C1) and (C2) may hold. In this section, we establish sufficient conditions on the Markov process that imply (C1) and (C2). Here, we mainly concentrate on reversible propagators P on $L^2(\mu)$ and refer for $L^1(\mu)$ and L_V^∞ to HUISINGA [2001] and MEYN and TWEEDIE [1993], respectively.

There are some sufficient conditions to guarantee the spectral properties of the propagator that are related to well studied stability properties of Markov processes. The $\mathcal{V}$-norm and total variation norm $\|\cdot\|_{\mathrm{TV}}$ stated in the next definition are defined in Appendix 9.1:

DEFINITION 5.2.1. Let p denote some stochastic transition function. Then

(a) p is called *geometrically ergodic* if

$$\| p^n(x, \cdot) - \mu \|_{\mathrm{TV}} \leqslant \mathcal{V}(x)q^n, \quad n \in \mathbf{N}, \tag{5.2.1}$$

for every $x \in \mathbf{X}$, some constant $q < 1$, and some integrable function $\mathcal{V} : \mathbf{X} \to \mathbf{R}$ satisfying $\mathcal{V} < \infty$ pointwise.

(b) p is called $\mathcal{V}$ *-uniformly ergodic* if

$$\| p^n(x, \cdot) - \mu \|_\mathcal{V} \leqslant C\mathcal{V}(x)q^n, \quad n \in \mathbf{N},$$

for every $x \in \mathbf{X}$, constants $q < 1$ and $C \leqslant \infty$, and some function $\mathcal{V} \in L^1(\mu)$ satisfying $1 \leqslant \mathcal{V}$ pointwise.

The relation between geometrical and $\mathcal{V}$-uniform ergodicity is as follows: By definition, $\mathcal{V}$-uniform ergodicity implies geometric ergodicity. On the other hand, for irreducible and aperiodic stochastic transition functions geometric ergodicity implies $\mathcal{V}$-uniform ergodicity according to (ROBERTS and ROSENTHAL [1997], Proposition 2.1). Either form of ergodicity implies the properties of interest:

THEOREM 5.2.1   (HUISINGA [2001]). *Let* $P: L^2(\mu) \to L^2(\mu)$ *denote a reversible propagator. Then* P *satisfies conditions* (C1) *and* (C2) *in* $L^2(\mu)$, *if its stochastic transition function is geometrically or* $\mathcal{V}$-*uniformly ergodic.*

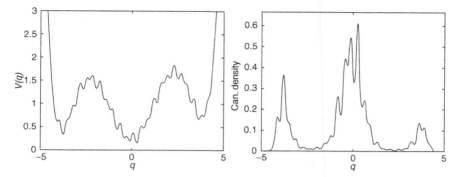

FIG. 5.1 Left: a perturbed three well potential V used for illustration. Right: canonical density corresponding to $\beta = 3.0$. Assume that the internal dynamics is given by the Smoluchowski system corresponding to $\gamma = 2.0$ and σ given by Eq. (3.2.6). The dominant spectrum of the propagator $P = P^\tau$ with $\tau = 1.0$ exhibits three eigenvalues close to 1. More precisely, we have $\lambda_1 = 1.0000$, $\lambda_2 = 0.9924$, $\lambda_3 = 0.9886$, which are separated from the remaining eigenvalues by a significant gap, since $\lambda_4 = 0.6634$. According to the key idea of the transfer operator approach we expect a decomposition of the state space into three metastable subsets, which is in agreement with our intuition for the above potential.

Conditions, under which Markov processes are geometrically or $\mathcal{V}$-uniformly ergodic are widely studied; for sufficient conditions w.r.t. the dynamical models introduced in Section 3.3, see, for example, HUISINGA [2001], Section 6, and cited references, or MATTINGLY, STUART and HIGHAM [2003], SCHÜTTE [1998].

5.3. Identification algorithm

The key idea of the transfer operator approach needs to be specified regarding the actual *algorithmic identification* of metastable subsets based on the most dominant eigenvectors. The basic idea is to reduce the problem of identifying a decomposition into metastable subsets to a clustering problem, which is done by incorporating dynamical information into the process of clustering. It is therefore different from statistical clustering that is solely based on geometrical information.

Following the key idea of the transfer operator approach we are aiming at a decomposition $\mathcal{D} = \{D_1, \ldots, D_m\}$ of the state space into m metastable subsets $D_1, \ldots, D_m$ such that the number of subsets m equals the number of dominant eigenvalues. Fig. 5.2 demonstrates the mechanism of the transfer operator approach to metastability.

Based on this mechanism, virtually almost every cluster algorithm can be used to identify a decomposition of the state space into metastable subsets, *if applied to the dynamically coded sampling points*. The identification procedure introduced in DEUFLHARD, HUISINGA, FISCHER and SCHÜTTE [2000] computes this decomposition from the *sign structure* of the coded sampling points as illustrated in Figs. 5.3 and 5.4. Given the m dominant eigenvectors $v_1, \ldots, v_m$, we can assign to every state $x \in \mathbf{X}$ a unique *sign structure*

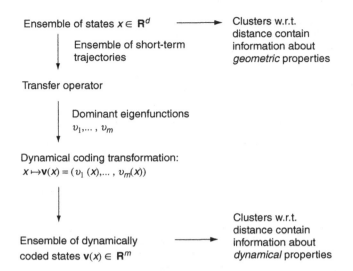

FIG. 5.2 Basic mechanism of incorporating dynamical information via dominant eigenfunctions of the transfer operator. Note that the dimension of the state space that have to be clustered reduces from d to m. In biomolecular applications, typically we have $m \ll d$.

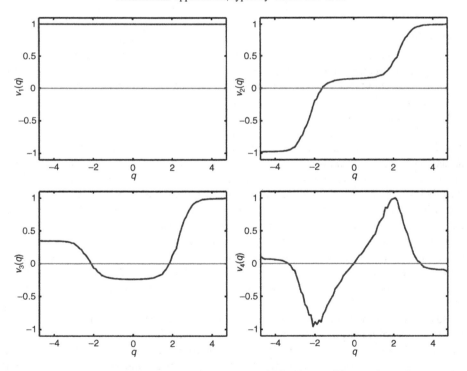

FIG. 5.3 Eigenfunction corresponding to propagator based on the Smoluchowski system in application to the perturbed three well potential V as illustrated in Fig. 5.1. Eigenfunctions corresponding to the largest eigenvalues 1.0000, 0.9924, 0.9886, 0.6634 (clockwise, starting top-left). As expected, the eigenfunction corresponding to $\lambda_1 = 1$ is constant. Note that the second and the third eigenfunction exhibit a very special structure: they are almost constant around the three wells (cf. potential in Fig. 5.1), while they show jumps near the saddle point regions.

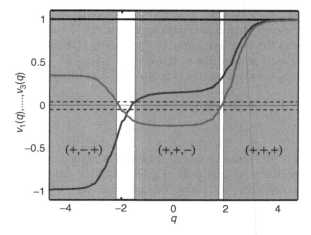

FIG. 5.4 Illustration of the identification procedure based on the perturbed three well potential presented in Fig. 5.1. The three gray shaded regions indicate the parts of the q-axis on which the three most dominant eigenvectors exhibit "unambiguous" sign structures (here $\theta = 0.05$ as indicated by dashed lines). To give an example: the gray shaded region on the very left has sign structure $(+, -, +)$ since $v_1(q), v_3(q) > \theta$ and $v_2(q) < -\theta$ for all states q on the very left.

$$s(x) = (s_1(x), \ldots, s_m(x)) = \{+, - \}^m,$$

where for some prescribed threshold value θ we define the "unambiguous" positive or negative sign $s_k(x)$ of $v_k(x)$ as

$$s_k(x) = \begin{cases} +, & v_k(x) > \theta, \\ -, & v_k(x) < \theta \end{cases}$$

as explained in Fig. 5.4. Denote by $S(\mathbf{X}) \subset \mathbf{X}$ the set of all states with at least one "unambiguous" sign s_k. If θ is large enough and the eigenfunctions are smooth, $S(\mathbf{X})$ decomposes into exactly m subsets each containing states of the same sign structure only (DEUFLHARD, HUISINGA, FISCHER and SCHÜTTE [2000]). These are the "core sets" of the metastable subsets. States with "unambiguous" sign structure are assigned to these core sets (DEUFLHARD, HUISINGA, FISCHER and SCHÜTTE [2000]) such that the resulting m metastable sets decompose the state space. This clustering algorithm has proved to be successful in many different situations; in the following subsection we give a mathematical justification of the algorithmic identification strategy from the point of view of the ensemble dynamics and the exit time approach.

5.4. Mathematical justification

We now give a mathematical justification of the key idea of the transfer operator approach, which illuminates the strong relation between the existence of a cluster of eigenvalues close to 1 and a possible decomposition of the state space into metastable subsets.

Justification within the ensemble dynamics approach. The investigation will be based on the following assumptions: The propagator $P{:}L^2(\mu) \rightarrow L^2(\mu)$ satisfies conditions (C1), (C2) and the underlying Markov process is reversible. As a consequence the propagator P is self-adjoint due to Proposition 5.1.1.

The close relation between transition probabilities and transfer operators becomes transparent in

$$p(B,C) = \frac{\langle P1_B, 1_C \rangle_\mu}{\langle 1_B, 1_B \rangle_\mu}. \tag{5.4.1}$$

Eq. (5.4.1) allows to give a mathematical statement relating dominant eigenvalues, the corresponding eigenfunctions and a decomposition of the state space into metastable subsets. For later reference we define the *metastability of a decomposition* $\mathcal{D}$ as the sum of the metastabilities of its subsets. The next result can be found in HUISINGA and SCHMIDT [2002]; a version for two subsets was published in HUISINGA [2001].

THEOREM 5.4.1 *Let $P{:} L^2(\mu) \rightarrow L^2(\mu)$ denote a reversible propagator satisfying (C1) and (C2). Then P is self-adjoint and the spectrum has the form*

$$\sigma(P) \subset [a,b] \cup \{\lambda_m\} \cup \cdots \cup \{\lambda_2\} \cup \{1\},$$

with $-1 < a \leqslant b < \lambda_m \leqslant \cdots \leqslant \lambda_1 = 1$ and isolated, not necessarily simple eigenvalues of finite multiplicity that are counted according to multiplicity. Denote by $v_m, \ldots, v_1$ the corresponding eigenfunctions, normalized to $\| v_k \|_2 = 1$. Let Q be the orthogonal projection of $L^2(\mu)$ onto span$\{1_{A_1}, \ldots, 1_{A_m}\}$. *The metastability of an arbitrary decomposition $\mathcal{D} = \{A_1, \ldots, A_m\}$ of the state space $\mathbf{X}$ can be bounded from above by*

$$p(A_1, A_1) + \cdots + p(A_m, A_m) \leqslant 1 + \lambda_2 + \cdots + \lambda_m,$$

while it is bounded from below according to

$$1 + \kappa_2 \lambda_2 + \cdots + \kappa_m \lambda_m + c \leqslant p(A_1, A_1) + \cdots + p(A_m, A_m),$$

where $\kappa_j = \| Q v_j \|^2_{L^2(\mu)}$ and $c = a(1 - \kappa_2) \cdots (1 - \kappa_n)$.

Theorem 5.4.1 highlights the strong relation between a decomposition of the state space into metastable subsets and dominant eigenvalues close to 1. It states that the metastability of an arbitrary decomposition $\mathcal{D}$ cannot be larger than $1 + \lambda_2 + \cdots + \lambda_m$, while it is at least $1 + \kappa_2 \lambda_2 + \cdots + \kappa_m \lambda_m + c$, which is close to the upper bound whenever the dominant eigenfunctions $v_2, \ldots, v_m$ are almost constant on the metastable subsets $A_1, \ldots, A_m$ implying $\kappa_j \approx 1$ and $c \approx 0$. The term c can be interpreted as a correction that is small, whenever $a \approx 0$ or $\kappa_j \approx 1$. It is demonstrated in HUISINGA and SCHMIDT [2002] that the lower and upper bounds are sharp and asymptotically exact.

In view of Theorem 5.4.1, it is natural to ask, whether there is an *optimal* decomposition with highest possible metastability. The answer is illustrated by Fig. 5.5: *Even if there exists an optimal decomposition, the problem of finding it might be ill-conditioned.* The graph shows the metastability of a family of decompositions. It is based on the propagator P corresponding to the Smoluchowski dynamics for the perturbed three-well potential. We identify a flat plateau of decompositions that are

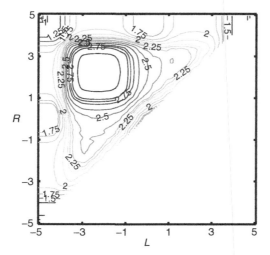

FIG. 5.5 Metastability of the decomposition of the state space $\mathbf{X} = \mathbf{R}$ into three subsets $\mathcal{D} = \{A, B, C\}$ with $A = (-\infty, L)$, $B = [L, R)$ and $C = [R, \infty)$ parameterized by $L < R \in \mathbf{R}$. The problem of finding the maximal value corresponding to the optimal decomposition is often ill-conditioned. There is a large region roughly characterized by $-3 < L < -1$ and $1 < R < 3$, in which the metastability of the corresponding decomposition is almost maximal. This region is more or less a flat plateau, for which the absolute maximum is hard to identify. The identification algorithm based on the sign structure identifies a decomposition with $L = -2.13$ and $R = 1.84$ with corresponding metastability 2.9558, which is quite close to the upper bound of 2.9916 resulting from Theorem 5.4.1. (Data based on Smoluchowski dynamics as illustrated in Fig. 5.1.)

nearly optimal. In this case the problem of finding the maximum is ill-conditioned. We also observe that the decomposition suggested by our identification algorithm is nearly optimal. The phenomenon illustrated by Fig. 5.5 is believed to be typical in our application context, which is due to the fact that the state space admits large regions corresponding to almost vanishing statistical weight (see also HUISINGA and SCHMIDT [2002]).

Justification within the exit time approach. The justification has been worked out in HUISINGA, MEYN and SCHÜTTE [2003] based on recent literature on Markov chains. The most important restriction of this approach is the restriction to Markov processes with continuous sample paths that admit a generator.

We will herein formulate the result for general Smoluchowski systems of type (3.2.7) from Section 3.2, because this allows us to remain within the framework of self-adjoint propagators with real-valued eigenvalues and eigenfunctions. The generalization to Langevin systems seems to be possible but requires immense technical effort.

THEOREM 5.4.2 *Assume that there is some continuous Lyapunov function* $\mathcal{V}$ *such that the Markov process is* $\mathcal{V}$-*uniformly ergodic, and that there exists a twice continuously differentiable eigenfunction* $v: \mathbf{X} \to \mathbf{R}$ *of the Smoluchowski generator* $\mathcal{A} = \mathcal{A}_{\text{Smo}}$. *Hence, there exists some eigenvalue* $\Lambda < 0$ *such that*

$$\mathcal{A}v = \Lambda v. \qquad\qquad (5.4.2)$$

Suppose moreover that the set C under consideration and the eigenfunction v satisfy the following conditions:

(1) $v(x) > 0$ *or* $v(x) < 0$ *for all* $x \in C$,
(2) $v(x) = 0$ *and* $(\nabla v(x))^T (\nabla v(x)) > 0$ *for* $x \in \partial C$,
(3) $K_n = \{x \in \mathbf{X} : \mathcal{V}(x) \leqslant nv(x)\}$ *is a compact subset of* $\mathbf{X}$ *for all* $n \geqslant 1$.

Then, the set C is metastable in the sense of definition (4.2.3) with exit rate $\Gamma(C) = -\Lambda$, *where* Λ *is the eigenvalue associated with* v.

This theorem has the following intriguing interpretation: if there is an eigenvalue $-\Gamma_0$ of the generator close to zero, then there is an eigenvalue $\exp(-\tau\Gamma_0)$ of the propagator P^τ close to $\lambda = 1$. The corresponding eigenfunction v of the generator is also the eigenfunction of the propagator. The set of zeros of this eigenfunction decomposes the state space into *open, connected subsets* C_k, restricted to each of which the eigenfunction is either positive or negative. Now Theorem 5.4.2 states that each of this subsets C_k is metastable with the same rate $\Gamma(C_k) = \Gamma_0$ which is considerably small since Γ_0 is.

The identification algorithm is based on the *collection* of the most dominant eigenfunctions. In seemingly contrast, Theorem 5.4.2 indicates that *each* single eigenfunction induces a metastable decomposition, in particular that different eigenfunctions might induce different metastable decompositions. We illustrate this for the three-well potential in Fig. 5.6. The second and third eigenfunctions v_2 and v_3 allow application of Theorem 5.4.2 and induce two different decompositions, namely $\{C_1, C_2\}$ and $\{D_1, D_2, D_3\}$. The rate of metastability of the first is superior ($\Gamma(C_k) = 0.022$); the latter reproduces the three-well structure of the potential but its metastable subsets show less significant metastability ($\Gamma(D_k) = 0.036$). Conclusively, the result is a hierarchy of metastable decompositions with decreasingly significant metastability of the subset.

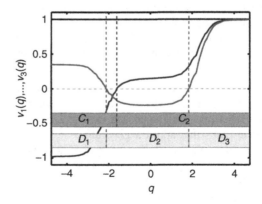

FIG. 5.6 Illustration of Theorem 5.4.2 for the perturbed three well potential illustrated in Fig. 5.1. The lowest eigenfunctions of the generator corresponding to the Smoluchowski dynamics are $\Lambda_1 = 0.0000$, $\Lambda_2 = -0.0076$, $\Lambda_3 = -0.0114$, while $\Lambda_4 = -0.4104$. The eigenfunctions v_2 and v_3 exhibit sign changes and allow to decompose the state space into subsets, on which the eigenfunction is either positive or negative. Since v_2 has exactly one zero a thereon based decomposition of the state space yields the subsets C_1 and C_2 with common exit rate $\Gamma_2 = \Lambda_2 = 0.0076$. Analogously, v_3 defines the subsets D_1, D_2 and D_3 with common exit rate $\Gamma_3 = -\Lambda_3 = 0.0114$.

However, the identification algorithm will result in a very similar hierarchy if iteratively applied to the set $\{v_1, v_2\}$ of eigenfunctions first and then to the set $\{v_1, v_2, v_3\}$.

6. Numerical realization

Identification of metastable subsets necessitates the computation of the most dominant
eigenfunctions of the propagator $P = P^\tau$ for some fixed observation time span $\tau > 0$.
In the following we describe the discretization procedure of the eigenvalue problem
$Pv = \lambda v$. Throughout this section we assume that P satisfies the conditions (C1)
and (C2) defined in Section 5.2. Part of this section follows from SCHÜTTE [1998],
SCHÜTTE, FISCHER, HUISINGA and DEUFLHARD [1999].

6.1. Galerkin discretization

Let $\chi = \{\chi_1, \ldots, \chi_n\} \subset L^2(\mu)$ denote a set of *nonnegative* functions with the property
to yield a partition of unity, that is,

$$\sum_{k=1}^{n} \chi k = 1_{\mathbf{X}}.$$

The *Galerkin projection* $\Pi_n : L^2(\mu) \to \mathcal{S}_n$ onto the associated finite-dimensional
ansatz space $\mathcal{S}_n = \mathrm{span}\{\chi_1, \ldots, \chi_n\}$ is defined by

$$\Pi_n v = \sum_{k=1}^{n} \frac{\langle v, \chi_k \rangle_\mu}{\langle \chi_k, \chi_k \rangle_\mu} \chi_k.$$

Application of the Galerkin projection to $Pv = \lambda v$ yields an eigenvalue problem for
the discretized propagator $\Pi_n P \Pi_n$ acting on the finite-dimensional space $\mathcal{S}_n$. The
matrix representation of this finite-dimensional operator is given by the $n \times n$ *transition matrix* $S = (S_{kl})$, whose entries are given by

$$S_{kl} = \frac{\langle P\chi_k, \chi_l \rangle_\mu}{\langle \chi_k, \chi_k \rangle_\mu} = \frac{\langle \chi_k, T\chi_l \rangle_\mu}{\langle \chi_k, \chi_k \rangle_\mu}, \tag{6.1.1}$$

where T denotes the adjoint transfer operator.

Properties of discretization matrix. Since P is a Markov operator and χ a partition
of unity, the Galerkin discretization S is a (row) stochastic matrix, that is, $S_{kl} \geqslant 0$ and
$\sum_{l=1}^{n} S_{kl} = 1$ for $k = 1, \ldots, n$. Consequently, all its eigenvalues λ satisfy $|\lambda| \leqslant 1$.
Moreover, we have the following three important properties (SCHÜTTE [1998],
SCHÜTTE, FISCHER, HUISINGA and DEUFLHARD [1999]):

(1) The row vector $\pi = (\pi_1, \ldots, \pi_n)$ with $\pi_k = \langle \chi_k, \chi_k \rangle_\mu = \| \chi_k \|_\mu^2$ represents the
   discretized invariant probability measure μ. It is a left eigenvector
   corresponding to the eigenvalue $\lambda = 1$, thus $\pi S = \pi$.
(2) S is *irreducible* and *aperiodic*. As a consequence, the eigenvalue $\lambda = 1$ is
   simple and *dominant*, hence $\lambda \in \sigma(S)$ implies $\lambda = 1$ or $|\lambda| < 1$. In particular,
   the discretized invariant density π is the *unique* invariant density of S.

(3) If P is reversible then S is *self-adjoint* w.r.t. the discrete scalar product $\langle u, v \rangle_\pi = \sum u_i \bar{v}_i \pi_i$. Equivalently, S satisfies the *detailed balance condition* $\pi_k S_{kl} = \pi_l S_{lk}$ for every $k, l \in \{1, \ldots, n\}$. Hence, all eigenvalues of S are real-valued and contained in the interval $(-1, 1]$.

Partitions of unity that are defined in terms of some decomposition $\mathcal{D} = \{D_1, \ldots, D_n\}$ of the state space are of particular interest. Any decomposition $\mathcal{D}$ defines a partition of unity $\chi = \{\chi_1, \ldots, \chi_n\}$ with $\chi_k = \mathbf{1}_{D_k}$. Galerkin discretization based on this *box discretization* then yields a transition matrix that in addition to the above properties has the advantageous property that its entries are one-step transition probabilities from D_k to D_l

$$S_{kl} = \frac{\langle P \mathbf{1}_{D_k}, \mathbf{1}_{D_l} \rangle_\mu}{\langle \mathbf{1}_{D_k}, \mathbf{1}_{D_k} \rangle_\mu} = p(D_k, D_l).$$

Another prominent example is given by partitions of unity χ whose elements $\chi_k \in \chi$ are so-called mollifies, that is, nonnegative C^∞-functions of compact support that approximate indicator functions. In this case, the discretization often is called *fuzzy set discretization*.

Summarizing, the discretization of the propagator can be interpreted as a *coarse graining* procedure, especially in the case of box discretization: Coarse graining the state space $\{x \in \mathbf{X}\} \to \{D_1, \ldots, D_n\}$ results in a coarse graining of the propagator $P \to S$ corresponding to a coarse graining of the Markov process $p(x, D) \to p(D_k, D_l)$ with invariant measures $\mu \to \pi$. In doing so, the discretization inherits the most important properties of the propagator.

REMARK. It is important to notice that – unless in very special situations – the discretization process does not commute with the semigroup property of the transfer operator P. Hence, in general we have

$$(S^2)_{kl} \neq p^2(D_k, D_l),$$

where S^2 denotes the square of the transition matrix S obtained from discretization P, and p^2 denotes the stochastic transition function corresponding to P^2.

6.2. *The eigenvalue problem*

We restrict our considerations to the important class of *reversible* propagators
$P: L^2(\mu) \to L^2(\mu)$ satisfying the conditions (C1) and (C2).

Under these assumptions, convergence results for the eigenvalues are simple consequences of, for example, the Rayleigh–Ritz theory (see SCHÜTTE [1998] for details). Obviously we have to require that the sequence of the Galerkin ansatz spaces $S_{n_1} \subset S_{n_2} \subset \cdots$ is dense in $L^2(\mu)$, and the corresponding partitions of unity $\chi^{(1)}$, $\chi^{(2)}, \ldots$ are getting gradually finer, for example, $\max_{\Phi \in \chi^{(k)}} \text{diam(supp } \Phi) \to 0$ as $n_k \to \infty$ with $k \to \infty$, where we assume that the functions in the kth partition of unity $\chi^{(k)}$ have compact support.

For the explicit numerical approximation of the dominant eigenvalues and eigenvectors different settings are available. Whenever one is interested in the dominant eigenvalues and eigenvectors of the transition matrix for a given discretization one may use iterative eigenvalue solvers, see, for example, LEHOUCQ, SORENSEN and YANG [1998]. These allow to compute the desired information even for matrices of size $10^6 \times 10^6$, for example. Whenever one is interested in subsequent refinements of the discretization to achieve approximations with high precision, one should apply multigrid techniques with optimal efficiency even for very fine grids like those constructed in DEUFLHARD, FRIESE and SCHMIDT [1997], for example. The mentioned preconditions for efficient convergence of the techniques perfectly fit to the scenario of metastability: it is required that the dominant eigenvalues are separated from the remainder of the spectrum by a significant gap. Then the convergence rate of those methods only depends on the spectral gap and is in principle independent of the size of the stochastic transition matrix. The numerical effort is given mainly by the effort of matrix-vector multiplications.

We emphasize, however, that already for small molecules the dimension of state space X is so high as to make the transfer operator approach computationally infeasible if naively applied directly in X. This problem is sometimes called the *curse of dimensionality*. To some extend it can be ameliorated by the use of adaptive algorithms such as those in DELLNITZ and JUNGE [1998] but, for many problems in high dimensions, combination of the transfer operator approach with clustering approaches, and/or with mathematical modeling such as the exploitation of fast/slow time-scale separation, is needed to circumvent the curse of dimensionality. We will address this problem in Section 6.5.

6.3. Evaluation of transition matrix

We consider the evaluation of the stochastic transition matrix S obtained from discretizing $P = P^\tau$. The corresponding discrete time Markov chain is denoted by $X_n = \{X_n\}_{n \in \mathbb{N}}$. Consider two elements χ_k, χ_l of the partition of unity used for discretization. Combining $T^\tau \chi_l(x) = E_x[\chi_l(X_\tau)]$ with (6.1.1) yields

$$S_{kl} = \frac{1}{\langle \chi_k, \chi_k \rangle_\mu} \int_X \chi_k(x) E_x[\chi_l(X_\tau)] \mu(dx),$$

which can be approximated within two steps:

(A1) approximation of the integral

$$\int_B g(x) \mu(dx) \approx \sum_{k=1}^N \alpha_k g(x_k)$$

by some deterministic or stochastic integration scheme with partition points or random variables $x_1, \ldots, x_N$, respectively, and weights $\alpha_1, \ldots, \alpha_N$ (DEUFLHARD and HOHMANN [1993], GILKS, RICHARDSON and SPIEGELHALTER [1997]);

(A2) approximation of the expectation value

$$E_x[\chi_l(X_\tau)] \approx \frac{1}{M} \sum_{j=1}^M \chi_l(X_\tau(\omega_j, x))$$

by relative frequencies, where $X_\tau(\omega_k, x)$ denotes a realization of the Markov process with initial distribution $X_0 = x$ (MEYN and TWEEDIE [1993], Chapter 17). A combination of (A1) and (A2) with $g(x) = \chi_k(x)E_x[\chi_l(X_\tau)]$ results in

$$S_{kl} \approx \frac{1}{M} \sum_{k=1}^{N} \sum_{j=1}^{M} \alpha_k \chi_k(x) \cdot \chi_l(X_\tau(\omega_{kj}, x_k));$$

hence, for each initial point x_k, the Markov process X_τ is realized M times. The approximation quality depends on the interplay between the two approximation steps (A1) and (A2). Numerical experiments in low dimensions show that it is even possible to choose small M, if the number of partition points N is chosen in such a way that the number of points in the support of any of the χ_k is reasonably large. For high-dimensional problems, we will in general be forced to use stochastic integration schemes such as Monte Carlo methods in order to approximate the integral in (A1). For the analysis of biomolecules Monte Carlo based techniques have been applied successfully (SCHÜTTE, FISCHER, HUISINGA and DEUFLHARD [1999], HUISINGA, BEST, ROITZSCH, SCHÜTTE and CORDES [1999]).

Whenever the discrete time Markov process $X_n = \{X_n\}_n \in \mathbb{N}$ is ergodic w.r.t. μ, and a statistically representative realization of X_n is available, we may also combine (A1) and (A2):

EXAMPLE 6.3.1. Let $x_0, \ldots, x_N$ denote a sequence of sampling points obtained from a realization of the discrete time Markov process X_n. Then

$$S_{kl} \approx S_{kl}^{(N)} = \frac{\sum_{j=1}^{N} \chi_k(x_j) \cdot \chi_l(x_{j+1})}{\sum_{j=1}^{N} \chi_k(x_j)^2}, \tag{6.3.1}$$

where convergence is guaranteed for μ-a.e. initial points x_0 by conditions (C1) and (C2) and the law of large numbers (MEYN and TWEEDIE [1993]).

6.4. Trapping problem

The *rate of convergence* of $S_{kl}^{(N)} \to S_{kl}$ depends on the smoothness of the partition functions χ_k as well as on the mixing properties of the Markov chain X_n (LEZAUD [2001]). The latter property is crucial here: The convergence is geometric with a rate constant $\lambda_1 - \lambda_2 = 1 - \lambda_2$ where λ_2 denotes the second largest eigenvalue (in modulus). That is, in case of metastability and thus λ_2 being very close to $\lambda_1 = 1$, we will have dramatically slow convergence. As this is the main problem for all approaches to biomolecular dynamics and statistics, this is also a bottleneck of the transfer operator approach. An entire bunch of the literature aims at tackling this problem that is often called the *trapping problem* (BERNE and STRAUB [1997], FERGUSON, SIEPMANN and TRUHLAR [1999]). In the framework of the transfer operator approach presented herein, A Fischer recently designed a hierarchical approach tailored to accelerate convergence called *uncoupling-coupling* (UC) (FISCHER [2000], FISCHER, SCHÜTTE,

DEUFLHARD and CORDES [2002]). Its *key assumption* is the following property of all stochastic systems designed to sample the canonical ensemble: Decreasing temperature induces stronger metastabilities, while metastability vanishes for large temperatures. Thus heating can help to reduce trapping which reappears if the system is annealed to lower temperatures. The uncoupling-coupling approach has been designed to circumvent this problems by combining the idea of domain decomposition with bridge sampling techniques.

In Section 5.3 a special bridge sampling technique called *Adaptive Temperature HMC* (ATHMC) (FISCHER, CORDES and SCHÜTTE [1998]) is used. Based on the Hybrid Monte Carlo procedure ATHMC reduces trapping problems by allowing to adapt the temperature during the sampling such that it can be increased to induce exits from metastable subsets. It was demonstrated that this procedure can be described by means of a generalized ensemble such that all parts of the sample (with different temperatures) can be reweighted to the temperature of interest (FISCHER, CORDES and SCHÜTTE [1998]).

6.5. Discretization in higher dimensions

Any discretization will suffer from the *curse of dimensionality* whenever it were based on uniform partition of all of the hundreds or thousands of degrees of freedom in a typical biomolecular system. Fortunately, chemical observations reveal that – even for larger biomolecules – the curse of dimensionality can be circumvented by exploiting the hierarchical structure of the dynamical and statistical properties of biomolecular systems: Firstly, only relatively few *conformational* or *essential degrees of freedom* are needed to describe the conformational transitions (AMADEI, LINSSEN and BERENDSEN [1993]). Furthermore, the canonical density has a rich spatial multiscale structure induced by the rich structure of the potential energy landscape. This structure induces a hierarchical cluster structure of the sampling data that can be identified and used to define a multilevel discretization adapted to the structures of the statistical data.

These observations give rise to a collection of approaches to the construction of *structure-adapted discretizations*:

Essential degrees of freedom. In the (low dimensional) subspace of essential degrees of freedom most of the positional fluctuations occur, while in the remaining degrees of freedom the motion can be considered as "physically constrained." Based on the available sampling, we may determine essential degrees of freedom either in the position space according to AMADEI, LINSSEN and BERENDSEN [1993] or in the space of internal degrees of freedom, for example, dihedral angles, by statistical analysis of circular data (HUISINGA, BEST, ROITZSCH, SCHÜTTE and CORDES [1999]). Either case is based on a principal component analysis of the sampling. As shown in HUISINGA, BEST, ROITZSCH, SCHÜTTE and CORDES [1999], this procedure may results in a enormous reduction of the number of degrees of freedom and, consequently, in a moderate number of subsets within the decomposition when discretizing the essential variables only. The principal component analysis is a linear approach to essential degrees of freedom. A characterization and identification of more general nonlinear essential degrees of freedom presently is a topic of further investigation.

Clustering algorithms. Another approach of decomposing the state space is based on clustering the sampling data by means of clustering algorithms (see, e.g., JAIN and DUBES [1988] and cited references). These methods cluster the sampling data according to structural similarity: The set of sampling points is partitioned into disjoint subsets with the property that two states belonging to the same subset are in some sense structural closer to each other than two states belonging to different subsets. A crucial question is the design of appropriate measures of structural similarity. In the biomolecular application context, these measures may either be based on the Cartesian coordinates or on the internal degrees of freedom. In contrast to the former the latter approach is invariant under rotations and translations of the entire molecule.

A novel promising approach to the above type of clustering problem uses self-organizing maps, a special kind of neural networks. Self-organizing maps allow to cluster the sampling data by assigning each sampling point to its nearest "neurons", each of them representing a subset of the decomposition. We have demonstrated its successful application to sampling data of biomolecular systems in GALLIAT, HUISINGA and DEUFLHARD [2000]. More advanced extensions, such as "box-neurons" and a hierarchical embedding, have recently been designed (GALLIAT and DEUFLHARD [2000], GALLIAT, DEUFLHARD, ROITZSCH and CORDES [2002]).

Whenever the statistical distribution allows to be clustered into a limited but significant number of clusters (e.g., a few thousand at most), these clusters can be used to define a statistics-adapted discretization as by fuzzy partitions of unity or by introducing discretization "boxes" such that each box contains a single cluster; for an application to biomolecular systems see HUISINGA, BEST, ROITZSCH, SCHÜTTE and CORDES [1999], WEBER and GALLIAT [2002].

7. Illustrative numerical experiments

We now want to illustrate the transfer operator approach to metastability in application to different dynamical descriptions that are based on the earlier introduced perturbed three-well potential (see Fig. 5.1, left). Further investigations, including dependence on parameters and discretization, can be found in HUISINGA [2001]. The theoretical justification of the transfer operator approach via conditions (C1) and (C2) for the below dynamical descriptions can be found in SCHÜTTE [1998], HUISINGA [2001] (Fig. 7.1).

Parameters. The perturbed three-well potential is defined by

$$V(q) = \frac{1}{100}(q^6 - 30q^4 + 234q^2 + 14q + 100 + 30\sin(17q) + 26\cos(11q)).$$

Below, we analyze the ensemble dynamics based on the Hamiltonian system with randomized momenta, the Langevin system and the Smoluchowski system. We choose $\beta = 2.0$ for the inverse temperature and $\gamma = 2.0$ for the friction constant. Then, σ is defined via the relation $\beta = 2\gamma/\sigma^2$ as stated in Eq. (3.2.6). For the observation time span we take $\tau = 1.0$.

Sampling. We sample the Markov processes given by the Hamiltonian system with randomized momenta with periodic boundary conditions on the positional state space

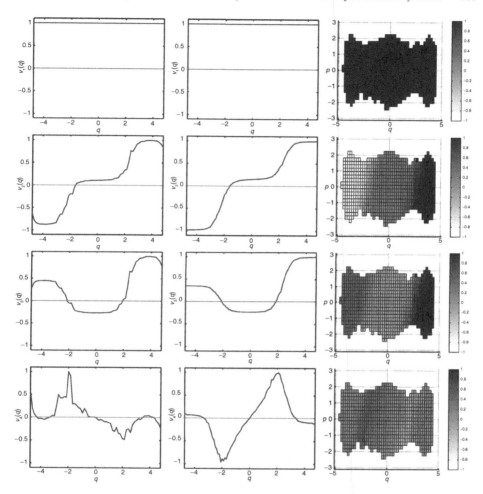

FIG. 7.1 The four dominant eigenfunctions of the propagator P_τ for different model systems. *Left*: Hamiltonian system with randomized momenta corresponding to the eigenvalues 1.0000, 0.9939, 0.9912, 0.6946 (from top to bottom). *Middle*: Smoluchowski equation for $\gamma = 2.0$ corresponding to the eigenvalues 1.0000, 0.9925, 0.9893, 0.6585. *Right*: Langevin equation for $\gamma = 2.0$ corresponding to the eigenvalues 1.0000, 0.9948, 0.9924, 0.6796.

$[-5, +5]$, the Langevin systems on $\mathbf{X} = \mathbf{R} \times \mathbf{R}$ and the Smoluchowski systems on $\mathbf{X} = \mathbf{R}$. The sampling length is $n = 300{,}000$ points.

Discretization. We discretize the positional state space into $n = 70$ equally sized intervals (boxes). For the Langevin dynamics, we additionally have to discretize the space of momenta. Examining the canonical density of momenta reveals that it is very unlikely to stay outside the interval $[-3, +3]$. Hence, we discretize the state space of momenta $\mathbf{R}$ by discretizing $[-3, +3]$ into 28 equally sized intervals and adding the two infinite intervals $(-\infty, -3)$ and $(3, \infty)$. Applying the discretization methods described in Section 5.1, we end up with an $n \times n$ stochastic transition matrix S with $n = 50 \times 30$ for the Langevin dynamics and $n = 70$ otherwise. For the Hamiltonian system with randomized momenta and the Smoluchowski dynamics, the transition matrix is self-adjoint.

Hamiltonian system with randomized momenta. Solving the eigenvalue problem for S yields:

λ_1	λ_2	λ_3	λ_4	λ_5	λ_6	...
1.000	0.9939	0.9912	0.6946	0.6481	0.5952	...

Application of our identification algorithm yields a decomposition of the state space $\mathcal{D} = \{C_1, C_2, C_3\}$ with $C_1 = \{q \le -2.13\}$, $C_2 = \{-2.13 < q \le 1.99\}$ and $C_3 = \{1.99 < q\}$. The statistical weights $\mu(C_k)$ within the positional canonical ensemble μ and the metastabilities $p(C_k, C_k)$ are given by the following table:

Metastable subset	C_1	C_2	C_3
statistical weight	0.2018	0.6979	0.1003
metastability	0.9883	0.9938	0.9800

The essential statistical behavior, that is, the probability of transitions between the metastable subsets, is described by the coupling matrix $C = (c_{jk})_{j,k=1,2,3}$ with $c_{jk} = p(C_j, C_k)$. For our example, we obtain

$$C = \begin{pmatrix} 0.9883 & 0.0117 & 0 \\ 0.0034 & 0.9938 & 0.0029 \\ 0 & 0.0200 & 0.9800 \end{pmatrix}$$

Langevin system. Solving the eigenvalue problem for S yields:

λ_1	λ_2	λ_3	λ_4	λ_5	...
1.0000	0.9948	0.9924	0.6796	0.5365	...

Application of our identification algorithm yields a decomposition of the state space $\mathcal{D} = \{C_1, C_2, C_3\}$. The statistical weights $\mu(C_k)$ within the canonical ensemble μ and the metastabilities $p(C_k, C_k)$ are given by the following table:

mMtastable subset	C_1	C_2	C_3
statistical weight	0.1918	0.7012	0.1070
metastability	0.9887	0.9946	0.9851

(7.1)

Calculating the coupling matrix yields

$$C = \begin{pmatrix} 0.9887 & 0.0113 & 0 \\ 0.0031 & 0.9946 & 0.0023 \\ 0 & 0.0149 & 0.9851 \end{pmatrix}.$$

(7.2)

Smoluchowski system.   Solving the eigenvalue problem for S yields:

λ_1	λ_2	λ_3	λ_4	λ_5	λ_6	...
1.0000	0.9925	0.9893	0.6585	0.5496	0.4562	...

Application of our identification algorithm yields a decomposition of the state space $\mathcal{D} = \{C_1, C_2, C_3\}$ with $C_1 = \{q \leqslant -2.04\}$, $C_2 = \{-2.04 < q \leqslant 1.94\}$ and $C_3 = \{1.94 < q\}$. The statistical weights $\mu(C_k)$ within the positional canonical ensemble μ and the metastabilities $p(C_k, C_k)$ are given by the following table:

metastable subset	C_1	C_2	C_3
statistical weight	0.2083	0.6936	0.0981
metastability	0.9863	0.9926	0.9770

The coupling matrix $C = (c_{jk})_{j,k=1,2,3}$ is given by

$$C = \begin{pmatrix} 0.9863 & 0.0137 & 0 \\ 0.0041 & 0.9926 & 0.0033 \\ 0 & 0.0230 & 0.9770 \end{pmatrix}.$$

Determining a decomposition of the state space following the exit time approach and partitioning according to the zeros of the second or third eigenfunction v_2 and v_3, respectively, we obtain:

Second eigenfunction: $\mathcal{D} = \{C_1, C_2\}$ with $C_1 = \{q \leqslant -1.32\}$ and $C_2 = \{-1.32 < q\}$ corresponding to the exit rate $\Gamma(C_i) = -\log(\lambda_2)/\tau = 0.0076$.
Third eigenfunction: $\mathcal{D} = \{D_1, D_2, D_3\}$ with $D_1 = \{q \leqslant -2.04\}$, $D_2 = \{-2.04 < q \leqslant 1.94\}$ and $D_3 = \{1.94 < q\}$ corresponding to the exit rate $\Gamma(D_i) = -\log(\lambda_3)/\tau = 0.0107$.

Hence, in this case the decomposition induced by h_3 is identical to the decomposition obtained via the identification algorithm.

8. Application to biomolecular systems

In this section we demonstrate that the algorithmic strategy presented in Section 5.2 can be applied to identify biomolecular conformations even for large systems as, for instance, small biomolecules with hundreds of atoms. For large systems, we have to face two particular problems:
   (1) How to generate a sample of the stationary distribution in a high-dimensional space?
   (2) How to decompose the highly-dimensional state space in order to discretize the propagator?
We will address these problems in the following.

Analyzing a small biomolecule.   This section illustrates the performance of the algorithmic approach to the tri-ribonucleotide adenylyl(3'-5')cytidylyl(3'-5')cytidin (r(ACC)) model system in vacuum, see Fig. 8.1. Its physical representation is based on the GROMOS96 extended atom force field (VAN GUNSTEREN, BILLETER, EISING,

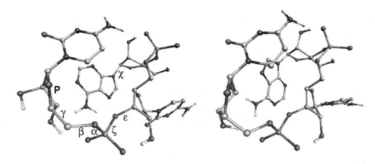

FIG. 8.1 Two representatives of different conformations of r(ACC). Left: The χ angle around the first gly-cosidic bond is in *anti* position (-175 degrees) and the terminal ribose pucker P is in C(3')endo C(2')exo conformation. Right: The χ angle is in *syn* position (19 degrees) and the terminal ribose in C(2')endo C(3') exo conformation. Visualization by AMIRA (KONRAD-ZUSE-ZENTRUM [2000]).

HÜNENBERGER, KRÜGER, MARK, SCOTT and TIRONI [1996]), resulting in $N = 70$ atoms, hence $\Omega = \mathbf{R}^{210}$ and $\Gamma = \mathbf{R}^{420}$. The internal fluctuations are modeled w.r.t. the Hamiltonian system with randomized momenta. For details see HUISINGA, BEST, ROITZSCH, SCHÜTTE and CORDES [1999].

The sampling of the canonical ensemble was generated using an adaptive temperature hybrid Monte Carlo method (FISCHER, CORDES and SCHÜTTE [1998]) at $T = 300$ K resulting in the sampling sequence $q_1, \ldots, q_{32000} \in \Omega$. The dynamical fluctuations within the canonical ensemble were approximated by integrating $M = 4$ short trajectories of length $\tau = 80$ fs starting from each sampling point. To facilitate transitions, analogous to the ATHMC sampling, the momenta were chosen according to the canonical ensemble of momenta corresponding to four different temperatures between 300–400 K and reweighted afterwards. This resulted in a total of $4 \times 32.000 = 128.000$ trajectories.

The configurational space was discretized using all four essential degrees of freedom, which were identified by means of a statistical analysis of the sampling data (see HUISINGA, BEST, ROITZSCH, SCHÜTTE and CORDES [1999]), resulting in $d = 36$ discretization subsets. Then the 36×36 stochastic transition matrix S was computed based on the 128.000 transitions taking the different weighting factors into account. The computation of the eigenvalues of S close to 1 yielded a cluster of eight eigenvalues with a significant gap to the remaining part of the spectrum, as shown in the following table:

k	1	2	3	4	5	6	7	8	9	$\ldots$
λ_k	1.00	0.99	0.98	0.97	0.96	0.95	0.93	0.90	0.81	$\ldots$

Finally, we computed conformations based on the corresponding eight eigenvectors of S via the identification algorithm presented in Section 5.3. We identified eight conformations; their statistical weights and metastabilities are shown in the following table:

conformations	C_1	C_2	C_3	C_4	C_5	C_6	C_7	C_8
statistical weight	0.11	0.01	0.12	0.03	0.32	0.04	0.29	0.10
metastability	0.99	0.94	0.96	0.89	0.99	0.95	0.98	0.96

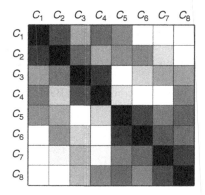

FIG. 8.2 Schematical visualization of the transition probabilities $p(\tau, C_i, C_j)$ between the conformation C_i (row) and C_j (column). The colors are chosen according to the logarithm of the corresponding entries: from $p \approx 0$ (light) to $p \approx 1$ (dark).

The transition probabilities between the different conformations are visualized schematically in Fig. 8.2. The matrix allows to define a hierarchy between the clusters: on the top level, there are two clusters, one consisting of the conformations $C_1, \ldots,$ C_4 and the other consisting of the conformations $C_5, \ldots, C_8$. This structure corresponds to the two 4×4 blocks on the diagonal. On the next level, each of these clusters splits up into two subclusters yielding four conformations $\{C_1, C_2\}, \{C_3, C_4\}, \{C_5, C_6\}, \{C_7, C_8\}$. On the bottom level, each cluster is further divided resulting in eight conformations.

9. Appendix

9.1. Some mathematical aspects of transfer operators

Consider the state space $\mathbf{X} \subset \mathbf{R}^m$ for some $m \in \mathbf{N}$ equipped with the Borel σ-algebra $\mathcal{A}$ on $\mathbf{X}$. The evolution of a single microscopic system is supposed to be given by a homogeneous Markov process $X_t = \{X_t\}_{t \in \mathbf{T}}$ in continuous or discrete time with $\mathbf{T} = \mathbf{R}_0^+$ or $\mathbf{T} = \mathbf{N}$, respectively. The motion of X_t is given in terms of a stochastic transition function p according to

$$p(t, x, A) = \mathbf{P}[X_{t+s} \in A | X_s = x],$$

for every $t, s \in \mathbf{T}$, $x \in \mathbf{X}$ and $A \subset \mathbf{X}$. The map $p : \mathbf{T} \times \mathbf{X} \times \mathcal{B}(\mathbf{X}) \to [0, 1]$ has the following properties
  (i) $x \mapsto p(t, x, A)$ is measurable for every $t \in \mathbf{T}$ and $\mathcal{A} \in \mathcal{B}(\mathbf{X})$;
  (ii) $A \mapsto p(t, x, A)$ is a probability measure for every $t \in \mathbf{T}$ and $x \in \mathbf{X}$;
  (iii) $p(0, x, \mathbf{X} \backslash \{x\}) = 0$ for every $x \in \mathbf{X}$;
  (iv) the Chapman–Kolmogorov equation

$$p(t + s, x, A) = \int_{\mathbf{X}} p(t, x, \mathrm{d}z) p(s, z, A)$$

holds for every $t, s \in \mathbf{T}$, $x \in \mathbf{X}$ and $A \subset \mathbf{X}$.

The relation between Markov processes and stochastic transition functions is one-to-one, that is, every homogeneous Markov process defines a stochastic transition functions satisfying properties (i) to (iv), and vice versa (MEYN and TWEEDIE [1993], Chapter 3).

To introduce the transfer operators consider the Banach spaces of equivalence classes of measurable functions

$$L^r(\mu) = \{u : \mathbf{X} \to \mathbf{C} : \int_{\mathbf{X}} |u(x)|^r \mu(\mathrm{d}x) < \infty\} \tag{9.1.1}$$

for $1 \le r < \infty$ and

$$L^\infty(\mu) = \{u : \mathbf{X} \to \mathbf{C} : \mu - \operatorname*{ess\,sup}_{x \in \mathbf{X}} |u(x)| < \infty\}$$

with corresponding norms $\|\cdot\|_r$ and $\|\cdot\|_\infty$, respectively. Due to Hölder's inequality we have $L^r(\mu) \subset L^s(\mu)$ for every $1 \le s \le r \le \infty$. The propagators or forward transfer operators $P^t : L^1(\mu) \to L^1(\mu)$ are defined by

$$P^t v(y) \mu(\mathrm{d}y) = \int_{\mathbf{X}} p(t, x, \mathrm{d}y) v(x) \mu(\mathrm{d}x),$$

while the backward transfer operators $T^t : L^\infty(\mu) \to L^\infty(\mu)$ are given by

$$T^t u(x) = \mathbf{E}_x[u(X_t)] = \int_{\mathbf{X}} u(y) p(t, x, \mathrm{d}y).$$

The assumed invariance of μ w.r.t. the stochastic transition function guarantees that the forward transfer operators is well-defined (REVUZ [1975], Chapter 4) and may be consider as acting on $L^r(\mu)$ for $1 \le r$. Moreover the backward transfer operator may be extended to any $L^r(\mu)$ with $1 \le r$.

Due to the properties of the stochastic transition function p both definitions define semigroups of Markov operators, that is, we especially have

$$T^t T^s = T^{t+s} \quad \text{and} \quad P^t P^s = P^{s+t},$$

and both operators, $A = T^t$ or $A = P^t$, conserve norm $\|Av\|_1 = \|A\|_1$ and positivity $Av \ge 0$ if $v \ge 0$. With the duality bracket defined in (5.1.6) the backward transfer operator is the adjoint of the propagator: $(P^t)^* = T^t$. Since $L^1(\mu)$ is a proper subset of the dual of $L^\infty(\mu)$, we have $P^t \subsetneq (T^t)$, hence P^t is not the adjoint of T^t. As a consequence, it is much easier to relate properties of P^t to T^t than vice versa.

Spectral properties. Consider a complex Banach space E with norm $\|\cdot\|$ and denote the spectrum[3] of a bounded linear operator $P : E \to E$ by $\sigma(P)$. For an eigenvalue $\lambda \in \sigma(P)$, the multiplicity of λ is defined as the dimension of the generalized eigenspace; see, for example, KATO [1995]. Eigenvalues of multiplicity 1 are called simple. The set of all eigenvalues $\lambda \in \sigma(P)$ that are isolated and of finite multiplicity

---

[3]For common functional analytical terminology see, e.g., DUNFORD and SCHWARTZ [1957], HEUSER [1986], KATO [1995], WERNER [1997].

is called the *discrete spectrum*, denoted by $\sigma_{\mathrm{discr}}(P)$. The *essential spectral radius* $r_{\mathrm{ess}}(P)$ of P is defined as the smallest real number, such that outside the ball of radius $r_{\mathrm{ess}}(P)$, centered at the origin, are only discrete eigenvalues, that is,

$$r_{\mathrm{ess}}(P) = \inf\{r \geq 0 : \lambda \in \sigma(\mathrm{P}) \text{ with } |\lambda| > r \text{ implies } \lambda \in \sigma_{\mathrm{discr}}(\mathrm{P})\}.$$

This definition of $r_{\mathrm{ess}}(P)$ is unusual in the sense that it does not involve any definition of the essential spectrum. This is owed to the surprising fact that although there are many different definitions of essential spectra, the associated spectral radii coincide (HUISINGA [2001], LEBOW and SCHECHTER [1971]) and are therefore somehow independent of the specific definition of essential spectra.

PROPOSITION 9.1.1 (HUISINGA [2001]).   *Consider the* Lebesgue decomposition *of the stochastic transition function*

$$p(x, \mathrm{d}y) = p_a(x, y)\mu(\mathrm{d}y) + p_s(x, \mathrm{d}y),$$

where p_a and p_s represent the absolutely continuous and the singular part w.r.t. μ, respectively. Assume that

(i) *the inequality*

$$\int_{\mathbf{X}} \int_{\mathbf{X}} p_a(x, y)^2 \mu(\mathrm{d}x)\mu(\mathrm{d}y) < \infty$$

holds, and

(ii) *there exists some $\eta < 1$ such that*

$$\eta = \sup p_s(x, \mathbf{X}) = 1 - \inf \int_{\mathbf{X}} \mathrm{p}_a(\mathrm{x}, \mathrm{y})\mu(\mathrm{d}y)$$

for μ-a.e. $x \in \mathbf{X}$.

Then, the essential spectrum is uniformly bounded away from 1, *more precisely, we have $r_{\mathrm{ess}}(P) < \sqrt{\eta} < 1$. In particular, condition* (C1) *is fulfilled.*

As stated in Section 5.2 the probabilistic interpretation of condition (C2) is that the Markov process admits a *unique* invariant probability measure and is *aperiodic*, hence does not show any periodic behavior. Consequently, the state space $\mathbf{X}$ can neither be decomposed into noninteracting (invariant) subsets nor into so-called cyclic subset such that the Markov process cycles with probability 1 along these cyclic subset. To interpret condition (C1), consider the Lebesgue decomposition of the stochastic transition function p. For simplicity assume that the invariant measure μ is absolutely continuous w.r.t. the Lebesgue measure as, for instance, the measure induced by the canonical ensemble. Then, as shown in Proposition 9.1.1, the essential spectral radius $r_{\mathrm{ess}}(P)$ is related to regularity conditions on the stochastic transition function. The essential spectral radius is close to one, if (a) the singular part p_s is close to one, or (b) the absolutely continuous part p_a shows a singularity like behavior, for example, grows too fast at infinity. There is a rapidly growing literature on testable conditions, which imply that neither (a) nor (b) are valid and thus the spectral radius is strictly bounded away from 1. We only want to make the following remarks: Firstly, for the special case of the deterministic Hamiltonian system, the absolutely continuous part

p_a vanishes such that $r_{ess}(P) = 1$ via condition (a). Then, for the case of Langevin or Smoluchowski dynamics with smooth potentials, the singular part vanishes and the validity of condition (C1) depends on the growths of p_a at infinity only; therefore Lyapunov conditions on p_a suffice to prove (C1) in these cases. Finally, we can safely exclude problems with condition (a) whenever the transition function p allows to reach an open set with positive measure w.r.t. μ from any point $x \in \mathbf{X}$. This shows that for the Hamiltonian system with randomized momenta we only have to exclude that there is an initial position q from which τ-trajectories with arbitrary initial momentum always end up in some discrete set of positions. This typically is not the case such that condition (a) will typically be no problem for the Hamiltonian system with randomized momenta whereas it is for the deterministic Hamiltonian system, see SCHÜTTE [1998] for details.

Implications of $\mathcal{V}$-uniform ergodicity. The main body of ergodic theory and spectral theory of transfer operators is based upon another vector space setting developed in MEYN and TWEEDIE [1993], Chapter 16. Let $\mathcal{V} : \mathbf{X} \to [1, \infty)$, finite a.e., be a given Lyapunov function, and denote by $L^\infty(\mathcal{V})$ the vector space of measurable functions $h: \mathbf{X} \to \mathbf{C}$ satisfying

$$\| h \|_{\mathcal{V}} = \sup_{x \in \mathbf{X}} \frac{|h(x)|}{\mathcal{V}(x)} < \infty.$$

Let $\| \cdot \|_{\mathcal{V}}$ also denote the operator norm induced by this $\mathcal{V}$-norm. The $\mathcal{V}$-norm on measures (as it is used in Section 5.3) is defined by

$$\| \nu \|_{\mathcal{V}} = \sup_{|v| \leqslant \mathcal{V}} \left| \int_{\mathbf{X}} v(x) \nu(dx) \right|,$$

where $|v| \leqslant \mathcal{V}$ is understood to hold pointwise for every measurable function v and every $x \in \mathbf{X}$; here $\mathcal{V}$ needs to be integrable. For the special case $\mathcal{V} \equiv 1$, the $\mathcal{V}$-norm coincides with the total variation norm $\| \cdot \|_{TV}$.

Consider the backward transfer operator acting on the function space $L^\infty(\mathcal{V})$. We have seen in Section 5.4 that the assumption of $\mathcal{V}$-ergodicity is crucial for our approach. Then, from Theorem 5.2 of DOWN, MEYN and TWEEDIE [1995] and the results of HUISINGA, MEYN and SCHÜTTE [2003] it follows

THEOREM 9.1.1. *If the stochastic transition function p is $\mathcal{V}$-uniformly ergodic then,*
  (1) *there is an invariant probability measure μ, and the semigroup T^t strongly converges in $L_{\mathcal{V}}^\infty$;*
  (2) *T^t admits a spectral gap in $L^\infty(\mathcal{V})$, that is, the set $\sigma(T^t) \cap \{z \in \mathbf{C} : |z - 1| \leqslant \varepsilon\}$ is finite for sufficiently small $\varepsilon > 0$;*
  (3) *for any $B \in \mathcal{B}$ with $\mu(B) > 0$, there exists $\bar{\Gamma}_B > 0$ and $b < \infty$ such that*

$$\mathbf{P}_x\{\%B \geqslant t\} \leqslant b\mathcal{V}(x)e^{-\bar{\gamma}_B t}, x \in \mathbf{X}. \tag{9.1.2}$$

The last properties nicely illustrates that there is a deep connection between $\mathcal{V}$-ergodicity, the existence of a spectral gap, and the exponential decay of the exit time distribution. This is important for the definition of exit rates in the next subsection.

9.2. Definition of exit rates

Consider the Markov process X_t with transition function p. As outlined above, p defines the semigroup of transfer operators via $T^t u(x) = E_x [u(X_t)]$. We now turn our attention to some open connected set B and define the *restricted process on B* induced by the process X_t via the semigroup of restricted transfer operators

$$T_B^t u(x) = E_x[u(X_t)1_X(\%_{B^c} \geq t)],$$

where $_{B^c}$ denotes the exit time from B^c as introduced in (4.2.1). Thus, the new process lives on B only; considering the original process exits from B are killed. The family $\{T_B^t\}$ is a semigroup of positive operators that in general are no longer Markov operators. We define the spectral radius for the family $\{T_B^t\}$ by

$$r(\{T_B^t\}) = \lim_{t \to \infty} (\| T_B^t \|_\mathcal{V})^{1/t},$$

and define the $\mathcal{V}$-*exit rate* from the set B by

$$\Gamma(B) = - \log r(\{T_B^t\}).$$

Now, one may raise the question whether the so-defined exit rate really measures the asymptotic decay of the distribution of exit times in the sense of Section 4.2. The answer is yes: If for the set B the exit time distribution decays due to (4.2.2) for some $\Gamma > 0$ that is constant in B, then $\| T^t \|_\mathcal{V}$ decays asymptotically as c exp $(-\Gamma t)$ such that $\Gamma(B) = - \log r(\{T_B^t\}) = \Gamma$. The other way around and in terms of a rigorous statement for the situation described in Section 5.4, we have:

THEOREM 9.2.1. *Suppose that the conditions of Theorem 5.4.2 hold implying that the subset C is metastable with exit rate* $\Gamma > 0$. *Then there exists* $\delta_0 > 0$ *such that for all s, T > 0 the conditional distribution of exit times satisfies*

$$F_x(s, T) = e^{-\Gamma s} \left[\frac{1 + O(G_x(s))}{1 + O(e^{-T\delta_0} G_x(s))} \right] (T \to \infty)$$

With $G_x(s) = e^{-\delta_0 s} \mathcal{V}(x)/h(x)$.

In HUISINGA, MEYN and SCHÜTTE [2003], δ_0 is some well-defined computable constant and it is shown that in typical situations we have $\delta_0 \gg \gamma$.

References

AMADEI, A., A.B.M. LINSSEN and H.J.C. BERENDSEN (1993), Essential dynamics of proteins, *Proteins* **17**, 412–425.

BERNE, B.J., CICCOTTI, G., AND COKER, D.F. eds., (1998). *Classical and Quantum Dynamics in Condensed Phase Simulations* (World Scientific, Singapore).

BERNE, B.J. and J.E. STRAUB (1997), Novel methods of sampling phase space in the simulation of biological systems, *Curr. Opin. Struct. Biol.* **7**, 181–189.

BOLHUIS, P.G., DELLAGO, C., CHANDLER, D. and GEISSLER, P. (2001), Transition path sampling: Throwing ropes over mountain passes, in the dark, *Ann. Rev. Phys. Chem.* (in press).

BOND, S.D., B.B.L. BENEDICT and J. LEIMKUHLER (1999), The Nosé–Poincaré method for constant temperature molecular dynamics, *JCP* **151**(1), 114–134.

BOVIER, A., M. ECKHOFF, V. GAYRARD and M. KLEIN (2001), Metastability in stochastic dynamics of disordered mean-field models, *Probab. Theor. Rel. Fields* **119**, 99–161.

CHANDLER, D. (1998), Finding transition pathways: Throwing ropes over rough mountain passes, in the dark, in: B. Berne, D. Ciccotti and Coker, eds., *Classical and Quantum Dynamics in Condensed Phase Simulations* (World Scientific, Singapore) pp. 51–66.

DAVIES, E.B. (1982), Metastable states of symmetric Markov semigroups I, *Proc. London Math. Soc.* **45**(3), 133–150.

DELLNITZ, M. and O. JUNGE (1998), An adaptive subdivision technique for the approximation of attractors and invariant measures, *Comput. Visual. Sci.* **1**, 63–68.

DELLNITZ, M. and O. JUNGE (1999), On the approximation of complicated dynamical behavior, *SIAM J. Num. Anal.* **36**(2), 491–515.

DEUFLHARD, P. and F. BORNEMANN (1994), *Numerische Mathematik II* (de Gruyter).

DEUFLHARD, P., DELLNITZ, M., JUNGE, O. and SCHÜTTE, C. (1999), Computation of essential molecular dynamics by subdivision techniques, in: P. Deuflhard, J. Hermans, B. Leimkuhler, A.E. Mark, S. Reich and R.D. Skeel, eds., *Computational Molecular Dynamics: Challenges, Methods, Ideas, Lecture Notes in Computational Science and Engineering*, Vol. 4 (Springer) pp. 98–115.

DEUFLHARD, P., T. FRIESE and F. SCHMIDT (1997), *A nonlinear multigrid eigenproblem solver for the complex Helmholtz equation* Preprint SC-97-55, (Konrad-Zuse-Zentrum, Berlin). Available via http://www.zib .de/bib/pub/pw/.

DEUFLHARD, P., HERMANS, J., LEIMKUHLER, B., MARK, A.E., REICH, S., AND SKEEL, R.D. eds., (1999). *Computational Molecular Dynamics: Challenges, Methods, Ideas, Lecture Notes in Computational Science and Engineering*, Vol. 4 (Springer).

DEUFLHARD, P. and A. HOHMANN (1993), *Numerische Mathematik I* (de Gruyter, Berlin).

DEUFLHARD, P., W. HUISINGA, A. FISCHER and C. SCHÜTTE (2000), Identification of almost invariant aggregates in reversible nearly uncoupled Markov chains, *Lin. Alg. Appl.* **315**, 39–59.

DOWN, D., S.P. MEYN and R.L. TWEEDIE (1995), Exponential and uniform ergodicity of Markov processes, *Ann. Prob.* **23**, 1671–1691.

DUNFORD, N. and J.T. SCHWARTZ (1957), *Linear Operators, Part I: General Theory, Pure and Applied Mathematics*, Vol. VII (Interscience, New York).

RAN, E.W.W. and E. VANDEN-EIJNDEN (2002), Probing multiscale energy landscapes using the string method, *Phys. Rev. Lett.* (submitted).

ELBER, R. and M. KARPLUS (1987), Multiple conformational states of proteins: A molecular dynamics analysis of Myoglobin, *Science* **235**, 318–321.

FERGUSON, D.M., SIEPMANN, J.I., and TRUHLAR, D.G. eds., (1999). *Monte Carlo Methods in Chemical Physics, Advances in Chemical Physics*, Vol. 105 (Wiley, New York).

FISCHER, A. (2000), *An Uncoupling–Coupling Technique for Markov chains Monte Carlo Methods* Report 00–04, (Konrad-Zuse-Zentrum, Berlin).

FISCHER, A., F. CORDES and C. SCHÜTTE (1998), Hybrid Monte Carlo with adaptive temperature in a mixed-canonical ensemble: Efficient conformational analysis of RNA, *J. Comput. Chem.* **19**, 1689–1697.

FISCHER, A., SCHÜTTE, C., DEUFLHARD, P. and CORDES, F. (2002), Hierarchial uncoupling–coupling of metastable conformations, in: Schlick, T. and H.H. Gan , eds., *Computational Methods for Macromolecules: Challenges and Applications, Proceedings of the 3rd International Workshop on Algorithms for Macromolecular Modelling, New York, Oct. 12–14, 2000, Lecture Notes in Computational Science and Engineering*, Vol. 24 (Springer).

FRAUENFELDER, H. and B.H. MCMAHON (2000), Energy landscape and fluctuations in proteins, *Ann. Phys.* (Leipzig) **9**(9–10), 655–667.

FRAUENFELDER, H., P.J. STEINBACH and R.D. YOUNG (1989), Conformational relaxation in proteins, *Chem. Soc.* **29A**, 145–150.

GALLIAT, T. and P. DEUFLHARD (2000), *Adaptive Hierarchical Cluster Analysis by Self-Organizing Box Maps* Report SC-00–13, (Konrad-Zuse-Zentrum, Berlin).

GALLIAT, T., P. DEUFLHARD, R. ROITZSCH and F. CORDES (2002), Automatic identification of metastable conformations via self-organized neural networks, In: T. Schlick and H.H. Gan, eds., *Computational Methods for Macromolecules: Challenges and Applications Proceedings of the 3rd International Workshop on Algorithms for Macromolecular Modelling, New York, Oct. 12–14, 2000, Lecture Notes in Computational Science and Engineering*, Vol. 24 (Springer).

GALLIAT, T., W. HUISINGA and P. DEUFLHARD (2000), Self-organizing maps combined with eigenmode analysis for automated cluster identification, In: H. Bothe and R. Rojas, eds., *Neural Computation* (ICSC Academic Press), pp. 227–232.

GILKS, W., RICHARDSON, S., and SPIEGELHALTER, D. eds., (1997). *Markov chain Monte-Carlo in Practice* (Chapman and Hall, London).

GRUBMÜLLER, H. (1995), Predicting slow structural transitions in macromolecular system: Conformational flooding, *Phys. Rev. E.* **52**, 2893–2906.

HEUSER, H. (1986), *Funktionalanalysis* (Teubner, Stuttgart).

HUBER, T., A.E. TORDA and W.F. VAN GUNSTEREN (1994), Local elevation: A method for improving the searching properties of molecular dynamics simulation, *JCAMD* **8**, 695–708.

HUISINGA, W. (2001), Metastability of Markovian systems: A transfer operator approach in application to molecular dynamics, PhD thesis, (Free University Berlin).

HUISINGA, W., C. BEST, R. ROITZSCH, C. SCHÜTTE and F. CORDES (1999), From simulation data to conformational ensembles: Structure and dynamic based methods, *J. Comp. Chem.* **20**(16), 1760–1774.

HUISINGA, W., MEYN, S. and SCHÜTTE, C. (2003), Phase transitions and metastability in Markovian and molecular systems, *Ann. Appl. Probab.* (accepted).

HUISINGA, W. and B. SCHMIDT (2002), *Metastability and Dominant Eigenvalues of Transfer Operators*, (in preparation).

JAIN, A.K. and R.C. DUBES (1988), *Algorithms for Clustering Data* (Prentice Hall, New Jersey), Advanced reference series edition.

KARATZAS, I. and S.E. SHREVE (1991), *Brownian Motion and Stochastic Calculus, Graduate Texts in Mathematics* (Springer, New York).

KATO, T. (1995), *Perturbation Theory for Linear Operators* (Springer, Berlin), reprint of the 1980 edn.

KLEEMAN, R. (2001), Measuring dynamical prediction utility using relative entropy, *J. Atmos. Sci.* (accepted).

KONRAD-ZUSE-ZENTRUM, (2000), Amira – advanced visualization, data analysis and geometry reconstruction, user's guide and reference manual (Konrad-Zuse-Zentrum für Informationstechnik Berlin (ZIB), (Indeed-Visual Concepts GmbH and TGS Template Graphics Software Inc).

LASOTA, A. and M.C. MACKEY (1994), *Chaos, Fractals and Noise, Applied Mathematical Sciences*, Vol. 97, 2nd edn. (Springer, New York).

LEBOW, A. and M. SCHECHTER (1971), Semigroups of operators and measures of noncompactness, *J. Funct. Anal.* **7**, 1–26.

LEHOUCQ, R.B., D.C. SORENSEN and C. YANG (1998), *ARPACK User's Guide: Solution of Large Eigenvalue Problems by Implicit Restartet Arnoldi Methods* (Rice University, Houston).

LEZAUD, P. (2001), Chernoff and Berry–Esséen inequalities for Markov processes, *ESIAM: P & S* **5**, 183–201.

MATTINGLY, J., A.M. STUART and D.J. HIGHAM (2003), Ergodicity for SDEs and approximations: Locally Lipschitz vector fields and degenerated noise, to appear in, *Stoch. Proc. Appl.*

MEYN, S. and R. TWEEDIE (1993), *Markov Chains and Stochastic Stability* (Springer, Berlin).

NIENHAUS, G.U., J.R. MOURANT and H. FRAUENFELDER (1992), Spectroscopic evidence for conformational relaxation in Myoglobin, *PNAS* **89**, 2902–2906.

OLENDER, R. and R. ELBER (1996), Calculation of classical trajectories with a very large time step: Formalism and numerical examples, *J. Chem. Phys.* **105**, 9299–9315.

REVUZ, D. (1975), *Markov Chains* (North-Holland, Amsterdam, Oxford).

RISKEN, H. (1996), *The Fokker–Planck Equation*, 2nd edn. (Springer, New York).

ROBERTS, G.O. and J.S. ROSENTHAL (1997), Geometric ergodicity and hybrid Markov chains, *Elect. Comm. in Probab.* **2**, 13–25.

SCHÜTTE, C. (1998), Conformational dynamics: modelling, theory, algorithm, and application to biomolecules, Habilitation Thesis, Fachbereich Mathematik und Informatik, Freie Universität Berlin.

SCHÜTTE, C., A. FISCHER, W. HUISINGA and P. DEUFLHARD (1999), A direct approach to conformational dynamics based on hybrid Monte Carlo, Special Issue on Computational Biophysics, *J. Comput. Phys.* **151**, 146–168.

SCHÜTTE, C. and W. HUISINGA (2000), On conformational dynamics induced by Langevin processes, in: B. Fiedler, K. Gröger and J. Sprekels, eds., *EQUADIFF 99 – International Conference on Differential Equations*, Vol. 2 (World Scientific, Singapore), pp. 1247–1262.

SCHÜTTE, C., W. HUISINGA and P. DEUFLHARD (2001), Transfer operator approach to conformational dynamics in biomolecular systems, in: B. Fiedler, ed., *Ergodic Theory, Analysis and Efficient Simulation of Dynamical Systems* (Springer), pp. 191–223.

SINGLETON, G. (1984), Asymptotically exact estimates for metastable Markov semigroups, *Quart. J. Math. Oxford* **35**(2), 321–329.

VAN GUNSTEREN, W.F., S.R. BILLETER, A.A. EISING, P.H. HÜNENBERGER, P. KRÜGER, A.E. MARK, W.R.P. SCOTT and I.G. TIRONI (1996), *Biomolecular Simulation: The GROMOS96 Manual and User Guide* (vdf Hochschulverlag AG, ETH Zürich).

WEBER, M. and T. GALLIAT (2002), *Characterization of Transition States in Conformational Dynamics Using Fuzzy Sets,* Technical Report 02–12, (Konrad-Zuse-Zentrum (ZIB), Berlin).

WERNER, D. (1997), *Funktionalanalysis*, 2nd edn. (Springer, Berlin).

ZEWAIL, A.H. (1995), Der Augenblick der Molekülbildung, *Spektrum der Wissenschaft, Digest 2: Moderne Chemie* 18–26.

ZEWAIL, A.H. (1996), Femtochemistry: Recent progress in studies of dynamics and control of reactions and their transition state, *J. Phys. Chem.* **100** (31), 12701–12724.

ZHOU, H.X., S.T. WLODEC and J.A. MCCAMMON (1998), Conformation gating as a mechanism for enzyme specificity, *Proc. Natl. Acad. Sci. USA* **95**, 9280–9283.

Numerical Methods for Molecular Time-Dependent Schrödinger Equations – Bridging the Perturbative to Nonperturbative Regime

André D. Bandrauk, Hui-Zhong Lu

Laboratoire de Chimie Théorique, Faculté des Sciences, Université de Sherbrooke, Canada

1. Introduction

Current laser technology has progressed significantly from ultrashort, that is, duration on the attosecond (10^{-18} s) time scale (HENTSCHEL et al. [2001]) and intensities exceeding that of electric fields internal to atoms and molecules. Recent technological advances in ultrafast optics have permitted the generation of light wavepackets comprising only a few oscillation cycles of electric and magnetic fields (BRABEC and KRAUSZ [2000]). This revolution in laser technology is expected to also bring a revolution in experimental sciences (HAROCHE [2000]). To appreciate this new technological revolution, one has to recall the scales in atomic units (where $h/2\pi = e = m_e = 1$) the magnitude of the following important physical parameters. The period of circulation of an electron in the ground state of the hydrogen (H) atom is:

$$\tau_H(\text{a.u.}) = 24.6 \times 10^{-18}\text{s} = 24.6 \text{ attoseconds (atts)}. \tag{1.1}$$

The electric field E_H and corresponding intensity I in that orbit are:

$$E_H(\text{a.u.}) = 5 \times 10^9 \text{ V/cm}, \quad I_H(\text{a.u.}) = cE^2/8\pi = 3 \times 10^{16} \text{ W/cm}^2. \tag{1.2}$$

Computational Chemistry
Special Volume (C. Le Bris, Guest Editor) of
HANDBOOK OF NUMERICAL ANALYSIS
P.G. Ciarlet (Editor)

Due to extreme temporal and spatial confinement using current compression techniques, pulse energies in peak intensities higher than 10^{15} W/cm^2 are readily available with table top lasers (BRABEC and KRAUSZ [2000]). The resulting electric field strength approach and can exceed that of the static Coulomb field experienced by outer-shell electrons in atoms and molecules. As a consequence the laser field is strong enough to suppress the binding Coulomb potential in atoms and trigger multiphoton optical-field ionization (KELDYSH [1965], CORKUM [1993]).

Much of our understanding of intense laser field-atom interaction physics has been documented in a book where new nonperturbative phenomena such as Above Threshold Ionization (ATI; the excess absorption of laser photons after ionization of an electron), high order harmonic generation, HOHG (the emission of a large number of laser photons after recollision of an electron with its parent or neighboring ion), have been treated theoretically and numerically from the perturbative to nonperturbative regime, as various approximations to appropriate time-dependent Schroedinger equations (TDSEs; GAVRILA [1992], EHLOTZKY [2001]). In the low frequency regime, especially with current Ti/Sap. lasers of wavelength $\lambda = 800$ nm, quasistatic tunneling models (KELDYSH [1965], CORKUM [1993], GAVRILA [1992], EHLOTZKY [2001]) have helped bridge the perturbative to nonperturbative regime in atom laser physics.

Molecules present a new challenge, due to the presence of extra degrees of freedom from nuclear motion. As an example, if one considers the frequency of vibration of the H_2 molecule, $\nu_{H_2} = 4000 \text{cm}^{-1}$ (HERTZBERG [1951]), this corresponds to a vibration period of:

$$\tau_{H_2} = (\nu_{H_2} c)^{-1} = 8.3 \times 10^{-15} \text{s} = 8.3 \text{ femtoseconds (fs)}, \qquad (1.3)$$

where $c = 3 \times 10^{10}$ cm/s is the velocity of light. This is to be compared to the Ti/Sap. laser ($\lambda = 800$ nm) frequency of 2.7 fs, CO_2 laser ($\nu = 1000 \text{ cm}^{-1}$) period of 30 fs. One finds therefore that the fastest atomic motion time, τ_p of a proton in molecules, is readily in the laser field frequency regime, that is,

$$3 \leq \tau_p \leq 30 \text{ fs}. \qquad (1.4)$$

The proton is indeed the most elusive particle with the shortest time scale for nuclear motion in molecules, yet it is probably the most important atom in chemistry and biology (e.g., DNA code). To date no one has been able to image its motion. Recent theoretical calculations based on exact TDSEs with intense ultrashort laser pulses, $\tau_L < 10$ fs, are proposing such pulses as the ideal tool for dynamic imaging of ultrafast nuclear motion (BANDRAUK and CHELKOWSKI [2001]).

Much of theoretical work in molecules has been focused in the past 50 years on *ab initio* electronic structure calculations, in the Born–Oppenheimer approximation, that is, assuming static nuclei (DEFRANCESCHI and LE BRIS [2001]). As indicated above, nuclear motion involves introducing dynamics into the structure calculations. Using Born–Oppenheimer or adiabatic molecular potentials obtained from *ab initio* calculations, considerable progress in describing nonperturbative multiphotons molecular processes has been achieved using a dressed state molecular representation. This has been summarized in various works (BANDRAUK [1994a], BANDRAUK [1994b]) and has led to the concept of laser induced avoided crossings of molecular potentials leading to laser induced molecular potentials (LIMPs), that is, new molecular potentials created by radiative interactions, recently confirmed experimentally (WUNDERLICH, KOBLER, FIGGER and HÄNSCH [1997]) and Above Threshold Dissociation (ATD; excess laser photon absorption after photodissociation), the analogue of ATI in atoms.

Intensities approaching the atomic unit of field strength (BRABEC and KRAUSZ [2000]), leads to considerable ionization. One thus has to treat in general dissociation-ionization processes where the ionization time scale, τ_I, the nuclear motion time scale τ_p and laser time scale τ_L become comparable, that is, one is in the few femtosecond regime where separability of electronic, nuclear and radiative processes is not possible. Thus the usual adiabatic (Born–Oppenheimer) methods of *ab initio* quantum chemistry (DEFRANCESCHI and LE BRIS [2001]) are no longer valid. Furthermore one has to deal with initial bound state (discrete spectra) to continuum state (continuum spectra) transitions. Such problems lead to the search for new numerical approaches, well beyond current quantum chemistry methods. To date only the one-electron, simplest molecule, H_2^+ has been solved exactly numerically using large grids and super-computers (CHELKOWSKI, FOISY and BANDRAUK [1998], CHELKOWSKI, CONJUSTEAU, ATABEK and BANDRAUK [1996]). Such simulations have helped discover new nonperturbative effects in molecules subjected to intense laser pulses such as Charge Resonance Enhanced Ionization (CREI; ionization induced by electron transfer; CHELKOWSKI and BANDRAUK [1995], BANDRAUK [2000]), Laser Induced Electron Diffraction (LIED; ZUO, BANDRAUK and CORKUM [1996]) and are leading the way to suggest laser control of dissociative-ionization (ZUO and BANDRAUK [1996], CHELKOWSKI, ZAMOJSKI and BANDRAUK [2001], BANDRAUK and YU [1998], BANDRAUK, FUJIMURA and GORDON [2002]).

In summary, we would like to emphasize once again the importance of the new ultrashort laser technology to the seminal problem of laser control and manipulation of molecules. Optimization of the laser parameters, that is, amplitude and phase, to optimize yields of interest to the chemical community has led to a new branch of photochemistry, *coherent control* of nuclear motion (see BRUMER and SHAPIRO [1994], RABITZ, TURINICI and BROWN [2003]). The present chapter focuses on a complete treatment of laser–molecule interaction, in order to arrive eventually at the laser control and manipulation of both *electronic* and *nuclear* states and their spaces, which become nonseparable, that is, nonadiabatic processes dominate, as laser intensities are increased. Of course a complete treatment of laser–molecule interaction will require investigating numerically solutions of the full coupled TDSE and Maxwell's equations (BANDRAUK [1994b]) and finally the Time-Dependent Dirac Equation (TDDE; THALLER [1992]). TDSEs are parabolic partial differential equations, PDEs, whereas Maxwell equations and TDDEs are hyperbolic PDEs. Clearly laser–molecule interaction problems offer new numerical challenges for the current accurate description of multiphotons processes in molecules from the perturbative to nonperturbative regimes. In the present work we shall focus on methods developed by us to treat exactly one electron problems beyond the standard adiabatic Born–Oppenheimer approximation. Extension to many-electrons problems will most probably require progress in *time-dependent density functional theory* (TDDFT) methods, (DREIZLER and GROSS [1990]), where *nonlocal* electron potentials occurring in *ab initio* quantum chemistry methods are replaced by effective local potentials. Then the one-electron numerical methods described in the present work will be easily implementable and applicable to describe completely laser–molecule interactions and the accompanying electron–nuclear dynamics for many electron systems.

2. Gauges and representations

Previous 3D calculations of intense laser–field matter interactions have relied on time-dependent Hartree–Fock (TDHF) approximations developed by KULANDER [1987] for atomic problems and applied also to H_2 (KRAUSE, SCHAFFER and KULANDER [1991]) using numerical grid methods. We have previously applied 3D Cartesian finite elements for TDHF calculations of H_2 and H_3^+ in intense laser fields (YU and BANDRAUK [1995]). Finite element bases are very flexible and are able to span bound and continuum states simultaneously. In order to go beyond TDHF method and 1D models, progress is required in exact nonperturbative numerical solution of complete TDSEs for multielectron non-Born–Oppenheimer simulations. Furthermore molecule-radiation interactions can be represented by Hamiltonians in different gauges (GAVRILA [1992], BANDRAUK [1994b]), with each gauge having advantages and disadvantages, both numerical and physical. Thus to solve molecular TDSEs in the field of an intense laser pulse, most current numerical methods use the length gauge $(\vec{E} \cdot \vec{r})$ or the velocity gauge (Coulomb, $\vec{A} \cdot \vec{p}$) with *Euclidean* stationary coordinates systems. In the length gauge, the molecule-field interaction appears as a *potential* term whereas in the velocity gauge, this interaction appears as a *convection* term. In classical fluid mechanics, it is customary to use a *Lagrangian* (moving) coordinate system which removes the convection term so that the conservation of physical relevant observables is automatically assured (GERHART, GROSS and HOCHSTEIN [1992], HAUSBO [2000]).

As we show below, the TDSE in the Lagrangian system is a generalization for molecules of the space translation (ST) method used in intense high frequency atom-laser interaction theories. This was originally introduced by PAULI and FIERZ [1938] for vacuum corrections to atom-field interactions. To our knowledge it was first used in quantum mechanics by HUSIMI [1953] to solve the complete time-dependent harmonic oscillator in an external time-dependent force. Thus in the Lagrangian or ST system, the laser field is included in the time-dependent coordinate displacements, reducing the TDSE to a time-independent equation in some simple cases (TRUSCOTT [1993]). We show that adopting such a reference system, one obtains TDSEs with moving coordinates thus requiring adaptive grid methods to solve these equations (LU and BANDRAUK [2001]). These new equations have the advantage that for large electron–nuclear separation as occurs in dissociative ionization processes, the grid moves exactly in phase with the electron wavefunction so that the grid in asymptotic limit is an *exact adaptive* grid. We show in the present work that the Lagrangian or ST method offers many advantages over other gauges and should facilitate exact computation for nonperturbative laser–molecule interactions. Of note is that for vanishing (zero) field, the ST method contains quantum corrections to vacuum fluctuations (WELTON [1948]).

Time-dependent fields in quantum mechanics can always be transformed to suitable equivalent (isospectral) representations by appropriate unitary transformations. Less well appreciated is the fact that one can eliminate time-dependent fields by a time-dependent coordinate transformation for particular cases. We start with the usual 1D TDSE in which the potential explicitly includes a spatially uniform field that is an arbitrary function of time $f(t)$ (we use atomic units a.u.: $e = m_e = h/2\pi = 1$),

$$i\frac{\partial\psi(x,t)}{\partial t} = -\frac{1}{2}\frac{\partial^2\psi(x,t)}{\partial x^2} + [V(x,t) - xf(t)]\psi(x,t). \tag{2.1}$$

Following the procedure of HUSIMI [1953], we transform from the x, t coordinates system to a new ξ, t coordinate system where:

$$\xi = x - q(t), \tag{2.2}$$

from which it follows:

$$\frac{\partial\xi}{\partial x} = 1, \quad \frac{\partial\xi}{\partial t} = -\dot{q}(t), \tag{2.3}$$

so that

$$\frac{\partial\psi}{\partial t} = -\dot{q}(t)\frac{\partial\psi}{\partial\xi}\bigg|_t + \frac{\partial\psi}{\partial t}\bigg|_\xi. \tag{2.4}$$

Introducing the transformation

$$\psi(\xi,t) = \exp(i\dot{q}\xi)\phi(\xi,t), \tag{2.5}$$

we obtain a new TDSE for $\phi(\xi,t)$:

$$i\frac{\partial\phi(\xi,t)}{\partial t} = -\frac{1}{2}\frac{\partial^2\phi(\xi,t)}{\partial\xi^2} + \left[V(\xi+q,t) + (\ddot{q}-f)\xi - \left(\frac{\dot{q}^2}{2}+fq\right)\right]\phi(\xi,t). \tag{2.6}$$

Setting

$$\dot{q} = \int f(s)\mathrm{d}s, \quad g(t) = \frac{\dot{q}^2}{2} + fq, \tag{2.7}$$

and performing another unitary transformation

$$\phi(\xi,t) = \exp\left[i\int g(s)\mathrm{d}s\right]\Phi(\xi,t), \tag{2.8}$$

one obtains a new TDSE for $\Phi(\xi,t)$:

$$i\frac{\partial\Phi(\xi,t)}{\partial t} = -\frac{1}{2}\frac{\partial^2\Phi(\xi,t)}{\partial\xi^2} + V(\xi+q(t))\Phi(\xi,t), \tag{2.9}$$

and the original solution of Eq. (2.1) is given by

$$\psi(x,t) = \exp[ix\dot{q}(t)]\exp\left[-\frac{i}{2}\int\dot{q}^2(t)\mathrm{d}t\right]\Phi(\xi,t). \tag{2.10}$$

We note that from Eq. (2.2), ξ is a time-dependent function, that is, it is a *moving* coordinate; however in Eq. (2.9) it is treated as a time-independent coordinate. In fact, for $V = V_0 = 0$, that is, a constant, $\Phi(\xi, t)$ is a plane wave $e^{ik\xi} = e^{ik(x - q(t))}$ and the phase factors in Eq. (2.10) contain the ponderomotive energy $U_p = E_0^2/4\omega^2$ due to field induced oscillation of a free particle in a sinusoidal field $f(t) = E(t) = E_0\sin(\omega t)$ (GAVRILA [1992], BANDRAUK [1994b]).

In the *length* gauge, using the total 3D electronic Hamiltonian of H_2^+ for fixed nuclei at distance R in a laser field $E(t)\cos(\omega t)$ parallel to the internuclear axis, the TDSE is written (CHELKOWSKI, CONJUSTEAU, ATABEK and BANDRAUK [1996]) as:

$$i\frac{\partial\psi(x,y,z,t)}{\partial t} = -\frac{2m_p+1}{4m_p}\left[\frac{\partial^2}{\partial x^2} + \frac{\partial^2}{\partial y^2} + \frac{\partial^2}{\partial z^2}\right]\psi(x,y,z,t) - \frac{\psi(x,y,z,t)}{\sqrt{x^2+y^2+(z\pm R/2)^2}}$$
$$+ \frac{2m_p+2}{2m_p+1}zE(t)\cos(\omega t)\psi(x,y,z,t).$$

$$(2.11)$$

Here, the Hamiltonian is the exact one-electron Hamiltonian for a molecule *aligned* with the z-axis by the intense laser field with fixed protons of mass m_p (we have used a.u.: $m_e = 1$).

To propagate the TDSE in the *velocity* (Coulomb) gauge, we now use the unitary (gauge) transformation:

$$\psi \to \psi\exp\left\{-i\frac{2m_p+1}{4m_p}\left[\int_0^t\left(\int_0^s E(\tau)d\tau\right)^2 ds\right] - iz\int_0^t E(s)ds\right\}, \qquad (2.12)$$

which gives the new TDSE:

$$i\frac{\partial\psi(x,y,z,t)}{\partial t} = -\frac{2m_p+1}{4m_p}\left[\frac{\partial^2}{\partial x^2} + \frac{\partial^2}{\partial y^2} + \frac{\partial^2}{\partial z^2}\right]\psi(x,y,z,t) - \frac{\psi(x,y,z,t)}{\sqrt{x^2+y^2+(z\pm R/2)^2}}$$
$$+ i\left[\frac{2m_p+1}{2m_p}\int_0^t E(s)ds\right]\frac{\partial\psi(x,y,z,t)}{\partial z}.$$

$$(2.13)$$

In this equation, we see clearly the *convection* term with velocity:

$$\text{Convection velocity} = \frac{2m_p+1}{2m_p}\int_0^1 E(s)ds. \qquad (2.14)$$

We note that Eq. (2.13) is similar to a *convection–diffusion* equation with *incompressible* convecting field but has an imaginary diffusion coefficient. Both of the two preceding formulations (2.11), (2.13) have used Eulerian (stationary) coordinates. To adopt the Lagrangian (moving) or ST coordinate system Eqs. (2.2)–(2.6), we change the variable z, as suggested by Eq. (2.2),

$$z \rightarrow u = z + \frac{2m_p + 1}{2m_p} \int_0^t \int_0^s E(\tau) \, d\tau \, ds = z + \alpha(t), \tag{2.15}$$

and this transformation which is called the ST method gives (GAVRILA [1992], BANDRAUK [1994b])

$$i \frac{\partial \psi(x, y, u, t)}{\partial t} = -\frac{2m_p + 1}{4m_p} \left[\frac{\partial^2}{\partial x^2} + \frac{\partial^2}{\partial y^2} + \frac{\partial^2}{\partial u^2} \right] \psi(x, y, u, t)$$
$$- \frac{\psi(x, y, u, t)}{\sqrt{x^2 + y^2 + (u - \alpha(t) \pm R/2)^2}}. \tag{2.16}$$

Comparing Eq. (2.16) with Eqs. (2.11), (2.13), we note that the former is simpler: the explicit laser pulse term is removed and its effect is included in the time-dependent potential. We note that the new coordinate system is moving with respect to the fixed protons and the movement is determined by the laser strength and its phase. It is clear that in the TDSE for the dynamics of the protons, we can use the same procedure to remove the convection term with respect to the nuclei (BANDRAUK [1994b]).

Previous applications of the ST method have been done in numerical 1D simulation of H_2 (WIEDEMANN and MOSTOWSKI [1994]) and recently also in a full 3D MCSCF calculation using a Floquet expansion (NGUYEN and NGUYEN-DANG [2000]). The latter corresponds to an adiabatic-field approach. Since in general, a laser field can be characterized by the set of parameters $E = \{E_0, \omega, \phi\}$ where E_0 is the instantaneous amplitude, ω is the frequency and ϕ is the phase, then the total time derivative for any TDSE thus becomes (CHU [1989])

$$\frac{d}{dt} = \frac{\partial}{\partial t} \Big|_E + \dot{E} \frac{\partial}{\partial E}, \tag{2.17}$$

where

$$\dot{E} \frac{\partial}{\partial E} = \dot{E}_0 \frac{\partial}{\partial E_0} + \dot{\omega} \frac{\partial}{\partial \omega} + \dot{\phi} \frac{\partial}{\partial \phi}. \tag{2.18}$$

This to be compared to Eq. (2.4) thus showing a close relation between the two approaches. All nonadiabatic field effects are contained in the time dependence of the potential via $q(t)$, Eq. (2.6). Thus accurate numerical solutions of the new TDSE, Eq. (2.6) are required to describe nonadiabatic effects expected in the short pulse limit where both amplitude E_0, frequency ω, and phase ϕ can be strongly time dependent (BRABEC and KRAUSZ [2000]). We address next this problem by investigating various discretized propagation schemes for the TDSE of H_2^+.

3. Numerical schemes

For simplicity, we describe all numerical methods using only two space variables (y, z) in the length gauge (Eulerian coordinates). All theses schemes are easily extendable to three variables and to the ST representation (Lagrangian coordinates).

A simple and efficient method is the Alternating Direction Implicit (ADI) method (see KAWATA and KONO [1999] for a recent application of the technique). The ADI method propagates the wavefunction by the ansatz,

$$\psi(t + \delta t) = \left[1 + i\frac{\delta t}{2}A_y\right]^{-1}\left[1 + i\frac{\delta t}{2}A_z\right]^{-1} \times \left[1 - i\frac{\delta t}{2}A_z\right]\left[1 - i\frac{\delta t}{2}A_y\right]\psi(t) \quad (3.1)$$

with the notations:

$$
\begin{aligned}
A_y &= -\frac{2m_p + 1}{4m_p}\frac{\partial^2}{\partial y^2} + \frac{V_c}{2}, \\
A_z &= -\frac{2m_p + 1}{4m_p}\frac{\partial^2}{\partial z^2} + \frac{V_c}{2} + \frac{2m_p + 2}{2m_p + 1}zE(t)\cos(\omega t).
\end{aligned}
\quad (3.2)
$$

Here, V_c is the Coulomb potential: In 2D, V_c is regularized to $V_c(y, z) = -1/\sqrt{0.5 + y^2 + (z \pm R/2)^2}$. In 3D, for Eulerian coordinates, the grid is chosen so that we don't need to evaluate $1/\sqrt{x^2 + y^2 + (z \pm R/2)^2}$ at its singularities but we conserve V_c in its originality. In 3D, for Lagrangian coordinates, when the laser is circularly/elliptically or linearly polarized we can adopt the same strategy as for Eulerian coordinates, but when the laser is a combination of a circular/elliptical laser and a perpendicular linear laser, we can regularize V_c by homogeneously distributing each proton's charge in a small sphere. Another way to regularize V_c in 3D is to distribute the proton's charge by a function which approximates the Dirac function and we discuss the problem of regularization next.

Regularization of Coulomb potentials to remove the singularity can be achieved by adding a positive constant to the electron–proton distance and has been used previously in 1D simulations of H_2 in the ST representation (WIEDEMANN and MOSTOWSKI [1994]). But this method has been found too inaccurate in our 3D calculation test and is abandoned. In order to conserve the 3D properties of real molecules, we investigate two regularization processes, averaging the singularity or discretizing the singularity in conformity with the grid structure. Averaging the singularity is very intuitive and is most easily achieved by distributing the charge over a small sphere of radius R_c which is generally chosen to be several times the grid cell's size and the Coulomb potential is replaced by

$$
V_c(\rho, z) = \begin{cases} -\dfrac{3}{2R_c} + \dfrac{\rho^2 + z^2}{2R_c^3} & \text{for } \sqrt{\rho^2 + z^2} \leqslant R_c, \\[2ex] -\dfrac{1}{\sqrt{\rho^2 + z^2}} & \text{for } \sqrt{\rho^2 + z^2} \geqslant R_c. \end{cases}
\quad (3.3)
$$

In the discretization procedure, we replace the Coulomb potential by the atomic relation for the Laplacian $\Delta = \nabla^2$,

$$-\frac{1}{\sqrt{\rho^2 + z^2}} = E_{1s} + \frac{\Delta\psi_{1s}}{\psi_{1s}}(\rho, z), \tag{3.4}$$

obtained from the Schrödinger equation of the H(1s) atomic orbital. The Laplacian is then discretized with the same finite difference schemes as in the TDSE. Alternatively, we can also use the exact relation:

$$\frac{1}{r} = \Delta\frac{r}{2}, \tag{3.5}$$

with $r = \sqrt{z^2 + \rho^2}$ and discretize again the Laplacian (Δ) in conformity with the grid discretization.

We remark that the discrete propagator Eq. (3.1) is not unitary because A_y (A_z) is dependent on z (y). Furthermore, the dependence of the operator on variables of other directions make its inversion difficult to be implemented in pipelines of parallel computers. One way to remove the intercoordinate dependence in the ADI method is to use the following modified propagator involving partial exponentiation as in general split-operator (SO) methods (see BANDRAUK and SHEN [1993] for a general Fourier SO method with higher order precision):

$$
\begin{aligned}
\psi(t+\delta t) = &\left[1 + i\frac{\delta t}{4}S_y^1\right]^{-1} \exp\left(\frac{i\delta t}{4}V_c\right)\left[1 - i\frac{\delta t}{4}S_y\right] \\
&\times \left[1 + i\frac{\delta t}{4}S_z\right]^{-1} \exp\left(\frac{i\delta t}{4}V_c\right)\left[1 - i\frac{\delta t}{4}S_z\right] \\
&\times \left[1 + i\frac{\delta t}{4}S_z\right]^{-1} \exp\left(\frac{i\delta t}{4}V_c\right)\left[1 - i\frac{\delta t}{4}S_z\right] \\
&\times \left[1 + i\frac{\delta t}{4}S_y\right]^{-1} \exp\left(\frac{i\delta t}{4}V_c\right)\left[1 - i\frac{\delta t}{4}S_y\right]\psi(t),
\end{aligned} \tag{3.6}
$$

with:

$$
\begin{aligned}
S_y &= -\frac{2m_p + 1}{4m_p}\frac{\partial^2}{\partial y^2}, \\
S_z &= -\frac{2m_p + 1}{4m_p}\frac{\partial^2}{\partial z^2} \quad \text{for(2.16)}, \\
S_z &= -\frac{2m_p + 1}{4m_p}\frac{\partial^2}{\partial z^2} + \frac{2m_p + 2}{2m_p + 1}zE(t)\cos(\omega t) \quad \text{for(2.11)}.
\end{aligned} \tag{3.7}
$$

This method (SO-1) is realizable on a pipeline of CPUs but still remains nonunitary.

To obtain a unitary propagator, one can combine the classical split operator (see BANDRAUK and SHEN [1993] for Fourier SO method with high precision) with the Crank–Nicholson method (HOFFMAN [1992]):

$$
\begin{aligned}
\psi(t + \delta t) = {} & \left[1 + i\frac{\delta t}{4}S_y\right]^{-1}\left[1 - i\frac{\delta t}{4}S_y\right]\left[1 + i\frac{\delta t}{4}S_z\right]^{-1}\left[1 - i\frac{\delta t}{4}S_z\right] \\
& \times \exp(i\delta t V_c) \\
& \times \left[1 + i\frac{\delta t}{4}S_z\right]^{-1}\left[1 - i\frac{\delta t}{4}S_z\right]\left[1 + i\frac{\delta t}{4}S_y\right]^{-1}\left[1 - i\frac{\delta t}{4}S_y\right]\psi(t).
\end{aligned}
\tag{3.8}
$$

This third method (SO-2) is *unitary* and implementable in a series of CPUs which are serially connected.

All three preceding methods ADI Eq. (3.1), SO-1 Eq. (3.6), SO-2 Eq. (3.8) are of second order precision in time (δt^2). We will also test a method of order one in δt (the simplest propagator), SO-3,

$$
\begin{aligned}
\psi(t + \delta t) = {} & \left[1 + i\frac{\delta t}{2}S_y\right]^{-1}\left[1 + i\frac{\delta t}{2}S_z\right]^{-1}\exp(i\delta t V_c) \\
& \times \left[1 - i\frac{\delta t}{2}S_z\right]\left[1 - i\frac{\delta t}{2}S_y\right]\psi(t).
\end{aligned}
\tag{3.9}
$$

Here, we have lost the unitarity of the propagator again as compared to SO-2, Eq. (3.8). For 3D calculations with the laser field parallel to the internuclear (z) axis, one replaces y by the cylindrical coordinate $\rho = \sqrt{x^2 + y^2}$ and $\partial^2/\partial y^2$ by $\partial^2/\partial\rho^2 + (1/\rho)(\partial/\partial\rho)$ in Eqs. (3.1)–(3.9).

For the spatial discretization, we have used a finite difference method with a *non-uniform* grid (see Fig. 3.1). Such adaptive grids are now popular in density functional, DFT, numerical calculations to treat large systems (GYGI and GALLI [1995]). The adaptive grid is obtained from a uniform grid as shown by KAWATA and KONO [1999]. We use a more general transformation: $y = T(\tilde{y}), z = T(\tilde{z})$. The following is our transformation scheme:

$$
T(u) = u\left(\frac{u^n + sp^n}{u^n + p^n}\right),
\tag{3.10}
$$

where n is an even integer, $p \geqslant 0$ representing the domain of grid refinement, and $0 \leqslant s \leqslant 1$ is set to be the minimum of DT/Du. Because $DT/Du = 1$ at infinity and $DT/Du = s$ for $T(u) = u = 0$, the generated grid is finest in the region of the singularities (neighborhood of $T(u) = 0$). In order to better conserve the wavefunction's norm, we transform it by: $\tilde{\psi} = \sqrt{T'(\tilde{y})T'(\tilde{z})}\psi$. With this spatial transformation, we use the following difference scheme to discretize the differential operator:

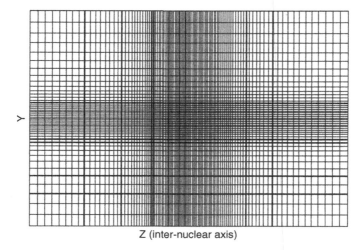

Z (inter-nuclear axis)

Fɪɢ. 3.1 A section of a nonuniform 3D grid.

$$\frac{\partial}{\sqrt{y'}\partial\tilde{y}}\frac{\partial}{y'\partial\tilde{y}}\frac{\tilde{\psi}}{\sqrt{y'}}(\tilde{y}_i)\simeq\frac{C_{i-3}+C_{i+3}}{90\delta\tilde{y}^2}-\frac{3(C_{i-2}+C_{i+2})}{20\delta\tilde{y}^2}+\frac{3(C_{i-1}+C_{i+1})}{2\delta\tilde{y}^2}$$

$$+\frac{1}{\delta\tilde{y}^2}\left[-\frac{D_{i+3}+D_{i-3}}{90}+\frac{3(D_{i+2}+D_{i-2})}{20}-\frac{3(D_{i+1}+D_{i-1})}{2}\right],$$

$$(3.11)$$

with the following notations:

$$C_j=\frac{\psi_j}{\sqrt{y'_i}y'_{(i+j)/2}\sqrt{y'_j}}, \quad D_j=\frac{\psi_i}{y'_iy'_{(i+j)/2}}.$$

In the Lagrangian–ST system, z is replaced by u, Eq. (2.15). The preceding scheme is of 5th order in space, that is, $\delta\tilde{y}^5$ and uses a stencil of 7 points. In the same way, we discretize the differential operators in other directions. The schemes for the four propagators (ADI, SO-1, SO-2, SO-3) can be easily adapted for this transformation of variable and wavefunction. We note that finite difference methods are necessary for adaptive grid methods since FFT methods as used in our previous SO methods are now difficult and inefficient (Wᴀʀᴇ [1998]).

All our 2D computations have used Cartesian coordinates with the idea of using this eventually for elliptically polarized light and the preceding discretization method without change. We have carried out a first series of 2D computations to compare different propagation methods. To show the performance of each discrete propagator, in Table 3.1 we give the CPU time of each time-step for the four methods used with a 304×688 ($y \times z$) grid on an IBM POWER3 system (375 MHz).

TABLE 3.1

CPU time/time-step for the four propagation methods

	ADI	SO-1	SO-2	SO-3
CPU (s)	0.35809	0.29747	0.28205	0.17772

Here, the CPU time for each propagator depends only on the grid point numbers, but is independent of gauge choices (Eqs. (2.11), (2.13), (2.16)). We note that the fastest method (SO-3), Eq. (3.9), needs only half of CPU time than the slowest method (ADI), Eq. (3.1), as it is of first order accuracy in δt whereas the three others, ADI, SO-1, SO-2 are of the second order accuracy. The superior performance of SO-2 (Eq. (3.8)) is to be noted due to its smaller number of exponentials and simplicity.

We next investigate numerically the above different regularization schemes (Eqs. (3.3)–(3.5)) to remove the singularities of the Coulomb potentials in a 3D simulation using cylindrical coordinates (ρ,z) for a laser field parallel to the internuclear axis. Our first calculation of ionization rates for H_2^+ in such coordinates removed the Coulomb singularity by expanding in eigenstates (Bessel functions) of the 2D cylindrical $(\rho = \sqrt{x^2 + y^2})$ Laplacian thus reducing the 3D electronic problem to 2D with large matrix for the effective Coulomb potentials (CHELKOWSKI, ZUO and BANDRAUK [1993]). The SO schemes, Eqs. (3.6)–(3.9) involve the exponentiation of the Coulomb potentials, resulting in undefined phases at the singularities. One can avoid the singularity problem partly by using nonuniform grids as in Eq. (3.10), thus enforcing that functions are zero at singularities in one direction. This results in complicated Laplacians and sometimes nonsymmetric finite difference representation (KONO et al. [1997]). We note that in the Lagrangian–ST representation, the singularity in the Coulomb potential V_c is now moving with time due to the replacement of z by u (see Eqs. (2.15)–(2.16)). In this case, regularization is not easily implemented and we use in the exponential of the propagators the exact time dependent V_c as given in Eq. (2.16).

We have compared the ionization rates Γ $(10^{12}$ s$^{-1})$ of 3D H_2^+ obtained from the time dependence of the norm of the wave function $N(t) = \|\psi(t)\|^2$,

$$N(t) = \|\psi(0)\|^2 \, e^{-\Gamma t}, \tag{3.12}$$

calculated using the SO-2 propagator (Eq. (3.8)) for the length gauge (ER) and the Lagrangian–ST representation. Note each time step is unitary but absorbing boundaries reduce the norm with time. Solutions of the corresponding TDSEs, Eqs. (2.11) and (2.16), were obtained by using the: (1) original Coulomb potential $(1/r)$; (2) a regularized potential averaged over a sphere; (3) a discretized potential using Eq. (3.5) and finally (4) a discretized potential using Eq. (3.4). As illustration, we show the result for regularized potential and discretized potential using Eq. (3.4) in Tables 3.2 and 3.3.

All calculations were performed for $I = 10^{14}$ W/cm^2, $\lambda = 1064$ nm ($\omega = 0.043$ a.u.), five cycle rise and then 20 cycle constant field. Table 3.2 shows results at $R = 4$ a.u., near equilibrium whereas Table 3.3 shows corresponding results at $R = 9.5$ a.u. where CREI, has been found to occur (CHELKOWSKI and BANDRAUK [1995], BANDRAUK [2000]). All time and space discrete steps δt, δz, $\delta \rho$ are in atomic units (a.u.). The protons are situated at the middle of the ρ-grid lines in order to avoid $\rho = 0$, but are situated either on z-grid lines, that is, with Z_p on-line, or off-line, in the latter case to avoid $z = 0$. Thus Z_p on-line results in the ER gauge are consistently poor due to the proximity of the Coulomb singularity at the proton positions Z_p.

TABLE 3.2

Ionization rates (10^{12} s^{-1}) at $R = 4$ a.u., $I = 10^{14}$ W/cm^2, $\lambda = 1064$ nm, Z_p is defined as the position of protons on the z direction grid

Gauge, Z_p	$\delta t = 0.05$ $\delta z = \delta \rho = 0.25$	$\delta t = 0.04$ $\delta z = \delta \rho = 0.20$	$\delta t = 0.03$ $\delta z = \delta \rho = 0.15$	$\delta t = 0.02$ $\delta z = \delta \rho = 0.10$
(a) Original potential				
ER, off-line	1.33423715973	1.31539308770	1.32842189536	1.34265241410
ER, on-line	0.89803248763	0.97239594933	1.07563578075	1.20825937841
ST, –	7.96097641265	4.13816007689	2.33437174921	1.53425672092
(b) Discrete potential ($\frac{1}{2} = E_{1s} - [\Delta \Psi_{1s}]/\Psi_{1s}$)				
ER, off-line	1.30362126727	1.32835165228	1.32600533461	1.34196387133
ER, on-line	1.30125816218	1.33284154864	1.32799130791	1.33980068003
ST, –	1.63386206026	1.38552395606	1.33912138036	1.34132620691

TABLE 3.3

Ionization rates (10^{12} s^{-1}) at $R = 9.5$ a.u., $I = 10^{14}$ W/cm^2, $\lambda = 1064$ nm, Z_p is defined as the position of protons on the z direction grid

Gauge, Z_p	$\delta t = 0.05$ $\delta z = \delta \rho = 0.25$	$\delta t = 0.04$ $\delta z = \delta \rho = 0.20$	$\delta t = 0.03$ $\delta z = \delta \rho = 0.15$	$\delta t = 0.02$ $\delta z = \delta \rho = 0.10$
(a) Original potential				
ER, off-line	31.4236573293	30.7844463378	29.4058863680	28.4667501641
ER, on-line	17.7299930008	19.9495445543	22.8293072187	25.2988645854
ST, –	30.1019935298	28.1778350731	27.0047899322	27.0457443269
(b) Discrete potential ($\frac{1}{2} = E_{1s} - [\Delta \Psi_{1s}]/\Psi_{1s}$)				
ER, off-line	28.4014692441	28.4290492609	28.3075801775	28.2110591261
ER, on-line	28.4066108394	28.4327977733	28.3066810375	28.2109847369
ST, –	28.3841322200	28.3170031848	28.1807871805	28.1880178017

All results converge to the same rate for the smallest time and space steps, $\delta t = 0.02$, $\delta z = \delta \rho = 0.1$, thus confirming the accuracy of the propagators for very fine grids. Both discrete potential methods show remarkable stability, as seen by the smallest sensitivity of the ionization rates as a function of discretization step size. This is indicative of the importance of the same discretization of all variables in grid-methods to ensure consistency of the numerical method.

The different numerical treatments of the Coulomb singularities in the present grid method have little effect on the absolute value of the wavefunction $|\psi(z, \rho, t)|$ calculated in the Lagrangian-ST and the Eulerian-length gauge, ER methods after propagation for 12.5 cycles (Figs. 3.2 and 3.4). Note that at $t = 12.5$ cycles, the field is zero and $z = u$ (Eq. (2.15)) for the Lagrangian–ST method.

The significant numerical differences occur in the phase of the function. In our previous careful studies of accuracies of various SO methods, we have shown that phase accuracy always lags that of amplitude accuracy by one order of magnitude in linear (BANDRAUK and SHEN [1993]) and nonlinear TDSEs (BANDRAUK [1994a]). We see this effect in Figs. 3.3 and 3.5 where using the direct ST/ER Coulomb potentials, that is, the original potentials (1) as compared to discretized Coulomb potentials (3) and (4) in conformity with the grid properties, produces functions at the nuclei which are of different phase for internuclear distance $R = 9.5$ a.u., that is, especially the real part ($R\ \psi$) in Fig. 3.3(a) are of opposite sign as the imaginary part ($I\ \psi$) in Fig. 3.5(b).

Comparing the ST function (Figs. 3.1 and 3.2) to those obtained in the ER gauge (Figs. 3.4 and 3.5) using off-line proton coordinates, we remark that:

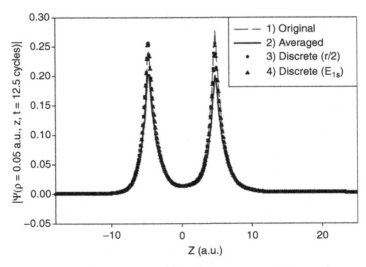

FIG. 3.2 Absolute electronic wavefunction $|\psi\ (\rho = 0.05$ a.u., $z, t = 12.5$ cycles)$|$ for H_2^+ at $I = 10^{14}$ W/cm^2, $\lambda = 1064$ nm, in the Lagrangian–Space Translation (ST) representation with (1) original V_c; (2) averaged V_c; (3) and (4) discretized V_c. Calculations were performed at $R = 9.5$ a.u. using SO-2 propagator (Eq. (3.8)) with steps: $\delta t = 0.02$, $\delta z = \delta \rho = 0.1$ a.u.

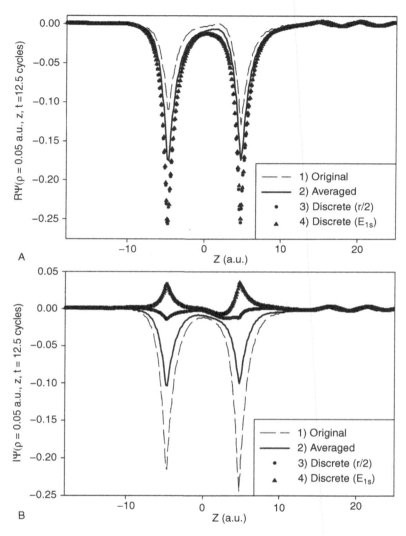

FIG. 3.3 (a) Real part $- R\psi$, (b) Imaginary part $- I\,\psi$ of ψ ($\rho = 0.05$ a.u., z, $t = 12.5$ cycles) for same parameters as Fig. 3.2, that is, in ST representation.

(1) In Fig. 3.5(a) and (b), the ER function obtained with the original potential has important numerical oscillation at $z = 10$ a.u. With the original potential, we have effectively approximated the wavefunction by a local polynomial which is regular in finite difference method, but using the discrete potentials the approximated function contain singular local basis. We know that the real wavefunction should be singular at the nuclei. This explains the appearance of the oscillation.

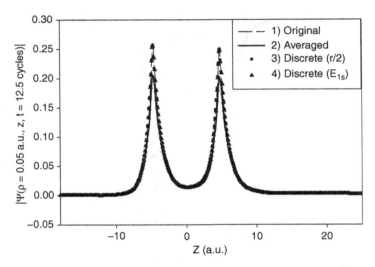

FIG. 3.4 Absolute electronic wavefunction $|\psi\,(\rho = 0.05$ a.u., z, $t = 12.5$ cycles)| for H_2^+ at $I = 10^{14}$ W/cm^2, $\lambda = 1064$ nm, in the length gauge (ER) with (1) original V_c; (2) averaged V_c; (3) and (4) discretized V_c. Calculations were performed at $R = 9.5$ a.u. using SO-2 propagator (Eq. (3.8)) with steps: $\delta t = 0.02$, $\delta z = \delta \rho = 0.1$ a.u.

(2) Note $\psi = re^{\phi}$ the wavefunction's value at a point, the difference between two functions at a point can be written as: $\psi_1 - \psi_2 = r_2[(r_1/r_2)e^{\phi_1-\phi_2} - 1]$. Figures 3.2 and 3.3 have shown that all the four potentials give almost the same absolute values for ST as for ER. In this case, we have

$$|e^{\phi_1-\phi_2} - 1| \simeq \sqrt{(R\psi_1 - R\psi_2)^2 + (I\psi_1 - I\psi_2)^2}. \tag{3.13}$$

According to this formula, Fig. 3.3 (ST) demonstrate an evident better convergence of wavefunction's phase than Fig. 3.5 (ER). We note furthermore that in the length gauge (ER) strong phase effects persist asymptotically and this has lead to a worse convergence of the wavefunction's phase. This is consistent with the fact that the ST wavefunction have the ponderomotive energy $U_p = E_0^2/4\omega^2$ phase effects already removed (Eq. (2.10)). These important laser-induced phase fluctuations remain in the length gauge (ER).

4. Boundary conditions and energy spectra

As shown in the previous section, two representations, the length gauge with ER radiative coupling or the ST method, give essentially the same accurate results for ionization rates. Nevertheless there are important differences numerically. The Ez coupling equation Eq. (2.11) diverges at large distance, especially at the grid boundaries,

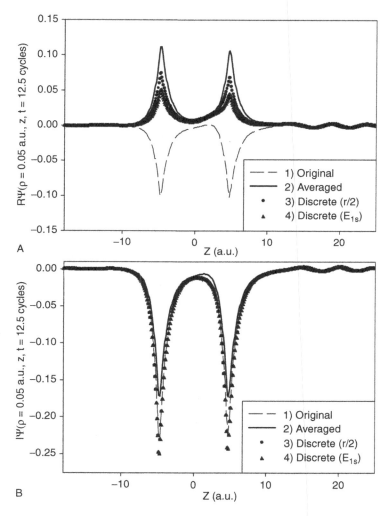

FIG. 3.5 (a) Real part $-R\psi$, (b) Imaginary part $-I\,\psi$ of ψ ($\rho = 0.05$ a.u., z, $t = 12.5$ cycles) for same parameters as Fig. 3.4, that is, in ER gauge.

$|z| \rightarrow \infty$, resulting in poor convergence of the phase in that gauge. In the ST representation, the wavefunction has the ponderomotive energy removed from the phase (Eq. (2.10)). These important laser-induced phase fluctuations remain in the length (ER) gauge.

An estimate of field induced electron displacements can be obtained from the classical equation of motion of a free ionized electron driven by a laser field E_0 $\cos(\omega t)$. Such classical models reproduce remarkably well ATI (electronic) spectra

(CORKUM [1993]) and HOHG spectra (BANDRAUK and SHEN [1994]). We start with the acceleration equation in the z-direction for the electron (see Fig. 3.1) and integrate twice to give the velocity $\dot{z}(t)$ and position $z(t)$, (or $\dot{q}(t)$ and $q(t)$, Eq. (2.7)):

$$\ddot{z}(t) = -E_0 \cos(\omega t + \phi), \tag{4.1}$$

$$\dot{z}(t) = -(E_0/\omega)[\sin(\omega t + \phi) - \sin(\phi)], \tag{4.2}$$

$$z(t) = -\alpha_0[\cos(\omega t + \phi) + \omega t \sin(\phi) - \cos(\phi)]. \tag{4.3}$$

ϕ is the initial laser phase at which the electron is ionized and we assume an initial velocity $\dot{z}(0) = 0$ in conformity with quasistatic tunneling models (CORKUM [1993]). α_0 is called the ponderomotive (quiver) radius and U_p is the corresponding ponderomotive energy:

$$\alpha_0(\text{a.u.}) = E_0/\omega^2, \quad U_p = \alpha_0^2 \omega^2/4 = E_0^2/4\omega^2. \tag{4.4}$$

These quantities correspond to $q(t)$ and $\dot{q}^2/2$ in the ST representation (Eq. (2.7)). These classical equations of motion can be used to predict the optimal condition for HOHG and ATI spectra. As a first hypothesis we shall look at maximum acceleration which in the classical limit should produce maximum radiation (BANDRAUK and YU [1999]). From Eq. (4.1) we obtain $\ddot{z}(\text{max})$ at phase $\omega t + \phi = \pi$. This yields the velocity $\dot{z} = E_0 \sin(\phi)/\omega$ with a maximum $\dot{z}(\text{max}) = E_0/\omega$ for $\phi = \pi/2$. The resulting energy is

$$E(\text{max}) = \dot{z}^2(\text{max})/2 = E_0^2/2\omega^2 = 2U_p. \tag{4.5}$$

The corresponding distance of the electron trajectory is

$$z(t) = \alpha_0(-1 + \omega t) = \alpha_0(\pi/2 - 1) = 0.57\alpha_0. \tag{4.6}$$

Thus the maximum acceleration criterion yields a maximum energy $E(\text{max}) = 2U_p$ or HOHG order $N_m = E(\text{max})/\omega = 2U_p/\omega$ if an electron collides with a neighboring nucleus at distance $R = 0.57\alpha_0$. The next scenario is *recombination* of the electron with the parent ion. For maximum velocity the phase requirement is $\omega t + \phi = 3\pi/2$, thus yielding the transcendental equation $(3\pi/2 - \phi)\sin(\phi) = \cos(\phi)$ for $z(t) = 0$, that is, recollision. The solution of this equation is $\phi = 0.08\pi$, $\dot{z}(\text{max}) = 1.26E_0/\omega$. The resulting maximum recollision energy is

$$E_m = 3.17U_p. \tag{4.7}$$

This simple classical model explains the maximum theoretical HOHG order obtained by recollision (CORKUM [1993]), that is,

$$N_m = (I_p + 3.17U_p)/\omega, \tag{4.8}$$

where I_p is the ionization potential. However in molecules, as we have shown in detail (BANDRAUK and YU [1999]), electron collision can occur with neighboring ions. Then the maximum velocity condition $\omega t + \phi = 3\pi/2$ results in this velocity being from Eq. (4.2), $\dot{z} = (8E_0/\omega)(1 + \sin\phi)$ or $2E_0/\omega$ at $\phi = \pi/2$. Thus the maximum energy obtained with no constraints is

$$E_m = 2E_0^2/\omega^2 = 8U_p. \tag{4.9}$$

This is gained during a half cycle $\omega t = \pi$ and the corresponding distance travelled by the electron is

$$z(\pi/\omega) = \pi\alpha_0. \tag{4.10}$$

In summary the classical model of an ionized electron driven by a laser field predicts maximum kinetic energies $2U_p$, $3.17U_p$ and $8U_p$ at distance $z = 0.57\alpha_0$, 0 and $\pi \alpha_0$ if the initial velocity of the electron $\dot{z} = 0$. This model allows us to estimate grid size necessary to capture accurately electron dynamics upon ionization. Thus at a laser intensity $I = 3 \times 10^{14}$ W/cm$^2 = 10^{-2}$ a.u., then $E_0 = 10^{-1}$ a.u. At the wavelength 1064 nm ($\omega = 0.043$ a.u.) one obtains readily $\alpha_0 = 54$ a.u. $= 28.6 \times 10^{-8}$ cm, whereas the corresponding ponderomotive energy $U_p = 1.35$ a.u. $= 36.8$ eV. Thus electron energies up to $8U_p = 300$ eV will appear in the energy spectrum of the ionized electron and electron trajectories $\pi \alpha_0 > 160$ a.u. need to be considered. The electron radiative coupling at these distances is at least $\pi E_0\alpha_0 \simeq 12U_p \simeq 440$ eV, clearly a nonperturbative regime.

In the ST representation, one clearly needs an adaptive moving grid due to the oscillatory motion $q(t)$ of the electron with maximum displacement greater even than α_0 (e.g., Eq. (4.4)). As an example, we give in Fig. 3.1 a section of a nonuniform grid which we use regularly in our simulations. This 3D grid (1064 × 512 × 512) contains 5.4×10^8 points. Introducing up to 10 discretization points in the z-laser direction between nearest neighbor points (i.e., $\delta z = 0.1$ a.u.) and also in the x and y directions results in about $10^{12} = 1$ terabyte of memory requirements per time iteration using any one of the propagation schemes described in the previous section. This does not include yet the nuclear motion which is usually restrained to 1D, that is, parallel to the laser field, since the strongest field–molecule interactions occur in this configuration, due to charge transfer effects between neighboring atoms (BANDRAUK [1994b]). Time iterations are performed for about 25 cycles $=$ 87.5 fs (here $\tau_{\text{cycle}} = 3.5$ fs $= 142$ a.u.) in the case of ionization, so that with time steps of $\delta t = 0.02$ a.u., one readily has $25 \times 142/0.02 \simeq 2 \times 10^5$ time iterations for calculating ionization rates.

In order to calculate ATI spectra, that is, electron kinetic energy spectra, one must wait long after the pulse is ended in order to capture slow (low energy) electrons which are propagating to the edge of the grids. To date we have only been able to obtain 1D ATI spectra for 1D H$_2^+$ (CHELKOWSKI, FOISY and BANDRAUK [1998], CHELKOWSKI, ZAMOJSKI and BANDRAUK [2001]) from the exact non-Born–Oppenheimer wavefunction $\psi(z, R, t)$ where z is the electronic coordinate and R the internuclear coordinate. Simultaneous calculation of the electron (ATI) kinetic energy and proton (ATD and CE (Coulomb Explosion, the rapid explosion of completely ionized nuclei from Coulomb repulsive forces)) spectra also allows for reconstructing initial nuclear wavefunctions, both static and dynamic in the long wavelength (800 nm) (BANDRAUK and YU [1999]) and UV (30 nm) laser regime (BANDRAUK and CHELKOWSKI [2001]). This is a new application, that is, *dynamic imaging* of nuclear motion with ultrashort intense laser pulses, which is expected to allow for time-dependent measurements of evolution of nuclear wavefunctions.

In order to obtain these kinetic energy spectra, both for electrons (ATI) and protons (ATD and CE), asymptotic projection onto exact solution of free particles are performed at large distances, z for the electron and R for the protons, that is, at the grid edges where interparticle, Coulomb forces, are negligible. In the calculations using the length ER gauge, the exact solutions of electrons in the laser field, called Volkov states (GAVRILA [1992], BANDRAUK [1994b]) are used (CHELKOWSKI, FOISY and BANDRAUK [1998], CHELKOWSKI, ZAMOJSKI and BANDRAUK [2001]). In the case of the ST representation (see previous section) one need only project onto field-free states. Comparison of the ER gauge and ST representation ATI spectra shows identical results with the ST method which requires smaller grids, thus allowing us to finally calculate ATI spectra for the two-electron systems H_2 and H_3^+ with *moving* nuclei (KAWATA and BANDRAUK [2002]).

The asymptotic projection method used here is a special application of *domain decomposition* for numerical solution of PDEs (QUARTERONI [1991]). The large grid is subdivided near the edge into a small internal and large external box. Exact solutions from the large box are then projected onto solutions of the small box. The latter contains the Volkov (ER gauge) or free (ST representation) solution. Propagation and projection is then continued well after the pulse is over until external probabilities defined as $|\psi(x, y, z, R, T)|^2$ become constant. This method is analogous to a single domain method where the size of the grid should be the sum of both internal and external boxes. Our projection step involves simple overlap of the exact internal solution onto the asymptotic Volkov or free particle states. It should be pointed out that our overlapping domain decomposition method does not iterate on the internal interface as the well-known classical Schwarz alternating domain decomposition method. Clearly, the iteration on the interface will assure a best convergence of the numerical solutions, but it is too expensive in computing time for problem in higher dimension (>3) aimed by our method. The numerical results of some domain decomposition methods with or without overlapping have demonstrated that the well chosen interface condition can offer very precise final results even if no iterations are made on the interface (GUERMOND, HUBERSON and SHEN [1993], GUERMOND and LU [2000]). In our approximation, we find that the wavepackets propagate from the internal box to the external box so we have used the transfer of population from internal box to the external one to replace the interface iteration. More sophisticated methods based on Schwarz method remain to be explored in order to permit better coupling between the sub-domains without iteration on the interface.

5. Beyond the dipole approximation

We have seen in Section 2 that there are three main approaches to treating molecules–laser field interaction in the nonlinear nonperturbative regime. The standard method, which relies on principles of electrostatics, is the length ER gauge, Eq. (2.11). In this gauge radiative interactions at grid boundaries become very large thus creating numerical instabilities. Another gauge, the velocity (Coulomb) gauge, Eq. (2.13) is similar to *convection–diffusion* equations of fluid dynamics. The convection velocity is in fact called the electromagnetic potential

$$A(t) = \int_0^t E(s)\mathrm{d}s, \tag{5.1}$$

and in the Coulomb gauge is *incompressible*, that is, $\vec{\nabla} \cdot \vec{A} = 0$. Thus both $A(t)$ and $E(t)$ have no spatial dependence and this is called the *dipole approximation* which is valid whenever $r/\lambda \ll 1$, where r is the atomic-molecular dimension ($\sim$1–10 Å) and λ is the wavelength of the radiation field. This implies the polarization (e.g., z) is always *orthogonal* to the propagation (e.g., x, y). Any spatially independent field $E(t)$ generates automatically a spatially independent potential $A(t)$ which automatically satisfies the Coulomb gauge or incompressibility condition.

In general, A is a vector spatial–temporal function $\vec{A}\,(\vec{r},t)$ so that the exact TDSE for a single particle in the velocity gauge is given by

$$i\frac{\partial \psi(\vec{r},t)}{\partial t} = \frac{1}{2}\left[i\vec{\nabla} + \frac{1}{c}\vec{A}\,(\vec{r},t)\right]^2 \psi(\vec{r},t) + V(\vec{r})\psi(\vec{r},t), \tag{5.2}$$

where $\vec{\nabla} = (\partial_x, \partial_y, \partial_z)$. The electromagnetic potential $\vec{A}\,(\vec{r},t)$ is related to the electric and magnetic fields by $\vec{E}\,(\vec{r},t) = (1/c)(\partial \vec{A}\,(\vec{r},t)/\partial t)$ and $\vec{B}\,(\vec{r},t) = \vec{\nabla} \wedge \vec{A}$ respectively (BANDRAUK [1994b]) so that $\vec{E} \mid \vec{A}$, $\vec{B} \perp \vec{A}$. We will assume that the laser field is polarized in the z direction and propagates in the y direction:

$$\vec{A}\,(\vec{r},t) = (0,0,A_z(y,t)); \quad \vec{E}\,(\vec{r},t) = (0,0,E_z(y,t)). \tag{5.3}$$

If one expands the potential $\vec{A}$ to first order in the space coordinate $\vec{r}\,(x,y,z)$ in term of the ratio r/λ or $r\omega/c$ this results in $A_z(y,t) = A_z(t) + y(\partial A_z(y,t)/\partial_y)$. This first order expansion of A corresponds to taking into account the magnetic dipole and electric quadrupole interactions (BANDRAUK [1994b]). Relativistic effects such as the mass correction and spin-orbit effects can be neglected up to $v/c \simeq 0.2$. This corresponds to an upper limit for a Ti:Sap ($\lambda = 800$ nm) laser peak intensity of $\sim 10^{17}$ W/cm^2 (WALSER, KEITEL, SCRINZI and BRABEC [2000]), since c is the vacuum light velocity = 137 a.u. (3×10^{10} cm/s) and $v \simeq E_0/\omega$, is the electron velocity in a field of peak amplitude E_0 and frequency ω (see Eq. (4.2)). Thus corrections to the dipole gauges, Section 2 must be considered whenever r/λ and/or $v/c \simeq 1$.

For higher intensities, and lower frequencies then one must consider corrections beyond the nonrelativistic limit ($v/c \sim 1$). Thus in the latter case, one expects modification of the electron–electron Coulomb interactions (BERGOU, VARRO and FEDOROV [1981]) whereas in the former, relativistic corrections will influence interference between direct and exchange electron–electron scattering (FEDOROV and ROSHCHUP-KIN [1984]).

We give next an example of such calculations, that is, beyond the dipole approximation for intense ($I > 10^{18}$ W/cm^2), short wavelength ($\lambda = 25$ nm, $\omega = 1.8$ a.u.). Such intensities in the VUV-X-Ray regime is expected to become available soon from Free-Electron lasers. We choose the electric field amplitude polarized in the z-direction but propagating only in the y-axis, where $T_{max} = 2n\pi/\omega$ is the number of cycles, and T is the pulse rise period,

$$
\begin{aligned}
E_z(y,t) &= E_0 \frac{t}{T} \sin(\omega t - \omega y/c), & t < T, \\
&= E_0 \sin(\omega t - \omega y/c), & T < t < T_{max} - T, \\
&= E_0 \frac{T_{max} - t}{T} \sin(\omega t - \omega y/c), & T_{max} - T < t.
\end{aligned} \tag{5.4}
$$

Due to the electromagnetic relation Eq. (5.1), that is, $\vec{E} = -(1/c)(\partial \vec{A}/\partial t)$, this implies the following corresponding form for the potential A,

$$
\begin{aligned}
A_z(y,t) &= \frac{cE_0}{\omega T}\left[t\cos\left(\omega t - \frac{\omega y}{c}\right) - \frac{1}{\omega}\sin\left(\omega t - \frac{\omega y}{c}\right) - \frac{1}{\omega}\sin\left(\frac{\omega y}{c}\right)\right], & t < T, \\
&= \frac{cE_0}{\omega}\cos\left(\omega t - \frac{\omega y}{c}\right), & T < t < T_{max} - T, \\
&= \frac{cE_0}{\omega T}\left[(T_{max} - t)\cos\left(\omega t - \frac{\omega y}{c}\right) + \frac{1}{\omega}\sin\left(\omega t - \frac{\omega y}{c}\right) + \frac{1}{\omega}\sin\left(\frac{\omega y}{c}\right)\right], & \\
& & T_{max} - T < t.
\end{aligned} \tag{5.5}
$$

The TDSE in the Coulomb (velocity) gauge is then given from Eq. (2.13),

$$
i\frac{\partial \psi(x,y,z,t)}{\partial t} = \frac{2m_p + 1}{4m_p}\left[-\Delta - i\frac{2}{c}\left(\frac{2m_p + 2}{2m_p + 1}\right)A_z\frac{\partial}{\partial_z}\right]\psi + \frac{2m_p + 1}{4m_p}\left(\frac{2m_p + 2}{2m_p + 1}\right)^2
$$

$$
A_z^2\psi(x,y,z,t) + V_c\psi(x,y,z,t). \tag{5.6}
$$

We now introduce a new ST, in analogy with the Husimi transformation Eq. (2.2),

$$
\frac{\partial \alpha}{\partial t} = \frac{1}{c}A_z(t, y = 0), \tag{5.7}
$$

or equivalently,

$$\alpha(t) = \frac{E_0}{\omega T}\left[\frac{t}{\omega}\sin(\omega t) - 2\frac{1 - \cos(\omega t)}{\omega^2}\right], \qquad t < T,$$

$$= \frac{E_0}{\omega^2}\sin(\omega t), \qquad\qquad T < t < T_{max} - T, \qquad (5.8)$$

$$= \frac{E_0}{\omega T}\left[\frac{T_{max} - t}{\omega}\sin(\omega t) + 2\frac{1 - \cos(\omega t)}{\omega^2}\right], \qquad T_{max} - T < t.$$

Proceeding further as in the ST representation, Eq. (2.15), we introduce a coordinate displacement,

$$z = u + 2\frac{2m_p + 1}{4m_p}\frac{2m_p + 2}{2m_p + 1}\alpha(t),$$

$$\frac{\partial u}{\partial t} = -\frac{2}{c}\frac{2m_p + 1}{4m_p}\frac{2m_p + 2}{2m_p + 1}A_z(t, y = 0). \qquad (5.9)$$

In the new ST coordinate system, the TDSE now becomes:

$$i\frac{\partial \psi(x, y, u)}{\partial t} = \frac{2m_p + 1}{4m_p}\left[-\Delta_{x,y,u} - 2iB_z(t, u)\frac{\partial}{\partial u}\right]\psi(x, y, u)$$

$$+ \frac{2m_p + 1}{4m_p}\left(\frac{2m_p + 2}{2m_p + 1}\right)^2\frac{1}{c^2}\vec{A}^2\psi + V_c(x, y, u)\psi, \qquad (5.10)$$

where

$$B_z(t, y) = \frac{1}{c}\frac{2m_p + 2}{2m_p + 1}[A_z(t, y) - A_z(t, y = 0)]. \qquad (5.11)$$

Such an equation is again an *imaginary convection–diffusion* equation in presence of a potential V_c and is a generalization of the long-wavelength, that is, dipole approximation, ST Eq. (2.16). Thus going beyond the dipole approximation does not allow us to eliminate the convection term as in Eq. (2.16). However due to the Coulomb gauge condition $\nabla \cdot A = 0$, the convection coefficient $B_z(t, y)$ remains *incompressible*, that is, $\nabla \cdot B = 0$.

The effect of the nondipole terms, that is, $(r / \lambda) \sim 1$, $(v / c) \sim 1$, will be to displace electrons perpendicular to the polarization direction z, the polarization of the electric field E_z, and the propagation direction y due to a magnetic component $B_x = (\nabla \wedge A)_x$ along the x-axis. Nevertheless the TDSEs Eq. (5.6) and Eq. (5.10) remain symmetric with respect to x, that is, $\psi(x) = \psi(-x)$. Due to this symmetry, one can employ a nonuniform grid Crank–Nicholson algorithm in this direction and thus reduce the number of discretization points. We use FFTs (see BANDRAUK

[1994a] for recent application of FFT in Fourier SO method) in the polarization z-direction and the propagation y-direction in order to maintain unitarity and high accuracy of the derivatives in both these directions. The optimum splitting for Eq. (5.10) is then called $FD(x) + FFT(yz) + ST(z)$, that is, finite difference in x, FFTs in with ST in z via the u transformation, Eq. (5.9):

$$
\begin{aligned}
\psi^{n+1} = \exp&\left\{i\frac{\delta t}{2}\frac{2m_p+1}{4m_p}\frac{\partial^2}{\partial y^2}\right\} \\
&\times \left[1 - i\frac{\delta t}{4}\frac{2m_p+1}{4m_p}\frac{\partial^2}{\partial x^2}\right]^{-1}\left[1 + i\frac{\delta t}{4}\frac{2m_p+1}{4m_p}\frac{\partial^2}{\partial x^2}\right] \\
&\times \exp\left\{-i\frac{\delta t}{2}\left[V_c + \frac{2m_p+1}{4m_p}\left(\frac{2m_p+2}{2m_p+1}\right)^2\frac{1}{c^2}\vec{A}^2\right]\right\} \\
&\times \exp\left\{\delta t\frac{2m_p+1}{4m_p}\left[i\frac{\partial^2}{\partial u^2} - 2B_z(t,y)\frac{\partial}{\partial u}\right]\right\} \\
&\times \exp\left\{-i\frac{\delta t}{2}\left[V_c + \frac{2m_p+1}{4m_p}\left(\frac{2m_p+2}{2m_p+1}\right)^2\frac{1}{c^2}\vec{A}^2\right]\right\} \\
&\times \left[1 - i\frac{\delta t}{4}\frac{2m_p+1}{4m_p}\frac{\partial^2}{\partial x^2}\right]^{-1}\left[1 + i\frac{\delta t}{4}\frac{2m_p+1}{4m_p}\frac{\partial^2}{\partial x^2}\right] \\
&\times \exp\left\{i\frac{\delta t}{2}\frac{2m_p+1}{4m_p}\frac{\partial^2}{\partial y^2}\right\}\psi^n.
\end{aligned}
\tag{5.12}
$$

Replacing u by z in Eq. (5.10) would give the equivalent Coulomb (velocity) gauge propagation scheme. In other words, in Eq. (5.10) which is the *new* nondipole ST equation, we have removed the dipole field (spatially independent) $A_z(t, y = 0)$ (see Eq. (5.11)). Thus in the dipole limit, $B_z \equiv 0$, $u(t) = z$, and the TDSE Eq. (5.10) reduce to the conventional ST TDSE, Eq. (2.16).

We have calculated the influence of the nondipole terms on the electron ionization of H_2^+ using the algorithm based on Eq. (5.12) which allows to calculate the electron density $|\psi(x, y, z, t)|^2$. We illustrate in Figs. 5.1 and 5.2 the density $|\psi(0, y, z, t)|^2$ in the yz plane, that is, in the direction of propagation (y) and polarization (z) for the intensities 2×10^{18} W/cm^2 and 10^{19} W/cm^2 at $\lambda = 25$ nm. Figure 5.1 correspond to 10 cycles whereas Fig. 5.2 are for 20 cycles. The z polarization of the E_z field is along the internuclear axis whereas the magnetic B_x component would be along the x-axis. Thus at the lower intensity (2×10^{18} W/cm^2) (Figs. 5.1 (a), 5.2(a)) one expects the electron to be emitted mainly along the z-axis as seen in both figures. The higher intensity (10^{19} W/cm^2) results, Figs. 5.1(a) and 5.2(b) show a marked difference for

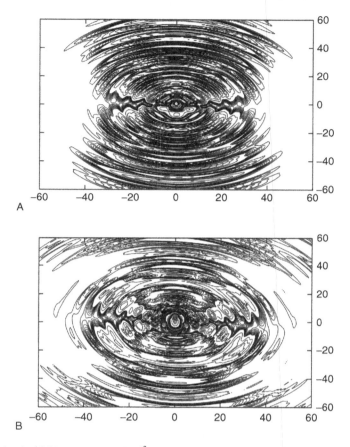

FIG. 5.1 The density $|\psi(0, y, z, t = 10 \text{ cycles})|^2$ iso-lines in the yz plane of the nondipole calculation for H_2^+ with $R = 2$ a.u. for the laser field polarized in the $z(R)$-internuclear direction and propagation in the y direction: (a) $I = 2 \times 10^{18}$ W/cm^2, $\lambda = 25$ nm; (b) $I = 10^{19}$ W/cm^2, $\lambda = 25$ nm.

longer times. At this intensity and wavelength the ponderomotive energy of the ionized electrons U_p (Eq. (4.4)) is 25 a.u. = 680 eV whereas at 2×10^{18} W/cm^2, $U_p = 5$ a.u. The remaining electrons (Fig. 5.2 (b)) are acquiring momentum in the direction of the propagation direction, that is, in the momentum direction of the field photons. Such effects do not appear at the lower intensity. Of further interest is that in all figures rings of electrons are created corresponding to a wavelength of about 6 a.u. $\simeq 3$ Å. As this corresponds to typical internuclear distances, one can foresee that such ionized electrons will interfere with the nuclei, leading to a new spectroscopy, LIED (ZUO, BANDRAUK and CORKUM [1996]).

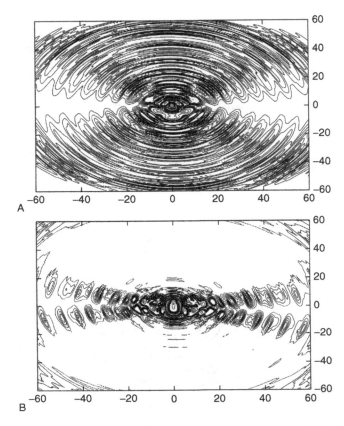

FIG. 5.2 The density $|\psi(0, y, z, t = 20 \text{ cycles})|^2$ iso-lines in the yz plane of the nondipole calculation for H_2^+ with $R = 2$ a.u. for the laser field polarized in the $z(R)$-internuclear direction and propagation in the y direction: (a) $I = 2 \times 10^{18}$ W/cm^2, $\lambda = 25$ nm; (b) $I = 10^{19}$ W/cm^2, $\lambda = 25$ nm.

6. Conclusion

We have focused on methods of solving accurately TDSEs to describe laser–molecule interactions from the perturbative to nonperturbative regime taking into consideration that modern laser technology is ever evolving towards controllable, more intense and shorter laser pulses. Our emphasis has been on the simplest one-electron H_2^+ molecule with a single nuclear degree of freedom. This simple system is the *benchmark* as it allows for exact numerical treatment of competing electron–nuclear dynamics, that is, including all non-Born–Oppenheimer corrections (see Section 4).

   In practice one has the liberty of choosing various gauges or representations, which are equivalent by unitary transformations (see Section 2) and thus will yield identical results only if the algorithms are exact. We have shown in fact that the most useful representation is the Space Transformation, ST, which we have generalized beyond the dipole approximation (see Section 5) in order to take into account retardation effects and to allow one to go to short wavelengths. The advantage of the ST method is that asymptotically, at the edge of the usual large grids necessary for high intensity calculations, radiative (laser–molecule) interactions vanish so that one can project the exact numerical solutions at grid boundaries onto well known analytic free particle states, either electronic and/or nuclear.

Many-body systems unfortunately require solving multidimensional PDEs, well beyond today's super-computer technology. A promising alternative is to reduce these problems to effective one-electron systems, via new methods such as TDDFT methods. The one-electron numerical methods discussed in the present work should be easily adaptable to these effective one-electron methods and these will allow for inclusion of nuclear motion, thus allowing for a complete dynamical description of molecules from the perturbative weak field limit to the nonperturbative intense field regime available with today's modern laser technology.

Acknowledgment

We wish to thank various colleagues who in the course of our work have pointed out many of the outstanding experimental and theoretical issues in laser–molecule science – P.B. Corkum (NRC, Ottawa), W. Kohn (U.C., Santa Barbara), C. Ulrich (Missouri), O. Atabek (Orsay), C. Le Bris (ENPC).

References

BANDRAUK, A.D. (1994a), *Int. Rev. Phys. Chem.* **13**, 123.

BANDRAUK, A.D. (1994b), *Molecules in Laser Fields* (M. Dekker Publication, New York).

BANDRAUK, A.D. (2000), In: Itikawa, Y. et al. (eds.), *The Physics of Electronic and Atomic Collision*, AIP Conf. Proc., vol. 500 (American Institute of Physics, New York), p. 102.

BANDRAUK, A.D. and S. CHELKOWSKI (2001), *Phys. Rev. Lett.* **87**, 273004.

Bandrauk, A.D., Y. Fujimura, and R.J. Gordon, (eds.) (2002). *Laser Control and Manipulation of Molecules*, ACS Symposium Series, vol. 821 (ACS, Washington)..

BANDRAUK, A.D. and H. SHEN (1993), *J. Chem. Phys.* **99**, 1185.

BANDRAUK, A.D. and H. SHEN (1994), *J. Phys. A* **27**, 7147.

BANDRAUK, A.D. and H. YU (1998), *Int. J. Mass. Spectrom.* **192**, 379.

BANDRAUK, A.D. and H. YU (1999), *Phys. Rev. A* **59**, 539.

BERGOU, J., S. VARRO and M.V. FEDOROV (1981), *J. Phys. A* **14**, 2305.

BRABEC, T. and F. KRAUSZ (2000), *Rev. Mod. Phys.* **72**, 545.

BRUMER, P. and M. SHAPIRO (1994), In: Bandrauk, A.D. (ed.) *Molecules in Laser Fields* (M. Dekker Pub., New York) Chapter 6.

CHELKOWSKI, S. and A.D. BANDRAUK (1995), *J. Phys. B* **28**, 1723.

CHELKOWSKI, S., A. CONJUSTEAU, O. ATABEK and A.D. BANDRAUK (1996), *Phys. Rev. A* **54**, 3235.

CHELKOWSKI, S., C. FOISY and A.D. BANDRAUK (1998), *Phys. Rev. A* **57**, 1176.

CHELKOWSKI, S., M. ZAMOJSKI and A.D. BANDRAUK (2001), *Phys. Rev. A* **63**, 023409.

CHELKOWSKI, S., T. ZUO and A.D. BANDRAUK (1993), *Phys. Rev. A* **48**, 3837.

CHU, S.I. (1989), In: Hirschfelder J.O., R.E. Wyatt and R.D. Coalson, (eds.), *Advances in Chemical Physics*, vol. 73 (Willy Interscience, New York) Chapter 17.

CORKUM, P.B. (1993), *Phys. Rev. Lett.* **71**, 1994.

Defranceschi, M. and C. Le Bris, (eds) (2001). *Mathematical Models and Methods for Ab-Initio Quantum Chemistry*, Lecture Notes in Chemistry, vol. 74 (Springer, New York).

DREIZLER, R.M. and E.K.U. GROSS (1990), *Density Functional Theory* (Springer, Berlin).

EHLOTZKY, F. (2001), *Phys. Rep.* **345**, 175.

FEDOROV, M.V. and S.P. ROSHCHUPKIN (1984), *J. Phys. A* **17**, 3143.

GAVRILA, M. (1992), *Atoms in Intense Laser Fields* (Academic Press, New York).

GERHART, P.M., R.J. GROSS and J.I. HOCHSTEIN (1992), *Fundamentals of Fluid Mechanics* (Addison-Wesley, New York).

GUERMOND, J.L., S. HUBERSON and W.Z. SHEN (1993), *J. Comp. Phys.* **108**, 557.

GUERMOND, J.L. and H.Z. LU (2000), *Comp. Fluids* **29**, 525.

GYGI, F. and G. GALLI (1995), *Phys. Rev. B* **52**, 2229.

HAROCHE, S. (2000), *Sciences aux Temps Ultracourts* (Académie des Sciences (Fr.)), Editions TEC.

HAUSBO, P. (2000), *J. Comp. Phys.* **159**, 274.

HENTSCHEL, M. et al. (2001), *Nature* **414**, 509.

HERTZBERG, G. (1951), *Spectra of Diatomic Molecules* (Van Nostrand, Amsterdam).

HOFFMAN, J.D. (1992), *Numerical Methods for Engineers and Scientists* (McGraw-Hill, New York).

HUSIMI, K. (1953), *Prog. Theor. Phys.* **9**, 381.

KAWATA, I. and A.D. BANDRAUK (2002), (in preparation).

KAWATA, I. and H. KONO (1999), *J. Chem. Phys.* **111**, 9498.

KELDYSH, L.V. (1965), *Sov. Phys. JETP* **20**, 1307.

KONO, H. et al. (1997), *J. Comput. Phys.* **130**, 148.

KRAUSE, J.L., K.J. SCHAFFER and K.C. KULANDER (1991), *Chem. Phys. Lett.* **178**, 573.

KULANDER, K. (1987), *Phys. Rev. A* **36**, 2726.

LU, H.Z. and A.D. BANDRAUK (2001), *J. Chem. Phys.* **115**, 1670.

NGUYEN, N.A. and T.T. NGUYEN-DANG (2000), *J. Chem. Phys.* **112**, 1229.

PAULI, W. and M. FIERZ (1938), *Nuovo Cimento* **15**, 167.

QUARTERONI, A. (1991), *Surv. Math. Ind.* **1**, 75.

RABITZ, H. and G. TURINICI, E. BROWN (2003). Ciarlet, P.G. and C. Le Bris, (eds.), *Computational Chemistry*, Handbook of Numerical Analysis, vol. X (Elsevier, Amsterdam), pp. 833–887.

THALLER, B. (1992), *The Dirac Equation* (Springer-Verlag, Berlin).

TRUSCOTT, W.S. (1993), *Phys. Rev. Lett.* **70**, 1900.

WALSER, M.W., C.H. KEITEL, A. SCRINZI and T. BRABEC (2000), *Phys. Rev. Lett.* **85**, 5082.

WARE, A.F. (1998), *SIAM Rev* **40**, 838.

WELTON, T.A. (1948), *Phys. Rev.* **74**, 1157.

WIEDEMANN, H. and T. MOSTOWSKI (1994), *Phys. Rev. A* **49**, 2719.

WUNDERLICH, C., E. KOBLER, H. FIGGER and T.W. HÄNSCH (1997), *Phys. Rev. Lett.* **78**, 2333.

YU, H. and A.D. BANDRAUK (1995), *J. Chem. Phys.* **102**, 1257.

ZUO, T. and A.D. BANDRAUK (1996), *Phys. Rev. A* **54**, 3254.

ZUO, T., A.D. BANDRAUK and P.B. CORKUM (1996), *Chem. Phys. Lett.* **259**, 313.

Control of Quantum Dynamics: Concepts, Procedures, and Future Prospects

Herschel Rabitz

Department of Chemistry, Princeton University, Princeton, NJ, USA.
E-mail: hrabitz@princeton.edu; URL: http://www.princeton.edu/~hrabitz

Gabriel Turinici

INRIA Rocquencourt, B.P. 105, Le Chesnay cedex, France
E-mail: Gabriel.Turinici@inria.fr; URL: http://www-rocq.inria.fr/Gabriel.Turinici

CERMICS–ENPC, Champs sur Marne, Marne la Vallée cedex, France. (*G. Turinici*)
E-mail: Gabriel.Turinici@inria.fr; URL: http://www-rocq.inria.fr/Gabriel.Turinici

Eric Brown

Program in Applied and Computational Mathematics, Princeton University, Princeton, NJ, USA. (E. Brown)
E-mail: ebrown@princeton.edu

1. Introduction

1.1. Historical perspective on quantum control

Along with the development of laser technology, a rising interest has naturally appeared in connection with using lasers to influence matter at the quantum dynamical level. This new field of research, commonly designated as "quantum control", has roots that go back to the earliest days of laser development in the 1960s. What distinguishes quantum control from the traditional means of chemical manipulation is the use of delicate quantum wave interferences to alter the outcome of molecular scale dynamics phenomena. Research in quantum control accelerated in the 1980s, with the first

Computational Chemistry
Special Volume (C. Le Bris, Guest Editor) of
HANDBOOK OF NUMERICAL ANALYSIS
P.G. Ciarlet (Editor)

successful results in SHI, WOODY and RABITZ [1988], TANNOR and RICE [1985], SHA-
PIRO and BRUMER [1989] driven by the recognition of the role of quantum interference
and the need for rigorous design tools. The introduction of the evolutionary paradigm
(see Section 4.5) in JUDSON and RABITZ [1992] and the accompanying experimental
successes of ASSION, BAUMERT, BERGT, BRIXNER, KIEFER, SEYFRIED, STREHLE and
GERBER [1998], LEVIS, MENKIR and RABITZ [2001], WEINACHT, AHN and BUCKSBAUM
[1999], BARDEEN, YAKOVLEV, WILSON, CARPENTER, WEBER and WARREN [1997],
BARDEEN, YAKOVLEV, SQUIER and WILSON [1998], HORNUNG, MEIER and MOTZKUS
[2000], KUNDE, BAUMANN, ARLT, MORIER-GENOUD, SIEGNER and KELLER [2000] in
the 1990s have brought an exciting new set of theoretical questions and laboratory possibilities to the field. The time scales involved lie mostly within the femtosecond
(10^{-15}) – picosecond (10^{-12}) range with spatial dimensions extending from one or
two atoms to large polyatomic molecules and solid state structures.

In practice the outcome of a control experiment is measured in terms of quantum
observables (e.g., selective dissociation of interatomic bonds) associated with Hermitian operators acting on the system at hand. The control process consists of steering
the appropriate observable or wavefunction from the initial state to a final desired
state. In the laboratory this is generally done by time-varying laser fields. We refer
the reader to Section 2 for a precise mathematical transcription of these concepts.

Although prospects for industrial applications motivated much of the early research
on quantum control, current applications span a wide range, from high harmonic
generation (BARTELS, BACKUS, ZEEK, MISOGUTI, VDOVIN, CHRISTOV, MURNANE
and KAPTEYN [2000]) and fast switching in semi-conductors (KUNDE, BAUMANN,
ARLT, MORIER-GENOUD, SIEGNER and KELLER [2000]) to Hamiltonian identification
(see Section 5.5).

1.2. Multidisciplinarity

Although research on quantum control was initiated within the field of physical chemistry, the subject has developed to involve researchers from myriad fields, including
engineering, mathematics, computer science and physics. The introduction of control
tools from the engineering contexts opened the field to the influence of (applied)
mathematicians; later, computer scientists also became involved through their interest
in quantum computing. Besides techniques employed by control specialists working
on controllability, observability and stabilization issues, the applied mathematics tools
involved include the resolution of the control problem through deterministic (direct
optimization or critical point equations) or stochastic (evolutionary algorithms)
approaches. The variety of research paradigms contributing to quantum control has
proved to be beneficial to the field and will likely be central to future advances.

1.3. Outline

This volume addresses numerical methods in various aspects of the chemical sciences
and the present chapter contributes in that fashion to the subject of quantum control.
However, in order to appreciate how numerical methods are relevant for quantum

control it is essential to understand how they fit into the overall subject including its laboratory implementations. Thus, the chapter attempts to give a full perspective on the subject including its theoretical foundations as well as its current state in the laboratory.

The balance of this review will proceed as follows: after an introduction of the fundamental concepts in Section 2 we present in Section 3 theoretical results regarding on the controllability of bilinear quantum systems. Then, in Section 4 we present the available numerical and experimental algorithmical approaches employed to control quantum phenomena. Present topics of interest and open questions are the object of Section 5. Concluding remarks are presented in Section 6. The text draws on many works in the literature and especially a recent prospectus (BROWN and RABITZ [2002]) on the field.

2. Basic principles

2.1. Mathematical formulation of the Hamiltonian and control law

Each particular control setting requires an appropriate quantum control model. The section will be mostly devoted to the description of the bilinear dipole coupling, which is often employed in practice. However, some other control paradigms will be presented and their domain of applicability and specific capabilities discussed. Controlling a quantum system requires the introduction of external interactions. Here, we will only consider control tools that act at the atomic/molecular level, with other, more classical tools such as temperature, pressure, catalysts, etc., being outside of our scope. The primary control for quantum systems discussed here will be the electric field of a laser.

Most of this paper will work under the assumption that the system to be controlled can be characterized by its state function $\psi(t)$. This is a proper representation for isolated systems starting in a pure state; the complementary case arises, for example, in collisional or condensed regimes, when a density operator $\rho(t)$ must be introduced to describe the statistical mixture of states making up the system. The density operator formulation will be discussed where relevant.

Consider a quantum system that evolves from the initial state $\psi(t = 0) = \psi_0$. The objective of quantum control can generally be expressed as the desire for the expectation value

$$\langle O(t) \rangle = \langle \psi(t) | O | \psi(t) \rangle \tag{2.1.1}$$

of some predefined observable operator O to be within a prescribed target set at the final time $t = T$. More general quantum control problem formulations may deal with several observable operators $O_1, \ldots, O_k$ $(k > 1)$ and additional intermediary times $0 \leqslant t_j \leqslant T$ $(j = 1, 2, \ldots)$.

In the absence of any external control influence, evolution of the state function $\psi(t)$ under the Schrödinger equation is determined by the free Hamiltonian H_0, which by assumption does not yield dynamics producing the desired expectation values. Quantum control theory considers the addition of a laboratory accessible control law term

$C(t)$ to the Hamiltonian in order to achieve these objectives, which makes $H = H_0 + C(t)$ the new Hamiltonian of the system and

$$i\hbar \frac{\partial \psi(t)}{\partial t} = [H_0 + C(t)]\psi(t) \tag{2.1.2}$$

the equation of motion. Appropriate regularity assumptions on the control law must be enforced so that the evolution of $\psi(t)$ is well defined. For example, a common control law for lasers has the form

$$C(t) = -\mu\varepsilon(t), \quad \varepsilon = (\varepsilon_i)_{i=1}^3, \quad \varepsilon_i(t) \in L^2[0,T], \quad i = 1, 2, 3,$$

where μ is the electric dipole operator, $\varepsilon(t)$ is the applied electric field, and the index i refers to spatial orientation.

REMARK 2.1. Depending on the problem, one may go beyond the first-order, bilinear term in Eq. (2.1.2) when describing the interaction between the laser and the system, cf. DION, BANDRAUK, ATABEK, KELLER, UMEDA and FUJIMURA [1999], DION, KELLER, ATABEK and BANDRAUK [1999].

Additional admissibility conditions may ensure that $\varepsilon(t)$ obeys laboratory limitations on the range of achievable laser frequencies, intensities, energy, or other criteria. In some applications, additional possibilities for $C(t)$ arise. These include
- the use of magnetic fields, in which case the control law becomes $-\mu_m B(t)$, where μ_m is the magnetic dipole operator and $B(t)$ is the magnetic field, and
- the use of materials whose design specifications themselves take the form of a control law, such as for quantum electron transport in semiconductors with variable material composition considered as the control.

Here, however, we will confine the discussion to time-dependent controls based on an external electric field $\varepsilon(t)$ coupled to the system through a dipole μ.

In some cases, it may be possible to obtain an adequate control description by replacing the Schrödinger equation with a classical representation of the system dynamics (see Section 2.6). This is especially true for interatomic phenomena, because the *de Broglie* wavelength associated with atoms is often short relative to interatomic length scales. While the relationships between classical and quantum models of molecular evolution have been extensively investigated, for example, DAVIS and HELLER [1984], KULANDER and OREL [1981], LEFORESTIER [1986], DARDI and GRAY [1982], the implications of these relationships for control are not completely understood and will be addressed later in the context of the quantum character of the control problem.

Assuming knowledge of H_0 and a well-defined control law $C(t)$, Eq. (2.1.2) or its classical equivalent represents a complete model of the system of interest. If $C(t)$ is given a priori, the solution of Eq. (2.1.2) is a standard numerical problem in time-dependent quantum mechanics. However, the essence of the control problem is to find $C(t)$ such that the objectives in Eq. (2.1.1) are met, and since at least one of the control objectives lies in the future, this task presents some additional challenges. In particular, the Hamiltonian depends on the future state of the system through the control objectives, as can be formally represented by the expression $C(t) = C(\psi(s))$:

$s \in [t, T]$). This noncausality introduces an entirely new set of mathematical issues which are not present in standard quantum or classical dynamics but are inherent to the theory and practice of temporal control in engineering and mathematical systems theory (cf. SONTAG [1998], BROCKETT [1970], KAILATH [1980], KHALIL [1996], NIJMEIHER and VAN DER SCHAFT [1990]). Their implications for the quantum regime are central issues. The formulation above is summarized in the following definition:

DEFINITION 2.2. The quantum control problem consists of determining a control law $C(t)$ that causes the system to optimally achieve the desired expectation values while possibly also satisfying auxiliary conditions. Quantum control theory encompasses methods of determining these control laws, their general properties, and their relationship to the underlying physical system and evolving quantum states.

2.2. Optimal control

Several approaches for determining control laws $C(t)$ will be discussed in this chapter. The present section introduces the concept of optimal control theory for this purpose; a more detailed algorithmic analysis will be given in Section 4. An extensive literature on optimal control theory can be found in classical engineering and mathematical systems theory (e.g., SONTAG [1998], LUENBERGER [1979], MOHLER [1983]) and increasingly in quantum mechanics (e.g., PEIRCE, DAHLEH and RABITZ [1988], ZHAO and RICE [1994], Chapter 6 of RICE and ZHAO [2000] and references therein). Considering control law design as an optimization problem is quite natural, as attaining the best possible final level of control is always the goal; further, optimization is essential when there are competing physical objectives that must simultaneously be met.

The *optimal control* approach seeks to optimize a cost functional J, which includes both terms that describe how well the objective has been met and terms that penalize undesirable effects. One simple example of a cost functional is

$$J(\varepsilon) = \langle \psi(T)|O|\psi(T)\rangle - \alpha \int_0^T \varepsilon^2(t)\mathrm{d}t, \qquad (2.2.1)$$

where $\alpha > 0$ is a parameter and O is the observable operator (positive semidefinite in this case) that specifies the goal. In mathematical terms, the observable O is a self-adjoint operator that acts on $\psi(T)$; in the case above, the goal is to achieve a large value $\langle \psi(T)|O|\psi(T)\rangle$. Note that in general attaining the maximum possible value of $\langle \psi(T)|O|\psi(T)\rangle$ comes at the price of a large laser fluence $\int_0^T \varepsilon^2(t)\mathrm{d}t$, so that the optimum evolution will strike a balance (weighted by α) between laser fluence and operator expectation values.

2.3. More nonlinear formulations

The quantum control problem becomes more complex when the control law and the free Hamiltonian cannot be treated independently. An important example of this phenomenon is intense-field laser control of molecular motion, where the electric field can directly alter the dipole operator through its manipulation of the electronic degrees of freedom:

$$C(t) = -\mu(\varepsilon(t))\varepsilon(t). \tag{2.3.1}$$

In simple cases, the relation in Eq. (2.3.1) may be expanded in terms of a low order polynomial in $\varepsilon(t)$ whose coefficients are the electric moments and polarizabilities of the system. Of special interest are situations in which the nonlinear structure may affect the controllability of the system (including the positive case in which this interaction makes a previously inaccessible target reachable).

REMARK 2.3. To date, very few mathematical studies exist to treat the situation where the control law has a nonlinear dependence on the control field; it is not clear whether successful approaches will draw upon existing methodology for proving, for example, quantum controllability (see Section 3) or upon new tools from mathematical systems theory.

The circumstances motivating Remark 2.3 can also be viewed from the larger perspective of a controllability analysis simultaneously including electronic and nuclear motion, for example, as in CANCÈS and LE BRIS [1999]. In the latter circumstance the control field will enter the Hamiltonian linearly, but at the expense of explicitly including the electronic degrees of freedom. The same comment also applies to the performance of optimal control designs in the strong-field regime. The practical importance of investigating the latter domain has recently been demonstrated experimentally (LEVIS, MENKIR and RABITZ [2001]).

A different formulation is necessary when the back action of the quantum medium upon the propagating control field is significant. In this case the medium is called *optically dense*. This scenario has been examined experimentally for a vapor of sodium (SHEN, SHI, RABITZ, LIN, LITTMAN, HERITAGE and WEINER [1993]), and the topic is of practical importance because the controlled medium will be dense in any application directed toward collecting large amounts of product. Optically dense media can interact with the electric field to alter its phase and/or amplitude structure as it propagates. In order to model this effect, the Schrödinger equation must be coupled with Maxwell's equations, for example, WANG and RABITZ [1996].

2.4. Density matrix formulation

In practical applications the medium will be at a finite temperature; in this case (and in any situation that concerns a statistical mixture of quantum states) the density operator formulation is necessary. In this formulation, the time evolution is given by the quantum Liouville equation (COHEN-TANNOUDJI, LIU and LALOË [1997]):

$$\frac{d\rho(t)}{dt} = \frac{1}{i\hbar}[H(t), \rho(t)], \tag{2.4.1}$$

and expectation values are calculated as

$$\langle O(t)\rangle = \text{Tr}(\rho(t)O).$$

Upon introducing a cost functional as in Eq. (2.2.1), quantum systems described by density operators can also be treated by optimal control theory.

2.5. Special control schemes: Pump–dump schemes, STIRAP

In addition to the general approach outlined below, special techniques for control law design have been developed for cases in which a priori specification of control mechanisms is possible. These methods include the weak-field regime, time-resolved "pump–dump" and simulated Raman adiabatic passage (STIRAP) schemes. Under particular quantum dynamics approximations and/or assumptions, these techniques allow for the derivation of closed form expressions for control laws that optimally accomplish certain control objectives under their specified conditions. The literature on related theoretical developments is extensive and will not be presented here (Chapters 3–5 of RICE and ZHAO [2000] and references therein, BROWN and SIBLEY [1998], SHAH and RICE [1999], MISHIMA and YAMASHITA [1999b], HOKI, OHTSUKI, KONO and FUJIMURA [1999], DE ARAUJO and WALMSLEY [1999], MISHIMA and YAMASHITA [1999a], CAO and WILSON [1997], BARDEEN, CHE, WILSON, YAKOVLEV, CONG, KOHLER, KRAUSE and MESSINA [1997], BARDEEN, CHE, WILSON, YAKOVLEV, APKARIAN, MARTENS, ZADOYAN, KOHLER and MESSINA [1997], AGARWAL [1997], GRØNAGER and HENRIKSEN [1998]). For experimental developments see Chapters 3–5 of RICE and ZHAO [2000] and references therein, BARDEEN, CHE, WILSON, YAKOVLEV, CONG, KOHLER, KRAUSE and MESSINA [1997], BARDEEN, CHE, WILSON, YAKOVLEV, APKARIAN, MARTENS, ZADOYAN, KOHLER and MESSINA [1997], MESHULACH and SILBERBERG [1998], PASTIRK, BROWN, ZHANG and DANTUS [1998].

2.6. Classical mechanics formulations

Classical modeling of quantum systems is a common and often successful technique, and it should have a level of applicability in molecular control. This section attempts to discuss its applicability, or the quantum character of molecular scale control. Nonclassical characteristics of dynamical behavior include tunneling, quantization of energy levels, and interference processes, but it is not clear which of these characterizations are most relevant to defining the quantum nature of a control problem. Understanding of this issue could be used to estimate the loss of reliability (i.e., defined upon comparison to the analogous quantum system response to the classically designed field) in resorting to the classical optimal control formulation.

Some aspects of this topic have been addressed in SCHWIETERS and RABITZ [1993], where quantum $C_q(t)$ and classical $C_c(t)$ control laws corresponding to equivalent representations of specific control problems are compared. The equations of motion analogous to Eq. (2.1.2) are:

$$\frac{dq_i^l}{dt} = \frac{\partial H}{\partial p_i^l}, \quad \frac{dp_i^l}{dt} = -\frac{\partial H}{\partial q_i^l}, \quad l = 1, \ldots, N_c, \tag{2.6.1}$$

and expectation values for the classical system are given by

$$\langle O_c \rangle = \sum_{l=1}^{N_c} \Gamma_l \bar{O}(\mathbf{q}^l, \mathbf{p}^l),$$

where H is the classical Hamiltonian of the system, i ranges over the particle co-ordinates, l indexes initial conditions $q^l(0)$, $p^l(0)$, and the weights Γ_l for the N_c initial conditions are chosen to mimic as best as possible the probability distribution function for the corresponding quantum system. Here, $\bar{O}$ is a classical observable corresponding to its quantum analog. It should be noted that the ordinary differential equations in Eq. (2.6.1) for some cases may be more expensive to solve than their quantum counterpart in Eq. (2.1.2). One motivation for considering classical optimal control design is for the physical insight possible with classical mechanics.

Optimal control theory has been used, for example, in SCHWIETERS and RABITZ [1993] to separately design a control field $\varepsilon(t)$ that minimizes the difference between $\langle O \rangle$ and $\langle O_c \rangle$ and the difference between each of these expectation values and the control objectives on $[0, T]$. For the example of a Morse oscillator, it was found that an optimal control law designed in this fashion produced very close agreement between $\langle O \rangle$ and $\langle O_c \rangle$. This result suggests that in some cases classically designed controls can also be successful as quantum controls. In related work SCHWIETERS and RABITZ [unpublished results], a method was developed for determining potentials under which evolving classical and corresponding quantum systems give similar values of classical and quantum observables; the approach met with considerable success for the control of dissociative flux and displacement. In general it is however not known for what classes of Hamiltonians and control objectives the quantum control problem can be adequately addressed using classical equations of motion. Because interference itself is a nonclassical phenomenon, this problem is related to the considerations of decoherence in Section 5.1.

3. Controllability of quantum mechanical systems

Prior to addressing the computation of the control law, it is natural to ask if the quantum control problem is well-posed such that a control law exists which will cause the objectives and auxiliary conditions to be (precisely) satisfied. Even if the answer to the latter question is negative, one may still be satisfied with achieving the control objectives (maybe only partially) through the techniques of optimal control. The fundamental importance of addressing controllability has long been recognized in engineering control applications; the broad literature on the classical aspects of the subject includes many comprehensive texts which cover linear (BROCKETT [1970], KAILATH [1980]) and nonlinear (SONTAG [1998], KHALIL [1996], NIJMEIHER and VAN DER SCHAFT [1990]) controllability. In addition, several works have considered various aspects of quantum controllability (e.g., RICE and ZHAO [2000], HUANG, TARN and CLARK [1983], JUDSON, LEHMANN, RABITZ and WARREN [1990], RAMAKRISHNA, SALAPAKA, DAHLEH, RABITZ and PIERCE [1995], TURINICI [2000c], TURINICI and

RABITZ [2001a], TURINICI and RABITZ [2001b], GIRARDEAU, SCHIRMER, LEAHY and KOCH [1998], GIRARDEAU, INA, SCHIRMER and GULSRUD [1997]).

Consider a quantum system (isolated from external influences for the moment) with internal Hamiltonian H_0 that is prepared in the initial state $\psi_0(x)$ described here in the coordinate representation; its dynamics obeys the time dependent Schrödinger equation. Denoting by $\psi(x, t)$ the state at the time t one can write the evolution equations for the free system:

$$
\begin{cases}
i\hbar \dfrac{\partial}{\partial t} \psi(x,t) = H_0 \psi(x,t), \\
\psi(x,t=0) = \psi_0(x), \quad \|\psi_0\|_{L^2(\mathbb{R}^\gamma)} = 1.
\end{cases} \tag{3.0.1}
$$

In the presence of external interactions that will be taken here as a control field amplitude $\varepsilon(t) \in \mathbb{R}, t \geqslant 0$, coupled to the system through a time-independent (e.g., dipole) operator μ, the (controlled) dynamical equations read:

$$
\begin{cases}
i\hbar \dfrac{\partial}{\partial t} \psi_\varepsilon(x,t) = H_0 \psi_\varepsilon(x,t) - \varepsilon(t)\mu\psi_\varepsilon(x,t) = H\psi_\varepsilon(x,t), \\
\psi_\varepsilon(x,t=0) = \psi_0(x).
\end{cases} \tag{3.0.2}
$$

REMARK 3.1. For the sake of simplicity we have chosen in this section to treat situations with only one control law present. Extensions to many control laws representing coupling via various other operators are also available (especially for finite dimensional settings); we refer the reader to RAMAKRISHNA, SALAPAKA, DAHLEH, RABITZ and PIERCE [1995], ALBERTINI and D'ALESSANDRO [2001], TURINICI, RAMAKHRISHNA, LI and RABITZ [2002] for details.

The L^2 norm of ψ_ε is conserved throughout the evolution:

$$
\| \psi_\varepsilon(x,t) \|_{L^2(\mathbb{R}^\gamma)} = \| \psi_0 \|_{L^2(\mathbb{R}^\gamma)}, \quad \forall t > 0, \tag{3.0.3}
$$

so the state (or wave-) function $\psi(t)$, evolves on the (complex) unit sphere $S = \{\psi \in L^2(\mathbb{R}^\gamma) : \| \psi \|_{L^2(\mathbb{R}^\gamma)} = 1\}$ according to the Schrödinger equation (3.0.2) from the initial state ψ_0 to some final state $\psi(T)$.

The study of the controllability of Eq. (3.0.2) is concerned with identifying the set of final states $\psi(T)$ that can be obtained from a given initial state ψ_0 for all admissible controls. To date, very different results are available for the infinite and finite dimensional settings: while controllability is reasonably well understood for finite dimensional systems, no positive results have been obtained for infinite dimensional systems.

3.1. Infinite dimensional controllability

Very few results are available concerning the controllability of Eq. (3.0.2) in its infinite dimensional form. Generic *negative* results have been obtained, as is the following (Theorem 1 from TURINICI [2000c]; see also BALL, MARSDEN and SLEMROD [1982], TURINICI [2000b]):

THEOREM 3.2. *Let S be the complex unit sphere of $L^2(\mathbb{R}^\gamma)$. Let μ be a bounded oper-ator from the Sobolev space $H_x^2(\mathbb{R}^\gamma)$ to itself and let H_0 generate a C^0 semigroup of bounded linear operators on $H_x^2(\mathbb{R}^\gamma)$. Denote by $\psi_\varepsilon(x, t)$ the solution of (3.0.2). Then the set of attainable states from ψ_0 defined by*

$$\mathcal{AS} = \bigcup_{T>0}\{\psi_\varepsilon(x, T); \varepsilon(t) \in L^2([0, T])\} \tag{3.1.1}$$

is contained in a countable union of compact subsets of $H_x^2(\mathbb{R}^\gamma)$. In particular its com-plement with respect to $S : \mathcal{N} = S\backslash\mathcal{AS}$ is everywhere dense on S. The same holds true for the complement with respect to $S \cap H_x^2(\mathbb{R}^\gamma)$.

The theorem implies that for any $\psi_0 \in H_x^2(\mathbb{R}^\gamma) \cap S$, within any open set around an arbi-trary point $\psi \in H_x^2(\mathbb{R}^\gamma) \cap S$ there exists a state unreachable from ψ_0 with L^2 controls.

REMARK 3.3. The lack of positive controllability results to complement Theorem 3.2 should be regarded as a failure of available control theory tools to provide insight into controllability rather than as an actual restriction. It is believed that new tools and concepts will make positive results possible, especially since such results are available for the finite dimensional setting (cf. Section 3.2).

Truncating an infinite-dimensional quantum control problem to a finite-dimensional problem (i.e., so that evolution takes place in a finite dimensional vector space) changes the nature of both the control and Hamiltonian operators and the set of states available as candidate members of reachable sets. The concern is to characterize these effects by asking how a controllability result obtained in a finite-dimensional space relates to the original infinite dimensional problem from which it was derived; there are also inherently finite dimensional quantum systems (as with spins) where the latter consideration does not arise. The controllability of finite dimensional systems is the subject of the next section.

3.2. Finite dimensional controllability

3.2.1. Introduction

Let $D = \{\psi_i(x); i = 1, \ldots, N\}$ be an orthonormal basis for a finite dimensional sub-space of $L^2(\mathbb{R}^\gamma)$ of interest (for instance the vector space spanned by the first N eigenstates of the internal Hamiltonian H_0 in Eq. (3.0.1)). Denote by M the linear space that D gener-ates, and let A and B be the matrices of the operators $-i H_0$ and $-i \mu$ respectively, with re spect to this basis. In order to exclude trivial control settings, it is supposed that $[A, B] \neq 0$ (the Lie bracket $[\cdot, \cdot]$ is defined as $[U, V] = UV - VU$). Note that since H_0 and μ are Hermitian operators, the matrices A and B are skew-Hermitian.

Let us denote by $c_\varepsilon(t) = (c_{\varepsilon i}(t))_{i=1}^N$ the coefficients of $\psi_i(x)$ in the expansion of the evolving state $\psi(t, x) = \sum_{i=1}^N c_{\varepsilon i}(t)\psi_i(x)$; then Eq. (3.0.2) reads

$$\begin{cases} \dfrac{\mathrm{d}}{\mathrm{d}t}c_\varepsilon(t) = Ac_\varepsilon(t) - \varepsilon(t)Bc_\varepsilon(t), \\ c_\varepsilon(t = 0) = c_0, \end{cases} \tag{3.2.1}$$

$$c_0 = (c_{0i})_{i=1}^N, c_{0i} = \langle\psi_0, \psi_i\rangle_{L^2(\mathbb{R}^\gamma)} \tag{3.2.2}$$

(where atomic units are used, i.e., we set $\hbar = 1$). Note as in Eq. (3.0.3) that the system (3.2.1) evolves on the unit sphere S_N of $L^2(\mathbb{R}^v) \cap M$ which reads

$$\sum_{i=1}^{N} |c_{\varepsilon i}(t)|^2 = 1, \forall t \geqslant 0. \tag{3.2.3}$$

Note also that the solution $c_\varepsilon(t)$ of Eq. (3.2.1) can be written

$$c_\varepsilon(t) = U_\varepsilon(t)c_0, \tag{3.2.4}$$

where the time evolution operator $U_\varepsilon(t)$ is the solution of the following

$$\begin{cases} \dfrac{\mathrm{d}}{\mathrm{d}t} U_\varepsilon(t) = AU_\varepsilon(t) - \varepsilon(t)BU_\varepsilon(t), \\[2mm] U_\varepsilon(t = 0) = I_{N \times N}. \end{cases} \tag{3.2.5}$$

The matrix $U(t)$ evolves in the Lie group of unitary matrices $U(N)$, or, if both the matrices A and B have zero trace, in the Lie group of special unitary matrices SU (N). Eq. (3.2.5) also prescribes the evolution of the density matrix operator: if the system starts in the mixture of states represented by the density operator ρ_0 then its evolution is given by the formula

$$\rho(t) = U(t)_{\rho_0} U(t)^\dagger, \tag{3.2.6}$$

where for any matrix X we denote by $X^\dagger$ its transpose-conjugate.

REMARK 3.4. Note that since $U(t)$ is unitary $\rho(t)$ has the same eigenvalues as $\rho(0)$.

We now introduce:

DEFINITION 3.5. Denote by $\mathcal{U}$ the set of all admissible control laws $\varepsilon(t)$. The system described by the state $c_\varepsilon(t)$ is called state-controllable if for any c_i, $c_f \in S_N$ there exists $0 < \tau < \infty$ and $\varepsilon \in \mathcal{U}$ such that $c_\varepsilon(t = \tau) = c_f$, where $c(t)$ satisfies Eq. (3.0.2) with $c_\varepsilon(t = 0) = c_i$.

Within the density matrix formulation the relevant definition of controllability becomes:

DEFINITION 3.6. The system described by the evolution of the density operator $\rho(t)$ satisfying Eqs. (3.2.5), (3.2.6) is called density-matrix-controllable if for any two density operators ρ_i and ρ_f compatible in the sense that there exists an unitary matrix U such that $\rho_f = U \rho_i U^\dagger$ there exists a control law $\varepsilon(t) \in \mathcal{U}$ such that given the initial condition $\rho(0) = \rho_i$ then $\rho(\tau) = \rho_f$ for some finite time τ.

REMARK 3.7. Because pure states my be represented by density matrices a system that is density-matrix-controllable is also state-controllable.

3.2.2. Lie group methods
Let us introduce the following
DEFINITION 3.8. A subset $\mathcal{T}$ of $U(N)$ (or $SU(N)$) is said to be *transitive (to act transitively)* on the sphere S_N if for any c_i, $c_f \in S_N$ there exists $g \in \mathcal{T}$ such that $c_f = gc_i$.

With this definition and considering Eq. (3.2.4) it follows that

- state-controllability of the wavefunction as in Definition 3.5 is equivalent to requiring that the set of all matrices attainable from identity $\{U_\varepsilon(t); 0 \leqslant t \leqslant \infty; \varepsilon \in \mathcal{U}, U$ verify Eq. (3.2.5)$\}$ be transitive on the sphere S_N, while
- density-matrix-controllability is equivalent to requiring that the set of all matrices attainable from identity be at least $SU(N)$.

Remarkable examples of transitive subgroups of $U(N)$ are $U(N)$ itself, $SU(N)$ and, when N is even, the symplectic matrices $Sp(N/2)$. It can be proven (see ALBERTINI and D'ALESSANDRO [2001]) that, except for some special values of N and up to an isomorphism, these are the only transitive subgroups arising in quantum control.

The important result that turns this remark into a powerful tool for studying the controllability of bilinear systems is that the set of all matrices attainable from the identity via Eq. (3.2.5) is given by the connected Lie subgroup e^L of the Lie algebra L generated by A and B (when this Lie group is compact) (cf. RAMAKRISHNA, SALAPAKA, DAHLEH, RABITZ and PIERCE [1995], ALBERTINI and D'ALESSANDRO [2001]). The group e^L is called the *dynamical Lie group* of the system.

Taking $\mathcal{U}$ to be the set of all piecewise continuous functions (unconstrained in magnitude) yields the following result:

THEOREM 3.9. (RAMAKRISHNA, SALAPAKA, DAHLEH, RABITZ and PIERCE [1995])
A sufficient condition for the density-matrix- (thus state-) controllability of the quantum system in Eq. (3.2.1) is that the Lie algebra L generated by A and B has dimension N^2 (as a vector space over the real numbers).

Furthermore, if both A and B are traceless then a sufficient condition for the density-matrix- (thus state-) controllability of quantum system is that the Lie algebra L has dimension $N^2 - 1$.

A following result builds on Theorem 3.9 to gives the necessary and sufficient condition of state controllability:

THEOREM 3.10. (ALBERTINI and D'ALESSANDRO [2001]) *The system is state controllable if and only if the Lie algebra L generated by A and B is isomorphic (conjugate) to $sp(\frac{N}{2})$ or to $su(N)$, if the dimension N is even, or to $su(N)$, if the dimension N is odd (with or without the iI, where I is the identity matrix).*

Theorem 3.9 lends itself to algorithmic (numerical) implementation: as soon as the matrices A and B that characterize the system are given, one can compute (numerically for instance) the Lie algebra they generate and obtain its dimension. However, except for small systems, this test becomes rapidly very computationally expensive; additional results are therefore required in order to shed some light on the relationship between controllability and the structure of the A and B matrices. Two studies in this direction are available: one, from SCHIRMER, FU and SOLOMON [2001], FU, SCHIRMER and SOLOMON [2001] is presented in this section and the second, the "connectivity graph" approach, is described in the next section (cf. TURINICI [2000c], TURINICI and RABITZ [2001a], TURINICI and RABITZ [2001b], TURINICI [2000b], TURINICI [2000a]).

When the basis $D = \{\psi_i(x); i = 1, \ldots, N\}$ is composed of eigenstates of the internal Hamiltonian H_0 the matrix A is diagonal with purely imaginary elements $-iE_k$, where E_k are the eigenvalues of H_0, $k = 1, \ldots, N$. Let $\delta_k = E_k - E_{k+1}$ for $n = 1, \ldots, N - 1$.

THEOREM 3.11. SCHIRMER, FU and SOLOMON [2001] *Suppose that the matrix B of the interaction operator $-i\mu$ with respect to the basis D is such that $B_{k,l} = 0$ for $|k - l| \neq 1$ and $B_{k,l} \neq 0$ for $|k - l| = 1$, $k, l = 1, \ldots, N$. Then if either*
(1) $\delta_1 \neq 0$ and $\delta_k \neq \delta_1$ for $k = 1, \ldots, N - 1$, or
(2) $\delta_{N-1} \neq 0$ and $\delta_{N-1} \neq \delta_k$ for $k = 1, \ldots, N - 1$
the dynamical Lie group of the system $A - \varepsilon(t) B$ with A and B as in Eq. (3.2.5) is at least $SU(N)$. If in addition $\operatorname{Tr} A \neq 0$ then the dynamical Lie group is $U(N)$. In both cases the system is density-matrix- (thus state-) controllable.

REMARK 3.12. Although the hypothesis of the theorem above are somewhat strong, it allows for the determination of controllability using only generic properties of the system under study, that is, without the need to know exactly the matrices A and B. We will see later in this section (see Remark 3.17) other results concerning the controllability of the wavefunction that do not require precise evaluation of the A and B matrices.

While the results above hold in the case that the control field amplitudes are not bounded, an open question suggested in this work is the extension of the result to stricter (and more realistic) admissibility conditions: can the Lie algebraic controllability conditions (e.g., RAMAKRISHNA, SALAPAKA, DAHLEH, RABITZ and PIERCE [1995]) be extended to treat the case where both the amplitude and the frequency of the control field are bounded from above and below? This issue has practical significance as it prescribes real laboratory conditions.

REMARK 3.13. Other open problems refer to the controllability within the product state space of the coupled Schrödinger–Maxwell equations for optically dense media. While the Schrödinger–Maxwell system has a product state space representing both $\rho(t)$ (the density operator) and $\varepsilon(t)$, expectation values depend only on $\rho(t)$, which is the usual focus of controllability studies. This inspires the question: in what, if any, cases is it possible for the Schrödinger–Maxwell system to be controllable in the state space of the quantum state but possibly not controllable in that of the electric field, or vice-versa? The latter case of controlling the electric field is of importance in the allied subject of optical field propagation (e.g., XIA, MERRIAM, SHARPE, YIN and HARRIS [1997]).

REMARK 3.14. To ease the assessment of controllability using the results of this section, an automatic tool that allows computation of the dimension of the Lie algebra generated by several (skew-Hermitian) matrices is available freely on the Internet (cf. TURINICI and SCHIRMER [2001]).

3.2.3. Controllability analysis via the connectivity graph
This section seeks to address controllability from an analysis of the kinematic structure of the Hamiltonian. Suppose that the basis $D = \{\psi_i(x); i = 1, \ldots, N\}$ is composed

of eigenstates of the internal Hamiltonian H_0, so that the matrix A is diagonal with purely imaginary elements $-iE_k$, where E_k are the eigenvalues of H_0, $k = 1,\ldots,N$, and that the diagonal elements of the matrix B are all zero (this is often the case in practice). We obtain the following structure:

$$A = -i \begin{pmatrix} E_1 & & & 0 \\ & E_2 & & \\ & & \ddots & \\ 0 & & & E_n \end{pmatrix}; \quad B = -i \begin{pmatrix} 0 & b_{12} & \cdots & b_{1N} \\ b_{12}^* & 0 & b_{ij} & \vdots \\ \vdots & b_{ij}^* & \ddots & \\ b_{1N}^* & \cdots & & 0 \end{pmatrix}.$$

Assume moreover that no degenerate transitions are present, that is,

$$|E_i - E_j| \neq |E_k - E_l|, \quad i,j,k,l = 1, \ldots, n, i \neq j, k \neq l, \{i,j\} \neq \{k,l\}. \tag{3.2.7}$$

The connectivity amongst the states $\{\psi_i; i = 1, \ldots, n\}$ provided by the elements b_{ij}, i, $j = 1, \ldots, n$, is central to issues of controllability. The structure in B can be conveniently expressed graphically by introducing a graph $G = (V, E)$ (see CHRISTOFIDES [1975] for an introduction to graph theory): let every state be a vertex (node) of the graph G so that the set of vertices $V = \{\psi_1, \ldots, \psi_N\}$, and let there be edges between every pair of nodes ψ_i and ψ_j with $b_{ij} \neq 0$ so that the set of edges $E = \{(\psi_i, \psi_i); b_{ij} \neq 0\}$. Two states ψ_i and ψ_j are said to be connected by a path if there exists a connected set of edges starting in ψ_i and ending in ψ_j. The graph G is called connected if there exists a path between every pair of vertices.

REMARK 3.15. Note that G being connected does *not* imply that any two states are necessarily *directly* connected (i.e., with a *direct* edge). One such example is the (connected) graph in Fig. 3.1 associated with the system

$$A = -i \begin{pmatrix} 1.1 & & & 0 \\ & 2.3 & & \\ & & 3.05 & \\ 0 & & & 4.6 \end{pmatrix}; \quad B = -i \begin{pmatrix} 0 & 0 & 1 & 0 \\ 0 & 0 & 1 & 0 \\ 1 & 1 & 0 & 1 \\ 0 & 0 & 1 & 0 \end{pmatrix}. \tag{3.2.8}$$

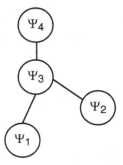

FIG. 3.1 The (connected) graph associated with the B matrix of the system in Eq. (3.2.8). Note that no direct path exits between, for example, ψ_1 and ψ_4.

These definitions allow formulating the following result (see also ALTAFINI [2002] for similar conclusions):

THEOREM 3.16. (TURINICI and RABITZ [2001a], TURINICI and RABITZ [2001b]) *Suppose that the graph G is connected and that the transitions of the internal Hamiltonian are nondegenerate in the sense of Eq.* (3.2.7). *Then the system in* (3.2.1) *is state-controllable.*

REMARK 3.17. The conditions of connectivity and nondegenerate transition frequencies involve only the eigenvalues of H_0 and the coefficients of B (which depend only upon which elements of B are nonzero, rather than their particular values) allowing conclusions about the controllability of the system in Eq. (3.2.1) even in the absence of quantitative information about this system.

REMARK 3.18. Controllability generally does not provide an actual control law that will achieve an objective of the form (2.1.1), but simply implies that at least one such control law exists. The proof of controllability in TURINICI [2000c] establishes an algorithm which constructively generates control laws for objective operators of the form $O_i = |\psi_i\rangle\langle\psi_i|$ in terms of sinusoidal electric fields of different fixed frequencies. Note that in this situation, the control objectives are populations of quantum states (e.g., $|\langle\psi|\psi_i\rangle|^2$).

This latter work, along with other work on special cases of constructive quantum control (e.g., HAREL and AKULIN [1999], RAMAKRISHNA [2000]), invites the question: can general constructive control solutions be developed for the quantum controllability of a broad class of objective expectation values? While the explicit construction of control solutions is generally an area of active research in control theory (NIJMEIHER and VAN DER SCHAFT [1990], SEPULCHRE, JANKOVIC and KOKOTOVIC [1998]), the case of quantum controllability of expectation values reduces this task to the specific structure of Schrödinger's equation that may be amenable to attack.

REMARK 3.19. The analysis of controllability in terms of state functions $\psi(t)$ reaches beyond what is necessary physically, as realistic objectives are expectation values of observable operators (cf. Eq. (2.1.1)). Since these quantities involve integrals of state functions, their control should generally be less demanding than that of the state itself. However, the quadratically nonlinear nature of the expectation values adds a level of additional complexity to the problem of determining controllability.

Having seen positive results for finite-dimensional wavefunction controllability we now investigate what phenomena prevent this controllability. Even if a final answer does not yet exist, available evidence suggests that conservation laws are responsible for the loss of controllability. Let us consider, as in TURINICI and RABITZ [2001a], the following simple 3-level system:

$$A = \begin{pmatrix} 1 & 0 & 0 \\ 0 & 2 & 0 \\ 0 & 0 & 3 \end{pmatrix}, \quad B = \begin{pmatrix} 0 & 1 & 0 \\ 1 & 0 & 1 \\ 0 & 1 & 0 \end{pmatrix}. \tag{3.2.9}$$

and the corresponding evolution equations

$$i\frac{d}{dt}c_{\varepsilon 1}(t) = c_{\varepsilon 1}(t) + \varepsilon(t)c_{\varepsilon 2}(t),$$

$$i\frac{d}{dt}c_{\varepsilon 2}(t) = 2c_{\varepsilon 2}(t) + \varepsilon(t)c_{\varepsilon 1}(t) + \varepsilon(t)c_{\varepsilon 3}(t),$$

$$i\frac{d}{dt}c_{\varepsilon 3}(t) = 3c_{\varepsilon 3}(t) + \varepsilon(t)c_{\varepsilon 2}(t).$$

This system has degenerate transitions as $E_2 - E_1 = E_3 - E_2$. Upon closer examination, a "hidden symmetry" is found for this system: more precisely it is easy to prove that for any $t > 0$ and $\varepsilon(t) \in L^2([0, t])$:

$$\left| c_{\varepsilon 1}(t)c_{\varepsilon 3}(t) - \frac{c_{\varepsilon 2}(t)^2}{2} \right| = \left| c_{\varepsilon 1}(0)c_{\varepsilon 3}(0) - \frac{c_{\varepsilon 2}(0)^2}{2} \right|. \qquad (3.2.10)$$

Therefore, any $\psi(t) = \sum_{i=1}^{3}c_{\varepsilon i}(t)\psi_i(x)$ that is reachable from $\psi(0)$ with $\psi(0) = \sum_{i=1}^{3}c_{\varepsilon i}(0)\psi_i(x)$ must satisfy the constraint (3.2.10). Let us consider a simple numerical example: suppose that the initial state is the ground state ψ_1 and the target is the first excited state ψ_2. We obtain for ψ_1:

$$\left| c_{\varepsilon 1}(0)c_{\varepsilon 3}(0) - \frac{c_{\varepsilon 2}(0)^2}{2} \right| = \left| 1 \cdot 0 - \frac{0^2}{2} \right| = 0$$

and for ψ_2:

$$\left| c_{\varepsilon 1}(t)c_{\varepsilon 3}(t) - \frac{c_{\varepsilon 2}(t)^2}{2} \right| = \left| 0 \cdot 0 - \frac{1^2}{2} \right| = \frac{1}{2}.$$

Since the two quantities are different, ψ_2 *is not reachable from* ψ_1 and therefore the system is not controllable, despite the fact that the connectivity assumption is satisfied.

The detailed analysis of the case $N = 3$ and a result that was communicated to us (cf. RAMAKRISHNA [2000]) suggest the conjecture that for general finite-dimensional systems, as long as no conservation laws appear (besides L^2 norm conservation) the system is controllable. This statement, if true, would have the merit of giving a result only in terms of the physical properties of the system under consideration (and independent of the mathematical transcription of the precise control situation).

REMARK 3.20. For the density matrix formulation, kinematical constraints on the controllability of systems in mixed states have been established (e.g., GIRARDEAU, SCHIRMER, LEAHY and KOCH [1998], GIRARDEAU, INA, SCHIRMER and GULSRUD [1997]), based on the eigenvalues of $\rho(0)$ and those of the objective operator.

3.2.4. Independent systems controllability and discrimination issues
This section introduces controllability results for independent quantum systems. Most of the text draws from TURINICI, RAMAKRISHNA, LI and RABITZ [2002].

Successful control may be expressed as a matter of high quality discrimination, whereby the control field steers the evolving quantum system dynamics out the desired channel, while diminishing competitive flux into other undesirable channels. A potential application of quantum control techniques is to the detection of specific molecules amongst others of similar chemical/physical characteristics. We call this procedure coherent molecular discrimination. Cases of special interest for discrimination include large polyatomic molecules of similar chemical nature, whose spectra can often mask each other.

Much work needs to be done to explore and develop the concept of coherent molecular discrimination, and a basic step in this direction is to establish the criteria for independently and simultaneously controlling the dynamics of several molecular species with the same control field. Discrimination of multiple molecules is a special case of the controllability concept, where the full system consists of a set of subsystems (i.e., molecules of different type). In the simplest circumstances, the molecules may be taken as independent and noninteracting, such that the initial state $\psi(0)$ is a product $\psi(0) = \prod_{\ell=1}^{L}\psi_\ell(0)$ of states $\psi_\ell(0)$ for each of the $L \geqslant 2$ molecular species. Full controllability would correspond to the ability to simultaneously and arbitrarily steer about each of initial states $\psi_\ell(0)$ to predefined targets $\psi_\ell(T) = \psi_\ell^{\text{target}}$ under the influence of a single control laser electric field $\varepsilon(t)$, where each molecule evolves under a separate Schrödinger equation

$$i\hbar \frac{\partial}{\partial t}\psi_\ell(t) = [H_0^\ell - \mu^\ell \cdot \varepsilon(t)]\psi_\ell(t). \tag{3.2.11}$$

Here, H_0^ℓ and μ^ℓ, respectively, are the free Hamiltonian and interaction operator (dipole) of the ℓth molecule. Other milder controllability criteria might also be specified.

The purpose of this section is to state theoretical criteria for the controllability of an ensemble of L separate quantum systems in the presence of a single electric field $\varepsilon(t)$. The criteria will refer to a *finite dimensional* setting where, for each $\ell, 1 \leqslant \ell \leqslant L$, the Hamiltonian H_0^ℓ and dipole operator μ^ℓ are expressed with respect to an eigenbasis of the internal Hamiltonians, as is often the case in applications. More precisely, let $D^\ell = \{\psi_i^\ell(x); i = 1, \ldots, N_\ell\}$ be the set of the first $N_\ell, N_\ell \geqslant 3$ eigenstates of the possibly infinite dimensional Hamiltonian H_0^ℓ, and let A^ℓ and B^ℓ be the matrices of the operators $-iH_0^\ell$ and $-i\mu^\ell$ respectively, with respect to this base. In order to exclude a trivial loss of controllability, it is supposed that $[A^\ell, B^\ell] \neq 0, \ell = 1, \ldots, L$. From the definition of the basis D^ℓ and the fact that H_0^ℓ and μ^ℓ are Hermitian operators, it follows that each A^ℓ is diagonal with purely imaginary elements and each B^ℓ is skew-Hermitian. We will suppose moreover that

For any $\ell = 1, \ldots, L : A^\ell$ has nonzero trace and B^ℓ has zero trace. (3.2.12)

With this notation, the wavefunction of the ℓth system can be written as $\psi_\ell(t) = \sum_{i=1}^{N_\ell} c_i^\ell(t)\psi_i^\ell$. The total eigenfunction $\prod_{\ell=1}^{L}\psi_\ell(t)$ will be represented as a column vector $c(t) = (c_1^1(t), \ldots, c_{N_1}^1(t), \ldots, c_1^L(t), \ldots, c_{N_L}^L(t))^T$. Denote $N = \sum_{\ell=1}^{L}N_\ell$, A be the $N \times N$ skew-Hermitian block-diagonal matrix obtained from

$A^\ell, \ell = 1, \ldots, L$, and B be the skew-Hermitian block-diagonal matrix obtained from $B^\ell, \ell = 1, \ldots, L$:

$$A = \begin{pmatrix} A^1 & 0 & \cdots & 0 \\ 0 & A^2 & \cdots & 0 \\ \vdots & \vdots & \ddots & \vdots \\ 0 & 0 & \cdots & A^L \end{pmatrix}; \quad B = \begin{pmatrix} B^1 & 0 & \cdots & 0 \\ 0 & B^2 & \cdots & 0 \\ \vdots & \vdots & \ddots & \vdots \\ 0 & 0 & \cdots & B^L \end{pmatrix}. \tag{3.2.13}$$

With the "atomic units" convention $\hbar = 1$, the dynamical equations read:

$$\frac{d}{dt} c(t) = Ac(t) - \varepsilon(t)Bc(t), \quad c(0) = c_0. \tag{3.2.14}$$

Recalling that each individual wavefunction $\psi_\ell(t) = \sum_{i=1}^{N_\ell} c_i^\ell(t)\psi_i^\ell$ is L^2 normalized to one, we obtain:

$$\sum_{i=1}^{N_\ell} |c_i^\ell(t)|^2 = 1, \quad \forall t \geq 0, \forall \ell = 1, \ldots, L. \tag{3.2.15}$$

Let $\mathcal{S}_{\mathbb{C}}^{k-1}$ be the complex unit sphere of $\mathbb{C}^k$. Then Eq. (3.2.15) gives:

$$c(t) \in \mathcal{S} = \prod_{\ell=1}^{L} \mathcal{S}_{\mathbb{C}}^{N_\ell-1}, \quad \forall t \geq 0. \tag{3.2.16}$$

Define the admissible control set $\mathcal{U}$ as the set of all piecewise continuous functions $\varepsilon(t)$. For every $\varepsilon \in \mathcal{U}$ Eq. (3.2.14) has an (unique) solution for all $t \geq 0$. The system $((A^\ell, B^\ell)_{\ell=1}^L, \mathcal{U})$ is said to be *controllable* if for any $c_i, c_f \in \mathcal{S}$ there exists an $t_f \geq 0$ (possibly depending on c_i, c_f) and $\varepsilon(t) \in \mathcal{U}$ such that the solution of Eq. (3.2.14) with initial data $c(0) = c_i$ satisfy $c(t_f) = c_f$. Of course, in order for the system $((A^\ell, B^\ell)_{\ell=1}^L, \mathcal{U})$ to be controllable each component system $(A^\ell, B^\ell, \mathcal{U}), \ell = 1, \ldots, N$, taken independently has to be controllable. However, requiring that all systems be controllable *at the same time and with the same laser field* is a more demanding condition. To illustrate this statement, we will consider the simple case of two ($L = 2$) three-level systems ($N_1 = 3, N_2 = 3$) of TURINICI, RAMAKRISHNA, LI and RABITZ [2002]:

$$A^1 = A^2 = -i \cdot \begin{pmatrix} 1 & 0 & 0 \\ 0 & 2 & 0 \\ 0 & 0 & 5 \end{pmatrix}, \quad B^1 = -i \cdot \begin{pmatrix} 0 & 1 & 0 \\ 1 & 0 & 2 \\ 0 & 2 & 0 \end{pmatrix}, \quad B^2 = -B^1. \tag{3.2.17}$$

Each system A^1, B^1 and A^2, B^2 is controllable, as can be checked by the Lie algebra criterion of RAMAKRISHNA, SALAPAKA, DAHLEH, RABITZ and PIERCE [1995] (the dimension of the Lie algebra is found to be 9).

However, denoting by (c_1^1, c_2^1, c_3^1) and by (c_1^2, c_2^2, c_3^2) the coefficients of the wavefunction of the first system and of the second system respectively, we obtain the following dynamical invariant (conservation law):

$$L(t) = \overline{c_1^1(t)}\, c_1^2(t) + \overline{c_3^1(t)}\, c_3^2(t) - \overline{c_2^1(t)}\, c_2^2(t) = \text{constant}, \quad \forall_\varepsilon \in \mathcal{U}. \tag{3.2.18}$$

The presence of this conservation law implies that the system is not controllable. For instance, starting with both systems in the ground state

$$\left(c_1^1(0), c_2^1(0), c_3^1(0)\right) = \left(c_1^2(0), c_2^2(0), c_3^2(0)\right) = (1, 0, 0)$$

one cannot steer both to their respective first excited state

$$\left(c_1^1(T), c_2^1(T), c_3^1(T)\right) = \left(c_1^2(T), c_2^2(T), c_3^2(T)\right) = (0, 1, 0)$$

since in the ground states the dynamical invariant takes the value $L(0) = 1 + 0 - 0 = 1$ while in the first excited states the value is $L(T) = 0 + 0 - 1 = -1$.
Defining the set of attainable states

$$\mathcal{A}(c_0, T) = \{c(t); c(t) \text{ solution of } (3.2.14), t \in [0, T], u \in \mathcal{U}\}, \qquad (3.2.19)$$

the system is controllable if (and only if) for any $c_0 \in \mathcal{S}$ the set of points attainable from c_0 : $\bigcup_{t \geqslant 0} \mathcal{A}(c_0, t)$ equals $\mathcal{S}$.
We are now ready to state the following controllability result:

THEOREM 3.21. TURINICI, RAMAKHRISHNA, LI and RABITZ [2002] *If the dimension (computed over the scalar field* $\mathbb{R}$*) of the Lie algebra* $\mathcal{L}$ *(A, B) generated by A and B equals* $1 + \sum_{\ell=1}^{L}(N_\ell^2 - 1)$ *then the system* $((A^\ell, B^\ell)_{\ell=1}^{L}, \mathcal{U})$ *is controllable. Moreover, when the system is controllable, there exists a time* $T > 0$ *such that all targets can be attained before or at time T, i.e., for any* $c_0 \in \mathcal{S}, \mathcal{A}(c_0, T) = \mathcal{S}$.

We refer to LI, TURINICI, RAMAKHRISHNA and RABITZ [2002], TURINICI, RAMAKHRISHNA, LI and RABITZ [2002] for more general results where the hypotheses of Eq. (3.2.12) are not satisfied or when multiple $s > 1$ external fields are considered (which can be expressed by introducing multiple dipole moment operators $\mu_i^\ell, i = 1, \ldots, s$).

3.3. Truncations

Little is known about the relationship between the controllability of the finite dimensional systems and that of infinite dimensional systems. We will discuss in this section some of the interesting aspects of this interplay through a list of open problems and questions.

Consider a quantum system that is controllable when its (truncated) equations of motion are expressed with respect to a particular n-dimensional basis which spans a finite dimensional space $\mathcal{H}_n$. According to Theorem 3.2, for every initial condition there must emerge a dense set of unreachable states in the limit n tends to infinity (depicted in Fig. 3.2(a)), assuming that the limiting process is well defined. In other words, for $n \to \infty$ the system may become uncontrollable in the strict sense defined above. This limit suggests the question: how does the controllability of a sequence of finite but increasingly higher dimensional quantum systems relate to the controllability of the corresponding infinite-dimensional quantum system in the limit $n \to \infty$ (if this limiting process exists)? How are the sets of unreachable states that emerge in this limit characterized?

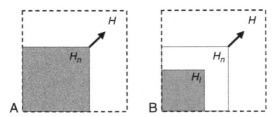

FIG. 3.2 Pictorial representation of the truncation problem. In both cases, controllability is of interest within the shaded subspace as the dimension of the truncated space H_n increases to infinity. Reprinted with permission from BROWN and RABITZ [2002, p. 23]; copyright 2002, Kluwer Academic Publishers

The analysis implied in the above questions can be subtle, as evident from a simple illustration involving the emergence or disappearance of unreachable states under finite increases in the dimensionality of $\mathcal{H}_n$. For example, in hydrogenic atoms the transitions due to emission or absorption of photons must satisfy the selection rules $\Delta l = \pm 1$ and $\Delta m = \pm 1$ or 0. If the step $\mathcal{H}_n \to \mathcal{H}_{n+1}$ of the limiting process adds a basis function to which there does not exist a sequence of allowed transitions from some function ψ_1 in $\mathcal{H}_n$, the additional dimension has caused a loss of system controllability. The converse situation may also arise where the additional basis function provides a "missing pathway" between states that were mutually unreachable in $\mathcal{H}_n$: in this case, the step $\mathcal{H}_n \to \mathcal{H}_{n+1}$ might cause an uncontrollable system to become controllable. It is an open question in quantum controllability to understand how such stepwise processes may be interpreted in the infinite limit.

Now consider the related issue of controllability within a "subspace of interest" $\mathcal{H}_I$ that is contained within $\mathcal{H}_n$ (as depicted in Fig. 3.2(b)). Let $\mathcal{H}_I$ be spanned by the first I elements of the set of basis functions $\{\psi_i: i = 1, \ldots, n\}$ spanning $\mathcal{H}_n$. Definition 3.5 may be modified to restrict analysis to the subspace of interest: controllability will be taken to mean that a system is controllable if the system can be steered between any two states ψ_1^I and ψ_2^I in $S \cap \mathcal{H}_I$. Controllability may be described as stationary within $\mathcal{H}_I$ if it remains unchanged as individual dimensions are added in any order to $\mathcal{H}_n$ until (if it exists) the limit $\lim_{n \to \infty, n \geqslant I} \mathcal{H}_n = \mathcal{H}$, is obtained. It is not known what characteristics of the Hamiltonian H_0, the dipole or other coupling coefficients, and the spaces $\mathcal{H}_I \subset \mathcal{H}_n \subset \mathcal{H}$ are required for stationary controllability within $\mathcal{H}_I$.

The discussion above does not address the effects on the evolution of states within the truncated space $\mathcal{H}_n$ arising from states that lie outside of $\mathcal{H}_n$. This consideration also has practical consequences. For example, suppose that controllability is satisfied within $\mathcal{H}_n$ or within $\mathcal{H}_I$ for some $\mathcal{H}_n$. A realizable laboratory control might inadvertently also access states lying outside of $\mathcal{H}_n$ which might even lift the controllability in the desired subspace. Techniques from optimal control theory would be the desirable way to handle the discovery of practical fields best satisfying the assumptions under an associated controllability analysis. Following upon the latter discussion, a new class of problems is introduced if a term is added to the Schrödinger equation to represent the interaction of the remainder states that are not explicitly modeled lying in $\mathcal{H}\backslash\mathcal{H}_n$. One such term introduced in BEUMEE and RABITZ [1992], cf. also SONTAG [1998], is a n-dimensional disturbance vector w:

$$\frac{\partial\psi}{\partial t} = [H_0 + C(t)]\psi + w,$$

where $\psi \in \mathcal{H}_n$. The form and magnitude of the disturbance term w is problem-dependent and assumed to be given a priori, and generally leads to a nonunitary evolution for $\psi(t)$. In min–max optimal control theory (cf. Section 4.3), w is selected to maximize the disruptive effect of the energy-bounded disturbance (e.g., fluctuations in the laboratory environment and apparatus). In another general context, w could represent coupling to a bath external to the dynamics described by $H_0 + C(t)$. This type of coupling is important for considerations of dynamical cooling (see BARTANA and KOSLOFF [1993], BARTANA and KOSLOFF [1997], TANNOR and BARTANA [1999], TANNOR, KOSLOFF and BARTANA [1999]) in the analogous density matrix formulation. It is not yet clear what general models of dynamics exterior to $\mathcal{H}_n \subset \mathcal{H}$ can cause controllable systems to become uncontrollable and vice versa.

4. Quantum control algorithms

As analytical solutions to the quantum optimal control equations cannot generally be found, iterative numerical algorithms must be employed. The time-dependent Schrödinger equation in multiple spatial variables is computationally very expensive to solve, and, although there appear to be many opportunities to develop special control design approximations, and while some work has already been done (e.g., RICE and ZHAO [2000]), there is room for much development.

REMARK 4.1. Since the Schrödinger equation must be solved at least once (and generally many times) in most optimal control and Hamiltonian identification methods, the numerical evolution of quantum systems with many degrees of freedom must be approximated in some fashion before control design techniques can be applied. Along these lines, broad classes of quantum dynamics approximations have been developed, and in principle any of them could be applied to quantum control design. While attempting to attain designs, it is worth investigating the effects on controllability of replacing the Schrödinger equation with its various quantum dynamical approximations. Significant influence of a dynamical approximation upon a system's controllability could have serious consequences for the reliability of any resultant control designs based on the approximation.

A very important feature of quantum control is the intricate relationship between theory and the laboratory implementations, where an optimization method, usually an evolutionary (e.g., genetic) algorithm, is used to drive the experimental work which, in its turn, feeds back the optimization algorithm with necessary data. Reflecting this connection, both numerical and experimental paradigms are presented in this section, the first in Section 4.2 (extending with connected topics through Section 4.4) and the latter in Section 4.5. Common to both, the construction of the cost functionals is the object of the next section.

4.1. Optimal control cost functional formulation

As explained in Section 2.2, the first step in formulating the general quantum optimal control problem is to define a "cost functional" whose minimization represents the balanced achievement of control and possibly other objectives. This cost functional is given by

$$J = \sum_k J_k(C(\psi(t); t \in [0, T])),$$

where the goal is to minimize (or maximize, as appropriately) J with respect to $C(t)$. While the specific form of the cost functional is flexible and problem-dependent, a term J_1 that addresses the achievement of the N_O optimal control objectives is always included:

$$J_1 = \sum_{j=1}^{N_O} \sum_{l=1}^{L_j} \begin{cases} \int_{\tau_j^l} W_{1,j}^l [\langle O_j(t) \rangle - \tilde{O}_j(t)]^\gamma dt, & \text{if } \tau_j^l \text{ is an interval,} \\ W_{1,j}^l [\langle O_j(\tau_j^l) \rangle - \tilde{O}_j(\tau_j^l)]^\gamma, & \text{if } \tau_j^l \text{ is a discrete time.} \end{cases} \quad (4.1.1)$$

Here the $W_{1,j}^l$ are positive design weights assigned to each of the objectives. The particular case $N_O = L_1 = \gamma = W_{1,1}^1 = 1, \tau_1^1 = T$ and $\tilde{O}_1(T) = 0$ has already been seen (cf. Eq. (2.2.1)) to give rise to a term of the form $J_1 = \langle \psi(T)|O_1|\psi(T) \rangle$.

For physically realistic control laws, the energy of the laboratory/molecular interaction must be bounded. This criteria is often included by adding the term

$$J_2 = \int_0^T W_2(t)\varepsilon(t)^2 dt \quad (4.1.2)$$

to the cost functional, which effectively limits the total electric field fluence. Here, $W_2(t)$ determines the time-dependent relative importance of minimizing the fluence. Note that the term J_2 does not prevent $\varepsilon(t)$ from being large in some small interval of time, although a cost on the local magnitude at any time could be introduced for this purpose.

Penalty terms may also be included, causing the minimization of the expectation values of $N_{O'}$ "undesirable" operators O_j' at the corresponding times $\tau_j'^l$:

$$J_3 = \sum_{j=1}^{N_O'} \sum_{l=1}^{L_j'} \begin{cases} \int_{\tau_j'^l} W_{3,j}^l |\langle O_j'(t) \rangle|^2 dt, & \text{if } \tau_j'^l \text{ is an interval,} \\ W_{3,j}^l |\langle O_j'(\tau_j'^l) \rangle|^2, & \text{if } \tau_j'^l \text{ is a discrete time.} \end{cases} \quad (4.1.3)$$

In addition to those explicitly given here, there are many other forms of J_k that could be incorporated into the cost functional. These terms could represent, for example, restrictions on the windowed Fourier transform of $\varepsilon(t)$ to a particular frequency band, minimization of sensitivity to small perturbations in the control law (as will be discussed below), or other characteristics of the desired optimal control solution.

One property of control law solutions $C(t)$ of practical import is simplicity. Several measures of simplicity could be used, such as the ability to decompose the control law into only a small number of spectral components with high accuracy (see GEREMIA, ZHU and RABITZ [2000]). However, the notion of field simplicity is best associated with the ease of stable and reliable generation in the laboratory, rather than any pre-conceived sense of simplicity associated with the presence of few field components. Design of simple control laws might be accomplished by introducing a term in the cost functional that favors solutions with suitable characteristics, or in a very ad hoc fashion by starting an iterative optimization algorithm with a simple control field and halting the process while some of this simplicity is still preserved but likely before complete convergence to the control objectives has been achieved. The latter suggestion follows from the observation that the final small fraction of progress toward the control objectives is often responsible for most of the complexity in the control field (e.g., see Table 6.1 of RICE and ZHAO [2000] or TERSIGNI, GASPARD and RICE [1990]). None of the above approaches has been subjected to a careful mathematical analysis, and further efforts to characterize the effects of these modifications on the optimal control process may be useful.

Thus far the terms in the cost functional have all been introduced to seek a control field that biases the objective or other goals in some specified direction. Under favorable circumstances one or more of these costs could be re-expressed as a hard demand by introducing a Lagrange multiplier. An example would be a requirement that the laser pulse energy be fixed at a specified laboratory accessible value; the reshaping would redistribute that energy as best as possible over a band of frequency components to meet the physical objective. Some absolute demands may lead to inconsistencies and resultant numerical design difficulties if the demand cannot be satisfied for some (often hidden) dynamical reason.

Once the cost functional has been defined, then the optimal control law is determined by minimizing the cost functional over the function space of admissible controls. Local or global, deterministic or stochastic optimization algorithms (e.g., gradient descent, genetic algorithms, etc.) may be used to find the minimum of the cost functional subject to satisfaction of the Schrödinger equation, possibly under suitable quantum dynamics approximations and assumptions (see KRAUSE and SCHAFER [1999], KRAUSE, REITZE, SANDERS, KUZNETSOV and STANTON [1998]). The existence of the minimum itself has been investigated in several works, including PEIRCE, DAHLEH and RABITZ [1988]; for a different formulation see CANCÈS, LE BRIS and MATHIEU [2000]. Alternatively, the Euler–Lagrange approach may be pursued, as explained below.

At the relevant minima of the cost functional, the first order variation with respect to the control law vanishes:

$$\frac{\delta J}{\delta C(t)} = 0. \qquad (4.1.4)$$

Eq. (4.1.4) is subject to the dynamical constraint that $\psi(t)$ satisfies the Schrödinger equation; this may be assured through the introduction of a Lagrange multiplier function $\lambda(t)$ (cf. SONTAG [1998]). The resultant variational problem produces Euler–Lagrange equations whose solutions define the controls $C(t)$ operative at each local extrema of J. To demonstrate some of the characteristics of these equations, consider as an example a quantum optimal control problem in which there is only one objective operator O whose expectation value is to be optimized at the (single target) time T. The control law is taken to be $C(t) = -\mu\varepsilon(t)$ and the cost functional $J = J_1 + J_2$ consists of the terms given in Eqs. (4.1.1) and (4.1.2) with weights set such that J is given by Eq. (2.2.1). The extended cost functional that includes the Lagrange multiplier λ, which will be called from now on the *adjoint state*, is

$$J(\varepsilon, \psi, \lambda) = \langle \psi(T)|O|\psi(T) \rangle - \alpha \int_0^T \varepsilon^2(t)\mathrm{d}t$$
$$- 2\mathrm{Re}\left[\int_0^T \left\langle \lambda(t)\left|\frac{\partial}{\partial t} + i[H_0 - \mu\varepsilon(t)]\right|\psi(t)\right\rangle\mathrm{d}t\right]. \tag{4.1.5}$$

With this definition we obtain first

$$\frac{\delta J(\varepsilon, \psi, \lambda)}{\delta\varepsilon} = -2\alpha\varepsilon(t) - 2\mathrm{Im}\{\langle\lambda(t)|\mu|\psi(t)\rangle\}. \tag{4.1.6}$$

The derivative of J with respect to the adjoint state yields (as expected) the equation of motion for the wavefunction. In order to compute the derivative of J with respect to $\psi(t)$ one integrates by parts in the last term of $J(\varepsilon, \psi, \lambda)$ and identifies the resulting terms. The Euler–Lagrange equations then become:

$$\begin{cases} i\dfrac{\partial\psi(t)}{\partial t} = [H_0 - \mu\varepsilon(t)]\psi(t), \\ \psi(0) = \psi_0, \end{cases} \tag{4.1.7}$$

$$\begin{cases} i\dfrac{\partial\lambda(t)}{\partial t} = [H_0 - \mu\varepsilon(t)]\lambda(t), \\ \lambda(T) = O\psi(T), \end{cases} \tag{4.1.8}$$

$$\alpha\varepsilon(t) = -\mathrm{Im}\{\langle\lambda(t)|\mu|\psi(t)\rangle\}. \tag{4.1.9}$$

If the field in Eq. (4.1.9) is substituted into Eqs. (4.1.7) and (4.1.8), the system becomes a pair of coupled nonlinear evolution equations.

REMARK 4.2. Under a mild set of assumptions the general quantum optimal control problem has been shown to possess a countable infinity of solutions, cf. DEMIRALP and RABITZ [1993]. This result has been obtained for cost functionals of the form $J = J_1 + J_2 + J_3$ having one objective operator O at a final time T and a single penalty operator O' evaluated over the entire control interval. In this work additional assumptions are that: (i) O and O' are bounded operators, (ii) O is either positive- or negative-definite, and (iii) $\mu \in (t: t \in [0, T])$ is bounded (although the proof can be extended for unbounded control terms).

4.2. Numerical algorithms

This section considers algorithms for determining the optimal controls based on the constrained variational problem in Eq. (4.1.4). Part of the text is drawn from the recent review ZHU and RABITZ [in press].

We will consider the quantum optimal control problem with the cost functional given in Eq. (2.2.1) and critical point Eqs. (4.1.7)–(4.1.9). The observable operator O is assumed to be positive semidefinite Hermitian. Note that Eq. (4.1.9) can also be written as

$$\begin{aligned}
\varepsilon(t) &= \frac{1}{\alpha}\mathrm{Re}\langle\psi(T,\{\varepsilon\})|O|\delta\psi(T,\{\varepsilon\})/\delta\varepsilon(t)\rangle \\
&= -\frac{1}{\alpha}\mathrm{Im}\langle\psi(T,\{\varepsilon\})|OU(T,t,\{\varepsilon\})\mu|\psi(t,\{\varepsilon\})\rangle
\end{aligned} \tag{4.2.1}$$

with U being the evolution operator of the system. The presence of the argument ε on the right hand side of Eq. (4.2.1) indicates that this is actually an implicit relation for determining the control field. To solve for the control field in Eq. (4.1.9) or Eq. (4.2.1), it is evident that some type of iteration algorithm needs to be employed.

Several numerical algorithms can be used; historically, the gradient-type methods (see SHI, WOODY and RABITZ [1988], COMBARIZA, JUST, MANZ and PARAMONOV [1991]) were the first to be used. In this approach,

(1) an initial guess ε_1 is set and the iteration count k is initialized to $k = 1$;
(2) the wavefunction $\psi_k(t)$ is propagated forward in time with the field ε_k by Eq. (4.1.7);
(3) the final data for the adjoint state $\lambda_k(T)$ is derived;
(4) the wavefunction $\psi_k(t)$ and the adjoint state $\lambda_k(t)$ are propagated backward in time with the field ε_k and a new electric field $\varepsilon_{k+1} = \varepsilon_k + \gamma\cdot(\delta J/\delta\varepsilon_k)$ is computed ($\delta J/\delta\varepsilon_k$ is given by Eq. (4.1.6)). The constant γ is found by a linear search to optimize J in the direction of the gradient $\delta J/\delta\varepsilon_k$;
(5) $k \to k + 1$; step (2) is returned to and the cycle is continued until convergence.

Note that in step (4) the propagation of the wavefunction $\psi_k(t)$ is required because the storage of its trajectory (already computed in step (2)) is usually too expensive. We refer the reader to TERSIGNI, GASPARD and RICE [1990] for a conjugated gradient version of the above algorithm.

Although this algorithm proved useful in some cases, its convergence is not guaranteed (here the setting is far from the quadratic cost functional that conjugated gradient type algorithms best optimize) and it may become slow; the algorithm used by Tannor (see TANNOR, KAZAKOV and ORLOV [1992], SOMLÓI, KAZAKOV and TANNOR [1993]) based on the Krotov method (KROTOV [1973], KROTOV [1974a], KROTOV [1974b]) was designed to correct this feature. The structure of this algorithm is as follows:

(1) an initial guess ε_1 is set and the iteration count k is initialized to $k = 1$;
(2) the wavefunction $\psi_1(t)$ is propagated with the field ε_1 by Eq. (4.1.7);
(3) the final data for the adjoint state $\lambda_k(T)$ is derived;
(4) the adjoint state $\lambda_k(t)$ is propagated backward in time with the field ε_k to give $\lambda_{k0} = \lambda_k(0)$;
(5) A new field is constructed by the simultaneous resolution of the following equations:

$$\begin{cases} i\dfrac{\partial \psi_{k+1}(t)}{\partial t} = [H_0 + \dfrac{1}{\alpha}\mu \mathrm{Im}\{\langle \lambda_k(t)|\mu|\psi_{k+1}(t)\rangle\}]\psi_{k+1}(t), \\ \psi_{k+1}(0) = \psi_0, \end{cases} \tag{4.2.2}$$

$$\begin{cases} i\dfrac{\partial \lambda_{k+1}(t)}{\partial t} = [H_0 - \mu\varepsilon_{k+1}(t)]\lambda_{k+1}(t), \\ \lambda_{k+1}(T) = O\psi_{k+1}(T), \end{cases} \tag{4.2.3}$$

$$\alpha\varepsilon_{k+1}(t) = -\mathrm{Im}\{\langle \lambda_k(t)|\mu|\psi_{k+1}(t)\rangle\}; \tag{4.2.4}$$

(6) $k \leftarrow k + 1$; step (3) is returned to and the cycle is continued until convergence. As in step (4) of the gradient descent algorithm, the propagation in Eq. (4.2.3) is motivated by memory storage considerations.

In order to analyze the numerical properties of this algorithm we evaluate the difference between the value of the cost functional between two successive iterations:

$$\begin{aligned} J(\varepsilon_{k+1}) - J(\varepsilon_k)) &= \langle \psi_{k+1}(T)|O|\psi_{k+1}(T)\rangle - \alpha\int_0^T \varepsilon_{k+1}(t)^2 \mathrm{d}t - \langle \psi_k(T)|O|\psi_k(T)\rangle \\ &+ \alpha\int_0^T \varepsilon_k(t)^2 \mathrm{d}t = \langle \psi_{k+1}(T) - \psi_k(T)|O|\psi_{k+1}(T) - \psi_k(T)\rangle \\ &+ 2\mathrm{Re}\langle \psi_{k+1}(T) - \psi_k(T)|O|\psi_k(T)\rangle + \alpha\int_0^T (\varepsilon_{k+1} - \varepsilon_k)(t)^2 \mathrm{d}t \\ &+ 2\alpha\int_0^T (\varepsilon_k - \varepsilon_{k+1})(t)\mathrm{d}t. \end{aligned} \tag{4.2.5}$$

Since we have also:

$$\begin{aligned} 2\mathrm{Re}\langle \psi_{k+1}(T) - \psi_k(T)|O|\psi_k(T)\rangle &= 2\mathrm{Re}\langle \psi_{k+1}(T) - \psi_k(T), O\psi_k(T)\rangle \\ &= 2\mathrm{Re}\langle \psi_{k+1}(T) - \psi_k(T), \lambda_k(T)\rangle \\ &= 2\mathrm{Re}\int_0^T \left\langle \frac{\partial(\psi_{k+1}(t) - \psi_k(t))}{\partial t}, \lambda_k(t)\right\rangle + \left\langle \psi_{k+1}(t) - \psi_k(t), \frac{\partial \lambda_k(t)}{\partial t}\right\rangle \mathrm{d}t \\ &= 2\mathrm{Re}\int_0^T \left\langle \frac{H_0 - \mu\varepsilon_{k+1}}{i}\psi_{k+1}(t) - \frac{H_0 - \mu\varepsilon_k}{i}\psi_k(t), \lambda_k(t)\right\rangle + \left\langle \psi_{k+1}(t) - \psi_k(t), \frac{H_0 - \mu\varepsilon_k}{i}\lambda_k(t)\right\rangle \\ &= 2\mathrm{Re}\int_0^T \varepsilon_{k+1}\left\langle \frac{-\mu}{i}\psi_{k+1}(t), \lambda_k(t)\right\rangle - \varepsilon_k\left\langle \frac{-\mu}{i}\psi_k(t), \lambda_k(t)\right\rangle + \varepsilon_k\left\langle \psi_{k+1}(t) - \psi_k(t), \frac{-\mu}{i}\lambda_k(t)\right\rangle \\ &= 2\int_0^T \varepsilon_{k+1}\cdot\alpha\varepsilon_{k+1} - \varepsilon_k\cdot\left\langle \frac{-\mu}{i}\psi_k(t), \lambda_k(t)\right\rangle - \varepsilon_k\cdot\alpha\varepsilon_{k+1} - \varepsilon_k\cdot\left\langle \psi_k(t), \frac{-\mu}{i}\lambda_k(t)\right\rangle \\ &= 2\alpha\int_0^T \varepsilon_{k+1}(t)\cdot(\varepsilon_{k+1} - \varepsilon_k)(t)\mathrm{d}t \end{aligned} \tag{4.2.6}$$

we obtain thus from (4.2.5) and (4.2.6)

$$
\begin{aligned}
J(\varepsilon_{k+1}) - J(\varepsilon_k) &= \langle \psi_{k+1}(T) - \psi_k(T)|O|\psi_{k+1}(T) - \psi_k(T)\rangle \\
&\quad + \alpha \int_0^T (\varepsilon_{k+1} - \varepsilon_k)(t)^2 \mathrm{d}t \geqslant 0
\end{aligned}
\tag{4.2.7}
$$

because the observable O is a positive semidefinite operator. Each step of this algorithm will therefore result in an increase of the value of the cost functional; this increase is expected to be important for initial steps where the critical point equations are not fulfilled and the difference between successive fields ε_k and ε_{k+1} will be important.

This algorithm was improved with the introduction of the monotonic quadratically convergent algorithm in ZHU, BOTINA and RABITZ [1998], ZHU and RABITZ [1998] that incorporates additional feedback from the adjoint state. The following material will present some aspects of the monotonically convergent algorithms and will analyze their convergence features. In the first step, a trial field $\varepsilon^{(0)}(t)$ is used to calculate the wavefunction $\psi(t, \{\varepsilon^{(0)}\})$, then Eq. (4.2.1) can be directly applied to obtain the first iteration to the control field $\varepsilon^{(1)}(t)$ by backward propagation of the evolution operator $U(T, t, \{\varepsilon^{(1)}\})$ from T to 0. In the second step, the new control field $\varepsilon^{(2)}(t)$ also can be directly obtained by forward propagation of the wavefunction $\psi(t, \{\varepsilon^{(2)}\})$ from 0 to T. A series of control fields can be repeatedly mapped out in this way. The algorithm as an iteration sequence has the following structure:

$$
\varepsilon^{(1)}(t) = -\frac{1}{\alpha}\mathrm{Im}\langle\psi(T, \{\varepsilon^{(0)}\})|OU(T, t, \{\varepsilon^{(1)}\})\mu|\psi(t, \{\varepsilon^{(0)}\})\rangle,
\tag{4.2.8}
$$

$$
\varepsilon^{(2)}(t) = -\frac{1}{\alpha}\mathrm{Im}\langle\psi(T, \{\varepsilon^{(0)}\})|OU(T, t, \{\varepsilon^{(1)}\})\mu|\psi(t, \{\varepsilon^{(2)}\})\rangle,
\tag{4.2.9}
$$

$$
\vdots
$$

$$
\varepsilon^{(2i+1)}(t) = -\frac{1}{\alpha}\mathrm{Im}\langle\psi(T, \{\varepsilon^{(2i)}\})|OU(T, t, \{\varepsilon^{(2i+1)}\})\mu|\psi(t, \{\varepsilon^{(2i)}\})\rangle,
\tag{4.2.10}
$$

$$
\varepsilon^{(2i+2)}(t) = -\frac{1}{\alpha}\mathrm{Im}\langle\psi(T, \{\varepsilon^{(2i)}\})|OU(T, t, \{\varepsilon^{(2i+1)}\})\mu|\psi(t, \{\varepsilon^{(2i+2)}\})\rangle,
\tag{4.2.11}
$$

The introduction of the adjoint state λ allows for rewriting the above algorithm in the simplified form

(1) an initial guess ε_1 is set and the iteration count k is initialized to $k = 1$;
(2) the wavefunction $\psi_0(t)$ is propagated with the field ε_1 by Eq. (4.1.7);
(3) the final data for the adjoint state $\lambda_k(T)$ is derived;
(4) the adjoint state $\lambda_k(t)$ is propagated backward in time

$$
\begin{cases}
i\dfrac{\partial\lambda_k(t)}{\partial t} = [H_0 + \dfrac{1}{\alpha}\mu\mathrm{Im}\langle\lambda_k(t)|\mu|\psi_{k-1}(t)\rangle]\lambda_k(t), \\
\lambda_k(T) = O\psi_{k-1}(T).
\end{cases}
\tag{4.2.12}
$$

Note: if the trajectory of the wavefunction $\psi_{k-1}(t)$ cannot be stored into memory, it is recomputed from the (stored) corresponding field.

(5) the wavefunction is evolved and a new field is constructed by the solution of the following equations:

$$
\begin{cases}
i\dfrac{\partial \psi_k(t)}{\partial t} = \left[H_0 + \dfrac{1}{\alpha}\mu\mathrm{Im}\langle \lambda_k(t)|\mu|\psi_k(t)\rangle \right]\psi_k(t), \\
\psi_k(0) = \psi_0,
\end{cases}
\tag{4.2.13}
$$

$$
\alpha\varepsilon_{k+1}(t) = -\mathrm{Im}\{\langle \lambda_k(t)|\mu|\psi_k(t)\rangle\};
\tag{4.2.14}
$$

As in step (4), if needed, the trajectory of the adjoint state $\lambda_k(t)$ is recomputed.

(6) $k \to k + 1$; step (3) is returned to and the cycle is continued until convergence. In order to analyze the convergence features of the above iteration sequence, we go back to the formulation in Eqs. (4.2.10) and (4.2.11) and consider the deviation of the objective functional between two neighboring steps (one step is defined as a pair of backward and forward propagations). Specifically, between the ith and $(i + 1)$st steps, the deviation is

$$
\Delta J_{i+1,i} = \left\{ \langle \psi(T, \{\varepsilon^{(2i+2)}\})|O|\psi(T, \{\varepsilon^{(2i+2)}\})\rangle - \alpha\int_0^T [\varepsilon^{(2i+2)}(t)]^2 dt \right\}
$$
$$
- \left\{ \langle \psi(T, \{\varepsilon^{(2i)}\})|O|\psi(T, \{\varepsilon^{(2i)}\})\rangle - \alpha\int_0^T [\varepsilon^{(2i)}(t)]^2 dt \right\}
\tag{4.2.15}
$$

Considering the dynamical equations of the wavefunctions $\psi(t, \{\varepsilon^{(2i + 2)}\})$ and $\psi(t, \{\varepsilon^{(2i)}\})$, and utilizing the field expressions in Eqs. (4.2.10) and (4.2.11), it can be derived (see ZHU and RABITZ [1998]) that:

$$
\Delta J_{i+1,i} = \alpha\int_0^T \left(\left[\varepsilon^{(2i+2)}(t) - \varepsilon^{(2i+1)}(t)\right]^2 + \left[\varepsilon^{(2i+1)}(t) - \varepsilon^{(2i)}(t)\right]^2 \right)dt
$$
$$
+ \langle \Delta\psi_{i+1,i}(T)|O|\Delta\psi_{i+1,i}(T)\rangle
\tag{4.2.16}
$$

where $\Delta\psi_{i + 1,i}(t) \equiv \psi(t, \{\varepsilon^{(2i + 2)}\}) - \psi(t, \{\varepsilon^{(2i)}\})$. Since O is a positive semidefinite operator,

$$
\langle \Delta\psi_{i+1,i}(T)|O|\Delta\psi_{i+1,i}(T)\rangle \geqslant 0,
\tag{4.2.17}
$$

then Eq. (4.2.16) satisfies

$$
\Delta J_{i+1,i} \geqslant 0,
\tag{4.2.18}
$$

where the equal sign occurs only if the initial guess for the control field happens to be an exact solution of Eq. (4.2.1). Eq. (4.2.18) indicates that regardless of the choice for the initial trial field, the objective functional will monotonically converge to a local maximum of J for the iterated sequence of control fields given in Eqs. (4.2.8)–(4.2.11).

It is straightforward to show that the total gain for the objective functional after N iteration steps will be

$$
\Delta J_{N,0} = \sum_{i=0}^{N-1} \langle \Delta\psi_{i+1,i}(T)|O|\Delta\psi_{i+1,i}(T)\rangle + \alpha\int_0^T \sum_{i=0}^{2N-1} [\Delta\varepsilon_{i+1,i}(t)]^2 dt,
\tag{4.2.19}
$$

where $\Delta\varepsilon_{i+1,i}(t) \equiv \varepsilon^{(i+1)}(t) - \varepsilon^{(i)}(t)$. Based on the above analysis, it can be concluded that a larger change of the field between neighboring steps will lead to faster convergence. As the initial guessed field usually will be far from the exact solution, the major contribution to the rapid convergence is expected to come from the first few iteration steps. An illustration of the monotonic convergence is shown in Fig. 4.1.

It is worthwhile to point out that the positive semidefiniteness of the operator O is not an intrinsic constraint on the monotonically convergent iteration algorithms: other types of monotonically convergent algorithms can lift this constraint. In the following, we will show one such procedure which is still monotonically convergent, but without the need to impose the positive semidefiniteness constraint on the operator. In the revised algorithm, the iteration sequence for the control field is slightly different from that shown in Eqs. (4.2.8)–(4.2.11). First, an initial guess for the field $\varepsilon^{(0)}(t)$ is needed to forward propagate the wavefunction $\psi(t, \{\varepsilon^{(0)}\})$. The modification of the algorithm just lies in the backward propagation. Specifically, the iteration sequence for determining the control field is (ZHU and RABITZ [1999a])

$$\varepsilon^{(1)}(t) = -\frac{1}{\alpha}\mathrm{Im}\langle\psi(t,\{\varepsilon^{(0)}\})|U^\dagger(T,t,\{\varepsilon^{(1)}\})OU(T,t,\{\varepsilon^{(1)}\})\mu \times |\psi(t,\{\varepsilon^{(0)}\})\rangle,$$

$$(4.2.20)$$

$$\varepsilon^{(2)}(t) = -\frac{1}{\alpha}\mathrm{Im}\langle\psi(t,\{\varepsilon^{(2)}\})|U^\dagger(T,t,\{\varepsilon^{(1)}\})OU(T,t,\{\varepsilon^{(1)}\})\mu \times |\psi(t,\{\varepsilon^{(2)}\})\rangle,$$

$$(4.2.21)$$

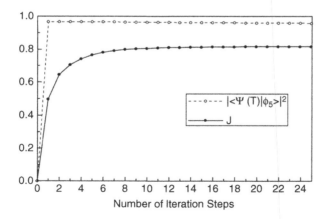

FIG. 4.1 Convergence of the cost functional to a (locally) optimal value in a numerical implementation of the quadratically convergent algorithm discussed in Section 4.2. The details of the calculation, which designs a minimum-fluence laser field to promote the $|3\rangle \rightarrow |5\rangle$ vibrational transition in an O–H bond, are given in ZHU and RABITZ [in press]. (Int. J. Quant. Chem., © copyright 2002 John Wiley & Sons Inc.)

$\vdots$

$$\varepsilon^{(2i+1)}(t) = -\frac{1}{\alpha}\text{Im}\langle\psi(t,\{\varepsilon^{(2i)}\})|U^\dagger(T,t,\{\varepsilon^{(2i+1)}\})OU(T,t,\{\varepsilon^{(2i+1)}\})\mu$$
$$\times|\psi(t,\{\varepsilon^{(2i)}\})\rangle, \tag{4.2.22}$$

$$\varepsilon^{(2i+2)}(t) = -\frac{1}{\alpha}\text{Im}\langle\psi(t,\{\varepsilon^{(2i+2)}\})|U^\dagger(T,t,\{\varepsilon^{(2i+1)}\})OU(T,t,\{\varepsilon^{(2i+1)}\})\mu$$
$$\times|\psi(t,\{\varepsilon^{(2i+2)}\})\rangle, \tag{4.2.23}$$

$\vdots$

REMARK 4.3. This second algorithm does not lend itself to an implementation in terms of direct and adjoint state only. The propagation of the wavefunction $\psi(t)$ and of the evolution operator $U(T, t, \{\varepsilon\})$ that is required is more costly than the propagation of $\psi(t)$ and $\lambda(t)$ of the previous monotonic convergent algorithm version because, in discrete form, it involves propagating a matrix and a vector as opposed to propagating two vectors.

To verify the monotonically convergent feature of the above algorithm, once again we evaluate the deviation of the objective functional between two neighboring steps as follows:

$$\Delta J_{i+1,i} = \left\{\langle\psi(T,\{\varepsilon^{(2i+2)}\})|O|\psi(T,\{\varepsilon^{(2i+2)}\})\rangle - \alpha\int_0^T\left[\varepsilon^{(2i+2)}(t)\right]^2\mathrm{d}t\right\}$$
$$- \left\{\langle\psi(T,\{\varepsilon^{(2i)}\})|O|\psi(T,\{\varepsilon^{(2i)}\})\rangle - \alpha\int_0^T[\varepsilon^{(2i)}(t)]^2\mathrm{d}t\right\}. \tag{4.2.24}$$

Considering the dynamical equations of the wavefunction $\psi(t, \{\varepsilon^{(2i+2)}\})$ and the operator $U(T, t, \{\varepsilon^{(2i+3)}\})$, and utilizing the field expressions in Eqs. (4.2.22) and (4.2.23), it can be proven (e.g., ZHU and RABITZ [1999a]) that:

$$\Delta J_{i+1,i} = \alpha\int_0^T([\varepsilon^{(2i+2)}(t) - \varepsilon^{(2i+1)}(t)]^2 + [\varepsilon^{(2i+1)}(t) - \varepsilon^{(2i)}(t)]^2)\mathrm{d}t. \tag{4.2.25}$$

Comparing Eq. (4.2.25) with Eq. (4.2.16), we see that the iteration algorithm given by Eqs. (4.2.20)–(4.2.23) is still monotonically convergent for the objective functional, but without the extra constraint on the operator O. The total monotonic gain of the objective functional after N iteration steps is simply

$$\Delta J_{N,0} = \alpha\int_0^T\sum_{i=0}^{2N-1}[\Delta\varepsilon_{i+1,i}(t)]^2\mathrm{d}t. \tag{4.2.26}$$

We will not go further here to explore other possible monotonically convergent algorithms to iteratively solve for optimal controls. It is important to keep in mind that monotonically convergent iteration algorithms may not exist for arbitrary objective functionals. However, extended work indicates that there exist monotonically convergent iteration algorithms for most common types of the objective functionals considered in the optimal control of quantum systems (ZHU, BOTINA and RABITZ [1998], SCHIRMER, GIRARDEAU and LEAHY [2000]), and that methods similar to those presented here may be applied to the density matrix formulation (OHTSUKI, ZHU and RABITZ [1999]).

REMARK 4.4. Besides these families of algorithms that aim to solve the critical point equations directly, an alternative approach has been proposed in SHEN, DUSSAULT and BANDRAUK [1994] that uses a penalization framework. Suppose that the objective can be expressed as $\langle \psi(T)|P|\psi(T)\rangle = 1$, where P is a projection operator; a sequence of cost functionals

$$J(\varepsilon, \gamma) = \int_0^T \varepsilon^2(t)\mathrm{d}t + \frac{1}{2\gamma} \parallel \langle \psi(T)|P|\psi(T)\rangle - 1\parallel^2, \tag{4.2.27}$$

(where $\psi(t)$ is the solution of Eq. (4.1.7)) is optimized with respect to ε for $\gamma \to 0$ by an inexact Newton method. The sequence of solutions of these optimization problems converges then to the solution of the initial control problem.

REMARK 4.5. The numerical resolution of the evolution equations as in Eqs. (4.1.7), (4.1.8), (4.2.2), (4.2.3), (4.2.12) or (4.2.13) requires a propagation scheme. Often used is the split-operator technique (e.g., ZHU and RABITZ [1998]) which can be written schematically: suppose that the equation to be solved is

$$i\frac{\partial}{\partial t}\chi(t) = (K + V(t))\chi(t), \tag{4.2.28}$$

where K is the kinetic energy operator and $V(t)$ is the total potential. Then, denoting by Δt the time step, the following recurrence is used:

$$\chi(t + \Delta t) = \mathrm{e}^{-iK\Delta t/2}\mathrm{e}^{-iV(t)\Delta t}\mathrm{e}^{-iK\Delta t/2}\chi(t) \tag{4.2.29}$$

which is known to be exact to second order in Δt. In order to apply the operators $e^{-iK\,\Delta t/2}$ and $e^{-iV(t)\Delta t}$ a dual *real space* $\leftrightarrow$ *Fourier(momentum)* representation is used; note that $V(t)$ is diagonal in real space while K is diagonal in momentum space; each operator is thus applied efficiently, with the transformation from one representation to the other realized by (fast) Fourier transforms.

4.3. Robust designs

Due to imperfect knowledge of system Hamiltonians and coupling operators as well as the limited precision and presence of background fluctuations inherent to any laboratory apparatus, it is impossible to perfectly reproduce either optimally designed control laws or the exact specifications under which they were designed. Hence, it is important to study the sensitivity of the control objective or cost functional to random variations or uncertainties in the operators and initial conditions describing the evolution of the system. There is extensive work on the general topic of robust optimal control in the engineering (DORATO [1987], HOSOE [1991], ACKERMAN [1985]) and quantum control literatures (DEMIRALP and RABITZ [1998], BEUMEE, SCHWIETERS and RABITZ [1990], ZHANG and RABITZ [1994]).

A general approach to assessing robustness and stability in quantum control has been considered in DEMIRALP and RABITZ [1998] based on introducing a stability operator S, the kernel of which is related to the curvature $\Delta^2 J[\varepsilon]/\Delta\varepsilon(t)\Delta\varepsilon(t')$ of the cost functional with respect to the control law. Considering the curvature is necessary as the null value of the first order variation $\Delta J[\varepsilon]/\Delta\varepsilon(t) = 0$ defines the optimal solution. Conditions for robustness and optimality of the control solutions can be expressed in terms of the spectrum of S, and this analysis can also reveal qualitative relationships between the various terms in the cost functional and the robustness/optimality features of the control solutions. Work is still needed in order to find a general relationship between the dominant characteristics of a system Hamiltonian, coupling operators, and the cost functional with respect to the eigenvalues of the stability operator S.

The introduction of a penalty term of the form

$$J_3 = \int_0^T W_3 |\langle \psi(t)|O'|\psi(t)\rangle|^2 dt, \tag{4.3.1}$$

where O' is an arbitrary positive definite operator, was observed (DEMIRALP and RABITZ [1998]) to improve the robustness of optimal control solutions. The presence of J_3 can bias the system to satisfy demands tangential to the true control objectives, causing an effective "drag" along the way to the goal. Hence, the effect of J_3 may be loosely interpreted as analogous to the presence of viscous drag in stabilizing a classical mechanical system about a weakly stable point in its phase space. However, the possible stabilization mechanisms have not been carefully studied or characterized and a complete mechanism to explain how the introduction of suitable ancillary objectives may stabilize the solutions to quantum optimal control problems remains to be found.

The robustness effects of penalty operators with more specific forms than that given by Eq. (4.3.1) may be easier to intuit. For example, the term

$$J_s = \int_0^T dt \left(\frac{\delta \langle O(T) \rangle}{\delta \varepsilon(t)} \right)^2$$

(or analogous expressions with higher derivatives) may be used (KOBAYASHI [1998]) to reduce the sensitivity of the achieved control objectives at the target time T to uncertainty in control fields. Analogs of this penalty term for the sensitivity of the target objective to uncertainty in other variables were found (see BEUMEE, SCHWIETERS and RABITZ [1990]) to be capable of reducing the sensitivity to errors in force constants and other model parameters.

Design of robust quantum optimal control solutions can be achieved through the min–max procedure, which involves simultaneously maximizing the effects of an energy-bounded disturbance and minimizing the objective functional. Solutions to such min–max problems represent the best possible control under the worst possible energy-bounded disturbances. For linear dynamical systems the min–max problem becomes H_∞ control, which has an exact solution through the Ricatti equations. This procedure is well-developed in engineering control theory (see, e.g., ATHANS and

FALB [1966]), and it has been applied to robust control designs for selective vibrational excitation in molecular harmonic oscillators (BEUMEE and RABITZ [1992]).

REMARK 4.6. In general, the min–max technique tends to give conservative robust solutions as it works against the worst possible bounded disturbance, and encountering this worst disturbance in practice is unlikely. This point suggests that consideration of a less extreme class of disturbances may also give useful solutions. The resulting analysis should give designs that are robust under more realistic conditions than those modeled in a worst case scenario.

The conclusions of min–max studies (ZHANG and RABITZ [1994]) reinforce the importance of this remark. For a diatomic molecule modeled as a Morse oscillator, the robustness properties of solutions to the min–max equations were compared with solutions to the standard Euler–Lagrange equations (cf. Eqs. (4.1.7)–(4.1.9)) derived without any robustness considerations. While the min–max controls performed better under the application of the worst-possible disturbance (for which they were designed), they did not necessarily outperform the standard Euler–Lagrange solutions under disturbances other than the worst case. For example, min–max control fields were demonstrated to be significantly less-robust than standard Euler–Lagrange control fields to sinusoidal perturbations with the same amplitude constraints as the worst-case disturbance. This underscores the importance of designing control laws that are robust to the particular class of disturbances most likely to occur.

Even in cases where the robustness properties of two designs are quite distinct, simulations have shown (see BEUMEE and RABITZ [1992]) that robust control designs may differ only slightly from nonrobust designs (i.e., the L^2 norm of the difference between the two control laws may be small). This similarity suggests that robustness properties in some cases may result from very subtle effects. It has also been noted. (BEUMEE and RABITZ [1992]) that the relationship between the robust field and the standard design (i.e., created without robustness considerations) can take two forms: the robust field can be either a scaled, self-similar version of the standard field (which may be described as achieving robustness by "speaking louder") or can have a qualitatively different form. At present, no means exists to predict in general when either of these two cases will occur. It is suggestive that self-similar robust fields will exist for weak disturbances, but there is presently no proof of this conjecture.

4.4. Tracking

As mentioned in Remark 4.2 of Section 4, there generally exist a multiplicity of solutions to the quantum optimal control equations, suggesting that it may be possible to predefine a selected path between the initial and final conditions satisfying the control objectives. The existence of such a path exactly matching the conditions at both ends assumes that the system is controllable. The path $y(t)$ can be implicitly defined by the expectation value of a tracking operator O_{tr}:

$$y(t) = \langle \psi(t)|O_{tr}|\psi(t)\rangle, \quad t \in [0, T]. \tag{4.4.1}$$

The quantum tracking control problem (e.g., GROSS, SINGH, RABITZ, HUANG and MEASE [1993], LU and RABITZ [1995], OHTSUKI, KONO and FUJIMURA [1998], ONG, HUANG, TARN and CLARK [1984]) may be viewed as a special case of optimal control theory with the target being the expectation value of O_{tr} over the entire time interval. (In some cases it may be physically attractive to only require that $\lim_{t \to T} O_{tr}(t) = O$, where O is the objective operator whose expectation value is desired at T.) Given the path defined in Eq. (4.4.1), the tracking algorithm for determining the control law may be derived from the Heisenberg equation of motion

$$i\hbar \frac{d\langle \psi(t)|O_{tr}|\psi(t)\rangle}{dt} = \langle \psi(t)|[H, O_{tr}]|\psi(t)\rangle + \left\langle \psi(t)\left|\frac{\partial O_{tr}}{\partial t}\right|\psi(t)\right\rangle. \qquad (4.4.2)$$

With a control law of the form $C(t) = -\mu\varepsilon(t)$ and the assumptions that O_{tr} is independent of time along with

$$[\mu, O_{tr}] \neq 0,$$

Eq. (4.4.2) can be rewritten to solve for the electric field:

$$\varepsilon(t) = \left(i\hbar \frac{dy}{dt} - \langle \psi(t)|[H_0, O_{tr}]|\psi(t)\rangle \right) \Big/ \langle \psi(t)|[\mu, O_{tr}]|\psi(t)\rangle. \qquad (4.4.3)$$

This equation may be substituted into the Schrodinger equation, which then can be numerically solved for $\psi(t)$; substituting $\psi(t)$ back into Eq. (4.4.3) gives an explicit expression for the required control law. One important feature of this technique is that it requires only a single numerical solution of the Schrödinger equation, as opposed to the iterative methods of standard optimal control.

Given the freedom in the selection of $y(t)$, one might unknowingly choose a track that generates one or more singularities, or events at which the denominator of the control field in Eq. (4.4.3) vanishes. This type of singularity may be classified as trivial (see ZHU, SMIT and RABITZ [1999]) if it exists for all $t \in [0,T]$. Trivial singularities may be removed by formulating a tracking equation analogous to Eq. (4.4.1) for control of the kth time-derivatives of $y(t)$. A rank index may be assigned to each tracking singularity by determining the smallest order k_r for which the corresponding tracking equation has no trivial singularity; if the rank index is infinite, then the track-system pair is uncontrollable. Otherwise, any remaining (isolated) singularities may be treated as nontrivial singularities of some relative order k_{nt} (as in ZHU, SMIT and RABITZ [1999]). The magnitude of the disturbance to the trajectory resulting from a nontrivial singularity depends inversely on the magnitude of the derivatives $\frac{\partial^i y}{\partial t^i}, i = 1, \ldots, k_{nt}$, evaluated at the singularity. This partially explains the effects of singularities on quantum tracking control and encourages the search for a noniterative algorithm to sense the occurrence of a forthcoming singularity and accordingly alter the path to avoid the momentary singularity while eventually reaching the objective.

Several extensions of exact inverse tracking which relax demands that could otherwise produce physically unreasonable fields have been developed in CHEN, GROSS, RAMAKRISHNA, RABITZ and MEASE [1995]. The first of these methods is local track generation, in which the problems associated with an a priori trajectory design are avoided by letting the track depend on the evolving quantum state: $y(t) = y(\psi(t))$. This approach is especially useful when the control objectives are not specifically defined by target operator expectation values as in Eq. (2.1.1), but rather can be expressed as the production of some qualitative change in a system. A second method is asymptotic tracking, in which the operator O_{tr} is modified to allow an asymptotic approach to possibly singular trajectories. Finally, in the competitive tracking technique a cost functional is defined whose minimization produces a solution optimally matching a number of trajectories for different tracking operators as well as minimizing the field fluence or satisfying other control objectives.

There is considerable room for further development of the tracking procedure guided by the attraction of performing only one solution of the Schrödinger equation to achieve a control design. Moreover, thus far tracking control has only been applied to the wavefunction formulation of quantum mechanics. A significant extension would be to treat mixed states in the density matrix formulation. In this context, the expectation value $\langle O_{tr}(t) \rangle = \text{Tr}(\rho(t)O_{tr})$ would be followed and the Schrodinger equation would be replaced by Eq. (2.4.1), with the possibility of additionally including decoherence processes.

4.5. Laboratory achievement of closed loop control

The design of control laws poses interesting theoretical challenges, and the practical motivation for such a task is to accomplish successful control in the laboratory. This section discusses the conceptual and theoretical aspects of laboratory operations in which information about the evolving quantum systems is used to improve or define effective control laws. We will cover the technique of quantum learning control, which is increasingly proving to be the most efficient method of practically achieving many control objectives, especially in complex quantum systems; we will also discuss aspects of feedback quantum control. Learning and feedback control are closed loop experimental procedures aimed at achieving control even in the presence of Hamiltonian uncertainties and laboratory disturbances.

The computational design of a control law to meet a physical objective requires (i) explicit knowledge of the system Hamiltonian and (ii) the ability to numerically solve the quantum control equations at least once (in the case of tracking control) or many times for convergence to an optimal solution. In practice, however, these requirements can rarely be met. If the system to be controlled is sufficiently complex (e.g., a polyatomic molecule), it is likely that the Hamiltonian will be only approximately known and the corresponding quantum design equations can only be solved under serious approximations.

In light of these limitations, a completely different and practical approach to the control of quantum dynamics phenomena has been developed (see JUDSON and RABITZ [1992]). In this quantum learning control technique, the laboratory quantum system in itself serves as an analog computer to guide its own control as indicated in Fig. 4.2. This approach addresses the requirements of (i) and (ii) above: a physical quantum system can solve its Schrödinger equation in real time and with exact knowledge of its own Hamiltonian, all without computational cost to the user. Hence, the

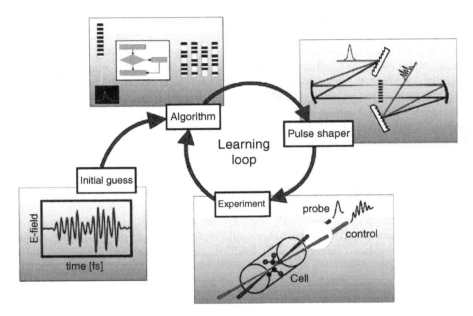

FIG. 4.2 A closed-loop process for teaching a laser to control quantum systems. The loop is entered with either an initial design estimate of even a random field in some cases. A current laser control field design is created with a pulse shaper and then applied to the sample. The action of the control is assessed, and the results are fed to a learning algorithm to suggest an improved field design for repeated excursions around the loop until the objective is satisfactory achieved. Reprinted with permission from RABITZ, DE VIVIE-RIEDLE, MOTZKUS, and KOMPA [2000, p. 824]. Copyright © 2000, American Association for the Advancement of Science.

burden of knowing the Hamiltonian and solving the Schrödinger equation is shifted over to a laboratory effort with a learning algorithm guiding the control experiments. The number of physical/chemical systems treated in this way is growing rapidly, and in many cases it is easier to do the experiments than to perform the designs. However, this approach can still benefit from even approximate control designs to start the laboratory learning process, and theory also has an important role to play in introducing the proper stable and reliable algorithms to make the experiments successful.

In summarizing the methodology of quantum learning control, we will consider a simple paradigm where the state of the system is to be optimized at the final time T only. The first step is to prepare the laboratory quantum system in a convenient initial state $\psi(0) = \psi_0$, or a mixed state or a distribution of incoherent states specified by $\rho(0)$. Next, the system is allowed to evolve under its Hamiltonian and some initial trial control law C_0 applied in the laboratory. At the time T, a measurement of the corresponding control objective(s) is made. The quantum system (perturbed by this measurement) is then discarded, and the control law may be updated to C_1 based on the information gained through this measurement. The method and the frequency with which the control law is updated depends on the specific learning algorithm being used (e.g., as described below, with a genetic algorithm the control law is updated after some multiple of N_{pop} experiments in each time interval). This updating continues until the learning algorithm has converged to some final control law C that achieves the objectives within the convergence bounds of the learning algorithm. The methods most widely used to accomplish the updating of learning control laws are evolutionary and genetic algorithms (GAs) (e.g., GOLDBERG [1989]), although other learning algorithms could be used.

A GA involves the evolution of successive generations of control laws from their parents, in some fashion mimicking biological evolution. Each trial control field is encoded to form a gene which is part of an overall population evolved in the laboratory during the search for an optimal control. Specifically, experiments are performed in which the quantum system evolves under each of the members of a control law "population". The "fitness" of the control law is evaluated based on the degree to which the control objective (s) are achieved, and the fittest control laws are preserved and/or modified in some prescribed fashion in the next generation. This procedure is continued until the fittest members of the control law population achieve the control objectives to the required extent.

REMARK 4.7. A more general learning control setting may include a discrete (yet finite) set of objective times in $\{t_i: i = 1,...,n\}$ or more generally unions of discrete and continuous time intervals. When more than one target is present the problem is divided into several independent subproblems solved sequentially in time. Note that all the experiments carried out thus far (e.g., ASSION, BAUMERT, BERGT, BRIXNER, KIEFER, SEYFRIED, STREHLE and GERBER [1998], BERGT, BRIXNER, KIEFER, STREHLE and GERBER [1999], WEINACHT, AHN and BUCKSBAUM [1999], BARDEEN, YAKOVLEV, WILSON, CARPENTER, WEBER and WARREN [1997], BARDEEN, YAKOVLEV, SQUIER and WILSON [1998], HORNUNG, MEIER and MOTZKUS [2000], LEVIS, MENKIR and RABITZ [2001]) have dealt with a single target goal.

The power of the GA and evolutionary algorithms lies in their ability to globally search the space of control laws and discover solutions that may be highly nonintuitive. This directed search takes advantage of the *ability to perform a great number of distinct control experiments in a short period of laboratory time*, and the closed loop technique has been demonstrated for a wide variety of quantum systems and control objectives. The method has also been shown to have interesting convergence properties. For example, it is observed in simulations (as in JUDSON and RABITZ [1992]) and experiments (e.g., ASSION, BAUMERT, BERGT, BRIXNER, KIEFER, SEYFRIED, STREHLE and GERBER [1998], BERGT, BRIXNER, KIEFER, STREHLE and GERBER [1999], WEINACHT, AHN and BUCKSBAUM [1999], BARDEEN, YAKOVLEV, WILSON, CARPENTER, WEBER and WARREN [1997], BARDEEN, YAKOVLEV, SQUIER and WILSON [1998], LEVIS, MENKIR and RABITZ [2001]) that the GA algorithm can converge for a set of randomly constructed initial control law populations. It is then interesting to know to what extent will the convergence properties of these algorithms be improved by incorporating trial designs into the search space.

REMARK 4.8. The choice of cost functionals (see Section 4.1) used in the experiments has the same freedom as for computational optimal control theory except that in

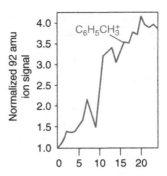

FIG. 4.3 An example the laboratory closed-loop control process. The objective, to maximize the flux of toluene (ions) from the dissociation rearrangement of acetophenone is achieved as the control law evolves over successive generations, see LEVIS, MENKIR and RABITZ [2001]. Reprinted with permission from LEVIS, MENKIR and RABITZ [2001, p. 709]. Copyright © 2001, American Association for the Advancement of Science.

laboratory learning control there is no direct access to the wavefunctions. At present the experiments have considered only the final target in the cost, but other criteria could be included giving rise to a possible enhancement to the procedure. Fig. 4.3 shows an application of the laboratory learning control concept.

Simulations have considered the effects of laboratory errors (modeled as distortions, or transformations, of true input and output data) and noise upon the learning control process (GROSS, NEUHAUSER and RABITZ [1993], TÓTH, LÖRINCZ and RABITZ [1994]). In TÓTH, LÖRINCZ and RABITZ [1994], input errors were modeled by performing various functional transformations on the control laws used in simulated experiments, while output errors were represented by transforming the expectation values of the control objectives corresponding to these experiments. In general, if the input errors are systematic and the output errors are random, they may not significantly affect the ability of the learning algorithm to find an optimal solution. The fitness of the final control laws found by the GA are also demonstrated to be reasonably insensitive to noise in the control fields. These results are based on very limited studies of simple model systems, and they invite further investigation. Such an analysis could give insight into how best to operate the laboratory experiments.

In principle, any optimization algorithm can be applied to quantum learning control. For example, gradient descent and simulated annealing algorithms have been explored in simulations (GROSS, NEUHAUSER and RABITZ [1993]), but the GA outperformed them in several test cases. However, this subject has not received a thorough examination and there may exist algorithms that converge with greater efficiency or robustness than the genetic algorithm for certain classes of quantum mechanical learning control problems. In treating this topic it is important to consider the ability to perform very large numbers of quantum control experiments, which may overcome certain algorithmic shortcomings found under more common conditions. This ability is almost unprecedented in other applications of learning algorithms.

Another approach to quantum learning control is provided by the use of input → output mapping techniques as in PHAN and RABITZ [1997], PHAN and RABITZ [1999], GEREMIA, WEISS and RABITZ [2001]. These methods develop an effective map between the inputs (i.e., the parameters or features defining the control laws) and the outputs (i.e., the expectation values of objective operators). A map from the control input space $\mathcal{C}$ to the space of possible expectation values may be determined directly from the laboratory input and output data; a series of these maps may be needed to cover a sufficiently large portion of $\mathcal{C}$. The control law that optimally satisfies the objectives can be identified from these maps using a suitable learning algorithm. A central issue is establishing the efficiency of mapping techniques as compared with eliminating the maps altogether in favor of having the learning algorithm directly interfaced with the laboratory experiments. Beyond issues of efficiency, mapping techniques may offer the additional benefit of providing physical insight into control mechanisms based on the observed map structure.

5. Challenges for the future

Previous sections have presented the general framework of quantum control, the central issue of controllability, the numerical algorithms for designing control laws and the algorithms for their laboratory discovery. Along the way, several unresolved questions and topics of current interest were introduced. The purpose of the current section is to highlight some emerging areas and unresolved questions that were not a part of this overview. Most of the material is drawn from BROWN and RABITZ [2002].

5.1. Coherence and control

The robustness of coherences for controlled quantum systems in mixed states (i.e., the robustness to decay of the off-diagonal terms in the density operator (cf. Section 2.4) is a topic of special interest. The effects associated with this decay are especially important in the quantum information sciences (see CHUANG, LAFLAMME, SHOR and ZUREK [1995]), where the development of methods to curtail decoherence in information processing algorithms is an active area of research (e.g., SHOR [1995], STEANE [1998]). A relevant contribution from control might be the combination of min–max optimal control techniques with the ideas of decoherence-free subspaces (LIDAR, CHUANG and WHALEY [1998], LIDAR, BACON and WHALEY [1999], BACON, LIDAR and WHALEY [1999]), in which dynamics are invariant to interference that would otherwise cause coherences to decay (this is related to the general notion of disturbance decoupling in mathematical systems theory, see SONTAG [1998]). The result would be control solutions that maximize coherences while simultaneously minimizing an objective cost functional, which could be of significance in developing a physical understanding of the mechanisms of decoherence and its suppression. We note that other schemes are also being considered for the dynamic manipulation of decoherence and control in the presence of dissipation, such as VIOLA and LLOYD [1998], VITALI and TOMBESI [1999], CAO, MESSINA and WILSON [1997], DUAN and GUO [1999].

When the persistence of coherence is not directly important to applications, it was shown in BARDEEN, CAO, BROWN and WILSON [1999] that coherence may not be operative in the control mechanism. In an n-dimensional space $\mathcal{H}_n$ and in the presence of rapid dephasing (which implies vanishing of the off-diagonal coherence terms of the density matrix), Eq. (2.4.1) reduces to a set of rate equations for the population in the n states. A case of special interest in this regard is the control of condensed phases. Under certain conditions, BARDEEN, CAO, BROWN and WILSON [1999] demonstrates that successful controls can be designed for quantum systems whose evolution is determined by these rate equations. This raises the question of whether quantitative measures of coherence can be developed to assess its role in any quantum control problem.

The application of control methods to the cooling of quantum systems is an active area of research (BARTANA and KOSLOFF [1993], BARTANA and KOSLOFF [1997], TANNOR and BARTANA [1999], TANNOR, KOSLOFF and BARTANA [1999]) and there are several means of defining cooling on the molecular scale. One typically utilized criterion (see BARTANA and KOSLOFF [1997]) aims to minimize the von Neumann entropy $\sigma = -\sum_k p_k \log pk$ corresponding to some observable O (such as the system Hamiltonian); here, p_k is the probability that the system is in the kth eigenstate of O. Another system cooling criterion is to increase in the Renyi entropy $\mathrm{Tr}(\rho^2)$ (as in BARTANA and KOSLOFF [1997]). With both measures, maximal cooling is achieved when all but one of the p_k are zero (i.e., achievement of a pure state). Thus, the ability to completely cool a molecular system is likely to be a challenging task in the presence of laser noise: for the purposes of molecular cooling, a laser control with noise fluctuations may be thought of as having an effective nonzero "temperature". Lower bounds related to a temperature beyond which a system cannot be cooled, if they exist, remain to be established.

5.2. *Can noise help attain control?*

In general, noise in $C(t)$ is thought of as harmful in the context of trying to achieve control objectives. However, hints from the subject of stochastic resonance (GAMMAITONI [1998]) suggest that under suitable conditions noise may possibly have beneficial effects, such as allowing the achievement of a particular level of control using smaller total field fluence than that required for the noise-free system. It remains to be demonstrated under what, if any, conditions the presence of noise can assist in the achievement of quantum control objectives, and to elucidate the possible physical mechanisms behind these effects. In addition, the process of seeking the optimal control will attempt to eliminate the deleterious influence of noise while attempting to reach the objectives.

5.3. *Qualitative behavior of optimal control solutions*

A large body of numerical studies provides examples of solutions to the quantum optimal control equations. However, none of this work has illuminated the general behavior, stability, and classes of solutions to the quantum optimal control equations (here, by classes of solutions we mean the qualitative notion of groups of solutions with particular properties, such as nondispersivity, periodicity, etc.). The possibility that unusual behavior can be expected is evident from one study which showed that the optimal control equations can be made equivalent to the standard nonlinear Schrödinger equation under suitable conditions (DEMIRALP and RABITZ [1997]) (see also in Section 5.4 the considerations on the deterministic feedback control).

5.4. Feedback quantum control

This section deals with the effects of real-time, laboratory measurements of the quantum system under two distinct experimental schemes, each described by a different type of feedback Schrödinger equation. The first scheme uses a sequence of repeated experiments to avoid the disturbance effects of measurement, while the second directly confronts these effects.

5.4.1. Deterministic feedback control

Here we consider the (deterministic) continuous-feedback Schrödinger equation

$$i\hbar \frac{\partial \psi(t)}{\partial t} = [H_0 + C(\langle \psi(t)|O_C|\psi(t)\rangle)]\psi(t), \qquad (5.4.1)$$

which conceptually follows from a sequence of laboratory experiments measuring some observable O_C at an increasing set of measurement times, in the limit that the intervals between these times are vanishingly small. In this approach, the probabilistic effect introduced by a measurement at each subsequent time is avoided by 'discarding' the quantum system after each measurement, using the measurement information to extend the control law over the following interval, and repeating the experiment up until the next measurement time.

Different choices of $C(\cdot)$ and O_C may result in Eq. (5.4.1) having qualitatively diverse behavior. An interesting case exists (DEMIRALP and RABITZ [1997]) under the assumptions (i) that $O_C = \delta(\mathbf{x} - \mathbf{x}')$ is the Dirac delta operator, and (ii) that the control law is $C(|\psi(\mathbf{x}, t)|^2) = -\gamma|\psi(\mathbf{x}, t)|^2$, where γ is a positive constant. The resulting equation

$$i\hbar \frac{\partial \psi(\mathbf{x}, t)}{\partial t} = -\left[\frac{\hbar^2}{2m}\nabla^2 + \gamma|\psi(\mathbf{x}, t)|^2\right]\psi(\mathbf{x}, t) \qquad (5.4.2)$$

admits dispersion free solutions (i.e., it preserves $|\psi(\mathbf{x} - \mathbf{v}t)|^2$) and also solitonic solutions under suitable conditions, cf. LAMB JR. [1980], DRAZIN and JOHNSON [1989]. These types of stable solutions may be significant in many applications of quantum control, including quantum information theory. Eq. (5.4.2) may also be derived from the quantum optimal control formalism under the assumptions stated above; thus, dispersion free control solutions are optimal under these same conditions. The existence of such a control law invites a search for other general classes of control Hamiltonians $H_0 + C(\langle O_C(t)\rangle)$ that exhibit nondispersive or other distinct types of qualitative behavior of practical interest.

5.4.2. Measurement disturbances and feedback control

This section is concerned with the effects of taking real-time measurements on a single quantum system while it is being controlled (here, "single" implies that the sequential "measure and discard" approach of the previous section is abandoned). Feedback may augment learning or optimal control methods by providing real-time information about the evolving quantum system for the stabilization of particularly sensitive objectives (e.g., locking a quantum system around an unstable point on its potential energy surface). This scenario naturally arises in the implementation of a feedback control law where measurements are taken at a discrete set of times $\{t_i\}$: the control law may be written as

$$C(t) = C(\langle \psi(t_i)|O_C|\psi(t_i)\rangle), \quad t_i \leqslant t. \tag{5.4.3}$$

There exist well-established procedures for determining feedback control laws based on measurements of evolving deterministic and stochastic classical systems in engineering control (e.g., SONTAG [1998], SOTINE and LI [1991]), and it is possible that many of these methods may be adapted to quantum mechanical control problems. Extensive consideration has been given to the effects of measurements on evolving quantum mechanical systems, including analysis in the contexts of continuous feedback and the control of quantum systems by homodyne detection (i.e., measurement of a component of the light field) cf. CARMICHAEL [1999a], CARMICHAEL [1999b], WISEMAN and MILBURN [1993b], WISEMAN and MILBURN [1993a], BRAGINSKY and KHALILI [1992], WISEMAN [1994], WISEMAN [1993], HOFMANN, MAHLER and HESS [1998], HOFMANN, HESS and MAHLER [1998], CARMICHAEL [1996]. These works generally treat the more difficult problem of random measurement times; here, we give only an elementary discussion of ideas relevant to feedback control with measurements taken at a deterministic, discrete set of times.

A postulate of quantum mechanics states that a perfectly precise measurement of an operator O_C (with nondegenerate spectrum) on a finite-dimensional Hilbert space must both yield one of the eigenvalues λ_j of the operator and result in a disturbance such that $\psi(t)$ collapses to the associated eigenstate ψ_j. The measurement process introduces a stochastic element into the evolution of the quantum system, with the probability of collapse into ψ_j being $|\langle \psi_j|\psi(t)\rangle|^2$. Each measurement in feedback quantum control therefore involves an information trade-off: the system is perturbed away from the deterministic Schrödinger equation, but a measurement is used to update the control law. These random transitions, and the evolution they determine on intervals between the t_i via the control laws Eq. (5.4.3), determine a "stochastic quantum map" (e.g., CARMICHAEL [1999b]) between states at these times. If the measurement process does not completely determine the quantum states (or if mixed states are present for other reasons), a formulation involving a stochastic map between conditioned density operators is required, cf. CARMICHAEL [1999b].

A crucial unresolved question is what general classes of quantum problems will be assisted by incorporating feedback measurements. More specifically, it may be possible to show that taking a certain number of measurements improves control (with reasonable assumptions about the problem-dependent frequency and timing of measurements to optimize the feedback quantum control problem). A related matter is the possibility of making "weak observations" that give useful information about a quantum system while introducing only minimal perturbations. The effects of measurements on the feedback control process for systems satisfactorily described semiclassically also remain to be characterized, and this domain may be especially amenable to performing weak measurements.

5.4.3. Closing the loop through machine feedback

The final topic on feedback control is the possibility of closing the control loop in laboratory hardware through machine feedback. Recent work in acoustics illustrates the capability of focusing reflected waves back upon their sources cf. FINK [1999], FINK and PRADA [1996] in an iterative fashion in order to enhance the intensity in the focal volume. An analogy of this technique relevant to quantum mechanics might be "reflection" through special measurement devices that could then send modified electromagnetic waves precisely back to an emitting quantum mechanical source to better achieve the control objectives. This process may be fully quantum mechanical if carried out in a suitable optical cavity, but in general the same closed loop observation/disturbance issues raised in the previous section must be considered here. At this juncture such a machine is only a gedanken process, but its potential strongly motivates an analysis of the concept.

5.5. Algorithms for the inversion of quantum dynamics data

Knowledge of the potential V and the dipole μ (or other coupling coefficients) is required for control law design and is of fundamental importance to many other applications in chemistry and physics. This section concerns dynamical algorithms that invert time-dependent laboratory data to identify these operators. This problem of determining $\mu(\mathbf{x})$ or $V(\mathbf{x})$ may be related (LU and RABITZ [1995]; see also BARGHEER, DIETRICH, DONOVANG and SCHWENTNER [1999] for a different approach) to the problem of determining the control law $C(t)$ (of the form (2.3.1)) that will cause a quantum system to follow a prescribed track (see Fig. 5.1). In particular, if the expectation values $y(t) \equiv \langle O_h(t) \rangle$ of a time-independent operator O_h are established from a series of observations of an evolving quantum system, the Schrödinger equation and the Heisenberg equation of motion form the pair of coupled (forward-inverse) equations

$$i\hbar\frac{d\psi(t)}{dt} = [H_0 - \mu\varepsilon(t)]\psi(t), \tag{5.5.1}$$

$$i\hbar\frac{dy(t)}{dt} = \langle\psi(t)|[H_0 - \mu\varepsilon(t), O_h]|\psi(t)\rangle. \tag{5.5.2}$$

The solution of these evolution equations may in principle be attempted for any two unknowns. As knowledge of $\psi(t)$ is not available in any physical problem, the wavefunction will always be considered as one of these unknowns; the other may be chosen from either $\varepsilon(t)$, $\mu(\mathbf{x})$, or $V(\mathbf{x})$ with the complementary pair assumed as known. The first of these possibilities was treated in Section 4.4, where the fact that the expectation value on the right hand side of Eq. (5.5.2) involves only spatial

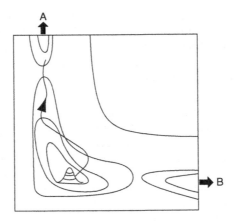

FIG. 5.1 Schematic illustration of a wavepacket track evolving on a potential surface (contours shown). For control a goal may be to find the field $\epsilon(t)$ that will steer the track out of product channel A or B, while for inversion the track is observed in the laboratory and the goal is to determine the potential in the regions traveled by the track, cf. LU and RABITZ [1995]. Reprinted with permission from LU and RABITZ [1995, p. 13731]. Copyright © 1995 American Chemical Society.

integration was exploited to write $\varepsilon(t)$ explicitly in the form of Eq. (4.4.3). We now turn to the solution for $\mu(\mathbf{x})$ and $V(\mathbf{x})$.

Eq. (5.5.2) may be rewritten (cf. LU and RABITZ [1995]) as a Fredholm integral equation of the second type, after regularization enforcing the physical criterion that $\mu(\mathbf{x})$ and $V(\mathbf{x})$ should decay as $x \to \infty$. The structure of this equation is

$$\int_{-\infty}^{\infty} \kappa(x, x') f(x') \mathrm{d}x' + \alpha f(x) = h(x), \tag{5.5.3}$$

where $f(x)$ is the unknown (μ or V), α is the regularization parameter, and the kernel $\kappa(x, x')$ and the inhomogeneity $h(x)$ involve the data $y(t)$, the operator O_h, and the solution $\psi(t)$ to the Schrödinger equation (5.5.1). Solution of the regularized pair of Eqs. (5.5.1) and (5.5.3) may be accomplished using the tracking procedure discussed in Section 4.4, with the role of $\varepsilon(t)$ replaced by $\mu(\mathbf{x})$ or $V(\mathbf{x})$. The procedure consists of formally solving Eq. (5.5.3) for $f(x)$ and substituting the result into the Schrödinger equation (5.5.1) to solve for $\psi(t)$, which is then used to determine $f(x)$ in a final inversion step.

In the special case that the data being inverted is the probability density function $|\psi(\mathbf{x}, t)|^2$ (formally, the expectation value of the Dirac delta operators $\delta(\mathbf{x} - \mathbf{x}')$) (see KRAUSE, SCHAFER, BEN-NUN and WILSON [1997]), a direct algorithm has been developed to identify $V(\mathbf{x})$ without the expensive requirement of numerically solving the Schrödinger equation cf. ZHU and RABITZ [1999b]. The algorithm relies on Ehrenfest's relation, and appears to be difficult to generalize to other measurement operators.

In general, the goal is to solve Eqs. (5.5.1) and (5.5.2) with minimum of distortion introduced by additional criteria; in the example above, the balance between this objective and stability requirements is set by α, whose optimal value depends on the details of the particular problem and solution method. Additionally, the evolution of the quantum system which determines $\kappa(x, x')$ in Eq. (5.5.3) is in turn governed by the applied field $\varepsilon(t)$ in Eq. (5.5.1). Hence, it should be possible to determine a control law which allows inversion with maximum stability to produce optimal dynamical regularization. Note that meaningful inversion of Eqs. (5.5.1) and (5.5.2) may only be expected if the control law $\varepsilon(t)$ steers the wavefunction to be nonzero in the domain in which μ or V is to be determined; the formulation of Eqs. (5.5.1) and (5.5.2) may be extended to incorporate multiple realizations of the control law $\varepsilon_j(t)$ (see LU and RABITZ [1995]) that, taken together, may provide the desired evolution over the entire spatial domain of interest. However, for dynamical reasons the kernel $\kappa(x, x')$ may still produce a singular operator in Eq. (5.5.3) where it is significantly nonzero. The additional conditions required to resolve this problem are not immediately apparent.

A complete laboratory device may be envisioned to function as an optimal dynamics inversion machine for the efficient and automatic discovery of μ or V for diverse quantum systems (cf. RABITZ and SHI [1991], RABITZ and ZHU [2000]). This machine would operate in a closed-loop mode to take advantage of the ability to perform a very large number of high throughput control-observation experiments, and would operate through the following steps, sketched in Fig. 5.2:

(1) Initial approximations for V and μ could be used to design an optimal control field aimed at causing the wavepacket to evolve in desired spatial areas where V and μ are being sought.

(2) Laboratory experiments using this control law would be performed to produce the data trajectory $\langle O_h(t) \rangle = y(t)$.

(3) An inversion would be performed to produce updated potential or dipole information.

(4) If the spatial domain of interest was not completely covered by the current trajectory or if the inversion quality is not adequate, the procedure would be repeated with the assistance of partial Hamiltonian information gained from step (3).

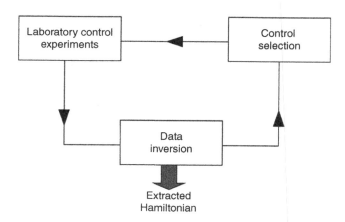

FIG. 5.2 The optimal dynamics identification machine. Reprinted with permission from BROWN and RABITZ [2002, p. 55]; copyright © 2002, Kluwer Academic Publishers.

5.6. Identification of quantum control mechanism and "rules of thumb"

A cornerstone of chemistry is that physically similar molecules tend to exhibit similar chemical behavior. The emphasis is on "similar" and in the context of quantum control the criteria for defining similarity is not known. From the rich behavior and information content in the design, closed loop, and dynamical inversion aspects of quantum control, one can anticipate using the emerging results to provide insight or estimates for the control laws for physically related, but as yet uninvestigated, problems. The body of relationships (as just yet beginning to be observed) between quantum systems, control objectives, and control laws may be called quantum control rules of thumb. A specialized example is the explanation of the timing of the pulses used in the STIRAP control method (cf. RICE and ZHAO [2000]). However, attempts to find general rules have proved much more difficult than was at first expected. Nevertheless, the implications of these rules for both the theory of quantum control and its practical implementation are substantial: resolution of this matter may be the most important challenge ahead for the field.

A natural strategy for identifying rules of thumb might ensue from a type of quantum mechanical reverse engineering: solutions $\{C(t), \psi(t), \lambda(t), \langle O \rangle\}$ to the optimal control equations, or $C(t)$ and $\langle O \rangle$ from closed loop experiments, could provide a physical basis for understanding the mechanisms and pathways leading the quantum system from initial conditions to final control objectives. However, there exist many examples in the literature in which the structure of the final control fields and the resulting control pathways are found to be highly nonintuitive, and judging the relevance of such solutions in terms of general rules of thumb is difficult in the presence of a possibly large number of (locally) optimal solutions. Further insight into the structure of these local minima may be gained by identifying the family of locally optimal control solutions and enumerating them based on their optimality. This problem might be partially alleviated by incorporating a global search procedure in the optimization algorithm (e.g., a genetic algorithm), both for theoretical design and laboratory control. If such techniques were developed, the existence of multiple solutions could possibly be exploited as a large body of data about control behavior.

In the context of closed-loop laboratory implementation of controls, the presence of multiple solutions to the quantum optimal control problem opens up several options (GEREMIA, ZHU and RABITZ [2000]). Given that there exist many possible solutions $C(t)$ from which identification of control mechanisms could be attempted, it is important to select solutions that contain a minimum of extraneous information that detract from this task. In addition, control rules of thumb would best be developed based on solutions that are robust to realistic laboratory noise. Both the suppression of extraneous structural components in $C(t)$ and the selection of robust control fields may be accomplished through the use of appropriate cost functionals (GEREMIA, ZHU and RABITZ [2000]). This "cleanup" of control laws is likely to assist in identifying rules of thumb for the control of quantum systems.

A first step toward identifying quantum control rules of thumb involves the effective classification of similarities and differences between molecules in a context relevant to the controls directing them to certain physical objectives. This type of classification is fundamental in many fields of chemistry and physics, in which the vast numbers of molecules are categorized according to their relevant behaviors or properties. However, presently the standard measures have not been able to consistently predict the structure of control fields for particular objectives.

A three-way classification structure will be necessary, relating (i) control laws, (ii) molecular Hamiltonians and coupling terms, and (iii) control objectives. Progress toward (i) will likely involve identification of the relevant properties of control laws $C(t)$. Experience thus far suggests that the most useful features will include, but extend beyond, description of spectral components and intensities. Progress may also follow from exploring the relation between the form of the Hamiltonian and (sets of) control solutions for a fixed particular control objective, possibly through numerical optimal control calculations for a series of quantum systems whose Hamiltonians differ by small increments, but collectively cover a broad sampling of physical systems.

6. Conclusion

This article aimed to present an overview of the current state of attempts to control quantum phenomena. Special emphasis was given to the conceptual, algorithmic and numerical aspects of the subject. in envisioning further advances along these lines it is very important to explicitly consider the special capabilities of performing massive numbers of control experiments over a short laboratory time. In many respects the field of control over quantum phenomena is a field which is young with the bulk of its developments lying ahead. It is hoped that this article provides some stimulus to push the field further.

Acknowledgments

H.R. and E.B. acknowledge support from the National Science Foundation and the Department of Defense. E.B. was supported by a NSF Graduate Fellowship.

G.T. would like to thank Wusheng Zhu from the Princeton University for helpful discussions and remarks.

References

ACKERMAN, J. (ed.) (1985). *Uncertainty and Control, Proceedings of an International Seminar, Bonn, Germany* (Springer, Berlin). May 1985.

AGARWAL, G. (1997), Nature of the quantum interference in electromagnetic-field-induced control of absorption, *Phys. Rev. A* **55**, 2467–2470.

ALBERTINI, F. and D. D'ALESSANDRO (2001), Notions of controllability for quantum mechanical systems Preprint arXiv:quant-ph/0106128.

ALTAFINI, C. (2002), Controllability of quantum mechanical systems by root space decompositions of $su(N)$, *J. Math. Phys.* **43**(5), 2051–2062.

ASSION, A., T. BAUMERT, M. BERGT, T. BRIXNER, B. KIEFER, V. SEYFRIED, M. STREHLE and G. GERBER (1998), Control of chemical reactions by feedback-optimized phase-shaped femtosecond laser pulses, *Science* **282**, 919–922.

ATHANS, M. and P.L. FALB (1966), *Optimal Control* (McGraw-Hill, New York).

BACON, D., D.A. LIDAR and K.B. WHALEY (1999), Robustness of decoherence-free subspaces for quantum computation, *Phys. Rev. A* **60**, 1944–1955.

BALL, J., J. MARSDEN and M. SLEMROD (1982), Controllability for distributed bilinear systems, *SIAM J. Contr. Optim.* **20**(4).

BARDEEN, C., V.V. YAKOVLEV, K.R. WILSON, S.D. CARPENTER, P.M. WEBER and W.S. WARREN (1997), Feedback quantum control of molecular electronic population transfer, *Chem. Phys. Lett.* **280**, 151–158.

BARDEEN, C.J., J. CAO, F.L.H. BROWN and K.R. WILSON (1999), Using time-dependent rate equations to describe chirped pulse excitation in condensed phases, *Chem. Phys. Lett.* **302**, 405–410.

BARDEEN, C.J., J. CHE, K.R. WILSON, V.V. YAKOVLEV, V.A. APKARIAN, C.C. MARTENS, R. ZADOYAN, B. KOHLER and M. MESSINA (1997), Quantum control of I_2 in the gas phase and in condensed phase solid Kr Matrix, *J. Chem. Phys.* **106**, 8486–8503.

BARDEEN, C.J., J. CHE, K.R. WILSON, V.V. YAKOVLEV, P. CONG, B. KOHLER, J.L. KRAUSE and M. MESSINA (1997), Quantum control of NaI photodissociation reaction product states by ultrafast tailored light pulses, *J. Phys. Chem. A* **101**, 3815–3822.

BARDEEN, C.J., V.V. YAKOVLEV, J.A. SQUIER and K.R. WILSON (1998), Quantum control of population transfer in green fluorescent protein by using chirped femtosecond pulses, *J. Am. Chem. Soc.* **120**, 13023–13027.

BARGHEER, M., P. DIETRICH, K. DONOVANG and N. SCHWENTNER (1999), Extraction of potentials and dynamics from condensed phase pump-probe spectra: Application to I_2 in Kr matrices, *J. Chem. Phys.* **111**, 8556–8564.

BARTANA, A. and R. KOSLOFF (1993), Laser cooling of molecular internal degrees of freedom by a series of shaped pulses, *J. Chem. Phys.* **99**, 196–210.

BARTANA, A. and R. KOSLOFF (1997), Laser cooling and internal degrees of freedom, II, *J. Chem. Phys.* **106**, 1435–1448.

BARTELS, R., S. BACKUS, E. ZEEK, L. MISOGUTI, G. VDOVIN, I.P. CHRISTOV, M.M. MURNANE and H.C. KAPTEYN (2000), Shaped-pulse optimization of coherent emission of high-harmonic soft X-rays, *Nature* **406**, 164–166.

BERGT, M., T. BRIXNER, B. KIEFER, M. STREHLE and G. GERBER (1999), Controlling the femtochemistry of Fe $(CO)_5$, *J. Phys. Chem. A* **103**, 10381–10387.

BEUMEE, J.G.B. and H. RABITZ (1992), Robust optimal control theory for selective vibrational excitation in molecules: A worst case analysis, *J. Chem. Phys.* **97**, 1353–1364.

BEUMEE, J., C. SCHWIETERS and H. RABITZ (1990), Optical control of molecular motion with robustness and application to vinylidene fluoride, *J. Opt. Soc. America B* 1736–1747.

BRAGINSKY, V. and F. KHALILI (1992), *Quantum Measurement* (Cambridge University Press, Cambridge).

BROCKETT, R. (1970), *Finite Dimensional Linear Systems* (Wiley, New York).

BROWN, E. and H. RABITZ (2002), Some mathematical and algorithmic challenges in the control of quantum dynamics phenomena, *J. Math. Chem.* **31**, 17–63.

BROWN, F. and R. SIBLEY (1998), Quantum control for arbitrary linear and quadratic potentials, *Chem. Phys. Lett.* **292**, 357–368.

CANCÈS, E. and C. LE BRIS (1999), On the time-dependent Hartree–Fock equations coupled with a classical nuclear dynamics, *Math. Mod. Methods Appl. Sci.* **9**(7), 963–990.

CANCÈS, E., C. LE BRIS and P. MATHIEU (2000), Bilinear optimal control of a Schrödinger equation, *C. R. Acad. Sci. Paris Sér. I Math.* **7**, 567–571.

CAO, J., M. MESSINA and K. WILSON (1997), Quantum control of dissipative systems: Exact solutions, *J. Chem. Phys.* **106**, 5239–5248.

CAO, J. and K. WILSON (1997), A simple physical picture for the quantum control of wave packet localization, *J. Chem. Phys.* **107**, 1441–1450.

CARMICHAEL, H. (1996), Stochastic Schrodinger equations: What they mean and what they can do, in: J. EBERLY, L. MANDEL and E. WOLF, eds., *Coherence and Quantum Optics VII* (p. 177) 177.

CARMICHAEL, H. (1999a), *Master Equations and Fokker Plank Equations* (Springer, Berlin).

CARMICHAEL, H. (1999b), *An Open Systems Approach to Quantum Optics* (Springer, Berlin).

CHEN, Y., P. GROSS, V. RAMAKRISHNA, H. RABITZ and K. MEASE (1995), Competitive tracking of molecular objectives described by quantum mechanics, *J. Chem. Phys.* **102**, 8001–8010.

CHRISTOFIDES, N. (1975), *Graph Theory: An Algorithmic Approach* (Academic Press, New York).

CHUANG, I.L., R. LAFLAMME, P.W. SHOR and W.H. ZUREK (1995), Quantum computers, factoring, and decoherence, *Science* **270**, 1633–1635.

COHEN-TANNOUDJI, C., B. LIU and F. LALOË (1997), *Quantum Mechanics* (Wiley, New York).

COMBARIZA, J., B. JUST, J. MANZ and G. PARAMONOV (1991), Isomerization controlled by ultrashort infrared laser pulses: Model simulations for the inversion of ligands (H) in the double-well potential of an organometallic compound [(C5H5)(CO)2FePH2], *J. Phys. Chem.* **95**, 10351.

DARDI, P.S. and S.K. GRAY (1982), Classical and quantum-mechanical studies of HF in an intense laser field, *J. Chem. Phys.* **77**, 1345–1353.

DAVIS, M.J. and E.J. HELLER (1984), Comparisons of classical and quantum dynamics for initially localized states, *J. Chem. Phys.* **80**, 5036–5048 (this and the following three references are from the list given in note 18 of Schwieters and Rabitz [1993]).

DE ARAUJO, L. and I. WALMSLEY (1999), Quantum control of molecular wavepackets: an approximate analytic solution for the strong-response regime, *J. Phys. Chem. A* **103**, 10409–10416.

DEMIRALP, M. and H. RABITZ (1993), Optimally controlled quantum molecular dynamics: A perturbation formulation and the existence of multiple solutions, *Phys. Rev. A* **47**, 809–816.

DEMIRALP, M. and H. RABITZ (1997), Dispersion-free wavepackets and feedback solitonic motion in controlled quantum dynamics, *Phys. Rev. A* **55**, 673–677.

DEMIRALP, M. and H. RABITZ (1998), Assessing optimality and robustness of control over quantum dynamics, *Phys. Rev. A* **57**, 2420–2425.

DION, C., A. BANDRAUK, O. ATABEK, A. KELLER, H. UMEDA and Y. FUJIMURA (1999), Two-frequency IR laser orientation of polar molecules, numerical simulations for HCN, *Chem. Phys. Lett.* **302**, 215–223.

DION, C., A. KELLER, O. ATABEK and A. BANDRAUK (1999), Laser-induced alignment dynamics of HCN: Roles of the permanent dipole moment and the polarizability, *Phys. Rev. A* **59**(2), 1382.

DORATO, P. (ed.) (1987). *Robust Control, IEEE Selected Reprint Series* (IEEE Press, New York).

DRAZIN, P. and R. JOHNSON (1989), *Solitons: An Introduction* (Cambridge University Press, Cambridge).

DUAN, L. and G. GUO (1999), Suppressing environmental noise in quantum computation through pulse control, *Phys. Lett. A* **261**, 139–144.

FINK, M. (1999), Time reversed acoustics, *Sci. Am.* 91–97.

FINK, M. and C. PRADA (1996), Ultrasonic focusing with time-reversal mirrors, in: BRIGGS, A. and W. ARNOLD, eds., *Advances in Acoustic Microscopy Series*, Vol. 2, 219–251.

Fu, H., S. Schirmer and A. Solomon (2001), Complete controllability of finite-level quantum systems, *J. Phys. A* **34**, 1679–1690.

Gammaitoni, L. (1998), Stochastic resonance, *Rev. Mod. Phys.* **70**, 223–287.

Geremia, J., E. Weiss and H. Rabitz (2001), Achieving the laboratory control of quantum dynamics phenomena using nonlinear functional maps, *Chem. Phys.* **267**, 209–222.

Geremia, J., W. Zhu and H. Rabitz (2000), Incorporating physical implementation concerns into closed loop quantum control experiments, *J. Chem. Phys.* **113**, 10841–10848.

Girardeau, M.D., M. Ina, S.G. Schirmer and T. Gulsrud (1997), Kinematical bounds on evolution and optimization of mixed quantum states, *Phys. Rev. A* **55**, R1565–R1568.

Girardeau, M.D., S.G. Schirmer, J.V. Leahy and R.M. Koch (1998), Kinematical bounds on optimization of observables for quantum systems, *Phys. Rev. A* **58**, 2684–2689.

Goldberg, D. (1989), *Genetic Algorithms in Search, Optimization and Machine Learning* (Addison-Wesley, New York).

Gross, P., D. Neuhauser and H. Rabitz (1993), Teaching lasers to control molecules in the presence of laboratory field uncertainty and measurement imprecision, *J. Chem. Phys.* **98**, 4557–4566.

Gross, P., J. Singh, H. Rabitz, G. Huang and K. Mease (1993), Inverse quantum-mechanical control: A means for design and a test of intuition, *Phys. Rev. A* **47**, 4593–4603.

Grønager, M. and N. Henriksen (1998), Real-time control of electronic motion: Application to NaI, *J. Chem. Phys.* **109**, 4335–4341.

Harel, G. and V.M. Akulin (1999), Complete control of Hamiltonian quantum systems: Engineering of Floquet evolution, *Phys. Rev. Lett.* **82**, 1–5.

Hofmann, H., O. Hess and G. Mahler (1998), Quantum control by compensation of quantum fluctuations, *Opt. Express* **2**, 339–346.

Hofmann, H.F., G. Mahler and O. Hess (1998), Quantum control of atomic systems by homodyne detection and feedback, *Phys. Rev. A* **57**, 4877–4888.

Hoki, K., Y. Ohtsuki, H. Kono and Y. Fujimura (1999), Quantum control of NaI predissociation in subpicosecond and several-picosecond time regimes, *J. Phys. Chem. A* **103**, 6301–6308.

Hornung, T., R. Meier and M. Motzkus (2000), Optimal control of molecular states in a learning loop with a parameterization in frequency and time domain, *Chem. Phys. Lett* **326**, 445–453.

Hosoe, S. (ed.) (1991). *Robust Control, Proceedings of a Workshop held in Tokyo, Japan, June 23–24, 1991* (Springer, Berlin).

Huang, G., T. Tarn and J. Clark (1983), On the controllability of quantum-mechanical systems, *J. Math. Phys.* **24**, 2608–2618.

Judson, R., K. Lehmann, H. Rabitz and W.S. Warren (1990), Optimal design of external fields for controlling molecular motion – Application to rotation, *J. Mol. Struct.* **223**, 425–456.

Judson, R. and H. Rabitz (1992), *Teaching lasers to control molecules, Phys. Rev. Lett.* .

Kailath, T. (1980), *Linear Systems* (Prentice Hall, Englewood Cliffs).

Khalil, H.K. (1996), *Nonlinear systems* (Macmillan, New York).

Kobayashi, M. (1998), Mathematics makes molecules dance, *SIAM News* **31**(9), 24.

Krause, J.L., D.H. Reitze, G.D. Sanders, A.V. Kuznetsov and C.J. Stanton (1998), Quantum control in quantum wells, *Phys. Rev. B* **57**, 9024–9034.

Krause, J. and K. Schafer (1999), Control of the emission from stark wave packets, *J. Phys. Chem. A* **103**, 10118–10125.

Krause, J.L., K.J. Schafer, M. Ben-Nun and K.R. Wilson (1997), Creating and detecting shaped Rydberg wave packets, *Phys. Rev. Lett.* **79**, 4978–4981.

Krotov, V. (1973), Optimization methods of control with minimax criteria, I, *Automat. Remote Control* **34**, 1863–1873 (translated from Avtomat. i Telemeh. 1973, no. 12, 5–17).

Krotov, V. (1974a), Optimization methods in control processes with minimax criteria, II, *Automat. Remote Control* **35**, (translated from Avtomat i Telemeh. 1974, no. 1, 5–15).

Krotov, V. (1974b), Optimization methods of control with minimax criteria, III, *Automat. Remote Control* **35**, 345–353 (translated from Avtomat. i Telemeh. 1974, no. 3, 5–14).

Kulander, K.C. and A.E. Orel (1981), Laser-collision induced chemical-reactions – A comparison of quantum-mechanical and classical-model results, *J. Chem. Phys.* **75**, 675–680.

KUNDE, J., B. BAUMANN, S. ARLT, F. MORIER-GENOUD, U. SIEGNER and U. KELLER (2000), Adaptive feedback control of ultrafast semiconductor nonlinearities, *Appl. Phys. Lett.* **77**, 924.

LAMB, G. (1980), *Elements of Soliton Theory* (Wiley, New York).

LEFORESTIER, C. (1986), Competition between dissociation and exchange processes in a collinear A + BC collision – Comparison of quantum and classical results, *Chem. Phys. Lett.* **125**, 373–377.

LEVIS, R.J., G. MENKIR and H. RABITZ (2001), Selective bond dissociation and rearrangement with optimally tailored, strong-field laser pulses, *Science* **292**, 709–713.

LI, B., G. TURINICI, V. RAMAKHRISHNA and H. RABITZ (2002), Optimal dynamic discrimination of similar molecules through quantum learning control, *J. Phys. Chem. B* **106**(33), 8125.

LIDAR, D., I.L. CHUANG and K.B. WHALEY (1998), Decoherence-free subspaces for quantum computation, *Phys. Rev. Lett.* **81**, 2594–2597.

LIDAR, D.A., D. BACON and K.B. WHALEY (1999), Concatenating decoherence-free subspaces with quantum error correcting codes, *Phys. Rev. Lett.* **82**, 4556–4559.

LU, Z.M. and H. RABITZ (1995), Unified formulation for control and inversion of molecular dynamics, *J. Phys. Chem.* **99**, 13731–13735.

LUENBERGER, D. (1979), *Introduction to Dynamic Systems: Theory, Models, and Applications* (Wiley, New York).

MESHULACH, D. and Y. SILBERBERG (1998), Coherent quantum control of two-photon transitions by a femtosecond laser pulse, *Nature* **396**, 239–241.

MISHIMA, K. and K. YAMASHITA (1999a), Quantum control of photodissociation wavepackets, *J. Mol. Struct.* **461–462**, 483–491.

MISHIMA, K. and K. YAMASHITA (1999b), Theoretical study on quantum control of photodissociation and photodesorption dynamics by femtosecond chirped laser pulses, *J. Chem. Phys.* **110**, 7756–7769.

MOHLER, R. (1983), *Bilinear Control Processes* (Academic Press, New York).

NIJMEIHER, H. and A. VAN DER SCHAFT (1990), *Nonlinear Dynamical Control Systems* (Springer, Berlin).

OHTSUKI, Y., H. KONO and Y. FUJIMURA (1998), Quantum control of nuclear wave packets by locally designed optimal pulses, *J. Chem. Phys.* **109**, 9318–9331.

OHTSUKI, Y., W. ZHU and H. RABITZ (1999), Monotonically convergent algorithm for quantum optimal control with dissipation, *J. Chem. Phys.* **110**, 9825–9832.

ONG, C.K., G.M. HUANG, T.J. TARN and J.W. CLARK (1984), Invertability of quantum-mechanical control systems, *Math. Syst. Theory* **17**, 335–350.

PASTIRK, I., E.J. BROWN, Q. ZHANG and M. DANTUS (1998), Quantum control of the yield of a chemical reaction, *J. Chem. Phys.* **108**, 4375–4378.

PEIRCE, A., M. DAHLEH and H. RABITZ (1988), Optimal control of quantum mechanical systems: Existence, numerical approximations, and applications, *Phys. Rev. A* **37**, 4950–4964.

PHAN, M.Q. and H. RABITZ (1997), Learning control of quantum-mechanical systems by laboratory identification of effective input-output maps, *Chem. Phys.* **217**, 389–400.

PHAN, M.Q. and H. RABITZ (1999), A self-guided algorithm for learning control of quantum-mechanical systems, *J. Chem. Phys.* **110**, 34–41.

RABITZ, H., R. DE VIVIE-RIEDLE, M. MOTZKUS and K. KOMPA (2000), Wither the future of controlling quantum phenomena? *Science* **288**, 824–828.

RABITZ, H. and S. SHI (1991), Optimal control of molecular motion: Making molecules dance, In: BOWMAN, (ed.), *Advances in Molecular Vibrations and Collision Dynamics*, vol. 1 Part A (JAI Press, Inc, Greenwich) pp. 187–214.

RABITZ, H. and W. ZHU (2000), Optimal control of molecular motion: Design, implementation, and inversion, *Acc. Chem. Res.* **33**, 572–578.

RAMAKRISHNA, V. (2000), *Private communication* .

RAMAKRISHNA, V., M. SALAPAKA, M. DAHLEH, H. RABITZ and A. PIERCE (1995), Controllability of molecular systems, *Phys. Rev. A* **51**(2), 960–966.

RICE, S. and M. ZHAO (2000), *Optical Control of Quantum Dynamics* (Wiley, New York) (many additional references to the subjects of this paper may also be found here).

SCHIRMER, S., M. GIRARDEAU and J. LEAHY (2000), Efficient algorithm for optimal control of mixed-state quantum systems, *Phys. Rev. A* **61**, 012101.

SCHIRMER, S.G., H. FU and A. SOLOMON (2001), Complete controllability of quantum systems, *Phys. Rev. A* **63**, 063410.

SCHWIETERS, C. and H. RABITZ (1993), Optimal control of classical systems with explicit quantum/classical difference reduction, *Phys. Rev. A* **48**, 2549–12457.

SCHWIETERS, C. and H. RABITZ Designing time-independent classically equivalent potentials to reduce quantum/classical observable differences using optimal control theory, (unpublished results).

SEPULCHRE, R., M. JANKOVIC and P. KOKOTOVIC (1998), *Constructive Nonlinear Control* (Springer, Berlin).

SHAH, S. and S. RICE (1999), Controlling quantum wavepacket motion in reduced-dimensional spaces: Reaction path analysis in optimal control of HCN isomerization, *Faraday Discuss.* **113**, 319–331.

SHAPIRO, M. and P. BRUMER (1989), Coherent chemistry: Controlling chemical reactions with lasers, *Acc. Chem. Res.* **22**, 407.

SHEN, H., J.P. DUSSAULT and A. BANDRAUK (1994), Optimal pulse shaping for coherent control by the penalty algorithm, *Chem. Phys. Lett.* **221**, 498–506.

SHEN, L., S. SHI, H. RABITZ, C. LIN, M. LITTMAN, J.P. HERITAGE and A.M. WEINER (1993), Optimal control of the electric susceptibility of a molecular gas by designed non-resonant laser pulses of limited amplitude, *J. Chem. Phys.* **98**, 7792–7803.

SHI, S., A. WOODY and H. RABITZ (1988), Optimal control of selective vibrational excitation in harmonic linear chain molecules, *J. Chem. Phys.* **88**, 6870–6883.

SHOR, P. (1995), Scheme for reducing decoherence in quantum computer memory, *Phys. Rev. A* **52**, R2493–R2496.

SOMLÓI, J., V. KAZAKOV and D. TANNOR (1993), Controlled dissociation of I_2 via optical transitions between the X and B electronic states, *Chem. Phys.* **172**, 85–98.

SONTAG, D. (1998), *Mathematical Control Theory* (Springer, Berlin), and references within.

SOTINE, J.J. and W. LI (1991), *Applied Nonlinear Control* (Prentice Hall, Englewood Cliffs).

STEANE, M. (1998), Introduction to quantum error correction, *Philos. Trans. R. Soc. Lond. A* **356**, 1739–1758.

TANNOR, D. and A. BARTANA (1999), On the interplay of control fields and spontaneous emission in laser cooling, *J. Chem. Phys. A* **103**, 10359–10363.

TANNOR, D., V. KAZAKOV and V. ORLOV (1992), Control of photochemical branching: Novel procedures for finding optimal pulses and global upper bounds, In: Broeckhove, J., Lathouwerse, L. (eds.), *Time Dependent Quantum Molecular Dynamics* (Plenum, New York), pp. 347–360.

TANNOR, D.J., R. KOSLOFF and A. BARTANA (1999), Laser cooling of internal degrees of freedom of molecules by dynamically trapped states, *Faraday Discuss.* **113**, 365–383.

TANNOR, D. and S. RICE (1985), Control of selectivity of chemical reaction via control of wave packet evolution, *J. Chem. Phys.* **83**, 5013–5018.

TERSIGNI, S.H., P. GASPARD and S. RICE (1990), On using shaped light-pulses to control the selectivity of product formation in a chemical-reaction – An application to a multiple level system, *J. Chem. Phys.* **93**, 1670–1680.

TÓÓTH, G., A. LÖRINCZ and H. RABITZ (1994), The effect of control field and measurement imprecision on laboratory feedback control of quantum systems, *J. Chem. Phys.* **101**, 3715–3722.

TURINICI, G. (2000a), *Analysis of numerical methods of simulation and control in Quantum Chemistry* PhD. thesis, (University of Paris VI).

TURINICI, G. (2000b), Controllable quantities for bilinear quantum systems, *Proceedings of the 39th IEEE Conference on Decision and Control, Sydney, Australia*, vol. 2, 1364–1369.

TURINICI, G. (2000c), On the controllability of bilinear quantum systems, In: Defranceschi, M., Le Bris, C. (eds.), *Mathematical models and methods for ab initio Quantum Chemistry*, Lecture Notes in Chemistry, vol. 74 (Springer, Berlin), pp. 75–92.

TURINICI, G. and H. RABITZ (2001a), Quantum wavefunction control, *Chem. Phys.* **267**, 1–9.

TURINICI, G. and H. RABITZ (2001b), *Wavefunction controllability in quantum systems, J. Phys. A* in press.

TURINICI, G., V. RAMAKHRISHNA, B. LI and H. RABITZ (2002), *Optimal discrimination of multiple quantum systems: Controllability analysis* in preparation.

TURINICI, G. and S. SCHIRMER (2001), *On-line controllability calculator* www-rocq.inria.fr/Gabriel. Turinici/ control/calculator.html.

VIOLA, L. and S. LLOYD (1998), Dynamical suppression of decoherence in two-state quantum systems, *Phys. Rev. A* **58**, 2733–2744.

VITALI, D. and P. TOMBESI (1999), Using parity kicks for decoherence control, *Phys. Rev. A* **59**, 4178–4185.

WANG, N. and H. RABITZ (1996), Optimal control of population transfer in an optically dense medium, *J. Chem. Phys.* 1173–1179.

WEINACHT, T., J. AHN and P. BUCKSBAUM (1999), Controlling the shape of a quantum wavefunction, *Nature* **397**, 233–235.

WISEMAN, H.M. (1993), Stochastic quantum dynamics of a continuously monitored laser, *Phys. Rev. A* **47**, 5180–5192.

WISEMAN, M. (1994), Quantum theory of continuous feedback, *Phys. Rev. A* **49**, 2133–2150.

WISEMAN, M. and G.J. MILBURN (1993a), Quantum-theory of field-quadrature measurements, *Phys. Rev. A* **47**, 643.

WISEMAN, M. and G.J. MILBURN (1993b), Quantum-theory of optical feedback via homodyne detection, *Phys. Rev. Lett.* **70**, 548–551.

XIA, H., A. MERRIAM, S. SHARPE, G. YIN and S.E. HARRIS (1997), Electromagnetically induced transparency in atoms with hyperfine structure, *Phys. Rev. A* **56**, R3362–R3365.

ZHANG, H. and H. RABITZ (1994), Robust optimal control of quantum molecular systems in the presence of disturbances and uncertainties, *Phys. Rev. A* **49**, 2241–2254.

ZHAO, M. and S. RICE (1994), Optimal control of product selectivity in reactions of polyatomic molecules: a reduced-space analysis, In: Hepburn, J. (ed.), *Laser Techniques for State-Selected and State-to-Chemistry II*, SPIE vol. 2124, pp. 246–257.

ZHU, W., J. BOTINA and H. RABITZ (1998), Rapidly convergent iteration methods for quantum optimal control of population, *J. Chem. Phys.* **108**, 1953–1963.

ZHU, W. and H. RABITZ (1998), A rapid monotonically convergent iteration algorithm for quantum optimal control over the expectation value of a positive definite operator, *J. Chem. Phys.* **109**, 385–391.

ZHU, W. and H. RABITZ (1999a), Noniterative algorithms for finding quantum optimal controls, *J. Chem. Phys.* **110**, 7142–7152.

ZHU, W. and H. RABITZ (1999b), Potential surfaces from the inversion of time dependent probability density data, *J. Chem. Phys.* **111**, 472–480.

ZHU, W. and H. RABITZ (to appear), Attaining optimal controls for manipulating quantum systems, *Int. J. Quant. Chem.* (in press).

ZHU, W., M. SMIT and H. RABITZ (1999), Managing dynamical singular behavior in the tracking control of quantum observables, *J. Chem. Phys.* **110**, 1905–1915.

Subject Index

Note: Page numbers followed by *f* indicate figures and followed by *t* indicate tables.

Printed in the United States
By Bookmasters